Modern Industrial
Microbiology and Biotechnology

Modern Industrial
Microbiology and Biotechnology

Nduka Okafor

Department of Biological Sciences
Clemson University, Clemson
South Carolina
USA

Science Publishers

Enfield (NH)   Jersey   Plymouth

SCIENCE PUBLISHERS
An imprint of Edenbridge Ltd., British Isles.
Post Office Box 699
Enfield, New Hampshire 03748
United States of America

Website: *http://www.scipub.net*

sales@scipub.net (marketing department)
editor@scipub.net (editorial department)
info@scipub.net (for all other enquiries)

Library of Congress Cataloging-in-Publication Data

Okafor, Nduka.
    Modern industrial microbiology and bitechnology/Nduka Okafor.
        p. cm.
    Includes bibliographical references and index.
    ISBN 978-1-57808-434-0 (HC)
    ISBN 978-1-57808-513-2 (PB)
      1. Industrial microbiology. 2. Biotechnology. I. Title

    QR53.O355 2007
    660.6'2--dc22

                                                                2006051256

ISBN   978-1-57808-434-0 (HC)
ISBN   978-1-57808-513-2 (PB)

Published by Science Publishers, Enfield, NH, USA
An imprint of Edenbridge Ltd.
Printed in India.

Dedication

This book is dedicated to the Okafor-Ozowalu family of Nri, Anambra State, Nigeria, and their inlaws.

Preface

The field of industrial microbiology has been undergoing rapid change in recent years. First, what has been described as the 'cook book' approach has been largely abandoned for the rational manipulation of microorganisms on account of our increased knowledge of their physiology. Second, powerful new tools and technologies especially genetic engineering, genomics, proteomics, bioinformatics and such like new areas promise exciting horizons for man's continued exploitation of microorganisms. Third, new approaches have become available for the utilization of some traditional microbial products such as immobilized enzymes and cells, site-directed mutation and metabolic engineering. Simultaneously, microbiology has addressed itself to some current problems such as the fight against cancer by the production of anti-tumor antibiotics; it has changed the traditional practice in a number of areas: for example the deep sea has now joined the soil as the medium for the search for new bioactive chemicals such as antibiotics. Even the search for organisms producing new products has now been broadened to include unculturable organisms which are isolated mainly on genes isolated from the environment. Finally, greater consciousness of the effect of fossil fuels on the environment has increased the call in some quarters for the use of more environmentally friendly and renewable sources of energy, has led to a search for alternate fermentation substrates, exemplified in cellulose, and a return to fermentation production of ethanol and other bulk chemicals. Due to our increased knowledge and changed approach, even our definitions of familiar words, such as antibiotic and species seem to be changing. This book was written to reflect these changes within the context of current practice.

This book is directed towards undergraduates and beginning graduate students in microbiology, food science and chemical engineering. Those studying pharmacy, biochemistry and general biology will find it of interest. The section on waste disposal will be of interest to civil engineering and public health students and practitioners. For the benefit of those students who may be unfamiliar with the basic biological assumptions underlying industrial microbiology, such as students of chemical and civil engineering, elements of biology and microbiology are introduced. The new elements which have necessitated the shift in paradigm in industrial microbiology such as bioinformatics, genomics, proteomics, site-directed mutation, metabolic engineering, the human genome project and others are also introduced and their relevance to industrial viii Modern Industrial Microbiology and Biotechnology

microbiology and biotechnology indicated. As many references as space will permit are included.

The various applications of industrial microbiology are covered broadly, and the chapters are grouped to reflect these applications. The emphasis throughout, however, is on the physiological and genomic principles behind these applications.

I would like to express my gratitude to Professors Tom Hughes and Hap Wheeler (Chairman) of the Department of Biological Sciences at Clemson University for their help and encouragement during the writing of the book. Prof Ben Okeke of Auburn University, Alabama, and Prof Jeremy Tzeng of Clemson University read portions of the script and I am deeply grateful to them.

My wife, Chinyelu was a source of constant and great support, without which the project might never have been completed. I cannot thank her enough.

Clemson, South Carolina   **Nduka Okafor**

Contents

SECTION F   PRODUCTION OF METABOLITES AS BULK CHEMICALS OR AS INPUTS IN OTHER PROCESSES

Section A

Introduction

CHAPTER 1

Introduction: Scope of Biotechnology and Industrial Microbiology

1.1 NATURE OF BIOTECHNOLOGY AND INDUSTRIAL MICROBIOLOGY

There are many definitions of biotechnology. One of the broadest is the one given at the United Nations Conference on Biological Diversity (also called the *Earth Summit*) at the meeting held in Rio de Janeiro, Brazil in 1992. That conference defined biotechnology as "any technological application that uses biological systems, living organisms, or derivatives thereof, to make or modify products or processes for specific use." Many examples readily come to mind of living things being used to make or modify processes for specfic use. Some of these include the use of microorganisms to make the antibiotic, penicillin or the dairy product, yoghurt; the use of microorganisms to produce amino acids or enzymes are also examples of biotechnology.

Developments in molecular biology in the last two decades or so, have vastly increased our understanding of the nucleic acids in the genetic processes. This has led to applications of biological manipulation at the molecular level in such technologies as genetic engineering. All aspects of biological manipulations now have molecular biology dimensions and it appears convenient to divide biotechnology into *traditional biotechnology* which does not directly involve nucleic acid or molecular manipulations and *nucleic acid biotechnology*, which does.

Industrial microbiology may be defined as the study of the large-scale and profit-motivated production of microorganisms or their products for direct use, or as inputs in the manufacture of other goods. Thus yeasts may be produced for direct consumption as food for humans or as animal feed, or for use in bread-making; their product, ethanol, may also be consumed in the form of alcoholic beverages, or used in the manufacture of perfumes, pharmaceuticals, etc. Industrial microbiology is clearly a branch of biotechnology and includes the traditional and nucleic acid aspects.

1.2   CHARACTERISTICS OF INDUSTRIAL MICROBIOLOGY

The discipline of microbiology is often divided into sub-disciplines such as medical microbiology, environmental microbiology, food microbiology and industrial microbiology. The boundaries between these sub-divisions are often blurred and are made only for convenience.

Bearing this qualification in mind, the characteristics of industrial microbiology can be highlighted by comparing its features with those of another sub-division of microbiology, medical microbiology.

1.2.1   Industrial vs Medical Microbiology

The sub-disciplines of industrial microbiology and medical microbiology differ in at least three different ways.

First is the immediate motivation: in industrial microbiology the immediate motivation is profit and the generation of wealth. In medical microbiology, the immediate concern of the microbiologist or laboratory worker is to offer expert opinion to the doctor about, for example the spectrum of antibiotic susceptibility of the microorganisms isolated from a diseased condition so as to restore the patient back to good health. The generation of wealth is of course at the back of the mind of the medical microbiologist but restoration of the patient to good health is the immediate concern.

The second difference is that the microorganisms per se used in routine medical microbiology have little or no direct economic value, outside the contribution which they make to ensuring the return to good health of the patient who may then pay for the services. In industrial microbiology the microorganisms involved or their products are very valuable and the *raison d'etre* for the existence of the industrial microbiology establishment.

The third difference between the two sub-disciplines is the scale at which the microorganisms are handled. In industrial microbiology, the scale is large and the organisms may be cultivated in fermentors as large as 50,000 liters or larger. In routine medical microbiology the scale at which the pathogen is handled is limited to a loopful or a few milliliters. If a pathogen which normally would have no economic value were to be handled on the large scale used in industrial microbiology, it would most probably be to prepare a vaccine against the pathogen. Under that condition, the pathogen would then acquire an economic value and a profit-making potential; the operation would properly be termed industrial microbiology.

1.2.2   Multi-disciplinary or Team-work Nature of Industrial Microbiology

Unlike many other areas of the discipline of microbiology, the microbiologist in an industrial establishment does not function by himself. He is usually only one of a number of different functionaries with whom he has to interact constantly. In a modern industrial microbiology organization these others may include chemical or production engineers, biochemists, economists, lawyers, marketing experts, and other high-level functionaries. They all cooperate to achieve the purpose of the firm, which is not philanthropy, (at least not immediately) but the generation of profit or wealth.

Despite the necessity for team work emphasized above, the microbiologist has a central and key role in his organization. Some of his functions include:

a. the selection of the organism to be used in the processes;
b. the choice of the medium of growth of the organism;
c. the determination of the environmental conditions for the organism's optimum productivity i.e., pH, temperature, aeration, etc.
d. during the actual production the microbiologist must monitor the process for the absence of contaminants, and participate in quality control to ensure uniformity of quality in the products;
e. the proper custody of the organisms usually in a culture collection, so that their desirable properties are retained;
f. the improvement of the performance of the microorganisms by genetic manipulation or by medium reconstitution.

1.2.3   Obsolescence in Industrial Microbiology

As profit is the motivating factor in the pursuit of industrial microbiology, less efficient methods are discarded as better ones are discovered. Indeed a microbiological method may be discarded entirely in favor of a cheaper chemical method. This was the case with ethanol for example which up till about 1930 was produced by fermentation. When cheaper chemical methods using petroleum as the substrate became available in about 1930, fermentation ethanol was virtually abandoned. From the mid-1970s the price of petroleum has climbed steeply. It has once again become profitable to produce ethanol by fermentation. Several countries notably Brazil, India and the United States have officially announced the production of ethanol by fermentation for blending into gasoline as gasohol.

1.2.4   Free Communication of Procedures in Industrial Microbiology

Many procedures employed in industrial microbiology do not become public property for a long time because the companies which discover them either keep them secret, or else patent them. The undisclosed methods are usually blandly described as 'know-how'. The reason for the secrecy is obvious and is designed to keep the owner of the secret one step ahead of his/her competitors. For this reason, industrial microbiology textbooks often lag behind in describing methods employed in industry. Patents, especially as they relate to industrial microbiology, will be discussed below.

1.3   PATENTS AND INTELLECTUAL PROPERTY RIGHTS IN INDUSTRIAL MICROBIOLOGY AND BIOTECHNOLOGY

All over the world, governments set up patent or intellectual property laws, which have two aims. First, they are intended to induce an inventor to disclose something of his/her invention. Second, patents ensure that an invention is not exploited without some reward to the inventor for his/her innovation; anyone wishing to use a patented invention would have to pay the patentee for its use.

The prerequisite for the patentability of inventions all over the world are that the claimed invention must be new, useful and unobvious from what is already known in 'the prior art' or in the 'state of the art'. For most patent laws an invention is patentable:

a. if it is new, results from inventive activity and is capable of industrial application, or

b. if it constitutes an improvement upon a patented invention, and is capable of industrial application.

For the purposes of the above:

a. an invention is new if it does not form part of the state of the art (i.e., it is not part of the existing body of knowledge);

b. an invention results from inventive activity if it does not obviously follow from the state of the art, either as to the method, the application, the combination of methods, or the product which is concerns, or as to the industrial result it produces, and

c. an invention is capable of industrial application if it can be manufactured or used in any kind of industry, including agriculture.

In the above, 'the art' means the art or field of knowledge to which an invention relates and 'the state of the art' means everything concerning that art or field of knowledge which has been made available to the public anywhere and at any time, by means of a written or oral description, or in any other way, before the date of the filing of the patent application.

Patents cannot be validly obtained in respect of:

a. plant or animal varieties, or essentially biological processes for the production of plants or animals (other than microbiological processes and their products), or

b. inventions, the publication or exploitation of which would be contrary to public order or morality (it being understood for the purposes of this paragraph that the exploitation of an invention is not contrary to public order or morality merely because its exploitation is prohibited by law).

Principles and discoveries of a scientific nature are not necessarily inventions for the purposes of patent laws.

It is however not always as easy as it may seem to show that an invention is 'new', 'useful', and 'unobvious'. In some cases it has been necessary to go to the law courts to decide whether or not an invention is patentable. It is therefore advisable to obtain the services of an attorney specializing in patent law before undertaking to seek a patent. The laws are often so complicated that the layman, including the bench-bound microbiologist may, without proper guidance, leave out essential details which may invalidate his claim to his invention.

The exact wording may vary, but the general ideas regarding patentability are the same around the world. The current Patent Law in the United States is the United States Code Title 35 – Patents (Revised 3 August, 2005), and is administered by the Patents and Trademarks Office while the equivalent UK Patent Law is the Patent Act 1977.

An examination of the patent laws of a number of countries will show that they often differ only in minor details. For example patents are valid in the UK and some other countries for a period of 20 years whereas they are valid in the United States for 17 years. International laws have helped to bridge some of the differences among the patent practices of various countries. The Paris Convention for the protection of Industrial Property has been signed by several countries. This convention provides that each country guarantees to the citizens of other countries the same rights in patent matters as their own citizens. The treaty also provides for the right of priority in case of dispute. Following from this, once an applicant has filed a patent in one of the member countries on a particular invention, he may within a certain time period apply for protection in all the other member countries. The latter application will then be regarded as having been filed on the same day as in the country of the first application. Another international treaty signed in Washington, DC came into effect on 1 June, 1968. This latter treaty, the Patent Cooperation Treaty, facilitates the filing of patent applications in different countries by providing standard formats among other things.

A wide range of microbiological inventions are generally recognized as patentable. Such items include vaccines, bacterial insecticides, and mycoherbicides. As will be seen below however, micro-organisms per se are not patentable, except when they are used as part of a 'useful' *process.*

On 16 June, 1980 a case of immense importance to the course of industrial microbiology was decided in the United States Court of Customs and Patent Appeals. In brief, the court ruled that "a live human-made micro-organism is patentable". Dr. Ananda Chakrabarty then an employee of General Electric Company had introduced into a bacterium of the genus *Pseudomonas* two plasmids (using techniques of genetic engineering discussed in Chapter 7) which enabled the new bacterium to degrade multiple components of crude oil. This single bacterium rather than a mixture of several would then be used for cleaning up oil spills. Claims to the invention were on three grounds.

a. Process claims for the method of producing the bacteria
b. Claims for an inoculum comprising an inert carrier and the bacterium
c. Claims to the bacteria themselves.

The first two were easily accepted by the lower court but the third was not accepted on the grounds that (i) the organisms are products of nature and (ii) that as living things they are not patentable. As had been said earlier the Appeals Court reversed the earlier judgment of the lower court and established the patentability of organisms imbued with new properties through genetic engineering.

A study of the transcript of the decision of the Appeals Court and other patents highlights a number of points about the patentability of microorganisms.

First, microorganisms by themselves are not patentable, being 'products of nature' and 'living things'. However they are patentable as part of a useful 'process' i.e. when they are included along with a chemical or an inert material with which jointly they fulfill a useful purpose. In other words it is the organism-inert material complex which is patented, not the organism itself. An example is a US patent dealing with a bacterium which kills mosquito larva granted to Dr L J Goldberg in 1979, and which reads thus in part:

What is claimed is:

A bacterial larvicide active against mosquito-like larvae *comprising* (this author's italics):

    a. an effective larva-killing concentration of spores of the pure biological strain of *Bacillus thuringiensis* var. WHO/CCBC 1897 as an active agent; and

    b. a carrier....

It is the combination of the bacterial larvicide and the carrier which produced a unique patentable material, not the larvicide by itself. In this regard, when for example, a new antibiotic is patented, the organism producing it forms part of the useful process by which the antibiotic is produced.

Second, a new organism produced by genetic engineering constitutes a 'manufacture' or 'composition of matter'. The Appeals Court made it quite clear that such an organism was different from a newly discovered mineral, and from Einstein's law, or Newton's law which are not patentable since they already existed in nature. Today most countries including those of the European Economic Community accept that the following are patentable: the creation of new plasmid vectors, isolation of new DNA restriction enzymes, isolation of new DNA-joining enzymes or ligases, creation of new recombinant DNA, creation of new genetically modified cells, means of introducing recombinant DNA into a host cell, creation of new transformed host cells containing recombinant DNA, a process for preparing new or known useful products with the aid of transformed cells, and novel cloning processes. Patents resulting from the above were in general regarded as process, not substance, patents. (The above terms all relate to genetic engineering and are discussed in Chapter 7.) The current US law specifically defines biotechnological inventions and their patentability as follows:

"For purposes of (this) paragraph .... the term 'biotechnological process' means:

  (A) a process of genetically altering or otherwise inducing a single- or multi-celled organism to-

     (i) express an exogenous nucleotide sequence,

    (ii) inhibit, eliminate, augment, or alter expression of an endogenous nucleotide sequence, or

   (iii) express a specific physiological characteristic not naturally associated with said organism;

  (B) cell fusion procedures yielding a cell line that expresses a specific protein, such as a monoclonal antibody; and

  (C) a method of using a product produced by a process defined by subparagraph (A) or (B), or a combination of subparagraphs (A) and (B)."

Third, the patenting of a microbiological process places on the patentee the obligation of depositing the culture in a recognized culture collection. The larvicidal bacterium, *Bacillus thuringiensis*, just mentioned, is deposited at the World Health Organization (WHO) International Culture depository at the Ohio State University Columbus Ohio, USA. The rationale for the deposition of culture in a recognized culture collection is to provide permanence of the culture and ready availability to users of the patent. The cultures must be pure and are usually deposited in lyophilized vials.

The deposition of culture solves the problems of satisfying patent laws created by the nature of microbiology. In chemical patents the chemicals have to be described fully and no need exists to provide the actual chemical. In microbiological patents, it is not very helpful to describe on paper how to isolate an organism even assuming that the isolate can be readily obtained, or indeed how the organism looks. More importantly, it is difficult to readily and accurately recognize a particular organism based on patent descriptions alone. Finally, since the organism is a part of the input of microbiological processes it must be available to a user of the patent information.

Culture collections where patent-related cultures have been deposited include the American Type Culture Collection, (ATCC), Maryland, USA, National Collection of Industrial Bacteria (NCIB), Aberdeen, Scotland, UK, Agricultural Research Service Culture Collection, Northern Regional Research Laboratory (NRRL), Peoria, Illinois, USA. A fuller list is available in the *World Directory of Cultures of Micro-organisms*. Culture collections and methods for preserving microorganisms are discussed in Chapter 8 of this book.

Fourth, where a microbiologist-inventor is an employee, the patent is usually assigned to the employer, unless some agreement is reached between them to the contrary. The patent for the oil-consuming *Pseudomonas* discussed earlier went to General Electric Company, not to its employee.

Fifth, in certain circumstances it may be prudent not to patent the invention at all, but to maintain the discovery as a trade secret. In cases where the patent can be circumvented by a minor change in the process without an obvious violation of the patent law it would not be wise to patent, but to maintain the procedure as a trade secret. Even if the nature of the compound produced by the microorganisms were not disclosed, it may be possible to discover its composition during the processes of certification which it must undergo in the hands of government analysts. The decision whether to patent or not must therefore be considered seriously, consulting legal opinion as necessary. It is for this reason that some patents sometimes leave out minor but vital details. As much further detail as the patentee is willing to give must therefore be obtained when a patent is being considered seriously for use.

In conclusion when all necessary considerations have been taken into account and it is decided to patent an invention, the decision must be pursued with vigor and with adequate degree of secrecy because as one patent law states:

> …. The right to patent in respect of an invention is vested in the statutory inventor, that is to say that person who whether or not he is the true inventor, is the first to file…(the) patent application.

1.4   THE USE OF THE WORD 'FERMENTATION' IN INDUSTRIAL MICROBIOLOGY

The word fermentation comes from the Latin verb *fevere*, which means to boil. It originated from the fact that early at the start of wine fermentation gas bubbles are released continuously to the surface giving the impression of boiling. It has three different meanings which might be confusing.

The first meaning relates to microbial physiology. In strict physiological terms, fermentation is defined in microbiology as the type of metabolism of a carbon source in which energy is generated by substrate level phosphorylation and in which organic molecules function as the final electron acceptor (or as acceptors of the reducing equivalents) generated during the break-down of carbon-containing compounds or catabolism. As is well-known, when the final acceptor is an inorganic compound the process is called respiration. Respiration is referred to as aerobic if the final acceptor is oxygen and anaerobic when it is some other inorganic compound outside oxygen e.g sulphate or nitrate.

The second usage of the word is in industrial microbiology, where the term 'fermentation' is any process in which micro-organisms are grown on a large scale, even if the final electron acceptor is not an organic compound (i.e. even if the growth is carried out under aerobic conditions). Thus, the production of penicillin, and the growth of yeast cells which are both highly aerobic, and the production of ethanol or alcoholic beverages which are fermentations in the physiological sense, are all referred to as fermentations.

The third usage concerns food. A fermented food is one, the processing of which micro-organisms play a major part. Microorganisms determine the nature of the food through producing the flavor components as well deciding the general character of the food, but microorganisms form only a small portion of the finished product by weight. Foods such as cheese, bread, and yoghurt are fermented foods.

1.5   ORGANIZATIONAL SET-UP IN AN INDUSTRIAL MICROBIOLOGY ESTABLISHMENT

The organization of a fermentation industrial establishment will vary from one firm to another and will depend on what is being produced. Nevertheless the diagram in Fig. 1.1 represents in general terms the set-up in a fermentation industry.

The culture usually comes from the firm's *culture collection* but may have been sourced originally from a public culture collection and linked to a patent. On the other hand it may have been isolated *ab initio* by the firm from soil, the air, the sea, or some other natural body. The *nutrients* which go into the medium are compounded from various raw materials, sometimes after appropriate preparation or modification including saccharification as in the case of complex carbohydrates such as starch or cellulose. An *inoculum* is first prepared usually from a lyophilized vial whose purity must be checked on an agar plate. The organism is then grown in shake flasks of increasing volumes until about 10% of the volume of the pilot fermentor is attained. It is then introduced into *pilot fermentor(s)* before final transfer into the *production fermentor(s)* (Fig. 1.2).

The *extraction* of the material depends on what the end product is. The methods are obviously different depending on whether the organism itself, or its metabolic product is the desired commodity. If the product is the required material the procedure will be dictated by its chemical nature. Quality control must be carried out regularly to ensure that the right material is being produced. *Sterility* is important in industrial microbiology processes and is maintained by various means, including the use of steam, filtration or by chemicals. Air, water, and steam and other *services* must be supplied and appropriately treated before use. The *wastes* generated in the industrial processes must also be disposed

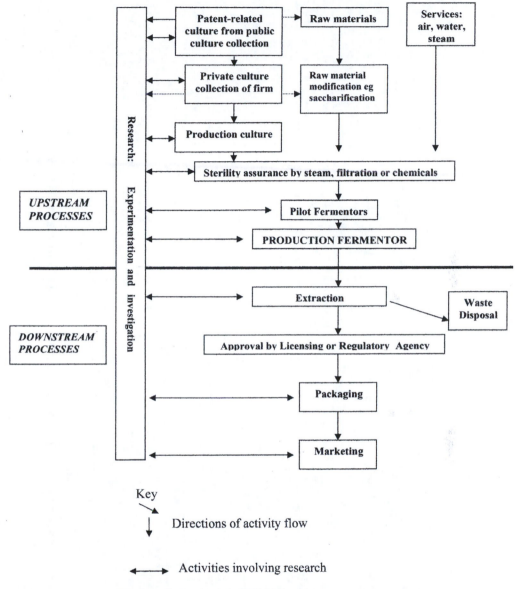

Fig. 1.1   Set-up in an Industrial Microbiology Establishment

off. Packaging and sales are at the tail end, but are by no means the least important. Indeed they are about the most important because they are the points of contact with the consumer for whose satisfaction all the trouble was taken in the first instance. The items in italics above are discussed in various succeeding chapters in this book.

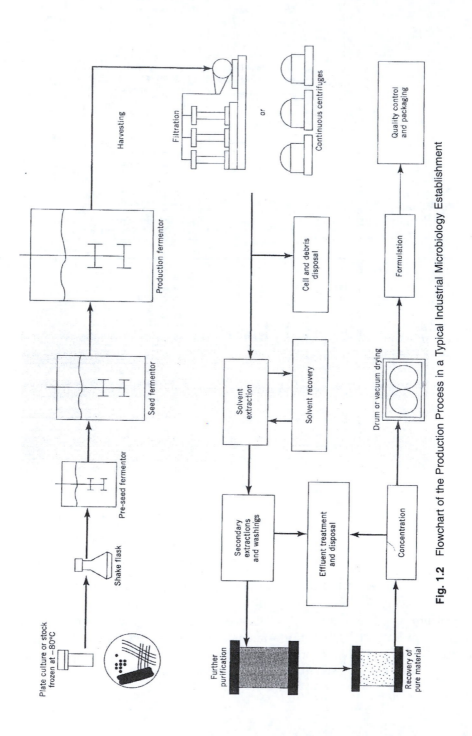

Fig. 1.2 Flowchart of the Production Process in a Typical Industrial Microbiology Establishment

SUGGESTED READINGS

Anon. 1979. General Information Concerning Patents. United States Government Printing Office. Washington, USA.

Anon. 1979. U.S. Patent 4,166,112 Goldberg, L.J., Mosquito larvae control using a bacterial larvicide, Aug. 28, 1979.

Anon. 1980. Supreme Court of the United States. Diamond, (Commissioner of Patents and Trademarks) v. Chakrabarty 447 US 303, 310, 206 USPQ 193, 197.

Anon. 1985. United States Patent Number 4,535,061 granted on August 13 1985 to Chakrabarty et al.: Bacteria capable of dissimilation of environmentally persistent chemical compounds. Washington, DC, USA.

Birch, R.G. 1997. Plant Transformation: Problems and Strategies for Practical Application Annual Review of Plant Physiology and Plant Molecular Biology 48, 297-326.

Bull, A.T., Ward, A.C., Goodfellow, M. 2000. Search and Discovery Strategies For Biotechnology:. The Paradigm Shift. Microbiology and Molecular Biology Reviews, 64, 573 - 548.

Dahod, S.K. 1999. Raw Materials Selection and Medium Development for Industrial Fermentation Processes. In: Manual of Industrial Microbiology and Biotechnology. A.L. Demain, J. E. Davies (eds) 2nd ed. American Society for Microbiology Press.

Doll, J.J. 1998. The patenting of DNA. Science 280, 689 -690.

Gordon, J. 1999. Intellectual Property. In: Manual of Industrial Microbiology and Biotechnology. A.L. Demain, J.E. Davies (eds) 2nd ed. American Society for Microbiology Press.

Kimpel, J.A. 1999. Freedom to Operate: Intellectual Property Protection in Plant Biology And its Implications for the Conduct of Research Annual Review of Phytopathology. 37, 29-51

Moran, K., King, S.R., Carlson, T.J. 2001. Biodiversity Prospecting: Lessons and Prospects. Annual Review of Anthropology, 30, 505-526.

Neijseel, M.O., Tempest, D.W. 1979. In: Microbial Technology; Current State, Future Prospects. A.T. Bull, D.C. Ellwood and C. Rattledge, (eds) Cambridge University Press, Cambridge, UK. pp. 53-82.

Biological Basis of Productivity in Industrial Microbiology and Biotechnology

CHAPTER 2

Some Microorganisms Commonly Used in Industrial Microbiology and Biotechnology

2.1 BASIC NATURE OF CELLS OF LIVING THINGS

All living things are composed of cells, of which there are two basic types, the **prokaryotic** cell and the **eucaryotic** cell. Figure 2.1 shows the main features of typical cells of the two types. The parts of the cell are described briefly beginning from the outside.

Cell wall: Procaryotic cell walls contain glycopeptides; these are absent in eucaryotic cells. Cell walls of eucaryotic cells contain chitin, cellulose and other sugar polymers. These provide rigidity where cell walls are present.

Procaryotic cell (*Bacillus* sp)	Eucaryotic cell (*Saccharomyces* sp)

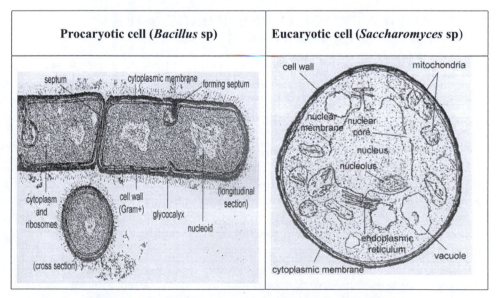

Fig. 2.1 Eucaryotic Cell (Yeast) and Procaryotic Cell (Bacillus)

Cell membrane: Composed of a double layer of phospholipids, the cell membrane completely surrounds the cell. It is not a passive barrier, but enables the cell to actively select the metabolites it wants to accumulate and to excrete waste products.

Ribosomes are the sites of protein synthesis. They consist of two sub-units. Procaryotic ribosomes are 70S and have two sub-units: 30S (small) and a 50S (large) sub-units. Eucaryotic ribosomes are 80S and have sub-units of 40S (small) and a 60S (large). (The unit S means Svedberg units, a measure of the rate of sedimentation of a particle in an ultracentrifuge, where the sedimentation rate is proportional to the size of the particle. Svedberg units are not additive–two sub-units together can have Svedberg values that do not add up to that of the entire ribosome). The prokaryotic 30S sub-unit is constructed from a 16S RNA molecule and 21 polypeptide chains, while the 50S sub-unit is constructed from two RNA molecules, 5S and 23S respectively and 34 polypeptide chains.

Mitochondria are membrane-enclosed structures where in aerobic eucaryotic cells the processes of respiration and oxidative phosphorylation occur in energy release. Procaryotic cells lack mitochondria and the processes of energy release take place in the cell membrane.

Nuclear membrane surrounds the nucleus in eukaryotic cells, but is absent in procaryotic cells. In procaryotic cells only one single circular macromolecule of DNA constitutes the hereditary apparatus or genome. Eucaryotic cells have DNA spread in several chromosomes.

Nucleolus is a structure within the eucaryotic nucleus for the synthesis of ribosomal RNA. Ribosomal proteins synthesized in the cytoplasm are transported into the nucleolus and combine with the ribosomal RNA to form the small and large sub-units of the eucaryotic ribosome. They are then exported into the cytoplasm where they unite to form the intact ribosome.

2.2   CLASSIFICATION OF LIVING THINGS: THREE DOMAINS OF LIVING THINGS

The classification of living things has evolved over time. The earliest classification placed living things into two simple categories, plants and animals. When the microscope was discovered in about the middle of the 16th century it enabled the observation of microorganisms for the first time. Living things were then divided into plants, animals and protista (microorganisms) visible only with help of the microscope. This classification subsisted from about 1866 to the 1960s. From the 1960s and the 1970s Whittaker's division of living things into five groups was the accepted grouping of living things. The basis for the classification were cell-type: procaryotic or eucaryotic; organizational level: single-celled or multi-cellular, and nutritional type: heterotrophy and autotrophy. On the basis of these characteristics living things were divided by Whitakker into five groups: Monera (bacteria), Protista (algae and protozoa), Plants, Fungi, and Animals.

The current classification of living things is based on the work of Carl R Woese of the University of Illinois. While earlier classifications were based to a large extent on morphological characteristics and the cell type, with our greater knowledge of molecular basis of cell function, today's classification is based on the sequence of ribosomal RNA (rRNA)in the 16S of the small sub-unit (SSU) of the procaryotic ribosome, and the 18S ribosomal unit of eucaryotes. The logical question to ask is, why do we use the rRNA sequence? It is used for the following reasons:

(i)   16S (or 18S) rRNA is essential to the ribosome, an important organelle found in all living things (i.e. it is universally distributed);

(ii) its function is identical in all ribosomes;

(iii) its sequence changes very slowly with evolutionary time, and it contains variable and stable sequences which enable the comparison of closely related as well as distantly related species.

The classification is evolutionary and attempts to link all livings things with evolution from a common ancestor. For this approach, an evolutionary time-keeper is necessary. Such a time-keeper must be available to, or used by components of the system, and yet be able to reflect differences and changes with time in other regions appropriate to the assigned evolutionary distances. The 16S ribosomal RNAs meet these criteria as ribosomes are involved in protein synthesis in all living things. They are also highly conserved (remain the same) in many groups and some minor changes observed are commensurate with expected evolutionary distances (Fig. 2.2).

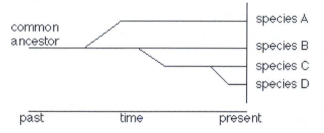

Fig. 2.2   Diagram Illustrating Evolutionary Relationship between Organisms with Time

According to the currently accepted classification living things are placed into three groups: Archae, Bacteria, and Eukarya. A diagram depicting the evolutionary relationships among various groups of living things is giving in Fig. 2.3, while the properties of the various groups are summarized in Table 2.1. Archae and Bacteria are procaryotic while Eucarya are eucaryotic.

2.3   TAXONOMIC GROUPING OF MICRO-ORGANISMS IMPORTANT IN INDUSTRIAL MICROBIOLOGY AND BIOTECHNOLOGY

The microorganisms *currently* used in industrial microbiology and biotechnology are found mainly among the bacteria and eukarya; the Archae are not used. However, as discussed in Chapter 1, the processes used in industrial microbiology and biotechnology are dynamic. Consequently, out-dated procedures are discarded as new and more efficient ones are discovered. At present organisms from Archae are not used for industrial processes, but that may change in future. This idea need not be as far fetched as it may

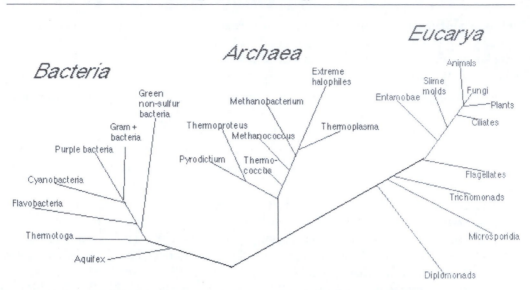

Fig. 2.3 The Three Domains of Living Things Based on Woese's Work

seem now. For as will be seen below, one of the criteria supporting the use of a microorganism for industrial purposes is the possession of properties which will enable the organism to survive and be productive in the face of competition from contaminants. Many organisms in Archae are able to grow under extreme conditions of temperature or salinity and these conditions may be exploited in industrial processes where such physiological properties may put a member of the Archae at an advantage over contaminants.

Plants and animals as well as their cell cultures are also used in biotechnology, and will be discussed in the appropriate sections below. Microorganisms have the following advantages over plants or animals as inputs in biotechnology:

i. Microorganisms grow rapidly in comparison with plants and animals. The generation time (the time for an organism to mature and reproduce) is about 12 years in man, about 24 months in cattle, 18 months in pigs, 6 months in chicken, but only 15 minutes in the bacterium, *E coli*. The consequence is that biotechnological products which can be obtained from microorganisms in a matter of days may take many months in animals or plants.

ii. The space requirement for growth microorganisms is small. A 100,000 litre fermentor can be housed in about 100 square yards of space, whereas the plants or animals needed to generate the equivalent of products in the 100,000 fermentor would require many acres of land.

iii. Microorganisms are not subject to the problems of the vicissitudes of weather which may affect agricultural production especially among plants.

iv. Microorganisms are not affected by diseases of plants and animals, although they do have their peculiar scourges in the form phages and contaminants, but there are procedure to contain them.

Despite these advantages there are occasions when it is best to use either plants or animals; in general however microorganisms are preferred for the reasons given above.

Table 2.1   Summary of differences among the three domains of living things, (from Madigan and Martimko, 2006)

S/No	Characteristic	Bacteria	Archae	Eukarya
Morphology and Genetics				
1	Prokaryotic cell structure	+	+	-
2	DNA present in closed circular form	+	+	-
3	[1]Histone proteins present	-	+	+
4	Nuclear membrane	-	-	+
5	Muramic acid in cell wall	+	-	-
6	Membrane lipids: Fatty acids or Branched hydrocarbons	Fatty acids	Branched hydrocarbons	Fatty acids
7	Ribosome size	70S	70S	80S
8	Initiator tRNA	Formyl-methionine	Methionine	Methionine
9	[2]Introns in most genes	-	-	+
10	[3]Operons	+	+	-
11	Plasmids	+	+	Rare
12	Ribosome sensitive to diphtheria toxin	-	+	+
13	Sensitivity to streptomycin, chloramphenicol, and kanamycin	+	-	-
14	[4]Transcription factors required	-	+	+
Physiological/Special Structures				
15	Methanogenesis	+	-	-
16	Nitrification	+	-?	-
17	Denitrification	+	+	-
18	Nitrogen fixation	+	+	-
19	Chlorophyll based photosynthesis	+	-	+ (plants)
20	Gas vesicles	+	+	-
21	Chemolithotrphy	+	+	-
22	Storage granules of poly-β-hydroxyalkanoates	+	+	-
23	Growth above 80°C	+	+	-
24	Growth above 100°C	-	+	-

[1]Histone proteins are present in eucaryotic chromosomes; histones and DNA give structure to chromosomes in eucaryotes. [2]Non-coding sequences within genes; [3]Operons: Typically present in prokaryotes, these are clusters of genes controlled by a single operator; [4]Transcription factor is a protein that binds DNA at a specific promoter or enhancer region or site, where it regulates transcription.

2.3.1   Bacteria

Bacteria are described in two compendia, *Bergey's Manual of Determinative Bacteriology* and *Bergey's Manual of Systematic Bacteriology*. The first manual (on *Determinative Bacteriology*) is designed to facilitate the identification of a bacterium whose identity is unknown. It was first published in 1923 and the current edition, published in 1994 is the ninth. The companion volume (on *Systematic Bacteriology)* records the accepted published descriptions of bacteria, and classifies them into taxonomic groups. The first edition was produced in four volumes and published between 1984 and 1989. The bacterial classification in the latest (second) edition of *Bergey's Manual of Sytematic Bacteriology* is based on 16S RNA sequences, following the work of Carl Woese, and organizes the Domain Bacteria into 18 groups (or *phyla;* singular, *phylum*) It is to be published in five volumes. Volume 1 which deals with the *Archae* and the deeply branching and phototrophic bacteria was published in 2001; Volume 2 published in 2005, deals with the *Proteobacteria* and has three parts while Volume 3 was published in 2006 and deals with the low G+C Gram-positive bacteria. The last two volumes, Volume 4 (the high C + C Gram-positive bacteria) and Volume 5 (The *Plenctomyces, Spirochaetes, Fibrobacteres, Bacteriodetes* and *Fusobacteria)* will be published in 2007. The manuals are named after Dr D H Bergey who was the first Chairman of the Board set up by the then Society of American Bacteriologists (now American Society for Microbiology) to publish the books. The publication of Bergey Manuals is now managed by the Bergey's Manual Trust.

Of the 18 phyla in the bacteria, (see Fig. 2.4) the Aquiflex is evolutionarily the most primitive, while the most advanced is the Proteobacteria. The bacterial phyla used in industrial microbiology and biotechnology are found in the Proteobacteria, the Firmicutes and the Actinobacteria.

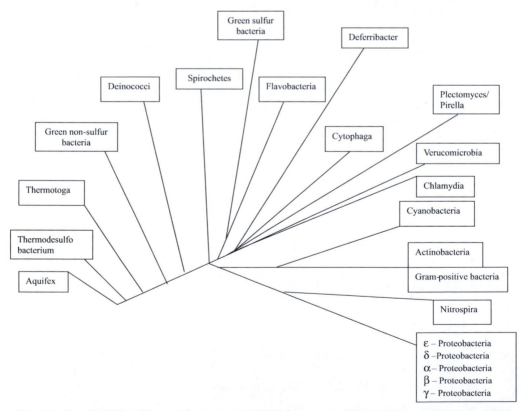

Fig. 2.4   The 18 Phyla of Bacteria Based on 16S RNA Sequences (After Madigan and Matinko, 2006)

2.3.1.1   The Proteobacteria

The **Proteobacteria** are a major group of bacteria. Due to the diversity of types of bacteria in the group, it is named after Proteus, the Greek god, who could change his shape. Proteobacteria include a wide variety of pathogens, such as *Escherichia, Salmonella, Vibrio* and *Helicobacter*, as well as free-living bacteria some of which can fix nitrogen. The group also includes the purple bacteria, so-called because of their reddish pigmentation, and which use energy from sun light in photosynthesis.

All Proteobacteria are Gram-negative, with an outer membrane mainly composed of lipopolysaccharides. Many move about using flagella, but some are non-motile or rely on bacterial gliding. There is also a wide variety in the types of metabolism. Most members are facultatively or obligately anaerobic and heterotrophic, but there are numerous exceptions.

Proteobacteria are divided into five groups: α (alpha), β (beta), γ (gamma), δ (delta), ε (epsilon). The only organisms of current industrial importance in the Proteobacteria are *Acetobacter* and *Gluconobacter*, which are acetic acid bacteria and belong to the Alphaproteobacteria. An organism also belonging to the Alphaproteobacteria, and which has the potential to become important industrially is *Zymomonas*. It produces copious amounts of alcohol, but its use industrially is not yet widespread.

2.3.1.1.1   The Acetic Acid Bacteria

The acetic acid bacteria are *Acetobacter* (peritrichously flagellated) and *Gluconobacter* (polarly flagellated). They have the following properties:

i. They carry out incomplete oxidation of alcohol leading to the production of acetic acid, and are used in the manufacture of vinegar (Chapter 14).

ii. *Gluconobacter* lacks the complete citric acid cycle and can not oxidize acetic acid; *Acetobacter* on the on the other hand, has all the citric acid enzymes and can oxidize acetic acid further to CO_2.

iii. They stand acid conditions of pH 5.0 or lower.

iv. Their property of 'under-oxidizing' sugars is exploited in the following:

    a. The production of glucoronic acid from glucose, galactonic aicd from galactose and arabonic acid from arabinose;

    b. The production of sorbose from sorbitol by acetic acid bacteria (Fig. 2.4), an important stage in the manufacture of ascorbic acid (also known as Vitamin C)

v. Acetic acid bacteria are able to produce pure cellulose when grown in an unshaken culture. This is yet to be exploited industrially, but the need for cellulose of the purity of the bacterial product may arise one day.

2.3.1.2   The Firmicutes

The **Firmicutes** are a division of bacteria, all of which are Gram-positive, in contrast to the Proteobacteria which are all Gram-negative. A few, the mycoplasmas, lack cell walls altogether and so do not respond to Gram staining, but still lack the second membrane found in other Gram-negative forms; consequently they are regarded as Gram-positive.

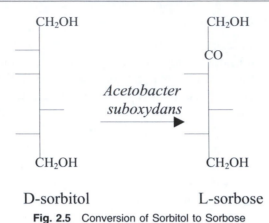

D-sorbitol   L-sorbose

Fig. 2.5   Conversion of Sorbitol to Sorbose

Originally the Firmicutes were taken to include all Gram-positive bacteria, but more recently they tend to be restricted to a core group of related forms, called the low G+C group in contrast to the Actinobacteria, which have high G+C ratios. The G+C ratio is an important taxonomic characteristic used in classifying bacteria. It is the ratio of Guanine and Cytosine to Guanine, Cytosine, Adenine, and Thymine in the cell. Thus the GC ratio = G+C divided by G+C+A+T x 100. It is used to classify Gram-positive bacteria: low G+C Gram-positive bacteria (ie those with G+C less than 50%) are placed in the Fermicutes, while those with 50% or more are in Actinobacteria. Fermicutes contain many bacteria of industrial importance and are divided into three major groups: i. spore-forming, ii. non-spore forming, and iii) wall-less (this group contains pathogens and no industrial organisms.)

2.3.1.2.1   Spore forming firmicutes

Spore-forming Firmicutes form internal spores, unlike the Actinobacteria where the spore-forming members produce external ones. The group is divided into two: *Bacillus* spp, which are aerobic and *Clostridium* spp which are anaerobic. *Bacillus* spp are sometimes used in enzyme production. Some species are well liked by mankind because of their ability to kill insects. *Bacillus papilliae* infects and kills the larvae of the beetles in the family *Scarabaeidae* while *B. thuringiensis* is used against mosquitoes (Chapter 17). The genes for the toxin produced by *B. thuringiensis* are also being engineered into plants to make them resistant to insect pests (Chapter 7). Clostridia on the other hand are mainly pathogens of humans and animals.

2.3.1.2.2   Non-spore forming firmicutes

The Lactic Acid Bacteria: The non-spore forming low G+C members of the firmicutes group are very important in industry as they contain the lactic acid bacteria.

The lactic acid bacteria are rods or cocci placed in the following genera: *Enterococcus, Lactobacillus, Lactococcus, Leuconostoc, Pediococcus* and *Streptococcus* and are among some of the most widely studied bacteria because of their important in the production of some foods, and industrial and pharmaceutical products. They lack porphyrins and cytochromes, do not carry out electron transport phosphorylation and hence obtain energy by substrate level phosphorylation. They grow anaerobically but are not killed by oxygen as is the case with many anaerobes: they will grow with or without oxygen. They obtain their energy from sugars and are found in environments where sugar is present. They have limited synthetic ability and hence are fastidious, requiring, when cultivated, the addition of amino acids, vitamins and nucleotides.

Lactic acid bacteria are divided into two major groups: The **homofermentative** group, which produce lactic acid as the sole product of the fermentation of sugars, and the **heterofermentative**, which besides lactic acid also produce ethanol, as well as CO_2. The difference between the two is as a result of the absence of the enzyme aldolase in the heterofermenters. Aldolase is a key enzyme in the E-M-P pathway and spits hexose glucose into three-sugar moieties. Homofermentative lactic acid bacteria convert the D-glyceraldehyde 3-phosphate to lactic acid. Heterofermentative lactic acid bacteria receive five-carbon xylulose 5 phosphate from the Pentose pathway. The five carbon xylulose is split into glyceraldehyde 3-phosphate (3-carbon), which leads to lactic acid, and the two-carbon acetyl phosphate which leads to ethanol (Fig. 2.6).

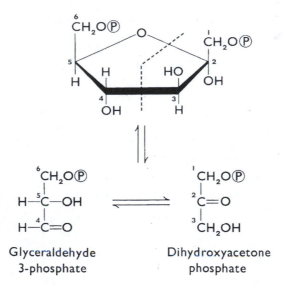

Glyceraldehyde
3-phosphate

Dihydroxyacetone
phosphate

Fig. 2.6 Splitting of 6-carbon Glucose into Three-carbon Compounds by the Enzyme Fructose Diphposphate Aldolase

Use of Lactic Acid Bacteria for Industrial Purposes:

The desirable characteristics of lactic acid bacteria as industrial microorganisms include

a. their ability to rapidly and completely ferment cheap raw materials,
b. their minimal requirement of nitrogenous substances,
c. they produce high yields of the much preferred stereo specific lactic acid
d. ability to grow under conditions of low pH and high temperature, and
e. ability to produce low amounts of cell mass as well as negligible amounts of other byproducts.

The choice of a particular lactic acid bacterium for production primarily depends on the carbohydrate to be fermented. *Lactobacillus delbreuckii* subspecies *delbreuckii* is able to ferment sucrose. *Lactobacillus delbreuckii* subspecies *bulgaricus* is able to use lactose while *Lactobacillus helveticus* is able to use both lactose and galactose. *Lactobacillus amylophylus* and *Lactobacillus amylovirus* are able to ferment starch. *Lactobacillus lactis* can ferment glucose, sucrose and galactose and *Lactobacillus pentosus* has been used to ferment sulfite waste liquor.

2.3.1.3 The Actinobacteria

The Acinobacteria are the Firmicutes with G+C content of 50% or higher. They derive their name from the fact that many members of the group have the tendency to form filaments or hyphae (*actinis*, Greek for ray or beam). The industrially important members

Table 2.2 Characteristics of the lactic acid bacteria

S/No	Group	Description	Habit	Importance
1	Streptococcus	Cocci in pairs or short chains	Some in respiratory tract, mouth, intestine; others found in fermenting vegetable and silage	Some cause sore throat; non-pathogenic strains used in yoghurt manufacture
2	Enterococcus	Cocco-bacilli usually in pairs; previously classified Streptococcus Lancefield Group D	Found as commensals in the human alimentary canal; sometimes cause urinary tract infections	Can be used to monitor water quality, (like *E. coli*)
3	Lactococcus	Coccoid, usually occuring in pairs; hardly form chains	Plant material and alimentary canals of animals	Used as starter in yoghurt manufacture; Used as probiotic for intestinal health; Produces copious amounts of lactic acid.
4	Pediococcus	Growth in tetrads	Found on plant materials	Spoils beer; but required in special beers such as lambic beer drunk in parts of Belgium
5	Leuconostoc	Cocco-bacili	Associated with plant materials	Tolerates high concentrations of salt and sugar and involved in the pickling of vegetables; produce dextrans from sucrose

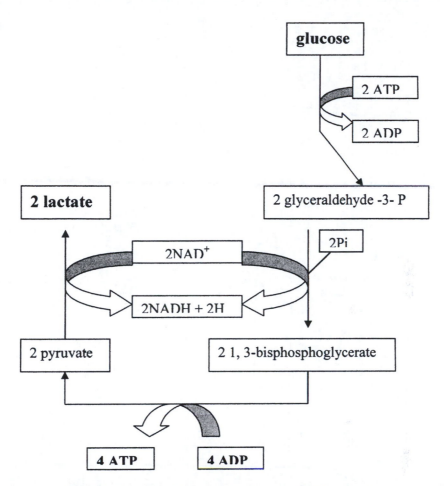

Fig. 2.7   Formation of lacttic acid by homofermentative bacteria

Table 2.3   Distinguishing characteristics of lactic acid bacteria

Character	Lactobacillus	Enterococcus	Lactocococcus	Leuconostoc	Pediococcus	Streptococcus
Tetrad formation	–	–	–	–	+	–
CO_2 from glucose	±	–	–	+	–	–
Growth at 10°C	±	+	+	+	±	–
Growth at 45°C	±	+	–	–	±	±
Growth at 6.5% NaCl	±	+	–	±	±	–
Growth at pH 4.4	±	+	±	±	+	–
Growth at pH 9.6	–	+	–	–	–	–
Lactic acid (optical orientation)	D, L, DL	L	L	D	L, DL	L

of the group are the Actinomycetes and *Corynebacterium*. *Corynebacterium* spp are important industrially as secreters of amino acids (Chapter 21). The rest of this section will be devoted to Actinomycetes.

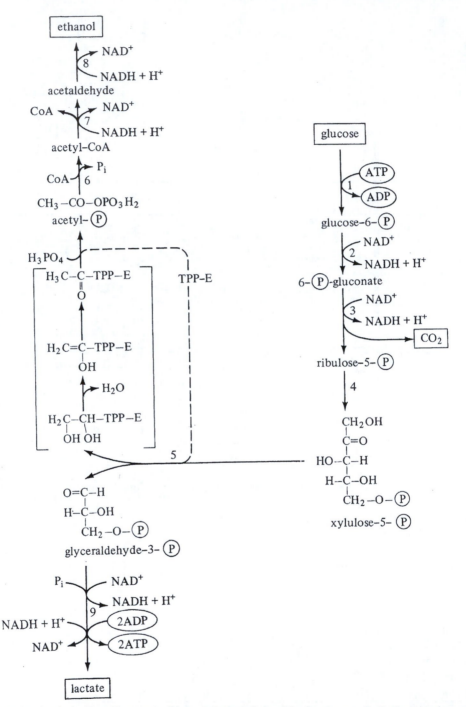

Enzymes involved: 1, Hexokinase; 2, Glucose-6-phosphate dehydrogenase; 3, 6-phosphogluconate dehydrogenase; 4, Ribulose-5-phosphate 3-epimerase; Phosphoketolase; 6, Phosphotransacetylase; 7, Acetaldehyde dehydrogenase; 8, Alcohol dehydrogense; 9, Enzymes of the homofermentative pathway

Fig. 2.8   Fermentation of Glucose by Heterofermentative Bacteria

Lactobacillus bulgaricus   **_Lactococcus lactis_**

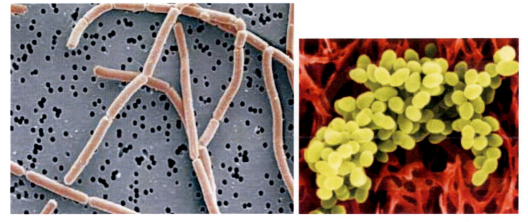

Fig. 2.9 Photomicrographs of Lactic Acid Bacteria

2.3.1.3.1   The Actinomycetes

They have branching filamentous hyphae, which somewhat resemble the mycelia of the fungi, among which they were originally classified. In fact they are unrelated to fungi, but are regarded as bacteria for the following reasons. First they have petidoglycan in their cell walls, and second they are about 1.0µ in diameter (never more than 1.5µ), whereas fungi are at least twice that size in diameter.

As a group the actinomycetes are unsurpassed in their ability to produce secondary metabolites which are of industrial importance, especially as pharmaceuticals. The best known genus is _Streptomyces_, from which many antibiotics as well as non-anti-microbial drugs have been obtained. The actinomycetes are primarily soil dwellers hence the temptation to begin the search for any bioactive microbial metabolite from soil.

2.3.2   Eucarya: Fungi

Although plants and animals or their cell cultures are used in biotechnology, microorganisms are used more often for reason which have been discussed. Fungi are members of the Eucarya which are commonly used in industrial production.

The fungi are traditionally classified into the four groups given in Table 2.4, namely Phycomycetes, Ascomycetes, Fungi Imprfecti, and Basidiomycetes. Among these the following are those currently used in industrial microbiology

Phycomycetes (Zygomycetes)
Rhizopus and **_Mucor_** are used for producing various enzymes

Ascomycetes
Yeasts are used for the production of ethanol and alcoholic beverages
Claviceps purperea is used for the production of the ergot alkaloids

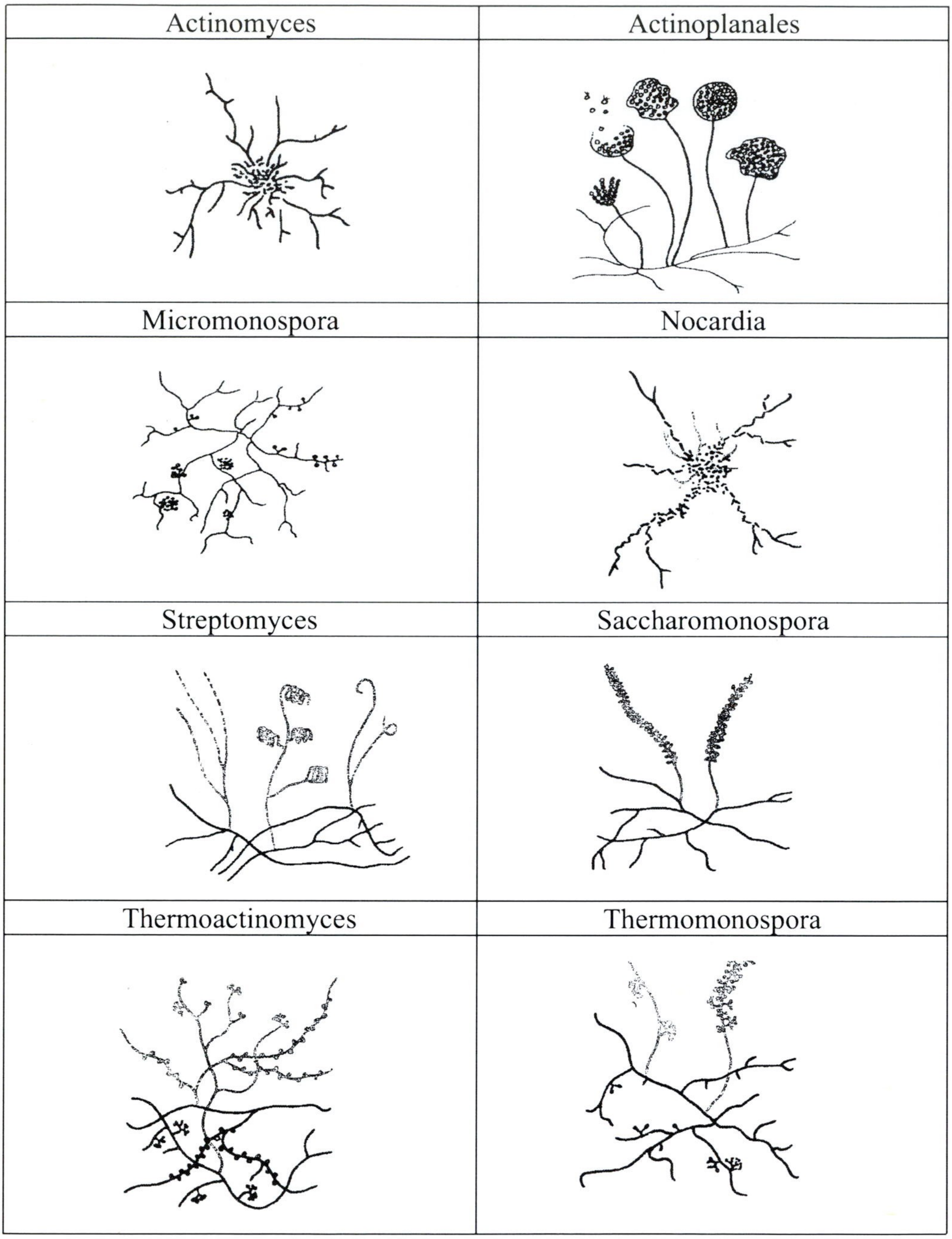

Actinomyces	Actinoplanales
Micromonospora	Nocardia
Streptomyces	Saccharomonospora
Thermoactinomyces	Thermomonospora

Fig. 2.10 Different Actinomycetes

Fungi Imperfecti

Aspergillus is important because it produces the food toxin, aflatoxin, while *Penicillium* is well-known for the antibiotic penicillin which it produces.

Basidiomycetes

Agaricus produces the edible fruiting body or mushroom

Numerous useful products are made through the activity of fungi, but the above are only a selection.

Table 2.4   Description of the various groups of fungi

Group	Ordinary Name	Septation of hyphae	Sexual Spores	Representative
Zygomycetes (Phycomycetes)	Bread molds	Non-septate	Zygospre	*Rhizopus, Mucor*
Ascomycetes	Sac fungi	Septate	Ascospore (in Perithecia)	*Neurospora, Saccharomyces (Yeasts)*
Basidiomycetes	Mushrooms	Septate	Basidiomycetes (Mushrooms)	*Agaricus*
Deuteromycetes	Fungi imperfecti	Septate	None	*Penicillium, Aspergillus*

2.4   CHARACTERISTICS IMPORTANT IN MICROBES USED IN INDUSTRIAL MICROBIOLOGY AND BIOTECHNOLGY

Microorganisms which are used for industrial production must meet certain requirements including those to be discussed below. It is important that these characteristics be borne in mind when considering the candidacy of any microorganism as an input in an industrial process.

  i. The organism must be able to grow in a simple medium and should preferably not require growth factors (i.e. pre-formed vitamins, nucleotides, and acids) outside those which may be present in the industrial medium in which it is grown. It is obvious that extraneous additional growth factors may increase the cost of the fermentation and hence that of the finished product.

 ii. The organism should be able to grow vigorously and rapidly in the medium in use. A slow growing organism no matter how efficient it is, in terms of the production of the target material, could be a liability. In the first place the slow rate of growth exposes it, in comparison to other equally effective producers which are faster growers, to a greater risk of contamination. Second, the rate of the turnover of the production of the desired material is lower in a slower growing organism and hence capital and personnel are tied up for longer periods, with consequent lower profits.

iii. Not only should the organism grow rapidly, but it should also produce the desired materials, whether they be cells or metabolic products, in as short a time as possible, for reasons given above.

 iv. Its end products should not include toxic and other undesirable materials, especially if these end products are for internal consumption.

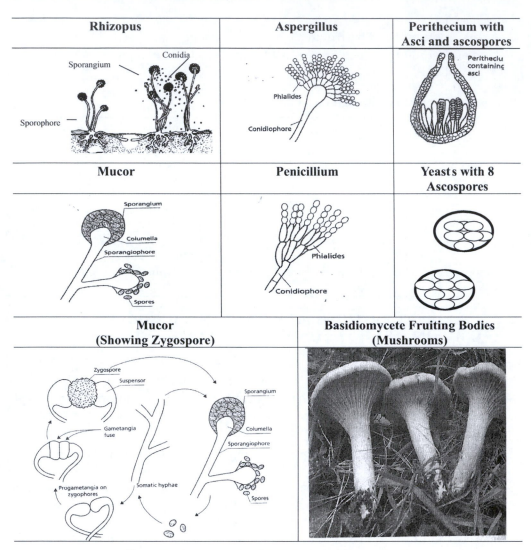

Fig. 2.11   Representative Structures from Different Fungi

   v. The organism should have a reasonable genetic, and hence physiological stability. An organism which mutates easily is an expensive risk. It could produce undesired products if a mutation occurred unobserved. The result could be reduced yield of the expected material, production of an entirely different product or indeed a toxic material. None of these situations is a help towards achieving the goal of the industry, which is the maximization of profits through the production of goods with predictable properties to which the consumer is accustomed.

  vi. The organism should lend itself to a suitable method of product harvest at the end of the fermentation. If for example a yeast and a bacterium were equally suitable for manufacturing a certain product, it would be better to use the yeast if the most appropriate recovery method was centrifugation. This is because while the bacterial diameter is approximately 1μ, yeasts are approximately 5μ. Assuming their densities are the same, yeasts would sediment 25 times more rapidly than bacteria. The faster sedimentation would result in less expenditure in terms of power, personnel supervision etc which could translate to higher profit.

vii. Wherever possible, organisms which have physiological requirements which protect them against competition from contaminants should be used. An organism with optimum productivity at high temperatures, low pH values or which is able to elaborate agents inhibitory to competitors has a decided advantage over others. Thus a thermophilic efficient producer would be preferred to a mesophilic one.

viii. The organism should be reasonably resistant to predators such as *Bdellovibrio* spp or bacteriophages. It should therefore be part of the fundamental research of an industrial establishment using a phage-susceptible organism to attempt to produce phage-resistant but high yielding strains of the organism.

ix. Where practicable the organism should not be too highly demanding of oxygen as aeration (through greater power demand for agitation of the fermentor impellers, forced air injection etc) contributes about 20% of the cost of the finished product.

x. Lastly, the organism should be fairly easily amenable to genetic manipulation to enable the establishment of strains with more acceptable properties.

SUGGESTED READINGS

Asai, T. 1968. The Acetic Acid Bacteria. Tokyo: The University of Tokyo Press and Baltimore: University Park Press.

Axelssson, L., Ahrne, S. 2000. Lactic Acid Bacteria. In: Applied Microbial Systematics, F.G. Priest, M. Goodfellow, (eds) A.H. Dordrecht, the Netherlands, pp. 367-388.

Barnett, J.A. , Payne, R.W., Yarrow, D. 2000. Yeasts: Characterization and Identification. 3rd Edition. Cambridge University Press. Cambridge, UK.

Garrity, G.M. 2001-2006. Bergey's Manual of Systematic Bacteriology. 2nd Ed. Springer, New York, USA.

Goodfellow, M., Mordaraski, M., Williams, S.T. 1984. The Biology of the Actinomycetes. Academic Press, London, UK.

Madigan, M., Martimko, J.M. 2006. Brock Biology of Microorganisms. Upper Saddle River: Pearson Prentice Hall. 11th Edition.

Major, A. 1975. Mushrooms Toadstools and Fungi: Arco New York, USA.

Narayanan, N., Pradip, K. Roychoudhury, P.K., Srivastava, A. 2004. L (+) lactic acid fermentation and its product polymerization. Electronic Journal of Biotechnology 7, Electronic Journal of Biotechnology [online]. 15 August 2004, 7, (3) [cited 23 March 2006]. Available from: http://www.ejbiotechnology.info/content/vol2/issue3/full/3/index.html. ISSN 0717-3458.

Samson, R., Pitt, J.I. 1989. Modern Concepts in Penicillium and Aspergillus Classification. Plenum Press New York and London.

Woese, C.R. 2002. On the evolution of cells Proceedings of the National Academy of Sciences of the United States of America 99, 8742-8747.

CHAPTER 3

Aspects of Molecular Biology and Bioinformatics of Relevance in Industrial Microbiology and Biotechnology

In recent times giant strides have been taken in harnessing our knowledge of the molecular basis of many biological phenomena. Many new techniques such as the polymerase chain reaction (PCR) and DNA sequencing have arrived on the scene. In addition major projects involving many countries such as the human genome project have taken place. Coupled with all these exciting technological developments, new vocabulary such as genomics has arisen. All this has transformed the approaches used in industrial microbiology. New approaches anchored on developments in molecular biology have been followed in many industrial microbiology processes and products such as vaccines, the search for new antibiotics, and the physiology of microorganisms. It therefore now appears imperative that any discussion of industrial microbiology and biotechnology must take these developments into account. This chapter will discuss only selected aspects of molecular biology in order to provide a background for understanding some of the newer directions of industrial microbiology and biotechnology. The discussion will be kept as simplified and as brief as possible, just enough in complexity and length needed to achieve the purpose of the chapter. The student is encouraged to look at many excellent texts in this field. In addition a glossary of some terms used in molecular biology is included at the end of the book.

3.1 PROTEIN SYNTHESIS

Proteins are very important in the metabolism of living things. They are in hormones for transporting messages around the body; they are used as storage such as in the whites of eggs of birds and reptiles and in seeds; they transport oxygen in the form of hemoglobin; they are involved in contractile arrangements which enable movement of various body parts, in contractile proteins in muscles; they protect the animal body in the form of antibodies; they are in membranes where they act as receptors, participate in membrane transport and antigens and they form toxins such as diphtheria and botulism. The most important function if it can be so termed is that form the basis of enzymes which catalyze all the metabolic activities of living things; in short proteins and the enzymes formed from them are the major engines of life.

In spite of the incredible diversity of living things, varying from bacteria to protozoa to algae to maize to man, the same 20 amino acids are found in all living things. On account of this, the principles affecting proteins and their structure and synthesis are same in all living things.

The genetic macromolecules (i.e. the macromolecules intimately linked to heredity) are deoxyribonucleic acid (DNA) and ribonucleic acid (RNA). The genetic information which determines the potential properties of a living thing is carried in the DNA present in the nucleus, except in some viruses where it is carried in RNA. DNA is also present in the organelles mitochondria and chloroplasts. (Just an interesting fact about mitochondrial DNA. Individuals inherit the other kinds of genes and DNA from both parents jointly. However, eggs destroy the mitochondria of the sperm that fertilize them. On account of this, the mitochondrial DNA of an individual comes exclusively from the mother. Due to the unique matrilineal transmission of mitochondrial DNA, data from mitochondrial DNA sequences is used in the study of genelogy and sometimes for forensic purposes).

DNA consists of four nucleotides, adenine, cytosine, guanine and thymine. RNA is very similar except that uracil replaces thymine (Fig. 3.1). RNA occurs in the nucleus and in the cytoplasm as well as in the ribosomes.

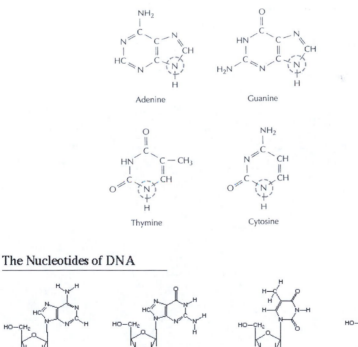

Adenine   Guanine

Thymine   Cytosine

The Nucleotides of DNA

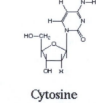

Adenine   Guanosine   Thymine   Cytosine

Purines   Pyrimidines

Fig. 3.1   The Nucleic Acid Bases

The processes of protein synthesis will be summarized briefly below. In protein synthesis, information flow is from DNA to RNA via the process of **transcription**, and thence to protein via **translation**. Transcription is the making of an RNA molecule from a DNA template. Translation is the construction of a polypeptide from an amino acid sequence from an RNA molecule (Fig. 3.2). The only exception to this is in retroviruses where reverse transcription occurs and where a single-stranded DNA is transcribed from a single-stranded RNA (the reverse of transcription); it is used by retroviruses, which includes the HIV/AIDS virus, as well as in biotechnology.

Transcription

An enzyme, RNA polymerase, opens the part of the DNA to be transcribed. Only one strand of DNA, the template or sense strand, is transcribed into RNA. The other strand,

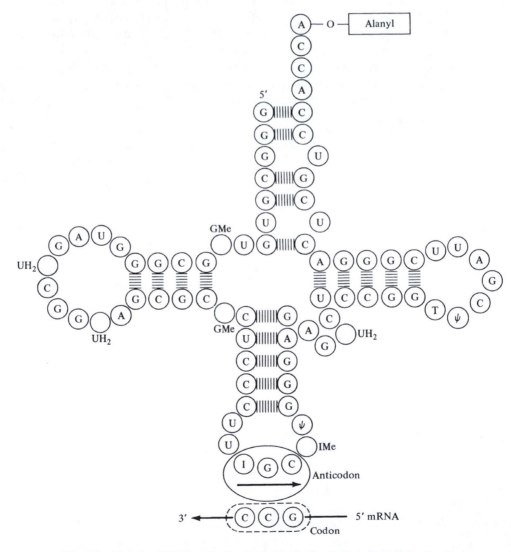

Fig. 3.2 Transfer RNA Transferring Amino Acids to mRNA in Protein Synthesis

the anti-sense strand is not transcribed. The anti-sense strand is used in making ripe tomatoes to remain hard. The RNA transcribed from the DNA is the messenger or mRNA (Fig. 3.3). As some students appear to be confused by the various types of RNA, it is important that we mention at this stage that there are two other types of RNA besides mRNA. These are ribosomal or rRNA and transfer or tRNA; they will be discussed later in this chapter. At this stage it is suffice to mention that in the analogy of a building, messenger RNA, mRNA is the blueprint or plan for construction of a protein (building); ribosomal RNA rRNA the construction site (plot of land) where the protein is made, while transfer RNA, tRNA, is the vehicle delivering the proper amino acid (building blocks) to the (building) site at the right time.

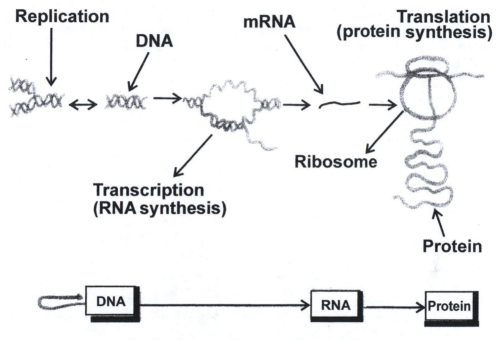

Fig. 3.3 Summary of Protein Synthesis Activities

When mRNA is formed, it leaves the nucleus in eukaryotes (there is no nucleus in prokaryotes!) and moves to the ribosomes.

Translation

In all cells, ribosomes are the organelles where proteins are synthesized. They consist of two-thirds of ribosomal RNA, rRNA, and one-third protein. Ribosomes consist of two sub-units, a smaller sub-unit and a larger sub-unit. In prokaryotes, typified by *E. coli* , the smaller unit is 30S and larger 50S. *S* is Svedberg units, the unit of weights determined from ultra centrifuge readings. The 30S unit has 16S rRNA and 21 different proteins. The 50S sub-unit consists of 5S and 23S rRNA and 34 different proteins. The smaller sub-unit has a binding site for the mRNA. The larger sub-unit has two binding sites for tRNA.

The *messenger RNA (mRNA)* is the 'blueprint' for protein synthesis and is transcribed from one strand of the DNA of the gene; it is translated at the ribosome into a polypeptide sequence. Translation is the synthesis of protein from amino acids on a template of messenger RNA in association with a ribosome. The bases on mRNA code for amino acids in triplets or codons; that is three bases code for an amino acid. Sometimes different triplet bases may code for the same amino acid. Thus the amino acid glycine is coded for by four different codons: GGU, GGC, GGA, and GGG. However, a codon usually codes for one amino acid. There are 64 different codons; three of these UAA, UAG, and UGA are stop codons and stop the process of translation. The remaining 61 code for the amino acids in proteins. (Table 3.1). Translation of the message generally begins at AUG, which also codes for methionine. For AUG to act as a start codon it must be preceded by a ribosome binding site. If that is not the case it simply codes for methionine.

Promoters are sequences of DNA that are the start signals for the transcription of mRNA. Terminators are the stop signals. mRNA molecules are long (500-10,000 nucleotides).

Ribosomes are the sites of translation. The ribosomes move along the mRNA and bring together the amino acids for joining into proteins by enzymes.

Table 3.1   The genetic code – codons

	U	C	A	G	
U	UUU = Phe	UCU = Ser	UAU = Tyr	UGU = Cys	U
	UUC = Phe	UCC = Ser	UAC = Tyr	UGC = Cys	C
	UUA = Leu	UCA = Ser	UAA = Stop	UAG = Stop	A
	UUG = Leu	UCG = Ser	UGA = Stop	UGG = Trp	G
C	CUU = Leu	CCU = Pro	CAU = His	CGU = Arg	U
	CUC = Leu	CCC = Pro	CAC = His	CGC = Arg	C
	CUA = Leu	CCA = Pro	CAA = Gln	CGA = Arg	A
	CUG = Leu	CCG = Pro	CAG = Gln	CGG = Arg	G
A	AUU = Ile	ACU = Thr	AAU = Asn	AGU = Ser	U
	AUC = Ile	ACC = Thr	AAC = Asn	AGC = Ser	C
	AUA = Ile	ACA = Thr	AAA = Lys	AGA = Arg	A
	AUG = Met	ACG = Thr	AAG = Lys	AGG = Arg	G
G	GUU = Val	GCU = Ala	GAU = Asp	GGU = Gly	U
	CUC = Val	GCC = Ala	GAC = Asp	GCG = Gly	C
	GUA = Val	GCA = Ala	GAA = Glu	GGA = Gly	A
	GUG = Val	GCG = Ala	GAG = Glu	GGG = Gly	G

AUG = start codon

UAA, UAG, and UGA = stop (nonsense) codons

Amino Acids

Phe = phenylalanine	Ser = serine	His = histidine	Glu = glutamic acid
Leu = leucine	Pro = proline	Gln = glutamine	Cys = cysteine
Ile = isoleucine	Thr = threonine	Asn = asparagine	Trp = tryptophan
Met = methionine	Ala = alanine	Lys = lysine	Arg = arginine
Val = valine	Tyr = tyrosine	Asp = aspartic acid	Gly = glycine

Transfer RNAs (tRNAs) carry amino acids to mRNA for linking and elongation into proteins. Transfer RNA is basically cloverleaf-shaped. (see Fig. 3.2) tRNA carries the proper amino acid to the ribosome when the codons call for them. At the top of the large loop are three bases, the anticodon, which is the complement of the codon. There are 61 different tRNAs, each having a different binding site for the amino acid and a different anticodon. For the codon UUU, the complementary anticodon is AAA. Amino acid linkage to the proper tRNA is controlled by the aminoacyl-tRNA synthetases. Energy for binding the amino acid to tRNA comes from ATP conversion to adenosine monophosphate (AMP).

Elongation terminates when the ribosome reaches a stop codon, which does not code for an amino acid and hence not recognized by tRNA.

After protein has been synthesized, the primary protein chain undergoes folding: secondary, tertiary and quadruple folding occurs. The folding exposes chemical groups which confer their peculiar properties to the protein.

Protein folding (to give a three-dimensional structure) is the process by which a protein assumes its functional shape or conformation. All protein molecules are simple unbranched chains of amino acids, but it is by coiling into a specific three-dimensional shape that they are able to perform their biological function. The three-dimensional shape (3D) conformation of a protein is of utmost importance in determining the properties and functions of the protein. Depending on how a protein is folded different functional groups may be exposed and these exposed group influence its properties.

The reverse of the folding process is protein denaturation, whereby a native protein is caused to lose its functional conformation, and become an amorphous, and non-functional amino acid chain. Denatured proteins may lose their solubility, and precipitate, becoming insoluble solids. In some cases, denaturation is reversible, and proteins may refold. In many other cases, however, denaturation is irreversible. Denaturation occurs when a protein is subjected to unfavorable conditions, such as unfavorable temperature or pH. Many proteins fold spontaneously during or after their synthesis inside cells, but the folding depends on the characteristics of their surrounding solution, including the identity of the primary solvent (either water or lipid inside the cells), the concentration of salts, the temperature, and molecular chaperones. Incorrect folding sometimes occurs and is responsible for prion related illness such as Creutzfeldt-Jakob disease and Bovine spongiform encephalopathy (mad cow disease), and amyloid related illnesses such as Alzheimer's Disease. When enzyme molecules are misfolded they will not function.

3.2 THE POLYMERASE CHAIN REACTION

The Polymerase Chain Reaction (PCR) is a technology used to amplify small amounts of DNA. The PCR technique was invented in 1985 by Kary B. Mullis while working as a chemist at the Cetus Corporation, a biotechnology firm in Emeryville, California. So useful is this technology that Muillis won the Nobel Prize for its discovery in 1993, eight years later. It has found extensive use in a wide range of situations, from the medical diagnosis to microbial systematics and from courts of law to the study of animal behavior.

The requirements for PCR are:

a. The DNA or RNA to be amplified
b. Two primers
c. The four nucleotides found in the nucleic acid,
d. A heat stable a thermostable DNA polymerase derived from the thermophilic bacterium, *Thermus aquaticus, Taq* polymerase

The Primer: A primer is a short segment of nucleotides which is complementary to a section of the DNA which is to be amplified in the PCR reaction.

Primers are anneal to the denatured DNA template to provide an initiation site for the elongation of the new DNA molecule. For PCR, primers must be duplicates of nucleotide sequences on either side of the piece of DNA of interest, which means that the exact order of the primers' nucleotides must already be known. These flanking sequences can be constructed in the laboratory or purchased from commercial suppliers.

The Procedure: There are three major steps in a PCR, which are repeated for 30 or 40 cycles. This is done on an automated cycler, which can heat and cool the tubes with the reaction mixture at specific intervals.

a. **Denaturation** at 94°C

The unknown DNA is heated to about 94°C, which causes the DNA to denature and the paired strands to separate.

b. **Annealing** at 54°C

A large excess of primers relative to the amount of DNA being amplified is added and the reaction mixture cooled to allow double-strands to anneal; because of the large excess of primers, the DNA single strands will bind more to the primers, instead of with each other.

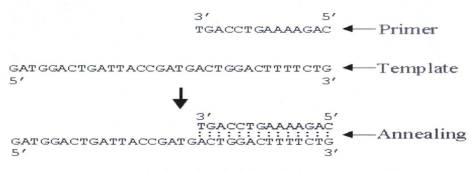

Fig. 3.4 Primer-Template Annealing

c. **Extension** at 72°C

This is the ideal working temperature for the polymerase. Primers that are on positions with no exact match, get loose again (because of the higher temperature) and donot give an extension of the fragment. The bases (complementary to the template) are coupled to the primer on the 3' side (the polymerase adds dNTP's from 5' to 3', reading the template from 3' to 5' side, bases are added complementary to the template).

d. **The Amplification:** The process of the amplification is shown in Fig. 3.3.

3.2.1 Some Applications of PCR in Industrial Microbiology and Biotechnology

PCR is extremely efficient and simple to perform. It is useful in biotechnology in the following areas:

(a) to generate large amounts of DNA for genetic engineering, or for sequencing, once the flanking sequences of the gene or DNA sequence of interest is known;

(b) to determine with great certainty the identity of an organism to be used in a biotechnological production, as may be the case when some members of a group of organisms may include some which are undesirable. A good example would be among the acetic acid bacteria where *Acetobacter xylinum* would produce slime rather acetic acid which *Acetobacter aceti* produces.

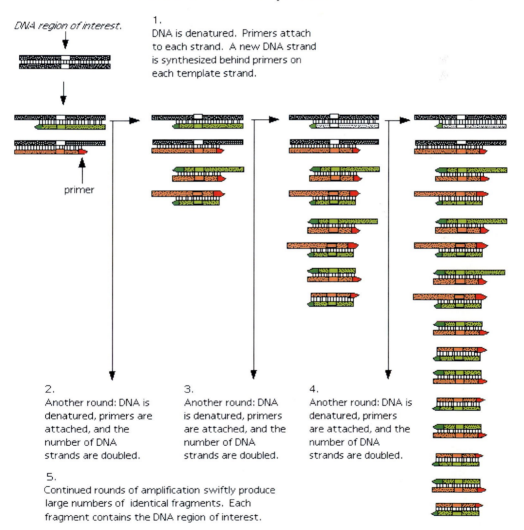

Fig. 3.5 Diagrammatic Representation of PCR

(c) PCR can be used to determine rapidly which organism is the cause of contamination in a production process so as to eliminate its cause, provided the primers appropriate to the contaminant is available.

3.3   MICROARRAYS

The availability of complete genomes from many organisms is a major achievement of biology. Aside from the human genome, the complete genomes of many microorganisms have been completed and are now available at the website of The Institue for Genomic Research (TIGR), a nonprofit organization located in Rockville, MD with its website at www.tigr.org. At the time of writing, TIGR had the complete genome of 294 microorganisms on its website (268 bacteria, 23 Archae, and 3 viruses). The major challenge is now to decipher the biological function and regulation of the sequenced genes. One technology important in studying functional microbial genomics is the use of DNA Microarrays.

Microarrays are microscopic arrays of large sets of DNA sequences that have been attached to a solid substrate using automated equipment. These arrays are also referred to as microchips, biochips, DNA chips, and gene chips. It is best to refer to them as microarrays so as to avoid confusing them with computer chips.

DNA microarrays are small, solid supports onto which the sequences from thousands of different genes are immobilized at fixed locations. The supports themselves are usually glass microscope slides; silicon chips or nylon membranes may also be used. The DNA is printed, spotted or actually directly synthesized onto the support mechanically at fixed locations or addresses. The spots themselves can be DNA, cDNA or oligonucleotides.

The process is based on hybridization probing. Single-stranded sequences on the microarray are labeled with a fluorescent tag or flourescein, and are in fixed locations on the support. In microarray assays an unknown sample is hybridized to an ordered array of immobilized DNA molecules of known sequence to produce a specific hybridization pattern that can be analyzed and compared to a given standard. The labeled DNA strand in solution is generally called the target, while the DNA immobilized on the microarray is the probe, a terminology opposite that used in Southern blot. Microarrays have the following advantages over other nucleic acid based approaches:

a. High through-put: thousands of array elements can be deposited on a very small surface area enabling gene expression to be monitored at the genomic level. Also many components of a microbial community can be monitored simultaneously in a single experiment.

b. High sensitivity: small amounts of the target and probe are restricted to a small area ensuring high concentrations and very rapid reactions.

c. Differential display: different target samples can be labeled with different fluorescent tags and then hybridized to the same microarray, allowing the simultaneous analysis of two or more biological samples.

d. Low background interference: non-specific binding to the solid surface is very low resulting in easy removal of organic and fluorescent compounds that attach to microarrays during fabrication.

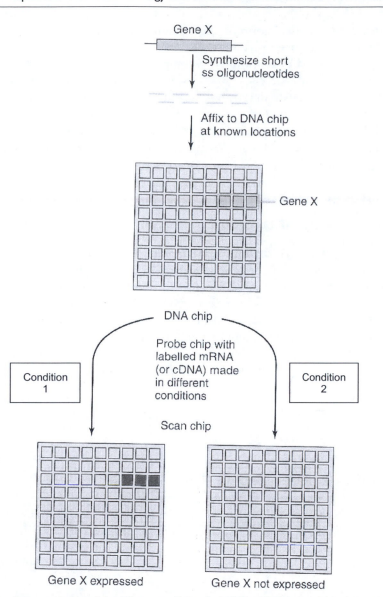

Gene X

Synthesize short
ss oligonucleotides

Affix to DNA chip
at known locations

Gene X

DNA chip

Probe chip with
labelled mRNA
(or cDNA) made
in different
conditions

Condition
1

Condition
2

Scan chip

Gene X expressed

Gene X not expressed

Fig. 3.6   Representation of the Microarray Procedure, (after Madigan and Martinko 2006)

e. Automation: microarray technology is amenable to automation making it
   ultimately cost-effective when compared with other nucleic acid technologies.

3.3.1   Applications of Microarray Technology

Microarray technology is still young but yet it has found use in a some areas which have
importance in microbiology in general as well as in industrial microbiology and
biotechnology, including disease diagnosis, drug discovery and toxicological research.

Microarrays are particularly useful in studying gene function. A microarray works by
exploiting the ability of a given mRNA molecule to bind specifically to, or hybridize to,
the DNA template from which it originated. By using an array containing many DNA

samples, it is possible to determine, in a single experiment, the expression levels of hundreds or thousands of genes within a cell by measuring the amount of mRNA bound to each site on the array. With the aid of a computer, the amount of mRNA bound to the spots on the microarray is precisely measured, generating a profile of gene expression in the cell. It is thus possible to determine the bioactive potential of a particular microbial metabolite as a beneficial material in the form of a drug or its deleterious effect.

When a diseased condition is identified through microarray studies, experiments can be designed which may be able to identify compounds, from microbial metabolites or other sources, which may improve or reverse the diseased condition.

3.4   SEQUENCING OF DNA

3.4.1   Sequencing of Short DNA Fragments

DNA sequencing is the determination of the precise sequence of nucleotides in a sample of DNA. Two methods developed in the mid-1970s are available: the Maxim and Gilbert method and the Sanger method. Both methods produce DNA fragments which are studied with gel electrophoresis. The Sanger method is more commonly used and will be discussed here. The Sanger method is also called the dideoxy method, or the enzymic method. The dideoxy method gets its name from the critical role played by synthetic analogues of nucleotides that lack the -OH at the 3' carbon atom (star position): dideoxynucleotide triphosphates (ddNTP) (Fig. 3.7). When (normal) deoxynucleotide

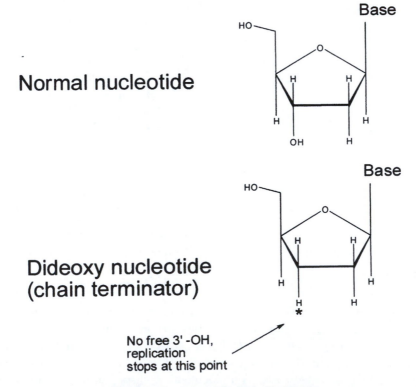

Fig. 3.7   Normal and Dedeoxy Nucleotides

triphosphates (dNTP) are used the DNA strand continues to grow, but when the dideoxy analogue is incorporated, chain elongation stops because there is no 3′ -OH for the next nucleotide to be attached to. For this reason, the dideoxy method is also called the chain termination method.

For Sanger sequencing, a single strand of the DNA to be sequenced is mixed with a primer, DNA polymerase I, an excess of normal nucleotide triphosphates and a limiting (about 5%) of the dideoxynucleotides labeled with a fluorescent dye, each ddNTP being labeled with a different fluorescent dye color. This primer will determine the starting point of the sequence being read, and the direction of the sequencing reaction. DNA synthesis begins with the primer and terminates in a DNA chain when ddNTP is incorporated in place of normal dNTP. As all four normal nucleotides are present, chain elongation proceeds normally until, by chance, DNA polymerase inserts a dideoxy nucleotide instead of the normal deoxynucleotide. The result is a series of fragments of varying lengths. Each of the four nucleotides is run separately with the appropriate ddNTP. The mix with the ddCTP produces fragments with C (cytosine); that with ddTTP (thymine) produces fragments with T terminals etc. The fluorescent strands are separated from the DNA template and electrophoresed on a polyacrilamide gel to separate them according their lengths. If the gel is read manually, four lanes are prepared, one for each of the four reaction mixes. The reading is from the bottom of the gel up, because the smaller the DNA fragment the faster it is on the gel. A picture of the sequence of the nucleotides can be read from the gel (Fig. 3.8). If the system is automated, all four are

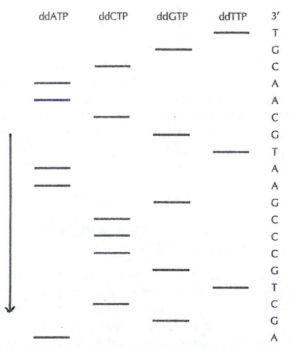

(Arrow shows direction of the electrophoresis. By convention the autoradiograph is read from bottom to the top).

Fig. 3.8 Diagram illustrating Autoradiograph of a Sequencing Gel of the Chain Terminating DNA Sequencing Method

mixed and electrphoresced together. As the ddNTPs are of different colors a scanner can scan the gel and record each color (nucleotide) separately. The sanger method is used for relatively short fragments of DNA, 700 -800 nucleotides. Methods for larger DNA fragments are described below.

3.4.2 Sequencing of Genomes or Large DNA fragments

The best example of the sequencing of a genome is perhaps that of the human genome, which was completed a few years ago. During the sequencing of the human genome, two approaches were followed: the use of bacterial artificial chromosomes (BACs) and the short gun approach.

3.4.2.1 Use of bacterial artificial chromosomes (BACs)

The publically-funded Human Genome Project, the National Institutes of Health and the National Science Foundation have funded the creation of 'libraries' of BAC clones. Each BAC carries a large piece of human genomic DNA of the order of 100-300 kb. All of these BACs overlap randomly, so that any one gene is probably on several different overlapping BACs. Those BACs can be replicated as many times as necessary, so there is a virtually endless supply of the large human DNA fragment. In the publically-funded project, the BACs are subjected to **shotgun sequencing** (see below) to figure out their sequence. By sequencing all the BACs, we know enough of the sequence in overlapping segments to reconstruct how the original chromosome sequence looks.

3.4.2.2 Use of the shot-gun approach

An innovative approach to sequencing the human genome was pioneered by a privately-funded sequencing project, Celera Genomics. The founders of this company realized that it might be possible to skip the entire step of making libraries of BAC clones. Instead, they blast apart the entire human genome into fragments of 2-10 kb and sequenced them. The challenge was to assemble those fragments of sequence into the whole genome sequence. It was like having hundreds of 500-piece puzzles, each being assembled by a team of puzzle experts using puzzle-solving computers. Those puzzles were like BACs - smaller puzzles that make a big genome manageable. Celera threw all those puzzles together into one room and scrambled the pieces. They, however, had scanners that scan all the puzzle pieces and used powerful computers to fit the pieces together.

3.5 THE OPEN READING FRAME AND THE IDENTIFICATION OF GENES

Regions of DNA that encode proteins are first transcribed into messenger RNA and then translated into protein. By examining the DNA sequence alone we can determine the putative sequence of amino acids that will appear in the final protein. In translation codons of three nucleotides determine which amino acid will be added next in the growing protein chain. The start codon is usually AUG, while the stop codons are UAA, UAG, and UGA. The open reading frame (ORF) is that portion of a DNA segment which will putatively code for a protein; it begins with a start codon and ends with a stop codon.

Once a gene has been sequenced it is important to determine the correct open reading frame. Every region of DNA has six possible reading frames, three in each direction because a codon consists of three nucleotides. The reading frame that is used determines which amino acids will be encoded by a gene. Typically only one reading frame is used in translating a gene (in eukaryotes), and this is often the longest open reading frame. Once the open reading frame is known the DNA sequence can be translated into its corresponding amino acid sequence.

For example, the sequence of DNA in Fig. 3.9 can be read in six reading frames. Three in the forward and three in the reverse direction. The three reading frames in the forward direction are shown with the translated amino acids below each DNA sequence. Frame **1** starts with the 'a', Frame **2** with the 't' and Frame **3** with the 'g'. Stop codons are indicated by an '*' in the protein sequence. The longest ORF is in Frame **1**.

```
  5'                                                                    3'
    atgcccaagctgaatagcgtagaggggttttcatcatttgaggacgatgtataa

1 atg ccc aag ctg aat agc gta gag ggg ttt tca tca ttt gag gac gat gta
                                                                      taa
    M   P   K   L   N   S   V   E   G   F   S   S   F   E   D   D   V
                                                          *
2 tgc cca agc tga ata gcg tag agg ggt ttt cat cat ttg agg acg atg tat
    C   P   S   *   I   A   *   R   G   F   H   H   L   R   T   M   Y
3  gcc caa gct gaa tag cgt aga ggg gtt ttc atc att tga gga cga tgt
```

Fig. 3.9 Sequence from a Hypothetical DNA Fragment

Genes can be identified in a number of ways, which are discussed below.

i. *Using computer programs*

As was shown above, the open reading frame (ORF) is deduced from the start and stop codons. In prokaryotic cells which do not have many extrons (intervening non-coding regions of the chromosome), the ORF will in most cases indicate a gene. However it is tedious to manually determine ORF and many computer programs now exist which will scan the base sequences of a genome and identify putative genes. Some of the programs are given in Table 3.2. In scanning a genome or DNA sequence for genes (that is, in searching for functional ORFs), the following are taken into account in the computer programs:

a. usually, functional ORFs are fairly long and are do not usually contain less than 100 amino acids (that is, 300 amino acids);

b. if the types of codons found in the ORF being studied are also found in known functional ORFs, then the ORF being studied is likely to be functional;

c. the ORF is also likely to be functional if its sequences are similar to functional sequences in genomes of other organisms;

d. in prokaryotes, the ribosomal translation does not start at the first possible (earliest 5') codon. Instead it starts at the codon immediately down stream of the Shine-Dalgardo binding site sequences. The Shine-Dalgardo sequence is a short sequence of nucleotides upstream of the translational start site that binds to

Table 3.2 Some Internet tools for the gene discovery in DNA sequence bases (modified from Fickett, (1996).

Category	Services	Organism(s)	Web address
Database search	BLAST; search sequence bases	Any	blast@ncbi.nlm.nih.gov
	FASTA; search sequence bases	Any	fasta@ebi.ac.uk
	BLOCKS; search for functional motifs	Any	blocks@howard.flicr.org
	Profilescan	Any	http://ulrec3.unil.ch.
	MotifFinder	Any	motif@genome.ad.jp
Gene Identification	FGENEH; integrated gene identification	Human	service@theory.bchs.uh.edu
	GeneID; integrated gene identification	Vetebrate	geneid@bircebd.uwf.edu
	GRAIL; integrated gene identification	Human	grail@ornl.gov
	EcoParse; integrated gene identification	*Escherichia coli*	

ribosomal RNA and thereby brings the ribosome to the initiation codon on the mRNA. The computer program searches for a Shine-Dalgardo sequence and finding it helps to indicate not only which start codon is used, but also that the ORF is likely to be functional.

e. if the ORF is preceded by a typical promoter (if consensus promoter sequences for the given organism are known, check for the presence of a similar upstream region)

f. if the ORF has a typical GC content, codon frequency, or oligonucleotide composition of known protein-coding genes from the same organism, then it is likely to be a functional ORF.

ii. *Comparison with Existing Genes*

Sometimes it may be possible to deduce not only the functionality or not of a gene (i.e. a functional ORF), but also the function of a gene. This can done by comparing an unknown sequence with the sequence of a known gene available in databases such as The Institute for Genomic Research (TIGR) in Maryland.

3.6 METAGENOMICS

Metagenomics is the genomic analysis of the collective genome of an assemblage of organisms or 'metagenome.' Metagenomics describes the functional and sequence-based analysis of the collective microbial genomes contained in an environmental sample (Fig. 3.10). Other terms have been used to describe the same method, including environmental DNA libraries, zoolibraries, soil DNA libraries, eDNA libraries, recombinant environmental libraries, whole genome treasures, community genome, whole genome shotgun sequencing. The definition applied here excludes studies that use PCR to amplify gene cassettes or random PCR primers to access genes of interest since these methods do not provide genomic information beyond the genes that are amplified.

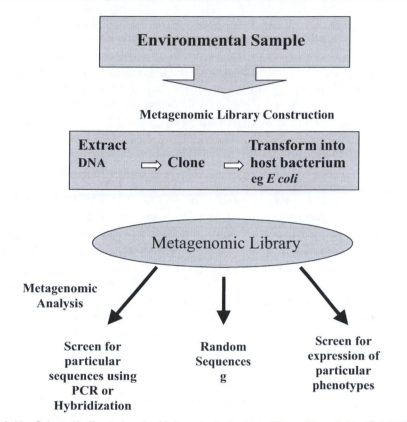

Fig. 3.10   Schematic Procedure for Metagenomic Analysis (From: Riesenfeld, et al (2004))

Many environments have been the focus of metagenomics, including soil, the oral cavity, feces, and aquatic habitats, as well as the hospital metagenome a term intended to encompass the genetic potential of organisms in hospitals that contribute to public health concerns such as antibiotic resistance and nosocomial infections.

Uncultured microorganisms comprise the majority of the planet earth's biological diversity. In many environments, as many as 99% of the microorganisms cannot be cultured by standard techniques, and the uncultured fraction includes diverse organisms that are only distantly related to the cultured ones. Therefore, culture-independent methods are essential to understand the genetic diversity, population structure, and ecological roles of the majority of microorganisms in a given environmental situation. Metagenomics, or the culture-independent genomic analysis of an assemblage of microorganisms, has potential to answer fundamental questions in microbial ecology. It can also be applied to determining organisms which may be important in a new industrial process still under study. Several markers have been used in metagenomics, including 16S mRNA, and the genes encoding DNA polymerases, because these are highly conserved (i.e., because they remain relatively unchanged in many groups). The marker most commonly used however is the sequence of 16S mRNA. The procedure in metagonomics is described in Fig. 3.10.

Its potential application in biotechnology and industrial microbiology is that it can facilitate the identification of uncultured organisms whose role in a multi-organism environment such as sewage or the degradation of a recalcitrant chemical soil may be hampered because of the inability to culture the organism. Indeed a method has been patented for isolating organisms of pharmaceutical importance from uncultured organisms in the environment. This is discussed in detail in Chapter 28 where approaches to drug discovery are discussed.

3.7 NATURE OF BIOINFORMATICS

Bioinformatics is a new and evolving science and may be defined as the use of computers to store, compare, retrieve, analyze, predict, or simulate the composition or the structure of the genetic macromolecules, DNA and RNA and their major product, proteins. Important research efforts in bioinformatics include sequence alignment, gene finding, genome assembly, protein structure alignment, protein structure prediction, prediction of gene expression and protein-protein interactions, and the modeling of evolution. Bioinformatics uses mathematical tools to extract useful information from a variety of data produced by high-throughput biological techniques. Examples of succesful extraction of orderly information from a 'forest' of seemingly chaotic information include the assembly of high-quality DNA sequences from fragmentary 'shotgun' DNA sequencing, and the prediction of gene regulation with data from mRNA microarrays or mass spectrometry.

The increased role in recent times of bioinformatics in biotechnology is due to a vast increase in computation speed and memory storage capability, making it possible to undertake problems unthinkable without the aid of computers. Such problems include large-scale sequencing of genomes and management of large integrated databases over the Internet. This improved computational capability integrated with large-scale miniaturization of biochemical techniques such as PCR, BAC, gel electrophoresis, and microarray chips has delivered enormous amount of genomic and proteomic data to the researchers. The result is an explosion of data on the genome and proteome analysis leading to many new discoveries and tools that are not possible in wet-laboratory experiments. Thus, hundreds of microbial genomes and many eukaryotic genomes including a cleaner draft of human genome have been sequenced raising the expectation of better control of microorganisms. Bioinformatics has been used in the following four areas:

a. *genomics* – sequencing and comparative study of genomes to identify gene and genome functionality;
b. *proteomics* – identification and characterization of protein related properties and reconstruction of metabolic and regulatory pathways;
c. cell visualization and simulation to study and model cell behavior; and
d. application to the development of drugs and anti-microbial agents.

The potential gains especially following from sequencing of the human genome and many microorganisms are greater understanding of the genetics of microorganisms and their subsequent improved control leading to better diagnosis of the diseases through the use of protein biomarkers, protection against diseases using cost effective vaccines and rational drug design, and improvement in agricultural quality and quantity. Some of these are discussed in Chapter 28 under the heading of drug discovery.

3.7.1   Some Contributions of Bioinformatics to Biotechnology

Some contributions made by bioinformatics to biotechnology include automatic genome sequencing, automatic identification of genes, identification of gene function, predicting the 3D structure modeling and pair-wise comparison of genomes.

i. Automatic genome sequencing

The major contribution of the bioinformatics in genome sequencing has been in the: (i) development of automated sequencing techniques that integrate the PCR or BAC based amplification, 2D gel electrophoresis and automated reading of nucleotides, (ii) joining the sequences of smaller fragments (contigs) together to form a complete genome sequence, and (iii) the prediction of promoters and protein coding regions of the genome.

PCR (Polymerase Chain Reaction) or BAC (Bacterial Artificial Chromosome)-based amplification techniques derive limited size fragments of a genome. The available fragment sequences suffer from nucleotide reading errors, repeats – very small and very similar fragments that fit in two or more parts of a genome, and chimera – two different parts of the genome or artifacts caused by contamination that join end-to-end giving a artifactual fragment. Generating multiple copies of the fragments, aligning the fragments, and using the majority voting at the same nucleotide positions solve the nucleotide reading error problem. Multiple experimental copies are needed to establish repeats and chimeras. Chimeras and repeats are removed before the final assembly of the genome-fragments. Using mathematical models, the fragments are joined. To join contigs, the fragments with larger nucleotide sequence overlap are joined first.

ii. Automated Identification of Genes

After the contigs are joined, the next issue is to identify the protein coding regions or ORFs (open reading frames) in the genomes. The identification of ORFs is based on the principles described earlier. The two programs which are used are GLIMMER and GenBank.

iii. Identifying gene function: searching and alignment

After identifying the ORFs, the next step is to annotate the genes with proper structure and function. The function of the gene has been identified using popular sequence search and pair-wise gene alignment techniques. The four most popular algorithms used for functional annotation of the genes are BLAST, BLOSUM, ClustalX, and SMART

iv. Three-dimensional (3D) structure modeling

A protein may exist under one or more conformational states depending upon its interaction with other proteins. Under a stable conformational state certain regions of the protein are exposed for protein-protein or protein-DNA interactions. Since the function is also dependent upon exposed active sites, protein function can be predicted by matching the 3D structure of an unknown protein with the 3D structure of a known protein. With bioinformatics it is possible to predict the possible conformations of the protein coded for by a gene and therefore the function of the protein.

v. Pair-wise genome comparison

After the identification of gene-functions, a natural step is to perform pair-wise genome comparisons. Pair-wise genome comparison of a genome against itself provides the details of paralogous genes – duplicated genes that have similar sequence with some variation in function. Pair-wise genome comparisons of a genome against other genomes have been used to identify a wealth of information such as orthololologous genes – functionally equivalent genes diverged in two genomes due to speciation, different types of gene-groups – adjacent genes that are constrained to occur in close proximity due to their involvement in some common higher level function, lateral gene-transfer – gene transfer from a microorganism that is evolutionary distant, gene-fusion/gene-fission, gene-group duplication, gene-duplication, and difference analysis to identify genes specific to a group of genomes such as pathogens, and conserved genes.

In conclusion, despite the recent emergence of bioinformatics it is already making big impacts on biotechnology. Except for the availability of bioinformatics techniques, the vast amount of data generated by genome sequencing projects would be unmanageable and would not be interpreted due to the lack of expert manpower and due to the prohibitive cost of sustaining such an effort. In the last decade bioinformatics has silently filled in the role of cost effective data analysis. This has quickened the pace of discoveries, the drug and vaccine design, and the design of anti-microbial agents. The major impact of bioinformatics in microbiology and biotechnology has been in automating microbial genome sequencing, the development of integrated databases over the Internet, and analysis of genomes to understand gene and genome function. Programs exist for comparing gene-pair alignments, which become the first steps to derive the gene-function and the functionality of genomes. Using bioinformatics techniques it is now possible to compare genomes so as to (i) identify conserved function within a genome family; (ii) identify specific genes in a group of genomes; and (iii) model 3D structures of proteins and docking of biochemical compounds and receptors. These have direct impact in the development of antimicrobial agents, vaccines, and rational drug design.

SUGGESTED READINGS

Bansal, K.A. 2005 Bioinformatics in microbial biotechnology – a mini review, Microbial Cell Factories 2005, 4, 19-30.

Dorrel, N., Champoin, O.L., Wren, B.W. 2002. Application of DNA Microarray for Comparative and Evolutionary Genomics In: Methods in Microbiology. Vol 33, Academic Press Amsterdam; the Netherlands pp. 83–99.

Handelsman, J., Liles, M., Mann, D., Riesenfeld, C., Goodman, R.M. 2002. In: Methods in Microbiology. Vol 33, Academic Press Amsterdam; the Netherlands pp. 242–255.

Hinds, J., Liang, K.G., Mangan, J.A., Butecer, P.D. 2002. Glass Slide Microarrays for Bacterial Genomes. In: Methods in Microbiology. Vol 33, Academic Press Amsterdam; the Netherlands 83–99.

Hinds, J., Witney, A.A., Vaas, J.K. 2002. Microarray Design for Bacterial Genomes. In: Methods in Microbiology. Vol 33, Academic Press Amsterdam; the Netherlands, 67-82.

Madigan, M., Martinko, J.M. 2006. Brock Biology of Microorganisms 11th ed. Pearson Prentice Hall, Upper Saddle River, USA.

Manyak, D.M., Carlson, P.S. 1999. Combinatorial Genomics™: New tools to access microbial chemical diversity In: Microbial Biosystems: New Frontiers, C.R. Bell, M. Brylinsky, P. Johnson-Green, (eds) Proceedings of the 8th International Symposium on Microbial Ecology Atlantic Canada Society for Microbial Ecology, Halifax, Canada, 1999.

Priest, F., Austin, B. 1993. Modern Bacterial Taxonomy. Chapman and Hall. London, UK.

Riesenfeld, C.S., Schloss, P.D., Handelsman, 2004. Metagenomics: Genomic Analysis of Microbial Communities. Annual Review of Genetics 38, 525-52.

Rogic, S., Mackworth, A.K., Ouellette, F.B.F. 2001. Evaluation of Gene-Finding Programs on Mammalian Sequences Genome Research 11, 817-832.

Whitford, D. 2005. Proteins: Structure and Function. John Wiley and Sons Chichester, UK.

Zhou, J. 2002. Microarrays: Applications in Environmental Microbiology. In: Encyclopedia of Environmental Microbiology Vol 4. Wiley Interscience, New York USA. pp. 1968-1979.

CHAPTER 4

Industrial Media and the Nutrition of Industrial Organisms

The use of a good, adequate, and industrially usable medium is as important as the deployment of a suitable microorganism in industrial microbiology. Unless the medium is adequate, no matter how innately productive the organism is, it will not be possible to harness the organism's full industrial potentials. Indeed not only may the production of the desired product be reduced but toxic materials may be produced. Liquid media are generally employed in industry because they require less space, are more amenable to engineering processes, and eliminate the cost of providing agar and other solid agents.

4.1 THE BASIC NUTRIENT REQUIREMENTS OF INDUSTRIAL MEDIA

All microbiological media, whether for industrial or for laboratory purposes must satisfy the needs of the organism in terms of carbon, nitrogen, minerals, growth factors, and water. In addition they must not contain materials which are inhibitory to growth. Ideally it would be essential to perform a complete analysis of the organism to be grown in order to decide how much of the various elements should be added to the medium. However, approximate figures for the three major groups of heterotrophic organisms usually grown on an industrial scale are available and may be used in such calculations (Table 4.1).

Carbon or energy requirements are usually met from carbohydrates, notably (in laboratory experiments) from glucose. It must be borne in mind that more complex carbohydrates such as starch or cellulose may be utilized by some organisms. Furthermore, energy sources need not be limited to carbohydrates, but may include hydrocarbons, alcohols, or even organic acids. The use of these latter substrates as energy sources is considered in Chapters 15 and 16 where single cell protein and yeast productions are discussed.

In composing an industrial medium the carbon content must be adequate for the production of cells. For most organisms the weight of organism produced from a given weight of carbohydrates (known as the yield constant) under aerobic conditions is about

Table 4.1   Average composition of microorganisms (% dry weight)

Component	Bacteria	Yeast	Molds
Carbon	48 (46-52)	48 (46-52)	48 (45-55)
Nitrogen	12.5 (10-14)	7.5 (6-8.5)	6 (4-7)
Protein	55 (50 –60)	40 (35-45)	32 (25-40)
Carbohydrates	9 (6-15)	38 (30-45)	49 (40-55)
Lipids	7 (5-10)	8 (5-10)	8 (5-10)
Nucleic Acids	23 (15-25)	8 (5-10)	5 (2-8)
Ash	6 (4-10)	6 (4-10)	4 (4-10)

Minerals (same for all three organisms)	
Phosphorus	1.0 - 2.5
Sulfur, magnesium	0.3 - 1.0
Potassium, sodium	0.1 - 0.5
Iron	0.01 - 0.1
Zinc, copper, manganese	0.001 – 0.01

0.5 gm of dry cells per gram of glucose. This means that carbohydrates are at least twice the expected *weight* of the cells and must be put as glucose or its equivalent compound.

Nitrogen is found in proteins including enzymes as well as in nucleic acids hence it is a key element in the cell. Most cells would use ammonia or other nitrogen salts. The quantity of nitrogen to be added in a fermentation can be calculated from the expected cell mass and the average composition of the micro-organisms used. For bacteria the average N content is 12.5%. Therefore to produce 5 gm of bacterial cells per liter would require about 625 mg N (Table 4.1).

Any nitrogen compound which the organism cannot synthesize must be added.

Minerals form component portions of some enzymes in the cell and must be present in the medium. The major mineral elements needed include P, S, Mg and Fe. Trace elements required include manganese, boron, zinc, copper and molybdenum.

Growth factors include vitamins, amino acids and nucleotides and must be added to the medium if the organism cannot manufacture them.

Under laboratory conditions, it is possible to meet the organism's requirement by the use of purified chemicals since microbial growth is generally usually limited to a few liters. However, on an industrial scale, the volume of the fermentation could be in the order of thousands of liters. Therefore, pure chemicals are not usually used because of their high expense, unless the cost of the finished material justifies their use. Pure chemicals are however used when industrial media are being developed at the laboratory level. The results of such studies are used in composing the final industrial medium, which is usually made with unpurified raw materials. The extraneous materials present in these unpurified raw materials are not always a disadvantage and may indeed be responsible for the final and distinctive property of the product. Thus, although alcohol appears to be the desired material for most beer drinkers, the other materials extraneous to the maltose (from which yeasts ferment alcohol) help confer on beer its distinctive flavor (Chapter 12).

4.2 CRITERIA FOR THE CHOICE OF RAW MATERIALS USED IN INDUSTRIAL MEDIA

In deciding the raw materials to be used in the production of given products using designated microorganism(s) the following factors should be taken into account.

(a) Cost of the material

The cheaper the raw materials the more competitive the selling price of the final product will be. No matter, therefore, how suitable a nutrient raw materials is, it will not usually be employed in an industrial process if its cost is so high that the selling price of the final product is not economic. Thus, although lactose is more suitable than glucose in some processes (e.g. penicillin production) because of the slow rate of its utilization, it is usually replaced by the cheaper glucose. When used, glucose is added only in small quantities intermittently in order to decelerate acid production. Due to these economic considerations the raw materials used in many industrial media are usually waste products from other processes. Corn steep liquor and molasses are, for example, waste products from the starch and sugar industries, respectively. They will be discussed more fully below.

(b) Ready availability of the raw material

The raw material must be readily available in order not to halt production. If it is seasonal or imported, then it must be possible to store it for a reasonable period. Many industrial establishments keep large stocks of their raw materials for this purpose. Large stocks help beat the ever rising cost of raw materials; nevertheless large stocks mean that money which could have found use elsewhere is spent in constructing large warehouses or storage depots and in ensuring that the raw materials are not attacked during storage by microorganisms, rodents, insects, etc. There is also the important implication, which is not always easy to realize, that the material being used must be capable of long-term storage without concomitant deterioration in quality.

(c) Transportation costs

Proximity of the user-industry to the site of production of the raw materials is a factor of great importance, because the cost of the raw materials and of the finished material and hence its competitiveness on the market can all be affected by the transportation costs. The closer the source of the raw material to the point of use the more suitable it is for use, if all other conditions are satisfactory.

(d) Ease of disposal of wastes resulting from the raw materials

The disposal of industrial waste is rigidly controlled in many countries. Waste materials often find use as raw materials for other industries. Thus, spent grains from breweries can be used as animal feed. But in some cases no further use may be found for the waste from an industry. Its disposal especially where government regulatory intervention is rigid could be expensive. When choosing a raw material therefore the cost, if any, of treating its waste must be considered.

(e) Uniformity in the quality of the raw material and ease of standardization

The quality of the raw material in terms of its composition must be reasonably constant in order to ensure uniformity of quality in the final product and the satisfaction of the customer and his/her expectations. In cases where producers are plentiful, they usually compete to ensure the maintenance of the constant quality requirement demanded by the user. Thus, in the beer industry information is available on the quality of the barley malt before it is purchased. This is because a large number of barley malt producers exist, and the producers attempt to meet the special needs of the brewery industry, their main customer. On the other hand molasses, which is a major source of nutrient for industrial microorganisms, is a by product of the sugar industry, where it is regarded as a waste product. The sugar industry is not as concerned with the constancy of the quality of molasses, as it is with that of sugar. Each batch of molasses must therefore be chemically analyzed before being used in a fermentation industry in order to ascertain how much of the various nutrients must be added. A raw material with extremes of variability in quality is clearly undesirable as extra costs are needed, not only for the analysis of the raw material, but for the nutrients which may need to be added to attain the usual and expected quality in the medium.

(f) Adequate chemical composition of medium

As has been discussed already, the medium must have adequate amounts of carbon, nitrogen, minerals and vitamins in the appropriate quantities and proportions necessary for the optimum production of the commodity in question. The demands of the microorganisms must also be met in terms of the compounds they can utilize. Thus most yeasts utilize hexose sugars, whereas only a few will utilize lactose; cellulose is not easily attacked and is utilized only by a limited number of organisms. Some organisms grow better in one or the other substrate. Fungi will for instance readily grow in corn steep liquor while actinomycetes will grow more readily on soya bean cake.

(g) Presence of relevant precursors

The raw material must contain the precursors necessary for the synthesis of the finished product. Precursors often stimulate production of secondary metabolites either by increasing the amount of a limiting metabolite, by inducing a biosynthetic enzyme or both. These are usually amino acids but other small molecules also function as inducers. The nature of the finished product in many cases depends to some extent on the components of the medium. Thus dark beers such as stout are produced by caramelized (or over-roasted) barley malt which introduce the dark color into these beers. Similarly for penicillin G to be produced the medium must contain a phenyl compound. Corn steep liquor which is the standard component of the penicillin medium contains phenyl precursors needed for penicillin G. Other precursors are cobalt in media for Vitamin B^{12} production and chlorine for the chlorine containing antibiotics, chlortetracycline, and griseofulvin (Fig. 4.1).

(h) Satisfaction of growth and production requirements of the microorganisms

Many industrial organisms have two phases of growth in batch cultivation: the phase of growth, or the **trophophase**, and the phase of production, or the **idiophase**. In the first phase cell multiplication takes place rapidly, with little or no production of the desired

Vitamin B^{12}	Chlortetracycline
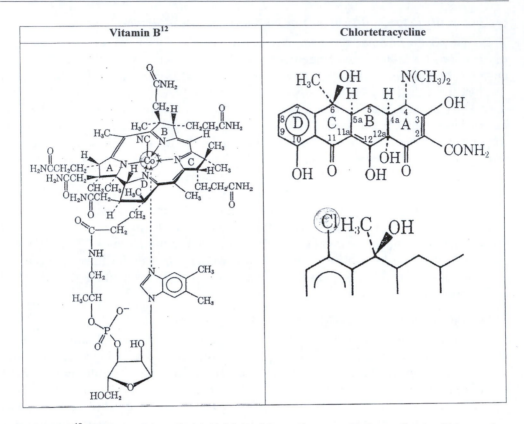	

Left: Vitamin B^{12}. Please note that cobalt is highlighted. It must be present in the medium in which organisms producing the vitamin are grown

Right: Top; The general structure of tetracyclines. Bottom; The structure of 7-Chlortetracycline; a chlorine atom is present in position 7. Chlorine must be present in the medium for producing chlortetracyline; note that chlorine is highlighted in position 7 in chlortetracycline.

Fig. 4.1 Vitamin B^{12} and Chlortetracycline Showing Location of Components Present as Precursors in the Medium

material. It is in the second phase that production of the material takes place, usually with no cell multiplication and following the elaboration of new enzymes. Often these two phases require different nutrients or different proportions of the same nutrients. The medium must be complete and be able to cater for these requirements. For example high levels of glucose and phosphate inhibit the onset of the idiophase in the production of a number of secondary metabolites of industrial importance. The levels of the components added must be such that they do not adversely affect production. Trophophase-idiophase relationship and secondary metabolites are discussed in detail in Chapter 5.

4.3 SOME RAW MATERIALS USED IN COMPOUNDING INDUSTRIAL MEDIA

The raw materials to be discussed are used because of the properties mentioned above: cheapness, ready availability, constancy of chemical quality, etc. A raw material which is cheap in one country or even in a different part of the same country may however not be cheap in another, especially if it has already found use in some other production process. In such cases suitable substitutes must be found if the goods must be produced in the new location. The use of local substitutes where possible is advantageous in reducing the transportation costs and even creating some employment in the local population. Prior experimentation may however be necessary if such new local materials differ substantially in composition from those already being used. Some well-known raw materials will now be discussed. In addition, some of potential useability will also be examined.

(a) Corn steep liquor

This is a by-product of starch manufacture from maize. Sulfur dioxide is added to the water in which maize is steeped. The lowered pH inhibits most other organisms, but encourages the development of naturally occurring lactic acid bacteria especially homofermentative thermophilic *Lactobacillus* spp. which raise the temperature to 38-55°C. Under these conditions, much of the protein present in maize is converted to peptides which along with sugars leach out of the maize and provide nourishment for the lactic acid bacteria. Lactic fermentation stops when the SO_2 concentration reaches about 0.04% and the concentration of lactic acid between 1.0 and 1.5%. At this time the pH is about 4. Acid conditions soften the kernels and the resulting maize grains mill better while the gel-forming property of the starch is not hindered. The supernatant drained from the maize steep is corn steep liquor. Before use, the liquor is usually filtered and concentrated by heat to about 50% solid concentration. The heating process kills the bacteria.

As a nutrient for most industrial organisms corn steep liquor is considered adequate, being rich in carbohydrates, nitrogen, vitamins, and minerals. Its composition is highly variable and would depend on the maize variety, conditions of steeping, extent of boiling etc. The composition of a typical sample of corn steep liquor is given in Table 4.2. As corn steep liquor is highly acidic, it must be neutralized (usually with $CaCO_3$) before use.

(b) Pharmamedia

Also known as proflo, this is a yellow fine powder made from cotton-seed embryo. It is used in the manufacture of tetracycline and some semi-synthetic penicillins. It is rich in

Table 4.2   Approximate composition of corn steep liquor (%)

Lactose	3.0-4.0
Glucose	0.-0.5
Non-reducing carbohydrates (mainly starch)	1.5
Acetic acid	0.05
Glucose lactic acid	0.5
Phenylethylamine	0.05
Amino aids (peptides, mines)	0.5
Total solids	80-90
Total nitrogen	0.15-0.2%

protein, (56% w/v) and contains 24% carbohydrate, 5% oil, and 4% ash, the last of which is rich in calcium, iron, chloride, phosphorous, and sulfate.

(c) Distillers solubles

This is a by-product of the distillation of alcohol from fermented grain. It is prepared by filtering away the solids from the material left after distilling fermented cereals (maize or barley) for whiskey or grain alcohol. The filtrate is then concentrated to about one-third solid content to give a syrup which is then drum-dried to give distillers soluble. It is rich in nitrogen, minerals, and growth factors (Table 4.3).

Table 4.3   Composition of maize distillers soluble

	%
Moisture	5
Protein	27
Lipid	9
Fibre	5
Carbohydrate	43
Ash (mainly K, Na, Mg, CO_3, and P)	11

(d) Soya bean meal

Soya beans (soja) (*Glycine max*), is an annual legume which is widely cultivated throughout the world in tropical, sub-tropical and temperate regions between 50°N and 40°S. The seeds are heated before being extracted for oil that is used for food, as an anti-foam in industrial fermentations, or used for the manufacture of margarine. The resulting dried material, soya bean meal, has about 11% nitrogen, and 30% carbohydrate and may be used as animal feed. Its nitrogen is more complex than that found in corn steep liquor and is not readily available to most microorganisms, except actinomycetes. It is used particularly in tetracycline and streptomycin fermentations.

(e) Molasses

Molasses is a source of sugar, and is used in many fermentation industries including the production of potable and industrial alcohol, acetone, citric acid, glycerol, and yeasts. It is a by-product of the sugar industry. There are two types of molasses depending on whether the sugar is produced from the tropical crop, sugar cane (*Saccharum officinarum*) or the temperate crop, beet, (*Beta alba*).

Four stages are involved in the manufacture of cane sugar. After *crushing*, a clear greenish dilute sugar solution known as 'mixed juice' is expressed from the canes. During the second stage known as *clarification* the mixed juice is heated with lime. Addition of lime changes the pH of the juice to alkaline and thus stops further hydrolysis (or inversion) or the cane sugar (sucrose), while heating coagulates proteins and other undesirable soluble portions of the mixed juice to form 'mud'. The supernatant juice is then *concentrated* (in the third stage) by heating under high vacuum and increasing low pressures in a series of evaporators. In the fourth and final stage of *crystallization*, sugar crystals begin to form with increasing heat and under vacuum, yielding a thick brown syrup which contains the crystals, and which is known as 'massecuite'. (In the beet industry it is known as 'fillmass'.) The massecuite is centrifuged to remove the sugar crystals and the remaining liquid is known as molasses. The first sugar so collected is 'A' and the liquid is 'A' molasses. 'A' molasses is further boiled to extract sugar crystals to yield 'B' sugar and 'B' molasses. Two or more boilings may be required before it is no longer profitable to attempt further extractions. This final molasses is known as 'blackstrap molasses'.

The sugar yielded with the production of black strap molasses is low-grade and brown in color, and known as raw sugar, cargo sugar, or refining sugar. This raw sugar is further refined, in a separate factory, to remove miscellaneous impurities including the brown color (due to caramel) to yield the white sugar used at the table. The heavy liquid discarded from the refining of sugar is known in the sugar refining industry as 'syrup' and corresponds to molasses in the raw sugar industry.

The above description has been of cane sugar molasses. In the beet sugar industry the processes used in raw refined sugar manufacture are similar, but the names of the different fractions recovered during purification differ. Cane and beet molasses differ slightly in composition (Table 4.4). Beet molasses is alkaline while cane molasses is acid.

Table 4.4   Average composition of beet and cane molasses

	Beet Molasses % (W/W)	Cane Molasses % (W/W)
Water	16.5	20.0
Sugars:	53.0	64.0
Sucrose	51.0	32.0
Fructose	1.0	15.0
Glucose	–	14.0
Raffinose	1.0	–
Non-sugar (nitrogeneous		
Materials, acids, gums, etc.)	19.0	10.0
Ash	11.5	8.0

Even within same type of molasses – beet or cane – composition varies from year to year and from one locality to another. The user industry selects the batch with a suitable composition and usually buys up a year's supply. For the production of cells the variability in molasses quality is not critical, but for metabolites such as citric acid, it is very important as minor components of the molasses may affect the production of these metabolites.

'High test' molasses (also known as inverted molasses) is a brown thick syrup liquid used in the distilling industry and containing about 75% total sugars (sucrose and reducing sugars) and about 18% moisture. Strictly speaking, it is not molasses at all but invert sugar, (i.e reducing sugars resulting from sucrose hydrolysis). It is produced by the hydrolysis of the concentrated juice with acid. In the so-called Cuban method, invertase is used for the hydrolysis. Sometimes 'A' sugar may be inverted and mixed with 'A' molasses.

(f) Sulfite liquor

Sulfite liquor (also called waste sulfite liquor, sulfite waste liquor or spent sulfite liquor) is the aqueous effluent resulting from the sulfite process for manufacturing cellulose or pulp from wood. Depending on the type, most woods contain about 50% cellulose, about 25% lignins and about 25% of hemicelluloses. During the sulfite process, hemicelluloses hydrolyze and dissolve to yield the hexose sugars, glucose, mannose, galactose, fructose and the pentose sugars, xylose, and arabinsoe. The acid reagent breaks the chemical bonds between lignin and cellulose; subsequently they dissolve the lignin. Depending on the severity of the treatment some of the cellulose will continue to exist as fibres and can be recovered as pulp. The presence of calcium ions provides a buffer and helps neutralize the strong lignin sulfonic acid. The degradation of cellulose yields glucose. Portions of the various sugars are converted to sugar sulfonic acids, which are not fermentable. Variable but sometimes large amounts of acetic, formic and glactronic acids are also produced.

Sulfite liquor of various compositions are produced, depending on the severity of the treatment and the type of wood. The more intense the treatment the more likely it is that the sugars produced by the more easily hydrolyzed hemicellulose will be converted to sulfonic acids; at the same time the more intense the treatment the more will glucose be released from the more stable cellulose. Hardwoods not only yield a higher amount of sugar (up to 3% dry weight of liquor) but the sugars are largely pentose, in the form of xylose. Hardwood hydrolyzates also contains a higher amount of acetic acid. Soft woods yield a product with about 75% hexose, mainly mannose.

Sulfite liquor is used as a medium for the growth of microorganisms after being suitably neutralized with $CaCO_3$ and enriched with ammonium salts or urea, and other nutrients. It has been used for the manufacture of yeasts and alcohol. Some samples do not contain enough assaimilable carbonaceous materials for some modern fermentations. They are therefore often enriched with malt extract, yeast autolysate, etc.

(g) Other Substrates

Other substrates used as raw materials in fermentations are alcohol, acetic acid, methanol, methane, and fractions of crude petroleum. These will be discussed under Single Cell Protein (Chapter 15). Barley will be discussed in the section dealing with the brewing of beer (Chapter 12).

4.4 GROWTH FACTORS

Growth factors are materials which are not synthesized by the organism and therefore must be added to the medium. They usually function as cofactors of enzymes and may be vitamins, nucleotides etc. The pure forms are usually too expensive for use in industrial media and materials containing the required growth factors are used to compound the medium. Growth factors are required only in small amounts. Table 4.5 gives some sources of growth factors.

4.5 WATER

Water is a raw material of vital importance in industrial microbiology, though this importance is often overlooked. It is required as a major component of the fermentation

Table 4.5 Some sources of growth factors

Growth factor	Source
Vitamin B	Rice polishing, wheat germ, yeasts
Vitamin B_2	Cereals, corn steep liquor
Vitamin B_6	Corn steep liquor, yeasts
Nicotinamide	Liver, penicillin spent liquor
Panthothenic Acid	Corn steep liquor
Vitamin B_{12}	Liver, silage, meat

medium, as well as for cooling, and for washing and cleaning. It is therefore used in rather large quantities, and measured in thousands of liters a day depending on the industry. In some industries such as the beer industry the quality of the product depends to some extent on the water. In order to ensure constancy of product quality the water must be regularly analyzed for minerals, color, pH, etc. and adjusted as may be necessary. Due to the importance of water, in situations where municipal water supplies are likely to be unreliable, industries set up their own supplies.

4.6 SOME POTENTIAL SOURCES OF COMPONENTS OF INDUSTRIAL MEDIA

The materials to be discussed are mostly found in the tropical countries, including those in Africa, the Caribbean, and elsewhere in the world. Any microbiological industries to be sited in these countries must, if they are not to run into difficulties discussed above, use the locally available substrates. It is in this context that the following are discussed.

4.6.1 Carbohydrate Sources

These are all polysaccharides and have to be hydrolyzed to sugar before being used.

(a) Cassava (manioc)

The roots of the cassava-plant *Manihot esculenta* Crantz serve mainly as a source of carbohydrate for human (and sometimes animal) food in many parts of the tropical world. Its great advantage is that it is high yielding, requires little attention when cultivated, and the roots can keep in the ground for many months without deterioration before harvest. The inner fleshy portion is a rich source of starch and has served, after hydrolysis, as a carbon source for single cell protein, ethanol, and even beer. In Brazil it is one of the sources of fermentation alcohol (Chapter 13) which is blended with petrol to form gasohol for driving motor vehicles.

(b) Sweet potato

Sweet potatoes *Ipomca batatas* is a warm-climate crop although it can be grown also in sub tropical regions. There are a large number of cultivars, which vary in the colors of the tuber flesh and of the skin; they also differ in the tuber size, time of maturity, yield, and sweetness. They are widely grown in the world and are found in South America, the USA, Africa and Asia. They are regarded as minor sources of carbohydrates in comparison with maize, wheat, or cassava, but they have the advantage that they do not require much agronomic attention. They have been used as sources of sugar on a semi-commercial basis because the fleshy roots contain saccharolytic enzymes. The syrup made from boiling the tubers has been used as a carbohydrate (sugar) source in compounding industrial media. Butyl alcohol, acetone and ethanol have been produced from such a syrup, and in quantities higher than the amounts produced from maize syrup of the same concentration. Since sweet potatoes are not widely consumed as food, it is possible that it may be profitable to grow them for use, after hydrolysis, in industrial microbiology media as well as for the starch industry. It is reported that a variety has been developed which yields up to 40 tonnes per hectare, a much higher yield than cassava or maize.

(c) Yams

Yams (*Dioscorea spp*) are widely consumed in the tropics. Compared to other tropical roots however, their cultivation is tedious; in any case enough of this tuber is not produced even for human food. It is therefore almost inconceivable to suggest that the crop should be grown solely for use in compounding industrial media. Nevertheless yams have been employed in producing various products such as yam flour and yam flakes. If the production of these materials is carried out on a sufficiently large scale it is to be expected that the waste materials resulting from peeling the yams could yield substantial amounts of materials which on hydrolysis will be available as components of industrial microbiological media.

(d) Cocoyam

Cocoyam is a blanket name for several edible members of the monocotyledonous (single seed-leaf) plant of the family *Araceae* (the aroids), the best known two genera of which are *Colocasia* (tano) and *Xanthosoma* (tannia). They are grown and eaten all over the tropical world. As they are laborious to cultivate, require large quantities of moisture and do not store well they are not the main source of carbohydrates in regions where they are grown. However, this relative unimportance may well be of significance in regions where for reasons of climate they can be suitably cultivated. Cocoyam starch has been found to be of acceptable quality for pharmaceutical purposes. Should it find use in that area, starchy by-products could be hydrolyzed to provide components of industrial microbiological media.

(e) Millets

This is a collective name for several cereals whose seeds are small in comparison with those of maize, sorghum, rice, etc. The plants are also generally smaller. They are classified as the minor cereals not because of their smaller sizes but because they generally do not form major components of human food. They are however hardy and will tolerate great drought and heat, grow on poor soil and mature quickly. Attention is being turned to them for this reason in some parts of the world. It is for this reason also that millets could become potential sources of cereal for use in industrial microbiology media. Millets are grown all over the world in the tropical and sub-tropical regions and belong to various genera: *Pennisetum americanum* (pearl or bulrush millet), *Setaria italica* (foxtail millet), *Panicum miliaceum* (yard millet), *Echinochloa frumentacea* (Japanese yard millet) and *Eleusine corcana* (finger millet). Millet starch has been hydrolyzed by malting for alcohol production on an experimental basis as far back as 50 years ago and the available information should be helpful in exploiting these grains for use as industrial media components.

(f) Rice

Rice, *Oryza sativa* is one of the leading food corps of the world being produced in all five continents, but especially in the tropical areas. Although it is high-cost commodity, it has the advantage of ease of mechanization, storability, and the availability of improved seeds through the efforts of the International Rice Research Institute, Philippines and other such bodies. The result is that this food crop is likely in the near future to displace, as a carbohydrate source, such other starch sources as yams, and to a lesser extent cassava in tropical countries. The increase in rice production is expected to become so efficient in many countries that the crop would yield substrates cheap enough for industrial microbiological use. Rice is used as brewing adjuncts and has been malted experimentally for beer brewing.

(g) Sorghum

Sorghum, *Sorghum bicolor,* is the fourth in term of quantity of production of the world's cereals, after wheat, rice, and corn. It is used for the production of special beers in various parts of the world. It has been mechanized and has one of the greatest potential among cereals for use as a source of carbohydrate in industrial media in regions of the world where it thrives. It has been successfully malted and used in an all-sorghum lager beer which compared favorably with barley lager beer (Chapter 12)

(h) Jerusalem artichoke

Jerusalem artichoke, *Helianthus tuberosus*, is a member of the plant family compositae, where the storage carbohydrate is not starch, but inulin (Fig. 4.2) a polymer of fructose into which it can be hydrolyzed. It is a root-crop and grows in temperate, semi-tropical and tropical regions.

4.6.2 Protein Sources

(a) Peanut (groundnut) meal

Various leguminous seeds may be used as a source for the supply of nitrogen in industrial media. Only peanuts (groundnuts) *Arachis hypogea* will be discussed. The nuts are rich in liquids and proteins. The groundnut cake left after the nuts have been freed of oil is often used as animal feed. But just as is the case with soya bean, oil from peanuts may be used as anti-foam while the press-cake could be used for a source of protein. The nuts and the cake are rich in protein.

(b) Blood meal

Blood consists of about 82% water, 0.1% carbohydrate, 0.6% fat, 16.4% nitrogen, and 0.7% ash. It is a waste product in abattoirs although it is sometimes used as animal feed. Drying is achieved by passing live steam through the blood until the temperature reaches about 100°C. This treatment sterilizes it and also causes it to clot. It is then drained, pressed to remove serum, further dried and ground. The resulting blood-meal is chocolate-colored and contains about 80% protein and small amounts of ash and lipids.

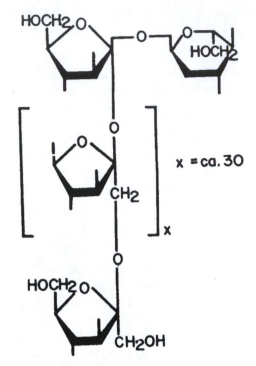

Fig. 4.2   Structure of Inulin. This Polymer of Fructose Replaces Starch in some Plants

Where sufficient blood is available blood meal could form an important source of proteins for industrial media.

(c) Fish Meal

Fish meal is used for feeding farm animals. It is rich in protein (about 65%) and, minerals (about 21% calcium 8%, and phosphorous 3.5%) and may therefore be used for industrial microbiological media production. Fish meal is made by drying fish with steam either aided by vacuum or by simple drying. Alternatively hot air may be passed over the fish placed in revolving drums. It is then ground into a fine powder.

4.7   THE USE OF PLANT WASTE MATERIALS IN INDUSTRIAL MICROBIOLOGY MEDIA: SACCHARIFICATION OF POLYSACCHARIDES

The great recommendation of plant agricultural wastes as sources of industrial microbiological media is that they are not only plentiful but that in contrast with petroleum, a major source of chemicals, they are also renewable. Serious consideration has therefore been given, in some studies, to the possibility of deriving industrial microbiological raw materials not just from wastes, but from crops grown deliberately for the purpose. However, plant materials in general contain large amounts of polysaccharides which are not immediately utilizable by industrial microorganisms and which will therefore need to be hydrolyzed or saccharified to provide the more available sugars. Thereafter the sugars may be fermented to ethyl alcohol for use as a chemical feed stock. The plant polysaccharides whose hydrolysis will be discussed in this section are starch, cellulose and hemicelluloses.

4.7.1 Starch

Starch is a mixture of two polymers of glucose: amylose and amylopectin. Amylose is a linear $(1 \rightarrow 4) \propto - D$ glucan usually having a degree of polymerization (D.P., i.e. number of glucose molecules) of about 400 and having a few branched residues linked with $(1 \rightarrow 6)$ bondings. Amylopectin is a branched D glucan with predominantly $\propto - D\ (1 \rightarrow 4)$ linkages and with about 4% of the $\propto - D\ (1 \rightarrow 6)$ type (Fig. 4.3). Amylopectin consists of amylose – like chains of D. P. 12 – 50 linked in a number of possible manners of which 3 in Fig. 4.4 seems most generally accepted. A comparison of the properties of amylose and amylopectin is given in Table 4.6.

Table 4.6 Some properties of amylose and amylopectin

Property	Amylose	Amylopectin
Structure	Linear	Branched
Behavior in water	Precipitates	
	Irreversibly	Stable
Degree of polymerization	10^3	10^4-10^5
Average chain length	10^3	20-25
Hydrolysis to maltose (%)		
(a) β - amylase	87	54
(b) β - amylase and		
debranching enzyme	98	79
Iodine Complex max (nm)	650	550

Starches from various sources differ in their proportion of amylopectin and amylose. The more commonly grown type of maize, for example, has about 26% of amylose and 74% of amylopectin (Table 4.7). Others may have 100% amylopectin and still others may have 80 – 85% of amylose.

4.7.1.1 Saccharification of starch

Starch occurs in discrete crystalline granules in plants, and in this form is highly resistant to enzyme action. However when heated to about 55°C – 82°C depending on the type, starch gelatinizes and dissolves in water and becomes subject to attack by various enzymes.

Before saccharification, the starch or ground cereal is mixed with water and heated to gelatinize the starch and expose it to attack by the saccharifying agents. The gelatinization temperatures of starch from various cereals is given in Table 12.1. The saccharifying agents used are dilute acids and enzymes from malt or microorganisms.

4.7.1.1.1 Saccharification of starch with acid

The starch-containing material to be hydrolyzed is ground and mixed with dilute hydrochloric acid, sulfuric acid or even sulfurous acid. When sulfurous acid is used it can be introduced merely by pumping sulfur dioxide into the mash.

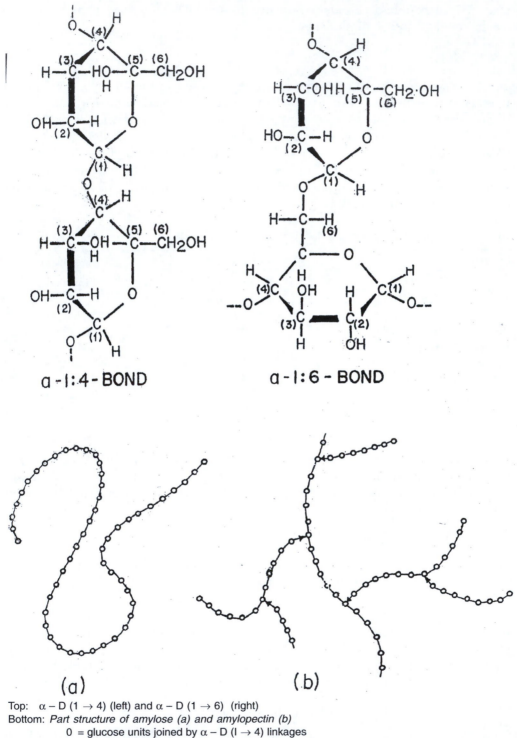

a - I:4 - BOND   a - I:6 - BOND

(a)   (b)

Top: α – D (1 → 4) (left) and α – D (1 → 6) (right)
Bottom: *Part structure of amylose (a) and amylopectin (b)*
        0 = glucose units joined by α – D (I → 4) linkages
        → = a – D (1-6) linkages

Fig. 4.3 Linkages of D-glucose

Table 4.7 Amylose contents of some starches

Source	Amylose Content %
Potato	20-22
Corn	20-27
Wheat	18-26
Oat	22-24
Waxymaize	0-5
Cassava	17-20
Sorghum	25-28
Rice	16-18

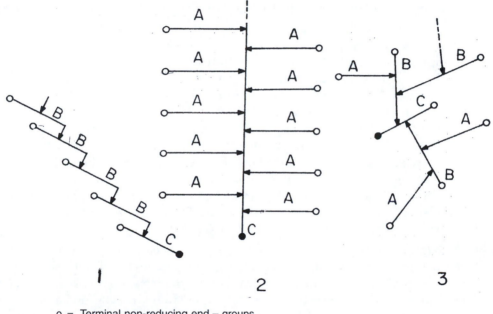

o = Terminal non-reducing end – groups
● = Reducing end group
→ = α - D - (1 → 6) linkage
— = Chain of 20 to 25 α - D - (1 → 4) linked D-glucose residues

Fig. 4.4 Diagrams Representing Three Proposed Structures of Amylopectin

The concentrations of the mash and the acid, length of time and temperature of the heating have to be worked out for each starch source. During the hydrolysis the starch is broken down from starch (about 2,000 glucose molecules) through compounds of decreasing numbers of glucose moieties to glucose. The actual composition of the hydrolysate will depend on the factors mentioned above. Starch concentration is particularly important: if it is too high, side reactions may occur leading to a reduction in the yield of sugar.

At the end of the reaction the acid is neutralized. If it is desired to ferment the hydrolysate for ethanol, yeast or single cell production, ammonium salts may be used as they can be used by many microorganisms.

4.7.1.1.2 Use of enzymes

Enzymes hydrolyzing starch used to be called collectively diastase. With increased knowledge about them, they are now called amylases. Enzymatic hydrolysis has several advantages over the use of acid: (a) since the pH for enzyme hydrolysis is about neutral, there is no need for special vessels which must stand the high temperature, pressure, and corrosion of acid hydrolysis; (b) enzymes are more specific and hence there are fewer side reactions leading therefore to higher yields; (c) acid hydrolysis often yields salts which may have to be removed constantly or periodically thereby increasing cost; (d) it is possible to use higher concentrations of the substrates with enzymes than with acids because of enzyme specificity, and reduced possibility of side reactions.

4.7.1.1.3 Enzymes involved in the hydrolysis of starch

Several enzymes are important in the hydrolysis of starch. They are divisible into six groups.

(i) *Enzymes that hydrolyse* $\alpha - 1, 4$ *bonds and by-pass* $\alpha - I, 6$ *bonding:* The typical example is α - amylase. This enzyme hydrolyses randomly the inner $(1 \rightarrow 4)$ - α - D - glucosidic bonds of amylose and amylopectin (Fig. 4.3). The cleavage can occur anywhere as long as there are at least six glucose residues on one side and at least three on the other side of the bond to be broken. The result is a mixture of branched α - limit dextrins (i.e., fragments resistant to hydrolysis and contain the α - D (1 6) linkage (Fig. 4.4) derived from amylopectin) and linear glucose residues especially maltohexoses, maltoheptoses and maltotrioses. α - Amylases are found in virtually every living cell and the property and substrate pattern of α - amylases vary according to their source. Thus, animal α - amylases in saliva and pancreatic juice completely hydrolyze starch to maltose and D-glucose. Among microbial α - amylases some can withstand temperatures near 100°C.

(ii) *Enzymes that hydrolyse the* $\alpha - 1, 4$ *bonding, but cannot by-pass the* $\alpha - 1,6$ *bonds: Beta amylase:* This was originally found only in plants but has now been isolated from micro-organisms. Beta amylase hydrolyses alternate $\alpha - 1,4$ bonds sequentially from the non-reducing end (i.e., the end without a hydroxyl group at the C – 1 position) to yield maltose (Figs. 4.3 and 4.5). Beta amylase has different actions on amylose and amylopectin, because it cannot by-pass the $\alpha - 1:6$ – branch points in amylopectin. Therefore, while amylose is completely hydrolyzed to maltose, amylopectin is only hydrolyzed to within two or three glucose units of the $\alpha - 1.6$ - branch point to yield maltose and a 'beta-limit' dextrin which is the parent amylopectin with the ends trimmed off. Debranching enzymes (see below) are able to open up the $\alpha - 1:6$ bonds and thus convert beta-limit dextrins to yield a mixture of linear chains of varying lengths; beta amylase then hydrolyzes these linear chains. Those chains with an odd number of glucose molecules are hydrolyzed to maltose, and one glucose unit per chain. The even numbered residues are completely hydrolyzed to maltose. In practice there is a very large population of chains and hence one glucose residue is produced for every two chains present in the original starch.

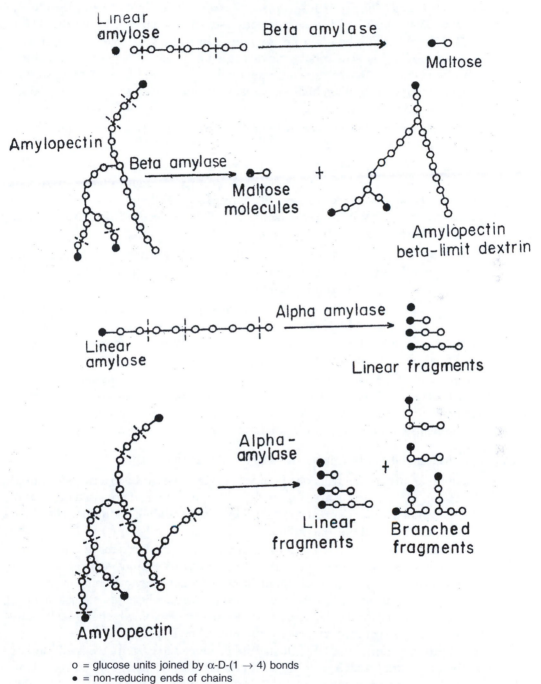

o = glucose units joined by α-D-(1 → 4) bonds
● = non-reducing ends of chains
— = point of attack by enzyme

Fig. 4.5   Pattern of Attack of Alpha Amylase and Beta-amylase on Amylose and Amylopectin Respectively

(iii) *Enzymes that hydrolyze (α —1, 4 and α — 1:6 bonds:* The typical example of these enzymes is amyloglucosidase or glucoamylase. This enzyme hydrolyzes α - D - (1 → 4) -D – glucosidic bonds from the non-reducing ends to yield D – glucose molecules. When the sequential removal of glucose reaches the point of branching in amylopectin, the hydrolysis continues on the (1 → 6) bonding but more slowly than on the (1 → 4) bonding. Maltose is attacked only very slowly. The end product is glucose.

(iv) *De-branching enzymes:* At least two de-branching enzymes are known: pullulanase and iso-amylase.

Pullulanase: This is a de-branching enzyme which causes the hydrolysis of α — D – (1 → 6) linkages in amylopectin or in amylopectin previsouly attacked by alpha-amylase. It does not attack α - D (1 → 4) bonds. However, there must be at least two glucose units in the group attached to the rest of the molecules through an α -D- (1 → 6) bonding.

Iso-amylase: This is also a de-branching enzyme but differs from pullulanase in that three glucose units in the group must be attached to the rest of the molecules through an α - D – (1 → 6) bonding for it to function.

(v) *Enzymes that preferentially attack α - 1, 4 linkages:* Examples of this group are glucosidases. The maltodextrins and maltose produced by other enzymes are cleaved to glucose by α - glucosidases. They may however sometime attack unaltered polysaccharides but only very slowly.

(vi) *Enzymes which hydrolyze starch to non-reducing cyclic D-glucose polymers known as cyclodextrins or Schardinger dextrins:* Cyclic sugar residues are produced by *Bacillus macerans*. They are not acted upon by most amylases although enzymes in Takadiastase produced by *Aspergillus oryzae* can degrade the residues.

4.7.1.1.4   Industrial saccharification of starch by enzymes

In industry the extent of the conversion of starch to sugar is measured in terms of dextrose equivalent (D.E.). This is a measure of the reducing sugar content, expressed in terms of dextrose, determined under defined conditions involving Fehling's solution. The D.E is calculated as percentage of the total solids.

For the saccharification of starch in industry acid is being replaced more and more by enzymes. Sometimes acid is used only initially and enzymes employed at a later stage. Acid saccharification has a practical upper limit of 55 D.E. Beyond this, breakdown products begin to accumulate. Furthermore, with acid hydrolysis reversion reactions occur among the sugar produced. These two deficiencies are avoided when enzymes are utilized. Besides, by selecting enzymes specific sugars can be produced.

Starch-splitting enzymes used in industry are produced in germinated seeds and by micro-organisms. Barley malt is widely used for the saccharification of starch. It contains large amounts of various enzymes notably β-amylase and α - glucosidase which further split saccharides to glucose.

All the enzymes discussed above are produced by different micro-organisms and many of these enzymes are available commercially. The most commonly encountered organisms producing these enzymes are *Bacillus* spp, *Streptomyeces* spp, *Aspergillus* spp, *Penicillium* spp, *Mucor* spp and *Rhizopus* spp.

4.7.2   Cellulose, Hemi-celluloses and Lignin in Plant Materials

4.7.2.1   Cellulose

Cellulose is the most abundant organic matter on earth. Unfortunately it does not exist pure in nature and even the purest natural form (that found in cotton fibres) contains about 6% of other materials. Three major components, cellulose, hemi-cellulose and lignin occur roughly in the ratio of 4:3:3 in wood. Before looking more closely at cellulose, the other two major components of plant materials will be briefly discussed.

4.7.2.2   Hemicelluloses

These are an ill-defined group of carbohydrates whose main and common characteristic is that they are soluble in, and hence can be extracted with, dilute alkali. They can then be precipitated with acid and ethanol. They are very easily hydrolyzed by chemical or biological means. The nature of the hemicellulose varies from one plant to another. In cotton the hemicelluloses are pectic substances, which are polymers of galactose. In wood, they consist of short (DP less than 200) branched heteropolymers of glucose, xylose, galactose, mannose and arabinose as well as uronic acids of glucose and galactose linked by $1-3$, $1-6$ and $1-4$ glycosidic bonding.

4.7.2.3   Lignin

Lignin is a complex three-dimensional polymer formed from cyclic alcohols. (Fig. 4.6). It is important because it protects cellulose from hydrolysis.

Cellulose is found in plant cell-walls which are held together by a porous material known as middle lamella. In wood the middle lamella is heavily impregnated with lignin which is highly resistant and thus protects the cell from attack by enzymes or acid.

4.7.2.4   Pretreatment of cellulose-containing materials before saccharification

In order to expose lignocellulosics to attack, a number of physical and chemical methods are in use, or are being studied, for altering the fine structure of cellulose and/or breaking the lignin-carbohydrate complex.

Table 4.8   Various pretreatment methods used in lignocellulose substrate preparation

Pretreatment type	Specific method
Mechanical	Weathering and milling-ball, fitz, hammer, roller
Irradiation	Gamma, electron beam, photooxidation
Thermal	Autohydrolysis, steam explosion, hydrothermolysis, boiling, pyrolysis, moist or dry heat expansion
Alkali	Sodium hydroxide, ammonium hydroxide
Acids	Sulfuric, hydrochloric, nitric, phosphoric, maleic
Oxidizing agents	Peracetic acid, sodium hypochlorite, sodium chlorite, hydrogen peroxide
Solvents Ethanol,	butanol, phenol, ethylamine, acetone, ethylene glycol
Gases	Ammonia, chlorine, nitrous oxide, ozone, sulfur dioxide
Biological	Ligninolytic fungi

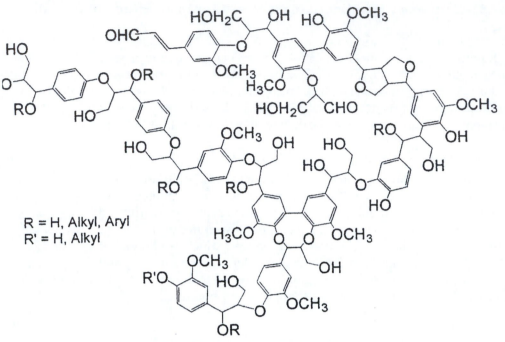

R = H, Alkyl, Aryl
R' = H, Alkyl

Fig. 4.6 Generalized Structure of Lignin

Chemical methods include the use of swelling agents such a NaOH, some amines, concentrated H_2SO_4 or HCl or proprietary cellulose solvents such as 'cadoxen' (tris thylene-diamine cadmium hydroxide). These agents introduce water between or within the cellulose crystals making subsequent hydrolysis, easier. Steam has also been used as a swelling agent. The lignin may be removed by treatment with dilute H_2SO_4 at high temperature.

Physical methods of pretreatment include grinding, irradiation and simply heating the wood.

4.7.2.5 Hydrolysis of cellulose

Following pretreatment, wood may be hydrolyzed with dilute HCl, H_2SO_4 or sulfites of calcium, magnesium or sodium under high temperature and pressure as described for sulfite liquor production in paper manufacture see section 4. above). When, however, the aim is to hydrolyze wood to sugars, the treatment is continued for longer than is done for paper manufacture.

A lot of experimental work has been done recently on the possible use of cellulolytic enzymes for digesting cellulose. The advantage of the use of enzymes rather than harsh chemicals methods have been discussed already. Fungi have been the main source of cellulolytic enzymes. *Trichoderma viride* and *T. koningii* have been the most efficient cellulase producers. *Penicillicum funiculosum* and *Fusarium solani* have also been shown to possess equally potent cellulases. Cellulase has been resolved into at least three components: C_1, C_x, and β-glucosidases. The C_1 component attacks crystalline cellulose and loosens the cellulose chain, after which the other enzymes can attack cellulose. The

C_x enzymes are β - $(1 \rightarrow 4)$ glucanases and hydrolyse soluble derivatives of cellulose or swoollen or partially degraded cellulose. Their attack on the cellulose molecule is random and cellobiose (2-sugar) and cellotroise (3-sugar) are the major products of their actions. There is evidence that the enzymes may also act by removing successive glucose units from the end of a cellulose molecule. β-glucosidases hydrolyze cellobiose and short-chain oligo-saccharides derived from cellulose to glucose, but do not attack cellulose. They are able to attack cellobiose and cellotriose rapidly. Many organisms described in the literature as 'cellulolytic' produce only C_x and β-glucosidases because they were isolated initially using partially degraded cellulose. The four organisms mentioned above produce all three members of the complex.

4.7.2.4.1   Molecular structure of cellulose

Cellulose is a linear polymer of D-glucose linked in the Beta-1, 4 glucosidic bondage. The bonding is theoretically as vulnerable to hydrolysis as the one in starch. However, cellulose – containing materials such as wood are difficult to hydrolyze because of (a) the secondary and tertiary arrangement of cellulose molecules which confers a high crystallinity on them and (b) the presence of lignin.

The degree of polymerization (D. P.) of cellulose molecule is variable, but ranges from about 500 in wood pulp to about 10,000 in native cellulose. When cellulose is hydrolyzed with acid, a portion known as the amorphous portion which makes up 15% is easily and quickly hydrolyzed leaving a highly crystalline residue (85%) whose DP is constant at 100-200. The crystalline portion occurs as small rod-like particles which can be hydrolyzed only with strong acid. (Fig. 4.7)

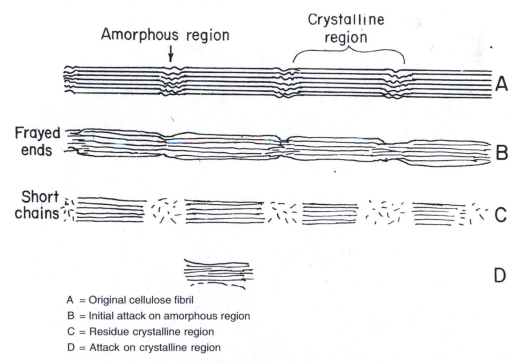

A = Original cellulose fibril
B = Initial attack on amorphous region
C = Residue crystalline region
D = Attack on crystalline region

Fig. 4.7 Diagram Illustrating Breakdown of Crystalline Cellulose

SUGGESTED READINGS

Barnes, A.C. 1974. *The Sugar cane.* Wiley, New York, USA.

Dahod, S.K. 1999. Raw Materials Selection and Medium Development for Industrial Fermentation Processes. In: Manual of Industrial Microbiology and Biotechnology. A. L. Demain and J. E. Davies (eds) American Society for Microbiology Press. 2nd Ed, Washington DC.

Demain, A.L. 1998. Induction of microbial secondary Metabolism International Microbiology 1, 259–264.

Flickinger, M.C., Drew, S.W. (eds) 1999. Encyclopedia of Bioprocess Technology - Fermentation, Biocatalysis, and Bioseparation, Vol 1-5. John Wiley, New York.

Ward, W.P., Singh, A. 2004. Bioethanol Technology: Developments and Perspectives *Advance in Applied Microbiology*, 51, 53 – 80.

CHAPTER 5

Metabolic Pathways for the Biosynthesis of Industrial Microbiology Products

5.1 THE NATURE OF METABOLIC PATHWAYS

In order to be able to manipulate microorganisms to produce maximally materials of economic importance to humans, but at minimal costs, it is important that the physiology of the organisms be understood as much as is possible. In this chapter relevant elements of the physiology of industrial organisms will be discussed.

A yeast cell will divide and produce CO_2 under aerobic conditions if offered a solution of glucose and ammonium salts. The increase in cell number resulting from the growth and the bubbling of CO_2 are only external evidence of a vast number of chemical reactions going on within the cell. The yeast cell on absorbing the glucose has to produce various proteins which will form enzymes necessary to catalyze the various reactions concerned with the manufacture of proteins, carbohydrates, lipids, and other components of the cell as well as vitamins which will form coenzymes. A vast array of enzymes are produced as the glucose and ammonium initially supplied are converted from one compound into another or metabolized. *The series of chemical reactions involved in converting a chemical (or a metabolite) in the organism into a final product is known as a metabolic pathway.* When the reactions lead to the formation of a more complex substance, that particular form of metabolism is known as *anabolism* and the pathway an anabolic pathway. When the series of reactions lead to less complex compounds the metabolism is described as *catabolism*. The compounds involved in a metabolic pathway are called *intermediates* and the final product is known as the end-product (see Fig. 5.1).

Catabolic reactions have been mostly studied with glucose. Four pathways of glucose breakdown to pyruvic acid (or glycolysis) are currently recognized. They will be discussed later. Catabolic reactions often furnish energy in the form of ATP and other high energy compounds, which are used for biosynthetic reactions. A second function of catabolic reactions is to provide the carbon skeleton for biosynthesis. Anabolic reactions lead to the formation of larger molecules some of which are constituents of the cell.

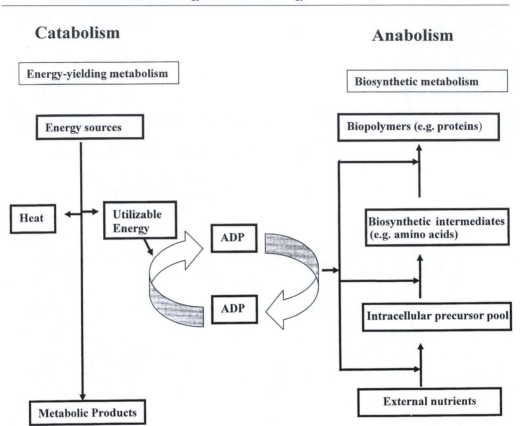

Fig. 5.1   Metabolism: Relationship between Anabolism and Catabolism in a Cell

Although anabolism and catabolism are distinct phenomena some pathways have elements of both kinds Metabolic intermediates which are derived from catabolism and which are also available for anabolism are known as amphibolic intermediates.

Methods for the study of metallic pathways are well reviewed in texts on microbial physiology and will therefore not be discussed here.

5.2   INDUSTRIAL MICROBIOLOGICAL PRODUCTS AS PRIMARY AND SECONDARY METABOLITES

Products of industrial microorganisms may be divided into two broad groups, those which result from primary metabolism and others which derive from secondary metabolism. The line between the two is not always clear cut, but the distinction is useful in discussing industrial products.

5.2.1   Products of Primary Metabolism

Primary metabolism is the inter-related group of reactions within a microorganism which are associated with growth and the maintenance of life. Primary metabolism is essentially the same in all living things and is concerned with the release of energy, and the synthesis of important macromolecules such as proteins, nucleic acids and other cell constituents. When primary metabolism is stopped the organism dies.

Products of primary metabolism are associated with growth and their maximum production occurs in the logarithmic phase of growth in a batch culture. Primary catabolic products include ethanol, lactic acid, and butanol while anabolic products include amino-acids, enzymes and nucleic acids. Single-cell proteins and yeasts would also be regarded as primary products (Table 5.1)

Table 5.1   Some industrial products resulting from primary metabolism

Anabolic Products	*Catabolic Products*
1. Enzymes	1. Ethanol and ethanol-containing products, e.g. wines
2. Amino acids	2. Butanol
3. Vitamins	3. Acetone
4. Polysaccharides	4. Lactic acid
5. Yeast cells	5. Acetic acid (vinegar)
6. Single cell protein	
7. Nucleic acids	
8. Citric acid	

5.2.2   Products of Secondary Metabolism

In contrast to primary metabolism which is associated with the growth of the cell and the continued existence of the organism, secondary metabolism, which was first observed in higher plants, has the following characteristics (i) Secondary metabolism has no *apparent* function in the organism. The organism continues to exist if secondary metabolism is blocked by a suitable biochemical means. On the other hand it would die if primary metabolism were stopped. (ii) Secondary metabolites are produced in response to a restriction in nutrients. They are therefore produced after the growth phase, at the end of the logarithmic phase of growth and in the stationary phase (in a batch culture). They can be more precisely controlled in a continuous culture. (iii) Secondary metabolism appears to be restricted to some species of plants and microorganisms (and in a few cases to animals). The products of secondary metabolism also appear to be characteristic of the species. Both of these observations could, however, be due to the inadequacy of current methods of recognizing secondary metabolites. (iv) Secondary metabolites usually have 'bizarre' and unusual chemical structures and several closely related metabolites may be produced by the same organism in wild-type strains. This latter observation indicates the existence of a variety of alternate and closely-related pathways. (v) The ability to produce a particular secondary metabolite, especially in industrially important strains is easily lost. This phenomenon is known as strain degeneration. (vi) Owing to the ease of the loss of the ability to synthesize secondary metabolites, particularly when treated with acridine dyes, exposure to high temperature or other treatments known to induce plasmid loss (Chapter 5) secondary metabolite production is believed to be controlled by plasmids (at least in some cases) rather than by the organism's chromosomes. A confirmation of the possible role of plasmids in the control of secondary metabolites is shown in the case of leupetin, in which the loss of the metabolite following irradiation can be reversed by conjugation with a producing parent. (vii) The factors which trigger secondary metabolism, the inducers, also trigger morphological changes (morphogenesis) in the organism.

Inducers of Secondary Metabolites

Autoinducers include the γ-butyrolactones (butanolides) of the actinomycetes, the N-acylhomoserine lactones (HSLs) of Gramnegative bacteria, the oligopeptides of Gram-positive bacteria, and B-factor [3'-(1-butylphosphoryl)adenosine] of rifamycin production in *Amycolatopsis mediterrane*. They function in development, sporulation, light emission, virulence, production of antibiotics, pigments and cyanide, plasmid-driven conjugation and competence for genetic transformation. Of great importance in actinomycete fermentations is the inducing effect of endogenous γ-butyrolactones, e.g. A-factor (2-S-isocapryloyl-3R-hydroxymethyl-γ-butyrolactone). A-factor induces both morphological and chemical differentiation in *Streptomyces griseus* and *Streptomyces bikiniensis*, bringing on formation of aerial mycelia, conidia, streptomycin synthases and streptomycin. Conidia can actually form on agar without A-factor but aerial mycelia cannot. The spores form on branches morphologically similar to aerial hyphae but they do not emerge from the colony surface. In *S. griseus*, A-factor is produced just prior to streptomycin production and disappears before streptomycin is at its maximum level. It induces at least 10 proteins at the transcriptional level. One of these is streptomycin 6-phosphotransferase, an enzyme which functions both in streptomycin biosynthesis and in resistance. In an A-factor deficient mutant, there is a failure of transcription of the entire streptomycin gene cluster. Many other actinomycetes produce A-factor, or related α-butyrolactones, which differ in the length of the side-chain. In those strains which produce antibiotics other than streptomycin, the γ-butyrolactones induce formation of the particular antibiotics that are produced, as well as morphological differentiation.

Secondary metabolic products of microorganism are of immense importance to humans. Microbial secondary metabolites include antibiotics, pigments, toxins, effectors of ecological competition and symbiosis, pheromones, enzyme inhibitors, immunomodulating agents, receptor antagonists and agonists, pesticides, antitumor agents and growth promoters of animals and plants, including gibbrellic acid, anti-tumor agents, alkaloids such as ergometrine, a wide variety of other drugs, toxins and useful materials such as the plant growth substance, gibberellic acid (Table 5.2). They have a major effect on the health, nutrition, and economics of our society. They often have unusual structures and their formation is regulated by nutrients, growth rate, feedback control, enzyme inactivation, and enzyme induction. Regulation is influenced by unique low molecular mass compounds, transfer RNA, sigma factors, and gene products formed during post-exponential development. The synthases of secondary metabolism are often coded for by clustered genes on chromosomal DNA and infrequently on plasmid DNA.

Unlike primary metabolism, the pathways of secondary metabolism are still not understood to a great degree. Secondary metabolism is brought on by exhausion of a nutrient, biosynthesis or addition of an inducer, and/or by a growth rate decrease. These events generate signals which effect a cascade of regulatory events resulting in chemical differentiation (secondary metabolism) and morphological differentiation (morphogenesis). The signal is often a low molecular weight inducer which acts by negative control, i.e. by binding to and inactivating a regulatory protein (repressor

Table 5.2 Some industrial products of microbial secondary metabolism

Product	Organism	Use/Importance
Antibiotics		
Penicillin	*Penicillium chrysogenum*	Clinical use
Streptomycin	*Streptomyces griseus*	Clinical use
Anti-tumor Agents		
Actinomyin	*Streptomyces antibioticus*	Clinical use
Bleomycin	*Streptomyces verticulus*	Clinical use
Toxins		
Aflatoxin	*Aspergiulus flavous*	Food toxin
Amanitine	*Amanita* sp	Food toxin
Alkaloids		
Ergot alkaloids	*Claviceps purpurea*	Pharmaceutical
Miscellaneous		
Gibberellic acid	*Gibberalla fujikuroi*	Plant growth hormone
Kojic acid	*Aspergillus flavus*	Food flavor
Muscarine	*Clitocybe rivalosa*	Pharmaceutical
Patulin	*Penicillium urticae*	Anti-microbial agent

protein/receptor protein) which normally prevents secondary metabolism and morphogenesis during rapid growth and nutrient sufficiency.

Thousands of secondary metabolites of widely different chemical groups and physiological effects on humans have been found. Nevertheless a disproportionately high interest is usually paid to antibiotics, although this appears to be changing. It would appear that the vast potential utility of microbial secondary metabolites is yet to be realized and that many may not even have been discovered. Part of this 'lopsided' interest may be due to the method of screening, which has largely sought antibiotics. The general topic of screening, especially of secondary metabolites, will be discussed in Chapters 7 and 28. In particular, an attempt will be made to discuss the screening of drugs outside antibiotics.

5.3 TROPHOPHASE-IDIOPHASE RELATIONSHIPS IN THE PRODUCTION OF SECONDARY PRODUCTS

From studies on *Penicillium urticae* the terms trophophase and idiophase were introduced to distinguish the two phases in the growth of organisms producing secondary metabolites. The trophophase (Greek, tropho = nutrient) is the feeding phase during which primary metabolism occurs. In a batch culture this would be in the logarithmic phase of the growth curve. Following the trophophase is the idio-phase (Greek, idio = peculiar) during which secondary metabolites peculiar to, or characteristic of, a given organism are synthesized. Secondary synthesis occurs in the late logarithmic, and in the stationary, phase. It has been suggested that secondary metabolites be described as 'idiolites' to distinguish them from primary metabolites.

5.4 ROLE OF SECONDARY METABOLITES IN THE PHYSIOLOGY OF ORGANISMS PRODUCING THEM

Since many industrial microbiological products result from secondary metabolism, workers have sought to explain the role of secondary metabolites in the survival of the organism. Due to the importance of antibiotics as clinical tools, the focus of many workers has been on antibiotics. This discussion while including antibiotics will attempt to embrace the whole area of secondary metabolites.

Some earlier hypotheses for the existence of secondary metabolism are apparently no longer considered acceptable by workers in the field. These include the hypotheses that secondary metabolites are food-storage materials, that they are waste products of the metabolism of the cell and that they are breakdown products from macro-molecules. The theories in currency are discussed below; even then none of these can be said to be water tight. The rationale for examining them is that a better understanding of the organism's physiology will help towards manipulating it more rationally for maximum productivity.

(i) *The competition hypothesis*: In this theory which refers to antibiotics specifically, secondary metabolites (antibiotics) enable the producing organism to withstand competition for food from other soil organisms. In support of this hypothesis is the fact that antibiotic production can be demonstrated in sterile and non-sterile soil, which may or may not have been supplemented with organic materials. As further support for this theory, it is claimed that the wide distribution of β-lactamases among microorganisms is to help these organisms detoxify the β-lactam antibiotics. The obvious limitation of this theory is that it is restricted to antibiotics and that many antibiotics exist outside Beta-lactams.

(ii) *The maintenance hypothesis*: Secondary metabolism usually occurs with the exhaustion of a vital nutrient such as glucose. It is therefore claimed that the selective advantage of secondary metabolism is that it serves to maintain mechanisms essential to cell multiplication in operative order when that cell multiplication is no longer possible. Thus by forming secondary enzymes, the enzymes of primary metabolism which produce precursors for secondary metabolism therefore, the enzymes of primary metabolism would be destroyed. In this hypothesis therefore, the secondary metabolite itself is not important; what is important is the pathway of producing it.

(iii) *The unbalanced growth hypothesis*: Similar to the maintenance theory, this hypothesis states that control mechanisms in some organisms are too weak to prevent the over synthesis of some primary metabolites. These primary metabolites are converted into secondary metabolites that are excreted from the cell. If they are not so converted they would lead to the death of the organism.

(iv) *The detoxification hypothesis*: This hypothesis states that molecules accumulated in the cell are detoxified to yield antibiotics. This is consistent with the observation that the penicillin precursor penicillanic acid is more toxic to *Penicillium chrysogenum* than benzyl penicillin. Nevertheless not many toxic precursors of antibiotics have been observed.

(v) *The regulatory hypothesis*: Secondary metabolite production is known to be associated with morphological differentiation in producing organisms. In the fungus *Neurospora crassa,* carotenoids are produced during sporulation. In *Cephalospoium acremonium,* cephalosporin C is produced during the idiophase when arthrospores are produced. Numerous examples of the release of secondary metabolites with some morphological differentiation have been observed in fungi. One of the most intriguing relationships between differentiation and secondary metabolite production, is that between the production of peptide antibiotics by *Bacillus* spp. and spore formation. Both spore formation and antibiotic production are suppressed by glucose; non-spore forming mutants of bacilli also do not produce antibiotics, while reversion to spore formation is accompanied by antibiotic formation has been observed in actinomycetes. Many roles have been assigned to antibiotics in spore formers but the most clearly demonstrated has been the essential nature of gramicidin in sporulation of *Bacillus* spp. The absence of the antibiotic leads to partial deficiencies in the formation of enzymes involved in spore formation, resulting in abnormally heat-sensitive spores. Peptide antibiotics therefore suppress the vegetative genes allowing proper development of the spores. In this theory therefore the production of secondary metabolites is necessary to regulate some morphological changes in the organism. It could of course be that some external mechanism triggers off secondary metabolite production as well as the morphological change.

(vi) ***The hypothesis of secondary metabolism as the expression of evolutionary reactions***: Zahner has put forth a most exciting role for secondary metabolism. To appreciate the hypothesis, it is important to bear in mind that both primary and secondary metabolism are controlled by genes carried by the organism. Any genes not required are lost. According to this hypothesis, secondary metabolism is a clearing house or a mixed bag of biochemical reactions, undergoing tests for possible incorporation into the cell's armory of primary reactions. Any reaction in the mixed bag which favorably affects any one of the primary processes, thereby fitting the organism better to survive in its environment, becomes incorporated as part of primary metabolism. According to this hypothesis, the antibiotic properties of some secondary metabolites are incidental and not a design to protect the microorganisms. This hypothesis is attractive because it implies that secondary metabolism must occur in all microorganisms since evolution is a continuing process. If that is the case, then the current range of secondary metabolites is limited only by techniques sensitive enough to detect them. That this is a possibility is shown by the increase in the number of antibiotics alone, since new methods were recently introduced in the processes used in screening for them. If therefore adequate methods of detection are devised it is possible that more secondary metabolites of use for humans could be found.

5.5 PATHWAYS FOR THE SYNTHESIS OF PRIMARY AND SECONDARY METABOLITES OF INDUSTRIAL IMPORTANCE

The main source of carbon and energy in industrial media is carbohydrates. In recent times hydrocarbons have been used. The catabolism of these compounds will be discussed briefly because they supply the carbon skeletons for the synthesis of primary as well as for secondary metabolites. The inter-relationship between the pathways of primary and the secondary metabolism will also be discussed briefly.

5.5.1   Catabolism of Carbohydrates

Four pathways for the catabolism of carbohydrates up to pyruvic acid are known. All four pathways exist in bacteria, actinomycets and fungi, including yeasts. The four pathways are the Embden-Meyerhof-Parmas, the Pentose Phosphate Pathways, the Entner Duodoroff pathway and the Phosphoketolase. Although these pathways are for the breakdown of glucose. Other carbohydrates easily fit into the cycles.

(i) *The Embden-Meyerhof-Parnas (EMP Pathways)*: The net effect of this pathway is to reduce glucose (C_6) to pyruvate (C_3) (Fig. 5.2). The system can operate under both aerobic and anaerobic conditions. Under aerobic conditions it usually functions with the tricarboxylic acid cycle which can oxidize pyruvate to CO_2 and H_2O. Under anaerobic conditions, pyruvate is fermented to a wide range of fermentation products, many of which are of industrial importance (Fig. 5.3).

(ii) *The pentose Phosphate Pathway (PP)*: This is also known as the Hexose Monophosphate Pathway (**HMP**) or the phosphogluconate pathway. While the **EMP** pathway provides pyruvate, a C_3 compound, as its end product, there is no end product in the PP pathway. Instead it provides a pool of triose (C_3) pentose (C_5), hexose (C_6) and heptose (C_7) phosphates. The primary purpose of the PP pathway, however, appears to be to generate energy in the form of **NADPA2** for biosynthetic and other purposes and pentose phosphates for nucleotide synthesis (Fig. 5.4)

(iii) *The Entner-Duodoroff Pathway (ED)*: The pathway is restricted to a few bacteria especially *Pseudomonas*, but it is also carried out by some fungi. It is used by some organisms in the enaerobic breakdown of glucose and by others only in gluconate metabolism (Fig. 5.5)

(iv) *The Phosphoketolase Pathway*: In some bacteria glucose fermentation yields lactic acid, ethanol and CO_2. Pentoses are also fermented to lactic acid and acetic acid. An example is *Leuconostoc mesenteroides* (Fig. 5.6).

Pathways used by microorganisms

The two major pathways used by microorganisms for carbohydrate metabolism are the **EMP** and the **PP** pathways. Microorganisms differ in respect of their use of the two pathways. Thus *Saccharomyces cerevisae* under aeaobic conditions uses mainly the **EMP** pathway; under anaerobic conditions only about 30% of glucose is catabolized by this pathway. In *Penicillium chrysogenum*, however, about 66% of the glucose is utilized via the PP pathway. The **PP** pathway is also used by *Acetobacter*, the acetic acid bacteria. Homofermentative bacteria utilize the EMP pathway for glucose breakdown. The **ED** pathway is especially used by *Pseudomonas*.

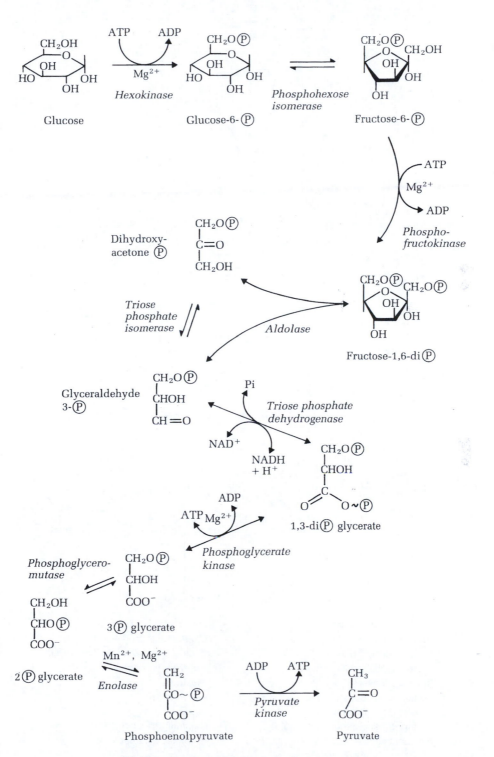

Fig. 5.2   The Embden-Meyerhof – Parnas Pathway

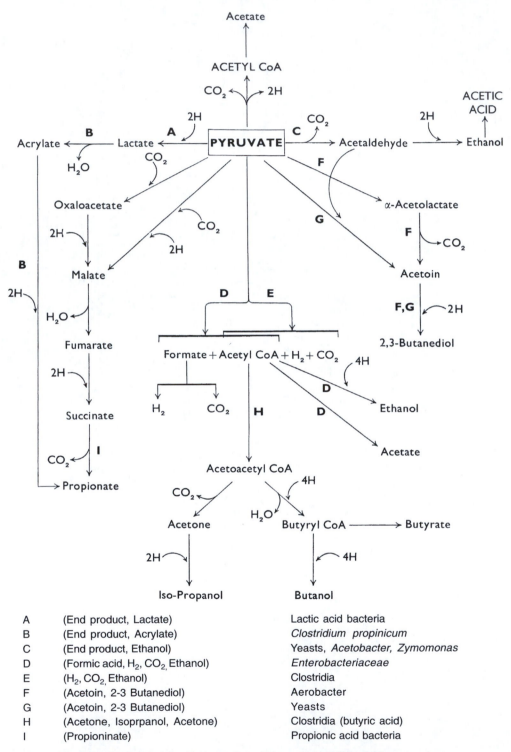

A	(End product, Lactate)	Lactic acid bacteria
B	(End product, Acrylate)	*Clostridium propinicum*
C	(End product, Ethanol)	Yeasts, *Acetobacter, Zymomonas*
D	(Formic acid, H_2, CO_2, Ethanol)	*Enterobacteriaceae*
E	(H_2, CO_2, Ethanol)	Clostridia
F	(Acetoin, 2-3 Butanediol)	Aerobacter
G	(Acetoin, 2-3 Butanediol)	Yeasts
H	(Acetone, Isoprpanol, Acetone)	Clostridia (butyric acid)
I	(Propioninate)	Propionic acid bacteria

Fig. 5.3 Products of the Fermentation of Pyruvate by Different Microorganisms

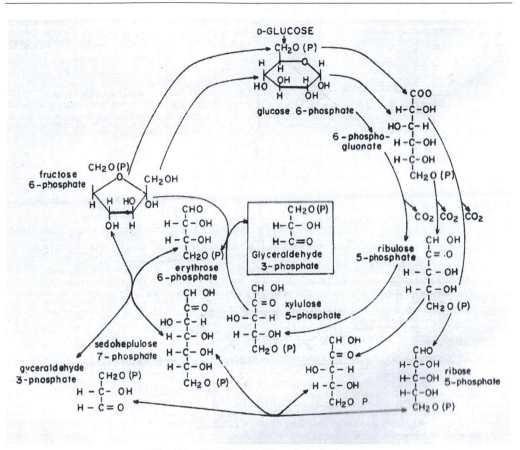

Fig. 5.4   The Pentose Phosphate Pathway

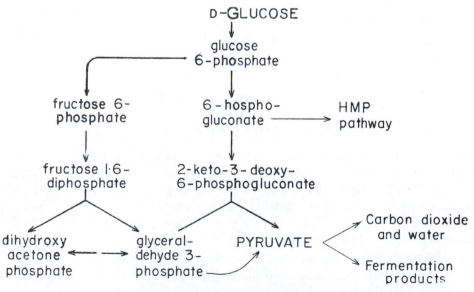

Fig. 5.5   The Enter-Doudoroff Pathway

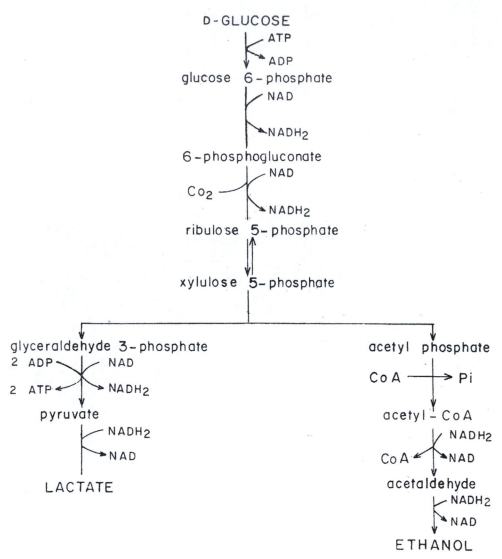

Fig. 5.6 The Phosphoketolase Pathway

5.5.2 The Catabolism of Hydrocarbons

Although the price of crude oil continues to rise, it is, along with other hydrocarbons still used in some fermentations as energy and carbon skeleton sources. Compared with carbohydrates, however, far fewer organism appear to utilize hydrocarbons. Hydrocarbons have been used in single cell protein production and in amino-acid production among other products. Their use by various organisms in industrial media is discussed more fully in Chapter 15.

  (i) *Alkanes*: Alkanes are saturated hydrocarbons that have the general formula $C_2 H_{n+2}$. When the alkanes are utilized, the terminal methyl group is usually oxidized to the corresponding primary alcohol thus:

$$R.CH_2CH_2CH_3 \rightarrow R.CH_2CH_2CH_2OH \rightarrow RCH_2CHO \rightarrow R.CH_2COOH$$

Alkane	Alcohol	Aldehyde	Fatty acid

The alcohol is then oxidized to a fatty acid, which then forms as ester with coenzyme A. Thereafter, it is involved in a series of β-oxidations (Fig. 5.7) which lead to the step-wise cleaving off of acetyl coenzyme A which is then further metabolized in the Tricarboxylic Acid Cycle.

(ii) *Alkenes*: The alkenes are unsaturated hydrocarbons and contain many double bonds. Alkenes may be oxidized at the terminal methyl group as shown earlier for alkanes. They may also be oxidized at the double bond at the opposite end of the molecule by molecular oxygen given rise to a diol (an alcohol with two –OH groups). Thereafter, they are converted to fatty acid and utilized as indicated above.

5.6 CARBON PATHWAYS FOR THE FORMATION OF SOME INDUSTRIAL PRODUCTS DERIVED FROM PRIMARY METABOLISM

The broad flow of carbon in the formation of industrial products resulting from primary metabolism may be examined under two headings: (i) catabolic products resulting from fermentation of pyruvic acid and (ii) anabolic products.

5.6.1 Catabolic Products

Industrial products which are catabolic products formed from carbohydrate fermentation are derived from pyruvic acid produced via the EMP, PP, or ED pathway. Those of importance are ethanol, acetic acid, 2, 3-butanediol, butanol, acetone and lactic acid. The general outline for deriving these from pyruvic acid has already been shown in Fig. 5.3. The nature of the products not only broadly depends on the species of organisms used but also on the prevailing environmental conditions such as pH, temperature, aeration, etc.

5.6.2 Anabolic Products

Anabolic primary metabolites of industrial interest include amino acids, enzymes, citric acid, and nucleic acids. The carbon pathways for the production of anabolic primary metabolites will be discussed as each product is examined.

5.7 CARBON PATHWAYS FOR THE FORMATION OF SOME PRODUCTS OF MICROBIAL SECONDARY METABOLISM OF INDUSTRIAL IMPORTANCE

The unifying features of the synthesis of secondary metabolic products by microorganisms can be summarized thus:

(i) conversion of a normal substrate into important intermediates of general metabolism;

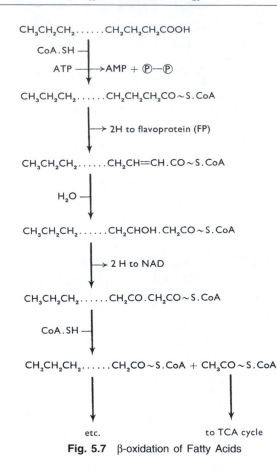

Fig. 5.7 β-oxidation of Fatty Acids

(ii) the assembly of these intermediates in an unusual way, by means of a combination of standard general mechanisms with a selection from a relatively small number of special mechanism;

(iii) these special mechanisms while being peculiar to secondary metabolism are not unrelated to general or primary mechanism;

(iv) the synthetic activity of secondary metabolism appears in response to conditions favorable for cell multiplication.

From the above, it becomes clear that although secondary metabolites are diverse in their intrinsic chemical nature as well as in the organism which produce them, they use only a few biosynthetic pathways which are related to, and use the intermediates of, the primary metabolic pathways. Based on the broad flow of carbon through primary metabolites to secondary metabolites, (depicted in Fig. 5.8) the secondary metabolites may then be classified according to the following six metabolic pathways.

(i) *Secondary products derived from the intact glucose skeleton*: The carbon skeleton of glucose is incorporated unaltered in many antibiotics and other secondary metabolites. The entire basic structure of the secondary product may be derived from glucose as in streptomycin or it may form the glycoside molecule to be combined with a non-sugar (aglycone portion) from another biosynthetic route.

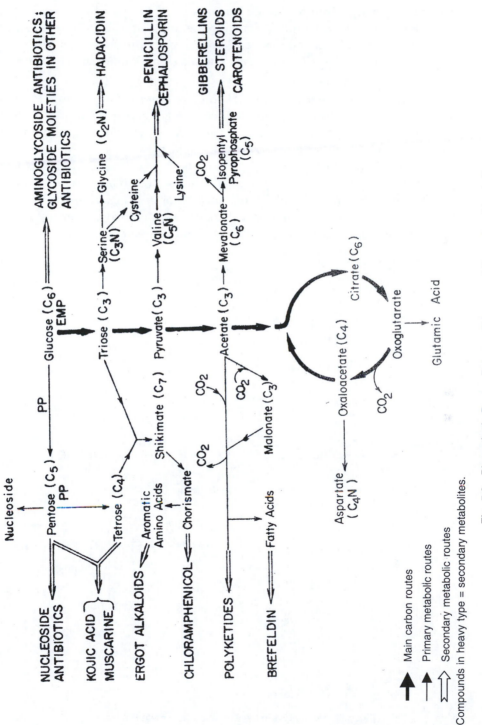

Fig. 5.8 Biosynthetic Routes Between Primary and Secondary Metabolites

The incorporation of the intact glucose molecule is more common among the actinomycetes than among the fungi.

(ii) *Secondary products related to nucleosides*: The pentose phosphate pathway provides ribose (5 carbon) for nucleoside biosynthesis. Many secondary metabolites in this group are antibiotics and are produced mainly by actinomycetes and fungi. Examples are nucleoside antibiotics such as bleomycin. (Chapter 21).

(iii) *Secondary products derived through the Shikimate-Chorismate Pathway*: Shikimic acid (C_7) is formed by the condensation of erythrose-4- phosphate (C_4) obtained from the PP pathway with phosphoenolypyruvate (C_3) from the EMP pathway. It is converted to chorismic acid which is a key intermediate in the formation of numerous products including aromatic aminoacids, such as phynylalamine, tryrosine and tryptophan. Chorismic acid is also a precursor for a number of secondary metabolites including chloramphenicol, p-amino benzoic acid, phenazines and pyocyanin which all have anticrobial properties (Fig. 5.9). The metabolic route leading to the formation of these compounds is therefore

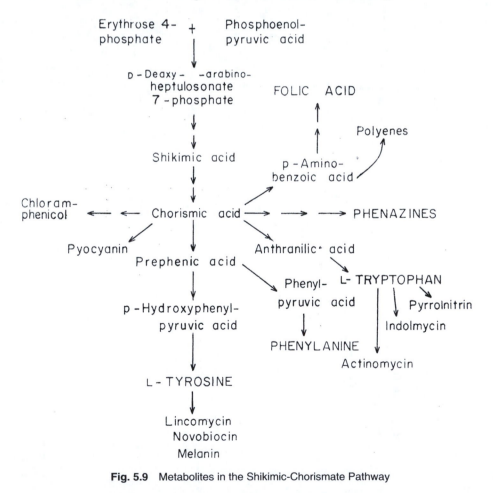

Fig. 5.9 Metabolites in the Shikimic-Chorismate Pathway

referred to as the shikimate pathway. In view of this central role of chorismic acid, however, the route is more widely known as the shikimate-chorismate route. The shikimate-chorismate route is an important route for the formation of aromatic secondary products in the bacteria and actinomycetes. Examples of such secondary products include chloramphenicol and novobiocin. The route is less used in fungi, where the polyketide pathway is more common for the synthesis of aromatic secondary products.

(iv) *The polyketide pathway*: polyketide biosynthesis is highly characteristic of the fungi, where more secondary metabolites are produced by it than by any other. Indeed most of the known polyketide-derived natural products have been obtained from the fungi, a much smaller number being obtained from bacteria and higher plants. The triose (C_3) derived from glucose in the EMP pathway is converted via pyruvic acid to acetate, which occupies a central position in both primary and secondary synthesis. The addition of CO_2 to an acetate group gives a malonate group. The synthesis of polyketides is very similar to that of fatty acids. In the synthesis of both groups of compounds acetate reacts with malonate with the loss of CO_2. By successive further linear reactions between the resulting compound and malonate, the chain of the final compound (fatty acid or polyketide) can be successively lengthened.

However, in the case of fatty acid the addition of each malonate molecule is followed by decarboxylation and reduction whereas in polyketides these latter reactions do occur. Due to this a chain of ketones or a β-polyketomethylene (hence the name polyketide) is formed (Fig. 5.10). The polyketide (β - poly-ketomethylene)

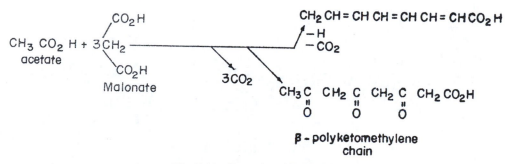

Fig. 5.10 Formation of Polyketides

chain made up of repeating C-CH$_2$ or 'C$_2$ units', is a reactive protein-bound intermediate which can undergo a number of reactions, notably formation into rings. Polyketides are classified as triketides, tetraketides, pentaketides, etc., depending on the number of 'C$_2$ units'. Thus, orsellenic acid which is derived from the straight chain compound in Fig. 5.11 with four 'C$_2$-units' is a tetraketide. Although the polyketide route is not common in actinomycetes, a modified polyketide route is used in the synthesis of tetracyclines by *Streptomyces griseus*.

(v) *Terpenes and steroids*: The second important biosynthetic route from acetate is that leading via mevalonic acid to the terpenes and steroids. Microorganisms

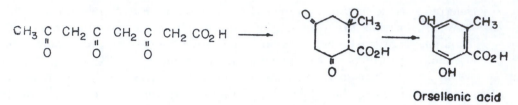

Fig. 5.11 Formation of the Triketide, Orsellenic Acid

especially fungi and bacteria synthesize a large number of terpenes, steroids, carotenoids and other products following the 'isoprene rule'. The central point of this rule is that these compounds are all derivatives of isoprene, the five-carbon compound.

Simply put the isoprene rules consist of the following (Fig. 5.12):

(i) Synthesis of mevalonate from acetate or leucine

(ii) Dehydratopm and decarboxylation to give isoprene followed by condensation to give isoprenes of various lengths.

(iii) Cyclization (ring formation) e.g., to give steroids (Chapter 26)

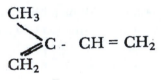

Isoprene

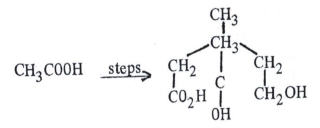

Mevalonic acid

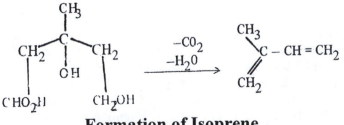

Formation of Isoprene

Fig. 5.12 Isoprene Derivatives

(iv) Further modification of the cyclised structure. The route leads to the formation of essential steroid hormones of mammals and to a variety of secondary metabolites in fungi and plants. it is not used to any extent in the actinomycetes.

(vi) ***Compounds derived from amino acids***: The amino acids are derived from various products in the catabolism of glucose. Serine (C_3N) and glycine (C_2N) are derived from the triose (C_3) formed glucose; valine (C_5N) is derived from acetate (C_3); aspartatic acid (C_4N) is derived from oxeloacetic acid (C_4) while glutamic acid (C_5N) is derived from oxoglutamic acid (C_5). The biosynthetic pathways for the formation of amino acids are shown in Fig. 5.13 from which it will be seen that aromatic amino acids are derived via the shikimic pathway.

Secondary products may be formed from one, two or more amino acids. an example of the first group (with one amino acid group) is hadacidin which inhibits plant tumors and

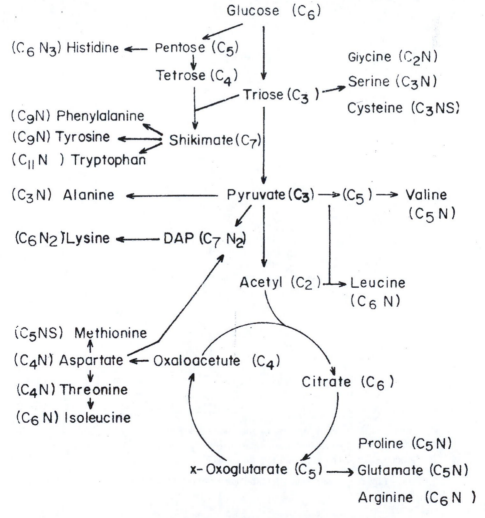

Fig. 5.13   Synthetic Routes of the Amino Acids

is produced from glycine and produced by *Penicillium frequentants* according to the formula shown below

$$H_2NCH_2CO_2H \rightarrow HN(OH)\,CH_2\,CO_2H \rightarrow OHCN(OH)\,CH_2\,CO_2H$$

Glycine   Hadacidin

Other examples are the insecticidal compound, ibotenic acid (Amanita factor C) produced by the mushroom *Amanita muscaria* and psilocybin, a drug which causes hallucinations and produced by the fungus *Psiolocybe* (Fig. 5.14), the ergot alkaloids (Chapter 25) produced by *Clavicepts purpureae* also belong in this group as does the antibiotic cycloserine.

Among the secondary products derived from two amino acids are gliotoxin which is produced by members of the Fungi Imperfecti, especially *Trichoderma* and which is a highly active anti-fungal and antibacterial (Fig. 5.14) and Arantoin, an antiviral drug also belongs to this group.

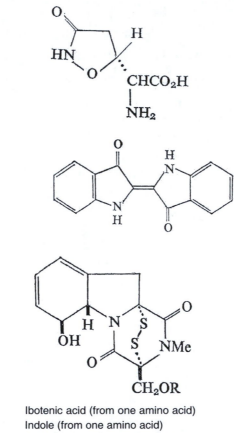

Top: Ibotenic acid (from one amino acid)
Middle: Indole (from one amino acid)
Bottom: Gliotoxin (from two amino acids)

Fig. 5.14 Secondary Metabolites Formed from Amino Acids

The secondary products derived from more than two amino acids include many which are of immense importance to man. These include many toxins from mushrooms e.g the *Aminita* toxins (Fig. 5.15) (phalloidin, amanitin) peptide antibiotics from *Bacillus* app and a host of other compounds.

An example of a secondary metabolite produced from three amino acids is malformin A (Fig. 5.15) which is formed by *Aspergillus* spp. It induces curvatures of beam shoots and maize seedlings. It is formed from L-leucine, D-leucine, and cysteine.

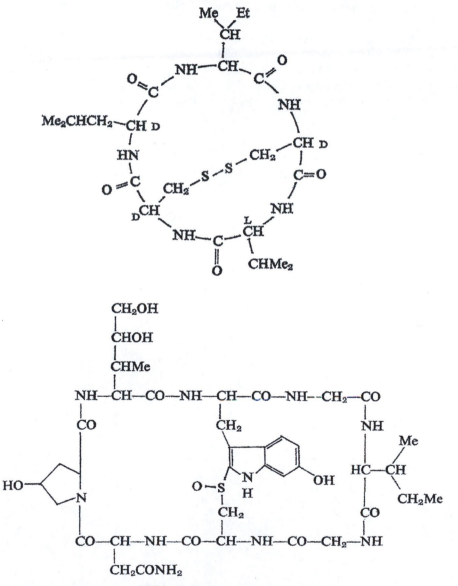

Top: Malformin A, a secondary metabolite formed from three amino acids
Bottom: Amanitin, a secondary metabolite formed from two amino acids

Fig. 5.15   Secondary Metabolites from Amino Acids

SUGGESTED READINGS

Bull, A.T., Ward, A.C., Goodfellow, M. 2000. Search and Discovery Strategies for Biotechnology: The Paradigm Shift Microbiology and Molecular Biology Reviews 64, 573 – 606.

Demain, A.L. 1998. Induction of microbial secondary Metabolism International Microbiology 1, 259–264.

Herrmann, K.H., Weaver, L.M. 1999. The Shikimate Pathway. Annual Review of Plant Physiology and Plant Molecular Biology. 50, 473–503.

Madigan, M.T., Martinko, J.M. 2006. Brock Biology of Micro-organisms. Pearson Prentice Hall Upper Saddle River, USA.

Meurer, G., Hutchinson, C.R. 1999. Genes for the Synthesis of Microbial Secondary M etabolites. In: Manual of Industrial Microbiology and Biotechnology. A.L. Demain and J.E. Davies, (eds). ASM Press. 2nd Ed. Washington, DC, USA pp. 740-758.

Zahner, H. 1978. In: Antibiotics and other Secondary Metabolites. R. Hutter, T. Leisenger, J. Nuesch, W. Wehrli (eds). Academic Press, New York, USA, pp. 1-17.

CHAPTER 6

Overproduction of Metabolites of Industrial Microorganisms

The complexity of the activities which go on within a cell was mentioned at the beginning of Chapter 5 when we discussed the metabolism of a yeast cell introduced into an aqueous solution of glucose and ammonium salts. The yeast cell must first permit the entry into itself of the glucose and ammonium salts. Under suitable environmental conditions such as pH and temperature it will grow by budding within about half an hour. For these buds to occur, hundreds of activities will have gone on within the cell. New proteins to be incorporated into enzymes and other structures will have been synthesized; nucleic acids for the chromosomes and carbohydrates for the cell walls will all have been synthesized. Hundreds of different enzymes will have participated in these synthetic activities. The organism must synthesize each of the compounds at the right time and in the appropriate quantities. If along side ammonium salts, amino acids were supplied, the yeast cells would stop absorbing the ammonium salt and instead utilize the supplied 'readymade' substrate.

A few yeasts can utilize starch. If our yeast belonged to this group and was supplied nothing but starch and ammonium salts, it would secrete extracellular enzyme(s) to breakdown the starch to sugars. These sugars would then be absorbed and would be used with ammonium salts, for the synthetic activities we described earlier.

Clearly therefore, while the organism's genetic apparatus determines in broad terms the organism's overall synthetic potentialities, what is actually synthesized depends on what is available in the environment. Most importantly, the organism is not only able to 'decide' when to manufacture and secrete certain enzymes to enable it to utilize materials in the environment, but it is able to decide to stop the synthesis of certain compounds if they are supplied to it. These sensing mechanisms for the switching on and off of the synthetic processes enable the organism to avoid the overproduction of any particular compound. If it did not have these regulatory mechanisms it would waste energy and resources (which are usually scarce in natural environments) in making materials it did not require.

An efficient or 'stringent' organism which does not waste its resources in producing materials it does not require will survive well in natural environments where competition is intense. Such an organism while surviving well in nature would not, however, be of much use as an industrial organism. The industrial microbiologist or biotechnologist prefers, and indeed, seeks, the wasteful, inefficient and 'relaxed' organism whose regulatory mechanisms are so poor that it will overproduce the particular metabolite sought. Knowledge of these regulatory mechanisms and biosynthetic pathways is essential, therefore, to enable the industrial microbiologist to derange and disorganize them so that the organism will overproduce desired materials.

In this chapter the processes by which the organism regulates itself and avoids over-production using enzyme regulation and permeability control will first be discussed. Then will follow a discussion of methods by which the microbiologist consciously deranges these two mechanisms to enable overproduction. Genetic manipulation of organisms will be discussed in the next chapter.

Regulatory methods and ways of disorganizing microorganisms for the over-production of metabolites are far better understood in primary metabolites than they are in secondary metabolites. Indeed for some time it was thought that secondary metabolites did not need to be regulated since the microorganisms had no apparent need for them. They are currently better understood and it is now known that they are also regulated.

In the discussions that follow, primary metabolites will first be considered. Only a minimum of examples will be given in respect of regulatory mechanisms of primary metabolites. Textbooks on microbial physiology may be consulted for the details.

6.1 MECHANISMS ENABLING MICROORGANISMS TO AVOID OVERPRODUCTION OF PRIMARY METABOLIC PRODUCTS THROUGH ENZYME REGULATION

Some of the regulatory mechanisms enabling organisms to avoid over-production are given in Table 6.1. Each of these will be discussed briefly.

Table 6.1 Regulatory mechanisms in microorganisms

---

1. Substrate Induction
2. Catabolite Regulation
    2.1 Repression
    2.2 Inhibition
3. Feedback Regulation
    3.1 Repression
    3.2 Inhibition
    3.3 Modifications used in branched pathways
        3.3.1 Concerted (multivalent) feedback regulation
        3.3.2 Cooperative feedback inhibition
        3.3.3 Cumulative feedback regulation
        3.3.4 Compensatory feedback regulation
        3.3.5 Sequential feedback regulation
        3.3.6 Isoenzyme feedback regulation
4. Amino acid Regulation of RNA synthesis
5. Energy Charge Regulation
6. Permeability Control

6.1.1   Substrate Induction

Some enzymes are produced by microorganisms only when the substrate on which they act is available in the medium. Such enzymes are known as *inducible* enzymes. Analogues of the substrate may act as the inducer. When an inducer is present in the medium a number of different inducible enzymes may sometimes be synthesized by the organism. This happens when the pathway for the metabolism of the compound is based on sequential induction. In this situation the organism is induced to produce an enzyme by the presence of a substrate. The intermediate resulting from the action of this enzyme on the substrate induces the production of another enzyme and so on until metabolism is accomplished. The other group of enzymes is produced whether or not the substrate on which they act, are present. These enzymes are known as *constitutive*.

Enzyme induction enables the organism to respond rapidly, sometimes within seconds, to the presence of a suitable substrate, so that unwanted enzymes are not manufactured.

Molecular basis for enzyme induction: The molecular mechanism for the rapid response of an organism to the presence of an inducer in the medium relates to protein synthesis since enzymes are protein in nature. Two models exist for explaining on a molecular basis the expression of genes in protein synthesis: one is a negative control and the other positive. The negative control of Jacob and Monod first published in 1961 is the better known and more widely accepted of the two and will be described first.

6.1.1.1   The Jacob-Monod Model of the (negative) control of protein synthesis

In this scheme (Fig. 6.1) the synthesis of polypeptides and hence enzymes protein is regulated by a group of genes known as the operon and which occupies a section of the chromosomal DNA. Each operon controls the synthesis of a particular protein. An operon includes a regulator gene (R) which codes for a repressor protein. The repressor can bind to the operator gene (O) which controls the activity of the neighboring structural genes (S). The production of the enzymes which catalyze the transcription of the message on the DNA into mRNA (namely, RNA polymerase) is controlled by the promoter gene (P). If the repressor protein is combined with the operator gene (O) then the movement of RNA polymerase is blocked and RNA complementary to the DNA in the structural genes (S) cannot be made. Consequently no polypeptide and no enzyme will be made. In the absence of the attachment of the repressor to the operator gene, RNA polymerase from the promoter can move to, and transcribe the structural genes, S.

Inducible enzymes are made when an inducer is added. Inducers inactivate or remove the repressor protein thus leaving the way clear for protein synthesis. Constitutive enzymes occur where the regulator gene (R) does not function, produces an inactive repressor, or produces a repressor to which the operator cannot bind. Often more than one structural gene may be controlled by a given operator.

Mutations can occur in the regulator (R) and operator (O) genes thus altering the nature of the repressor or making it impossible for an existing repressor to bind onto the operator. Such a mutation is called constitutive and it eliminates the need for an inducer. The structural genes of inducible enzymes are usually repressed because of the

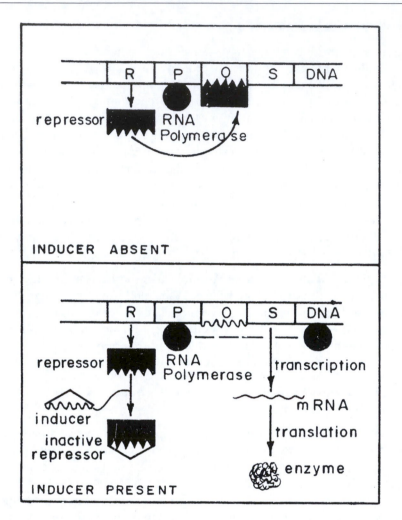

Fig. 6.1   Diagram Illustrating Negative Control of Protein Synthesis According to the Jacob and Monod Model

attachment of the repressor to the operator. During induction the repressor is no longer a hindrance, hence induction is also known as de-repression. In the model of Jacob and Monod gene expression can only occur when the operator gene is free. (i.e., in the absence of the attachment of the repressor protein the operator gene O. For this reason the control is said to be negative.

6.1.1.2   Positive control of protein synthesis

Positive control of protein synthesis has been less well studied but has been established in at least one system, namely the *ara* operon, which is responsible for L-arabinose utilization in *E. coli*. In this system the product of one gene (ara C) is a protein which combines with the inducer arabinose to form an activator molecule which in turn initiates action at the operon. In the scheme as shown in Fig. 6.2, 'C' protein combines with arabinose to produce an arabinose – 'C' protein complex which binds to the

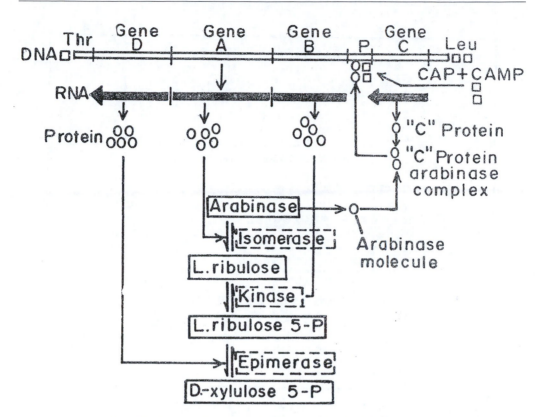

Fig. 6.2 Diagram Illustrating Positive Control of Protein Synthesis

Promoter P and initiates the synthesis of the various enzymes isomerase, kinase. epimerase) which convert L-arabinose to D-xylulose-5-phosphate, a form in which it can be utilized in the Pentose Phosphate pathway (Chapter 5). Positive control of protein synthesis also operates during catabolite repression (see below).

6.1.2 Catabolite Regulation

The presence of carbon compounds other than inducers may also have important effects on protein synthesis. If two carbon sources are available to an organism, the organism will utilize the one which supports growth more rapidly, during which period enzymes needed for the utilization of the less available carbon source are repressed and therefore will not be synthesized. As this was first observed when glucose and lactose were supplied to E. coli, it is often called the 'glucose effect', since glucose is the more available of the two sugars and lactose utilization is suppressed as long as glucose is available. It soon became known that the effect was not directly a glucose effect but was due to some catabolite. The term *catabolite repression* was therefore adopted as more appropriate. It must be borne in mind that other carbon sources can cause repression (see later) and that sometimes it is glucose which is repressed.

The active catabolite involved in catabolite repression has been found to be a nucleotide cyclic 3'5'-adenosine monophosphate (cAMP), (Fig. 6.3). In general, less

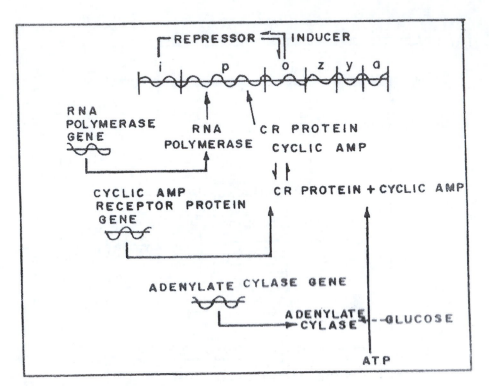

Fig. 6.3   Action of Cyclic Amp on the Lac Operon

c-AMP accumulates in the cell during growth on carbon compounds supporting rapid growth of the organism, vice versa.

During the rapid growth that occurs on glucose, the intracellular concentration of cyclic AMP is low. C-AMP stimulates the synthesis of a large number of enzymes and in necessary for the synthesis of the mRNA for all the inducible enzymes in *E.coli*. When it is low as a result of growth on a favorable source the enzymes which need to be induced for the utilization of the less available substrate are not synthesized.

Unlike the negative control of Jacob and Monod, c-Amp exerts a positive control. Another model explains the specific action in catabolite repression of glucose. In this model an increased concentration of c-AMP is a signal for energy starvation. When such a signal is given, c-Amp binds to an intracellular protein, c-AMP-receptor protein (CRP) for which it has high affinity. The binding of this complex to the promoter site of an operon stimulates the initiation of operon transcription by RNA polymerase (Fig. 6.3). The presence of glucose or a derivative of glucose inhibits adenylate cyclase the enzyme which converts ATP to c-AMP. Transcription by susceptible operons is inhibited as a result. In short, therefore, catabolite repression is reversed by c-AMP.

In recent times, for instance, it has been shown that c-AMP and CRP are not the only mediators of catabolite repression. It has been suggested that while catabolite repression in enterobacteria at least is exerted by the catabolite(s) of a rapidly utilized glucose source it is regulated in a two-fold manner: positive control by c-AMP and a negative control by a catabolite modulation factor (CMF) which can interfere with the operation of operons senstitive to catbolite repression. In *Bacillus* c-AMP has not been observed, but an analogue of c-AMP is probably involved.

6.1.3   Feedback Regulation

Feedback or end-product regulations control exerted by the end-product of a metabolic pathway, hence its name. Feedback regulations are important in the control over anabolic or biosynthetic enzymes whereas enzymes involved in catabolism are usually controlled by induction and catabolite regulation. Two main types of feedback regulation exist: feedback inhibition and feedback repression. Both of them help adjust the rate of the production of pathway end products to the rate at which macro-molecules are synthesized (see Fig. 6.4).

6.1.3.1   Feedback inhibition

In feedback inhibition the final product of metabolic pathway inhibits the action of earlier enzymes (usually the first) of that sequence. The inhibitor and the substrate need not resemble each other, hence the inhibition is often called *allosteric* in contrast with the *isosteic* inhibition where the inhibitor and substrate have the same molecular conformation. Feedback inhibition can be explained on an enzymic level by the structure of the enzyme molecule. Such enzymes have two type of protein sub-units. The binding site on the sub-unit binds to the substrate while the site on the other sub-unit binds to the feedback inhibitor. When the inhibitor binds to the enzyme the shape of the enzymes is changed and for this reason, it is no longer able to bind on the substrate. The situation is known as the allosteric effect.

6.1.3.2   Feedback Repression

Whereas feedback inhibition results in the reduction of the activity of an already synthesized enzyme, feedback repression deals with a reduction in the rate of synthesis of the enzymes. In enzymes that are affected by feedback repression the regulator gene (R) is said to produce a protein aporepressor which is inactive until it is attached to corepressor, which is the end-product of the biosynthetic pathway. The activated repressor protein then interacts with the operator gene (O) and prevents transcription of the structural genes (S) on to mRNA. A derivative of the end-product may also bring about feedback repression. It is particularly active in stopping the over production of vitamins, which are required only in small amounts (see Fig. 6.1).

While feedback inhibition acts rapidly, sometimes within seconds, in preventing the wastage of carbon and energy in manufacturing an already available catabolite, feedback repression acts more slowly both in its introduction and in its removal. About two generations are required for the specific activity of the repressed enzymes to rise to its maximum level when the repressing metabolite is removed; about the same number of generations are also required for the enzyme to be repressed when a competitive metabolite is introduced.

6.1.3.3 Regulation in branched pathway

In a branched pathway leading to two or more end-products, difficulties would arise for the organism if one of them inhibited the synthesis of the other. For this reason, several patterns of feedback inhibition have been evolved for branched pathways of which only six will be discussed. Each type of applicable to either feedback inhibition or feedback repression The descriptions below refer to Fig. 6.4

(i) *Concerted or multivalent feedback regulation*: Individual end-products F and H have little or no negative effect, on the first enzyme, E_1, but together they are potent inhibitors. It occurs in *Salmonella* in the branched sequence leading to valine, leucine, isoleucine and pantothenic acid.

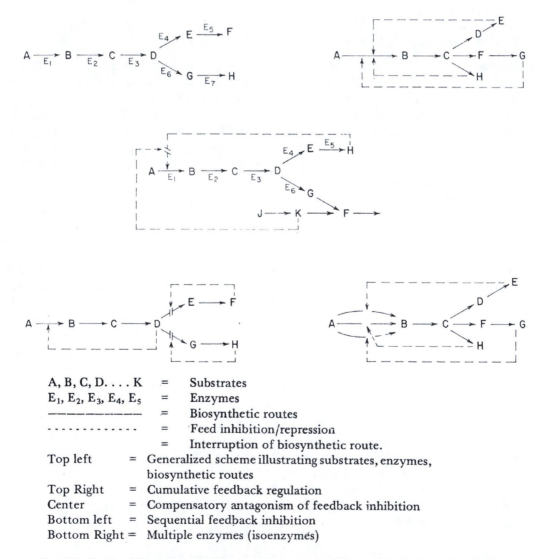

A, B, C, D . . . . K	=	Substrates
E_1, E_2, E_3, E_4, E_5	=	Enzymes
——————	=	Biosynthetic routes
- - - - - - - - - - - -	=	Feed inhibition/repression
	=	Interruption of biosynthetic route.
Top left	=	Generalized scheme illustrating substrates, enzymes, biosynthetic routes
Top Right	=	Cumulative feedback regulation
Center	=	Compensatory antagonism of feedback inhibition
Bottom left	=	Sequential feedback inhibition
Bottom Right =		Multiple enzymes (isoenzymes)

Fig. 6.4 Feedback Regulation (Inhibition and Repression) of Enzymes in Branched Pathways

(ii) *Cooperative feedback regulation*: In this case the end-products F and H are individually weakly inhibiting to the primary enzyme, E_1, but together they act synergistically, exerting an inhibition exceeding the sum of their individual activities.

(iii) *Cumulative feedback regulation*: In this system an end-product for example (H), inhibits the primary enzyme E_1 to a degree which is not dependent on other inhibitors. A second inhibitor further increases the total inhibition but not synergistically. Complete inhibition occurs only when all the products (E, G, H in Fig. 6.4) are present.

(iv) *Compensatory antagonism of feedback regulation*: This system operates where one of the end-products, F, is an intermediate in another pathway J, K, F (Fig. 6.4). In order to prevent the other end-product, H, of the original pathway from inhibiting the primary Enzyme E_1, and thus ultimately causing the accumulation of H, the intermediate in the second pathway J, K is able to prevent its own accumulation by decreasing the inhibitory effect of H on the primary enzyme E_1.

(v) *Sequential feedback regulation*: Here the end-products inhibit the enzymes at the beginning of the bifurcation of the pathways. This inhibition causes the accumulation of the intermediate just before the bifurcation. It is the accumulation of this intermediate which inhibits the primary enzyme of the pathway.

(vi) *Multiple enzymes (isoenzymes) with specific regulatory effectors*: Multiple primary enzymes are produced each of which catabolyzes the same reaction from A to B but is controlled by a different end-product. Thus if one end-product inhibits one primary enzyme, the other end products can still be formed by the mediation of one of the remaining primary enzymes.

6.1.4   Amino Acid Regulation of RNA Synthesis

Both protein synthesis and RNA synthesis stop when an amino acid requiring mutant exhausts the amino acid supplied to it in the medium. In this way the cell avoids the overproduction of unwanted RNA. Such economical strains are 'stringent'. Certain mutant strains are however 'relaxed' and continue to produce RNA in the absence of the required amino acid. The stoppage of RNA synthesis in stringent strains is due to the production of the nucleotide guanosine tetraphosphate (PpGpp) and guanosine pentaphosphate (ppGpp) when the supplied amino acid becomes limiting. The amount of ppGpp in the cell is inversely proportional to the amount of RNA and the rate of growth. Relaxed cells lack the enzymes necessary to produce ppGpp from guanosine diphosphate and ppGpp from guanosine triphosphate.

6.1.5   Energy Charge Regulation

The cell can also regulate production by the amount of energy it makes available for any particular reaction. The cell's high energy compounds adenosine triphosphate, (ATP), adenosine diphosphate (ADP), and adenosine monophosphate (AMP) are produced during catabolism. The amount of high energy in a cell is given by the adenylate charge or energy charge. This measures the extent to which ATP-ADP-AMP systems of the cell contains high energy phosphate bonds, and is given by the formula.

$$\text{Energy charge} = \frac{(ATP) + 1/2\,(ADP)}{(ATP) + (ADP) + AMP}$$

Using this formula, the charge for a cell falls between 0 and 1.0 by a system resembling feedback regulation, energy is denied reactions which are energy yielding and shunted to those requiring it. Thus, at the branch point in carbohydrate metabolism phosphoenolpyruvate is either dephosphorylated to give pyruvate or carboxylated to give oxalocetate. A high adenylate charge inhibits dephosphorylation and so leads to decreased synthesis of ATP. A high energy charge on the other hand does not affect carboylation to oxoloacetate. It may indeed increase it because of the greater availability of energy.

6.1.6   Permeability Control

While metabolic control prevents the overproduction of essential macromolecules, permeability control enables the microorganisms to retain these molecules within the cell and to selectively permit the entry of some molecules from the environment. This control is exerted at the cell membrane.

A solute molecule passes across a lipid-protein membrane only if there is driving force acting on it, and some means exists for the molecule to pass through the membrane. Several means are available for the transportation of solutes through membranes, and these can be divided into two: (a) passive diffusion, (b) active transport via carrier or transport mechanism.

6.1.6.1   Passive transport

The driving force in this type of transportation is the concentration gradient in the case of non-electrolytes or in the case of ions the difference in electrical charge across the membrane between the internal of the cell and the outside. Yeasts take up sugar by this method. However, few compounds outside water pass across the border by passive transportation.

6.1.6.2   Transportation via specific carriers

Most solutes pass through the membrane via some specific carrier mechanism in which macro-molecules situated in the cell membrane act as ferryboats, picking up solute molecules and helping them across the membrane. Three of such mechanisms are known:

   (i) *Facilitated diffusion*: This is the simplest of the three, and the driving force is the difference in concentration of the solute across the border. The carrier in the membrane merely helps increase the rate of passage through the membrane, and not the final concentration in the cell.

  (ii) *Active transport*: This occurs when material is accumulated in the cell against a concentration gradient. Energy is expended in the transportation through the aid of enzymes known as permeases but the solute is not altered. The permeases act on specific compounds and are controlled in many cases by induction or repression so that waste is avoided.

(iii) *Group translocation*: In this system the solute is modified chemically during the transport process, after which it accumulates in the cell. The carrier molecules act like enzymes catalysing group-transfer reactions using the solute as substrate. Group translocation can be envisaged as consisting of two separate activities: the entrance process and the exit process. The exit process increases in rate with the accumulation of cell solute and is carrier-mediated, but it is not certain whether the same carriers mediate entrance and efflux.

Carrier-mediated transportation is important because it is selective, and also because it is the rate-limiting step in the metabolism of available carbon and energy sources. As an increased rate of accumulation of metabolisable carbon source can increase the extent of catabolite repression of enzyme synthesis, the rate of metabolisable carbon transport may have widespread effects on the metabolism of the entire organism.

6.2   DERANGEMENT OR BYPASSING OF REGULATORY MECHANISMS FOR THE OVER-PRODUCTION OF PRIMARY METABOLITES

The mechanisms already discussed by which microorganisms regulate their metabolism ensure that they do not overproduce metabolites and hence avoid wastage of energy or building blocks. From the point of view of the organism an efficient organism such as *Escherichia coli* is one which does not permit any wastage: it switches on and off its synthetic mechanisms only as they are required and makes no concessions to the need of the industrial microbiologist to keep his job through obtaining excess metabolites from it!

The interest of the biotechnologist, the industrial microbiologist, and the biochemical engineer and indeed the entire industrial establishment and even the consuming public, is to see that the microorganism over produces desirable metabolites. If the microorganism is highly efficient and economical about what it makes, then the adequate approach is to disorganize its armamentarium for the establishment of order and thus cause it to overproduce. In the previous section we discussed methods by which the organism avoids overproduction. We will now discuss how these control methods are disorganized. First the situation concerning primary metabolites will be discussed, and later secondary metabolites will be looked at.

The methods used for the derangement of the metabolic control of primary metabolites will be discussed under the following headings: (1) Metabolic control; (a) feedback regulation, (b) restriction of enzyme activity; (2) Permeability control.

6.2.1   Metabolic Control

6.2.1.1   Feedback control

Feedback control is the major means by which the overproduction of amino acids and nucleotides is avoided in microorganisms. The basic ingredients of this manipulation are knowledge of the pathway of synthesis of the metabolic product and the manipulation of the organism to produce the appropriate mutants (methods for producing mutants are discussed in Chapter 7).

(i) **Overproduction of an intermediate in an unbranched pathway**: The accumulation of an intermediate in an unbranched pathway is the easiest of the various manipulations to be considered. Consider the production of end-product E following the series in Fig. 6.5.

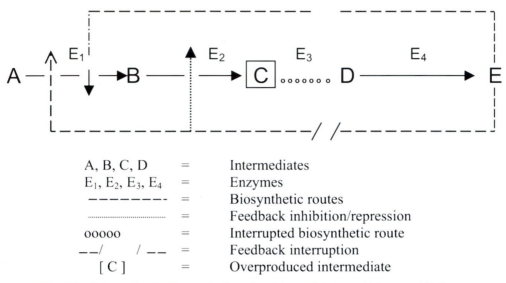

$$A \longrightarrow B \longrightarrow C \cdots\cdots\cdots D \longrightarrow E$$

A, B, C, D	=	Intermediates
E_1, E_2, E_3, E_4	=	Enzymes
– – – – – – – –	=	Biosynthetic routes
··············	=	Feedback inhibition/repression
ooooo	=	Interrupted biosynthetic route
– –/ / – –	=	Feedback interruption
[C]	=	Overproduced intermediate

Fig. 6.5 Scheme for the Overproduction of an Intermediate in an Unbranched Pathway

End-product E inhibits Enzyme 1 and represses Enzymes 2, 3, and 4. An auxotrophic mutant is produced (Chapter 7) which lacks Enzyme 3. Such a mutant therefore requires E for growth. If limiting (low levels) of E are now supplied to the medium, the amount in the cell will not be enough to cause inhibition of Enzyme 1 or repression of Enzyme 2 and C will therefore be over produced, and excreted from the cells. This principle is applied in the production of ornithine by a citrulline-less mutant (citrulline auxotroph) of *Corynebacterium glutamicum* to which low level of arginine are supplied (Fig. 6.6).

(ii) **Overproduction of an intermediate of a branched pathway; Inosine –5-monophosphate (IMP) fermentation**: This is a little more complicated than the previous case. Nucleotides are important as flavoring agents and the over-production of some can be carried out as shown in Fig. 6.7. In the pathway shown in Fig. 6.7 end-products adenosine 5- monophosphate (AMP) and guanosine –5-monophsophate (GMP) both cumulatively feedback inhibit and repress the primary enzyme [1].

Furthermore, AMP inhibits enzyme [11] which coverts IMP to xanthosine-5-monophosphate (XMP). By feeding low levels of adenine to an auxotrophic mutant of *Corynebacterium glutamicum* which lacks enzyme [11] (also known as adenineless because it cannot make adenine) IMP is caused to accumulate. The conversion of IMP to XMP is inhibited by GMP at [13]. When the enzyme [14] is removed by mutation, a strain requiring both guanine and adenine is obtained. Such a strain will excrete high amounts of XMP when fed limiting concentrations of guanine and adenine.

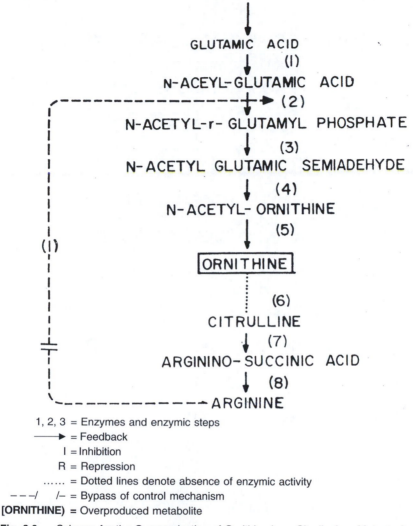

1, 2, 3 = Enzymes and enzymic steps
⟶ = Feedback
I = Inhibition
R = Repression
...... = Dotted lines denote absence of enzymic activity
‒ ‒ ‒/ /‒ = Bypass of control mechanism
[ORNITHINE) = Overproduced metabolite

Fig. 6.6   Scheme for the Overproduction of Ornithine by a Citrulineless Mutant of
Corynebacterium glutamicum

(iii) *Overproduction of end-products of a branched pathway*: The overproduction of end production of end-products is more complicated than obtaining intermediates. Among end-products themselves the production of end-products of branched pathways is easier than in unbranched pathways. Over-production of end-products of branched pathways will be discussed in this section; unbranched pathway will be dealt with later.

This is best illustrated (Fig. 6.8) using lysine, an important amino acid lacking in cereals and therefore added as a supplement to cereal foods especially in animal foods. It is produced using either *Corynebacterium glutamicum* or *Brevibacterium flavum*. Lysine is produced in these bacteria by a branched pathway that also produces methionine, isoleucine, and threonine. The initial enzyme in this pathway aspartokinase is regulated by concerted feedback inhibition of threonine

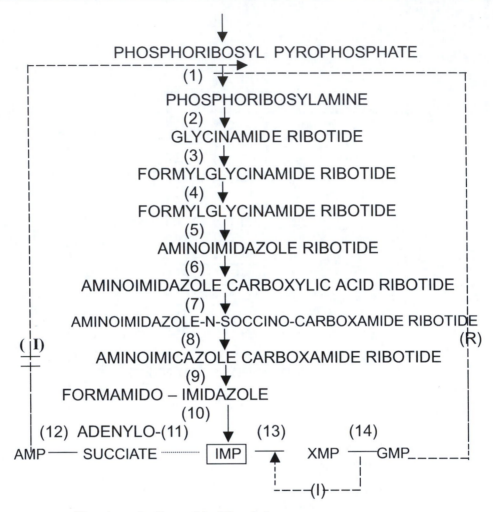

Key is as indicated in Fig. 6.6

Fig. 6.7 Scheme for the Overproduction of Inosinic Acid by an Adenine Auxotroph of *Corynebacterium glutamicum*

and lysine. By mutational removal of the enzyme which converts aspartate semi-aldehyde to homoserine, namely homoserine dehydrogenase, the mutant cannot grow unless methionine and threonine are added to the medium. As long as the threonine is supplied in limiting quantities, the intracelluar concentration of the amino acid is low and does not feed back inhibit the primary enzyme, aspartokinase. The metabolic intermediates are thus moved to the lysine branch and lysine accumulates in the medium (Fig. 6.8).

(iv) *Overproduction of end-product of an unbranched pathway*: Two methods are used for the overproduction of the end-product of an unbranched pathway. The first is the use of a toxic analogue of the desired compound and the second is to back-mutate an auxotrophic mutant.

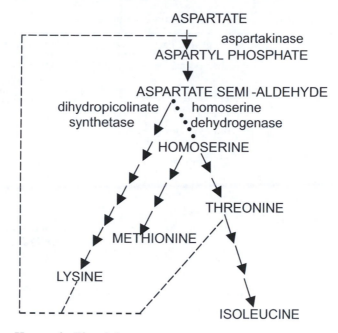

Key as in Fig. 6.6

Fig. 6.8   Lysine Overproduction Using a Mutant of *Corynebacterium glutamicum* Lacking
the Enzyme Homoserine Dehydrogenase

Use of toxic or feedback resistant analogues: In this method the organism (bacterial or yeast cells, or fungal spores) are first exposed to a mutagen. They are then plated in a medium containing the analogue of the desired compound, which is however also toxic to the organism. Most of the mutagenized cells will be killed by the analogue. Those which survive will be resistant to the analogue and some of them will be resistant to feedback repression and inhibition by the material whose overproduction is desired. This is because the mutagenized organism would have been 'fooled' into surviving on a substrate similar to, but not the same as offered after mutagenesis. As a result it may exhibit feedback inhibition in a medium containing the analogue but may be resistant to feed back inhibition from the material to be produced, due to slight changes in the configuration of the enzymes produced by the mutant. The net effect is to modify the enzyme produced by the mutant so that it is less sensitive to feedback inhibition. Alternatively the enzyme forming system may be so altered that it is insensitive to feedback repression. Table 6.2 shows a list of compounds which have been used to produce analogue-resistant mutants.

Use of reverse Mutation: A reverse mutation can be caused in the structural genes of an auxotrophic mutant in a process known as reversion. Enzymes which differ in structure from the original enzyme, but which are nevertheless still active, often result. It has been reported that the reversion of auxotrophic mutants lacking the primary enzyme in a metabolic pathway often results in revertants which excrete the end-product of the pathway. The enzyme in the revertant is active but differs from the original enzyme in being insensitive to feedback inhibition.

Table 6.2 Excretion of end-products by analogue-resistant mutants

Analogue	Compound Excreted	Organism
p–Fluorophenylalanin	Phenylalanine	*Pseudomonas* sp.
		Mycobacterium sp.
p–fluorophenylalanine	Tyrosine	*Escherichia coli*
Thienylalanine	Tyrosin + phenylalanine	*E. coli*
Thienylalanine	Phenylalanine	*E. coli*
Ethionine	Methionine	*E. coli, candida utilis*
		Neurospora crassa
Norleucine	Methinonine	*E. coli*
6-Methyltryptophan	Tryptophan	*Salmonella typhimurium*
5-Methyltryptophan	Tryptophan	*E. coli, escherichia*
		animdolica
Canavanine	Arginine	*E. coli*
Trifluoroleucine	Leucine	*S. typhimurium*
Valine	Isoleucine	*E. coli*
2–Thiazolealanine	Histidine	*E. coli, S. typhimurium*
3,4 – Dehydroproline	Proline	*E. coli*
2, 6 – Diaminopurine	Adenine	*S. typhimurium*

6.2.1.2 Restriction of enzyme activity

In the tricarboxylic acid cycle the accumulation of citric acid can be encouraged in *Aspergillus niger* by limiting the supply to the organism of phosphate and the metals which form components of co-enzymes. These metals are iron, manganese, and zinc. In citric acid production the quantity of these is limited, while that of copper which inhibits the enzymes of the TCA cycle is increased (Chapter 20).

6.2.2 Permeability

Ease of permeability is important in industrial microorganisms not only because it facilitates the isolation of the product but, more importantly, because of the removal of the product from the site of feedback regulation. If the product did not diffuse out of the cell, but remained cell-bound, then the cell would have to be disrupted to enable the isolation of the product, thereby increasing costs. The importance of permeability is most easily demonstrated in glutamic acid producing bacteria. In these bacteria, the permeability barrier must be altered in order that a high level of amino acid is accumulated in the broth. This increased permeability can be induced by several methods:

(i) *Biotin deficiency*: Biotin is a coenzyme in carboxylation and transcarboxylation reactions, including the fixation of CO_2 to acetate to form malonate. The enzyme which catalyses this is rich in biotin. The formation of malonyl COA by this enzyme (acetyl-COA carboxylase) is the limiting factor in the synthesis of long chain fatty acids. Biotin deficiency would therefore cause aberrations in the fatty acid produced and hence in the lipid fraction of the cell membrane, resulting in leaks in the membrane. Biotin deficiency has been shown also to cause aberrant forms in *Bacillus polymax, B. megaterium,* and in yeasts.

(ii) **Use of fatty acid derivatives**: Fatty acid derivatives which are surface-acting agents e.g. polyoxylene-sorbitan monostearate (tween 60) and tween 40 (-mono-palmitate) have actions similar to biotin and must be added to the medium before or during the log phase of growth. These additives seem to cause changes in the quantity and quality of the lipid components of the cell membrane. For example they cause a relative increase in saturated fatty acids as compared to unsaturated fatty acids.

(iii) **Penicillin**: Penicillin inhibits cell-wall formation in susceptible bacteria by interfering with the crosslinking of acetylmuranmic-polypeptide units in the mucopeptide. The cell wall is thus deranged causing glutamate excretion, probably due to damage to the membrance, which is the site of synthesis of the wall.

6.3 REGULATION OF OVERPRODUCTION IN SECONDARY METABOLITES

The physiological basis of secondary metabolite production is much less studied and understood than primary metabolism. Nevertheless there is increasing evidence that controls similar to those discussed above for primary metabolism also occur in secondary metabolites. Some examples will be given below:

6.3.1 Induction

The stimulatory effect of some compounds in secondary metabolite fermentation resembles enzyme induction. A good example is the role of tryptophan in ergot alkaloid fermentation by *Claviceps* sp. Although the amino acid is a precursor, its role appears to be more important as an inducer of some of the enzymes needed for the biosynthesis of the alkaloid. This is because analogues of tryptophan while not being incorporated into the alkaloid, also induce the enzymes used for the biosynthesis of the alkaloid. Furthermore, tryptophan must be added during the growth phase otherwise alkloid formation is severely reduced. This would also indicate that some of the biosynthetic enzymes, or some chemical reactions leading to alkaloid transformation take place in the trophophase, thereby establishing a link between idiophase and the trophophase. A similar induction appears to be exerted by methionine in the synthesis of cephalosporin C by *Cephalosporium ocremonium.*

6.3.2 Catabolite Regulation

Catabolite regulation as seen earlier can be by repression or by inhibition. It is as yet not possible to tell which of these is operating in secondary metabolism. Furthermore, it should be noted that catabolite regulations not limited to carbon catabolites and that the recently discovered nitrogen catabolite regulation noted in primary metabolism also occurs in secondary metabolism

6.3.2.1   Carbon catabolite regulation

The regulation of secondary metabolism by carbon has been known for a long time. In penicillin production it had been known for a long time that penicillin is not produced in a glucose-containing medium until after the exhaustion of the glucose, when the idiophase sets in; the same effect has been observed with cephalosporin production. Indeed the 'glucose effect' in which production is suppressed until the exhaustion of the sugar is well known in a large number of secondary products. Although the phenomenon where an easily utilizable source is exhausted before a less available is used has been described as glucose effect, it is clearly a misleading term because other carbon sources may be preferred in two-sugar systems when glucose is absent. Thus, β-carotene production by *Mortierella* sp. is best on fructose even though galoctose is a better carbon-source for growth. Carbon sources which have been found suitable for secondary metabolite production include sucrose (tetracycline and erythromycin), soyabena oil (kasugamycin), glycerol (butirosin) and starch and dextrin (fortimicin). Table 6.3 shows a list of secondary metabolites whose production is suppressed by glucose as well as non- interfering carbon sources.

It is fairly easy to decide whether the catabolite is *repressing or inhibiting the synthesis.* In *catabolite* repression the synthesis of the enzymes necessary for the synthesis of the metabolite is repressed. It is tested by the addition of the test substrate just prior to the initiation of secondary metabolite synthesis where upon synthesis is severely repressed. To test for catabolite inhibition by glucose or other carbon source it is added to a culture already producing the secondary metabolite and any inhibition in the synthesis noted.

Table 6.3   Secondary metabolites whose production is suppressed by glucose

Secondary Metabolite	Organism	Non-interfering Carbon Sources
Actinomycin	*Streptomyces antibioticus*	Galoactose
Indolmycin	*Streptomyces griseus*	Fructose
Kanamycin	*Streptomyces kanamyceticus*	Galactose
Mitomycin	*Streptomyces verticillatus*	Low glucose
Neomycin	*Streptomyces fradiae*	Maltose
Puromycin	*Streptomyces alboniger*	Glycerol
Siomycin	*Streptomyces sioyaensis*	Maltose
Streptomycin	*Steptomyces griseus*	Mannan
Bacitracin	*Bacillus licheniformis*	Citrate
Prodigiosin	*Seratia marcescens*	Galactose
Violacein	*Chromobacterium violaceum*	Maltose
Cephalosporin C	*Cephalosporium acremonium*	Sucrose
Ergot alkaloids	*Claviceps purperea*	-
Enniatin	*Fusarium sambucinum*	Lactose
Gibberellic acid	*Fusarium monoliforme*	-
Penicillin	*Penicillium chrysogenum*	Lactose

6.3.2.2 Nitrogen catabolite regulation

Nitrogen catabolite regulation has also been observed in primary metabolism. It involves the suppression of the synthesis of enzymes which act on nitrogen-containing substances (proteases, ureases, etc.) until the easily utilizable nitrogen sources e.g., ammonia are exhausted. In streptomycin fermentation where soyabean meal is the preferred substrate as a nitrogen source the advantage may well be similar to that of lactose in penicillin, namely that of slow utilization. Secondary metabolites which are affected by nitrogen catabolite regulation include trihyroxytoluene production by *Aspergillus fumigatus*, bikaverin by *Gibberella fujikuroi* and cephamycins by *Streptomyces* spp.

In all these cases nitrogen must be exhausted before production of the secondary metabolite is initiated.

6.3.3 Feedback Regulation

That feedback regulation exists in secondary metabolism is shown in many examples in which the product inhibits its further synthesis. An example is penicillin inhibition by lysine. Penicillin biosynthesis by *Penicillium chrysogenum* is affected by feedback inhibition by L-lysine because penicillin and lysine are end-products of a brack pathway (Fig. 6.9). Feedback by lysine inhibits the primary enzyme in the chain, homocitrate synthetase, and inhibits the production of α-aminoadipate. The addition of α-aminoadipate eliminats the inhibitory effect of lysine.

Self-inhibition by secondary meabolites: Several secondary products or even their analogues have been shown to inhibit their own production by a feedback mechanism. Examples are audorox, an antibiotic active against Gram-positive bacteria, and used in poultry feeds, chloramphenicol, penicillin, cycloheximids, and 6-methylsallicylic acid (produced by *Penicillium urticae*). Chloramphenicol repression of its own production is shown in Fig. 6.10, which also shows chorismic acid inhibition by tryptophan.

6.3.4 ATP or Energy Charge Regulation of Secondary Metabolites

Secondary metabolism has a much narrower tolerance for concentrations of inorganic phosphate than primary metabolism. A range of inorganic phosphate of 0.3-30 mM permits excellent growth of procaryotic and eucaryotic organisms. On the other hand the average highest level that favors secondary metabolism is 1.0 mM while the average lower quantity that maximally suppresses secondary process is 10 mM High phosphate levels inhibit antibiotic formation hence the antibiotic industry empirically selects media of low phosphate content, or reduce the phosphate content by adding phosphate-complexing agents to the medium. Several explanations have been given for this phenomenon. One of them is that phosphate stimulates high respiration rate, DNA and RNA synthesis and glucose utilization, thus shifting the growth phase from the idiophase to the trophophase. This shift can occur no matter the stage of growth of the organisms. Exhaustion of the phosphate therefore helps trigger off idiophase. Another hypothesis is that a high phosphate level shifts carbohydrate catabolism ways from

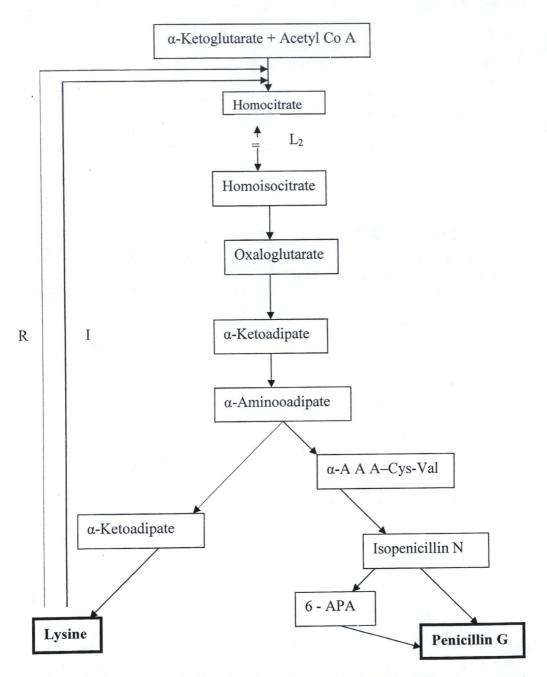

Penicillin and lysine are synthesized by a branched pathway in a mutant of *Penicillium chrysogenum*, L_2. α-AAA is the branching intermediate. Mutant L_2 is blocked before homoisocitrate and therefore accumulates homocitrate. The first enzyme is repressed (R) and and inhibited (I) by L-lysine, but not by penicillin G. 6-APA = 6-amino penicillanic acid.

Fig. 6.9 Penicillin Synthesis by a Mutant of *Penicillium chrysogenum*, L_2

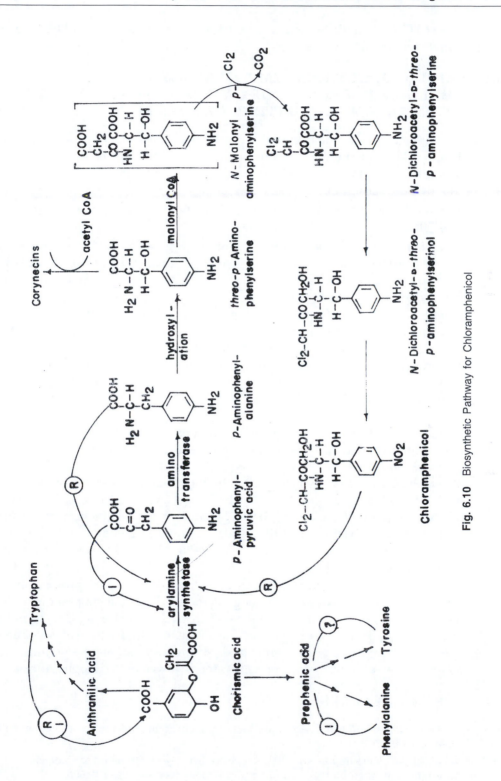

Fig. 6.10 Biosynthetic Pathway for Chloramphenicol

HMP to the EMP pathway favoring glycolysis. If this is the case then NADPH would become limiting of ridiolite synthesis.

6.4 EMPIRICAL METHODS EMPLOYED TO DISORGANIZE REGULATORY MECHANISMS IN SECONDARY METABOLITE PRODUCTION

Metabolic pathways for secondary metabolites are becoming better known and more rational approaches to disrupting the pathways for overproduction are being employed. More work seems to exist with regard to primary metabolites. Methods which are used to induce the overproduction of secondary metabolites are in the main empirical. Such methods include mutations and stimulation by the manipulation of media components and conditions.

   (i) *Mutations*: Naturally occurring variants of organisms which have shown evidence of good productivity are subjected to mutations and the treated cells are selected randomly and tested for metabolite overproduction. The nature of the mutated gene is often not known.

  (ii) *Stimulatory effect of precursors*: In many fermentations for secondary metabolites, production is stimulated and yields increased by the addition of precursors. Thus penicillin production was stimulated by the addition of phenylacetic acid present in corn steep liquor in the early days of penicillin fermentation. For the experimental synthesis of aflatoxin by *Aspergillus parasiticus*, methionine is required. In mitomycin formation by *Streptomyces verticillatus*, L-citurulline is a precursor.

 (iii) *Inorganic compounds*: Two inorganic compounds which have profound effects of fermentation for secondary metabolites are phosphate and maganese. The effect of inorganic phosphate has been discussed earlier. In summary, while high levels of phosphate encourage growth, they are detrimental to the production of secondary metabolites. Manganese on the other hand specifically encourages idiophase production particularly among bacilli, including the production of bacillin, bacitracin, mycobacillin, subtilin, D-glutamine, protective antigens and endospores. Surprisingly, the amount needed are from 20 to several times the amount needed for growth.

 (iv) *Temperature*: While the temperature range that permits good growth (in the trophophase) spans about 25°C among microorganisms, the temperature range within which secondary metabolites are produced is much lower, being in the order of only 5-10°C. Temperatures used in the production of secondary metabolites are therefore a compromise of these situations. Sometimes two temperatures – a higher for the trophophase and a lower for the idiophase are used.

SUGGESTED READINGS

Betina, V. 1995. Differentiation and Secondary Metabolism in Some Prokaryotes and Fungi Folia Microbiologica 40, 51–67.

Bull, A.T., Ward, A.C., Goodfellow, M. 2000. Search and Discovery Strategies for Biotechnology: the Paradigm Shift. Microbiology and Molecular Biology Reviews, 64, 573–606.

Demain, A.L. 1998. Induction of microbial secondary metabolism. International Microbiology, 1: 259–264.

Martin, J.F., Demain, A.L. 1980. Control of antibiotic biosynthesis. Microbiological Reviews 44, 230–251.

Krumphanzl, V., Sikyta, B., Vanck, Z. 1982. Overproduction of Microbial Products. Academic Press, London and New York.

Spizek, J.J., Tichy, P. 1995. Some Aspects of Overproduction of Secondary Metabolites. Folia Microbiologica 40, 43–50.

Vinci, A.V., Byng, G. 1999. Strain Improvement by Nono-recombinant Methods. In: Manual of Industrial Microbiology and Biotechnology. A S M Press, 2nd Ed. Washington, DC, USA, pp. 103–113.

Watts, J.E.M., Huddleston-Anderson, A.S., Wellington, E.M.H. 1999. Bioprospecting. In: Manual of Industrial Microbiology and Biotechnology. 2nd ed. A S M Press, Washington, DC, USA, pp. 631–641.

CHAPTER 7

Screening for Productive Strains and Strain Improvement in Biotechnological Organisms

In the last several decades, a group of microbial secondary metabolites, the antibiotics, has emerged as one of the most powerful tools for combating disease. So important are antibiotics as chemotherapeutic agents that much of the effort in searching for useful bioactive microbial products has been directed towards the search for them. Thousands of secondary metabolites are, however, known and they include not only antibiotics, but also pigments, toxins, pheromones, enzyme inhibitors, immunomodulating agents, receptor antagonists and agonists, pesticides, antitumor agents and growth promoters of animals and plants. When appropriate screening has been done on secondary metabolites, numerous drugs outside antibiotics have been found. Some of such non-antibiotic drugs are shown in Table 7.1. It seems reasonable from this to conclude that the exploitation of microbial secondary metabolites, useful to man outside antibiotics, has barely been touched. A special effort is made in this book to discuss method for assaying microbial metabolites for drugs outside of antibiotics (Chapter 28).

This section will therefore discuss in brief general terms the principles involved in searching for microorganisms producing metabolites of economic importance. More detailed procedures will be examined when various products are discussed. The genetic improvement of strains of organisms used in biotechnology, including microorganisms, plants and animals is also discussed.

7.1 SOURCES OF MICROORGANISMS USED IN BIOTECHNOLOGY

7.1.1 Literature Search and Culture Collection Supply

If one was starting from scratch and had no idea which organism produced a desired industrial material, then perhaps a search on the web and in the literature, including patent literature, accompanied by contact with one or more of the established culture collections (Chapter 8) and the regulatory offices dealing with patents (Chapter 1) may

Table 7.1 Some microbial metabolites with non-antibiotic pharmacological activity

Compound	Activity	Producing Microorganism
Aspergillic acid	Antihypertensive	*Aspergillus* sp
Astromentin	Sommoth muscle relaxant	*Monascus* sp
Siolipn	Acceleration of fibrin clot	*Streptomyces sioyaensis*
Azaserine	Antidiuretic, antitumor	*Streptomyces fragilis*
Ovalicin	Immunosuppressive, antitumor	*Pseudeurotum ovalis*
Candicidin (and other polyene Macrolides)	Cholesterol lowering	*Streptomyces noursei*
Streptozotocin	Hyperglycemic, antitumor	*Streptomyces achromogenes*
Zygosporin A	Anti-inflammatory	*Cephalosporium acremonium*
Fusaric acid	Hypotensive	*Fusarium oxysporum*
Leupeptin family	Plasmin inhibitor	*Bacillus* sp
Pepstatin	Pepsin inhibitor	*Aspergiluus niger*
Oosponol	Dopamine β-hydroxlyase inhibitor	*Oospora adringens*
Fumagallin	Angiogenesis inhibitor	*Aspergillus fumigatus*

provide information on potentially useful microbial cultures. The cultures may, however, be tied to patents, and fees may be involved before the organisms are supplied, along with the right to use the patented process for producing the material. Generally, cultures are supplied for a small fee from most culture collections irrespective of whether or not the organism is part of a process patent.

7.1.2   Isolation *de novo* of Organisms Producing Metabolites of Economic Importance

Although the well-known ubiquity of microorganism implies that almost any natural ecological entity–water, air, leaves, tree trunks – may provide microorganisms, the soil is the preferred source for isolating organisms, because it is a vast reservoir of diverse organisms. Indeed microorganisms capable of utilizing virtually any carbon source will be found in soil if adequate screening methods are used. In recent times, other 'new' habitats, especially the marine environment, have been included in habitats to be studied in searches for bioactive microbial metabolites or 'bio-mining'. Some general screening methods are described below. Detailed methods for the discovery of new antibiotics and other bioactive metabolites will be discussed in Chapter 21 and Chapter 28.

7.1.2.1   Enrichment with the substrate utilized by the organism being sought

If the organism being sought is one which utilizes a particular substrate, then soil is incubated with that substrate for a period of time. The conditions of the incubation can also be used to select a specific organism. Thus, if a thermophilic organism attacking the substrate is required, then the soil is incubated at an elevated temperature. After a period of incubation, a dilution of the incubated soil is plated on a medium containing the substrate and incubated at the previous temperature (i.e., elevated for thermopile search). Organisms can then be picked out especially if some means has been devised to select them. Selection could, for instance, be based on the ability to cause clear zones in an agar plate as a result of the dissolution of particles of the substrate in the agar. In the search for α-amylase producers, the soil may be enriched with starch and subsequently suitable soil dilutions are plated on agar containing starch as the sole carbon source. Clear halos form around starch-splitting colonies against a blue background when iodine is introduced in the plate.

Continuous culture (Chapter 9) methods are a particularly convenient means of enriching for organisms from a natural source. The constant flow of nutrients over material from a natural habitat such as soil will encourage, and after a time, select for organisms able to utilize the substrate in the nutrient solution. Conditions such as pH, temperature, etc., may also be adjusted to select the organisms which will utilize the desired substrate under the given conditions. Agar platings of the outflow from the continuous culture setup are made at regular intervals to determine when an optimum population of the desired organism has developed.

7.1.2.2   Enrichment with toxic analogues of the substrate utilized by the organism being sought

Toxic analogues of the material where utilization is being sought may be used for enrichment, and incubated with soil. The toxic analogue will kill many organisms which utilize it. The surviving organisms are then grown on the medium with the non-toxic substrate. Under the new conditions of growth many organisms surviving from exposure to toxic analogues over-produce the desired end-products. The physiological basis of this phenomenon was discussed earlier in Chapter 6.

7.1.2.3   Testing microbial metabolites for bioactive activity

(i) Testing for anti-microbial activity

For the isolation of antibiotic producing organisms the metabolites of the test organism are tested for anti-microbial activity against test organisms. One of the commonest starting point is to place a soil suspension or soil particles on agar seeded with the test organism(s). Colonies around which cleared zones occur are isolated, purified, and further studied. This method is discussed more fully in Chapter 21 where discussion of the search for antibiotics is included.

(ii) Testing for enzyme inhibition

Microorganisms whose broth cultures are able to inhibit enzymes associated with certain disease may be isolated and tested for the ability to produce drugs for combating the disease. Enzyme inhibition may be determined using one of the two methods among those discussed by Umezawa in 1982. In the first method the product of the reaction between an enzyme and its substrate is measured using spectroscopic methods. The quantity of the inhibitor in the test sample is obtained by measuring (a) the product in the reaction mixture without the inhibitor and (b) the product in the mixture with the inhibitor (i.e., a broth or suitable fraction of the broth whose inhibitory potency is being tested). The percentage inhibition (if any) is calculated by the formula

$$\frac{(a - b)}{a} \times 100$$

The second method determines the quantity of the unreacted substrate. For this determination the following measurements of the substrate are made: (a) with the enzyme and without the inhibitor (i.e., broth being tested); (b) with the enzyme and with the inhibitor and; (c) without the enzyme and without the inhibitor. Percentage inhibition (if any) is determined by *(c-a) – (c-b)* x 100. The results obtained above enable the assessment of the existence of enzyme inhibitors and facilitate the comparison of the inhibitory ability of broths from several sources.

(iii) Testing for morphological changes in fungal test organisms
The effect on spore germination or change in hyphal morphology may be used to detect the presence of pharmacologically active products in the broth of a test organism. This method does not rely on the death or inhibition of microbial growth, which has been so widely used for detecting antibiotic presence in broths.

(iv) Conducting animal tests on the microbial metabolites
The effect of broth on various animal body activities such as blood pressure, immunosuppressive action, anti-coagulant activity are carried out in animals to determine the content of potentially useful drugs in the broth. This method is discussed extensively in Chapter 21, which discusses details of the search for the production of bioactive metabolites from microorganisms.

7.2 STRAIN IMPROVEMENT

Several options are open to an industrial microbiology organization seeking to maximize its profits in the face of its competitors' race for the same market. The organization may undertake more aggressive marketing tactics, including more attractive packaging while leaving its technical procedures unchanged. It may use its human resources more efficiently and hence reduce costs, or it may adopt a more efficient extraction system for obtaining the material from the fermentation broth. The operations in the fermentor may also be improved by its use of a more productive medium, better environmental conditions, better engineering control of the fermentor processes, or it may genetically improve the productivity of the microbial strain it is using. Of all the above options, strain improvement appears to be the one single factor with the greatest potential for contributing to greater profitability.

While realizing the importance of strain improvement, it must be borne in mind that an improved strain could bring with it previously non-existent problems. For example, a more highly yielding strain may require greater aeration or need more intensive foam control; the products may pose new extraction challenges, or may even require an entirely new fermentation medium. The use of a more productive strain must therefore be weighed against possible increased costs resulting from higher investments in extraction, richer media, more expensive fermentor operations and other hitherto non-existent problems. This possibility not withstanding, strain improvement is usually part of the program of an industrial microbiology organization.

To appreciate the basis of strain improvement it is important to remember that the ability of any organism to make any particular product is predicated on its capability for the secretion of a particular set of enzymes. The production of the enzymes, themselves depends ultimately on the genetic make-up of the organisms. Improvement of strains can therefore be put down in simple term as follows:

(i)   regulating the activity of the enzymes secreted by the organisms;
(ii) in the case of metabolites secreted extracellularly, increasing the permeability of the organism so that the microbial products can find these way more easily outside the cell;
(iii) selecting suitable producing strains from a natural population;
(iv) manipulation of the existing genetic apparatus in a producing organism;
(v)   introducing new genetic properties into the organism by recombinant DNA technology or genetic engineering.

Items (i) and (ii) above have been discussed in Chapter 6. The other possible procedures, namely selection from natural variants, modification of the genetic apparatus without the introduction of foreign DNA and the use of foreign DNA will be discussed below (Table 7.1).

7.2.1   Selection from Naturally Occurring Variants

In selection of this type, naturally occurring variants which over-produce the desired product are sought. Strains which were encountered but not selected should not be automatically discarded; the better ones are usually kept as stock cultures in the organization's culture collection for possible use in future genetic manipulations. Selection from natural variants is a regular feature of industrial microbiology and biotechnology. For example, in the early days of antibiotic production the initial increase in yield was obtained in both penicillin and griseofulvin by natural variants producing higher yields in submerged rather than in surface culture. Another example is lager beer manufacture where the constant selection of yeasts that flocculate eventually gave rise to strains which are now used for the production of the beverage. Similarly in wine fermentation yeasts were repeatedly taken from the best vats until yeasts of suitable properties were obtained.

Selection of this type is not only slow but its course is largely outside the control of the biotechnologist, an intolerable condition in the highly competitive world of modern industry. Strain improvement is therefore mostly achieved by other means described below.

7.2.2   Manipulation of the Genome of Industrial Organisms in Strain Improvement

The manipulation of the genome for increased productivity may be done in one of two general procedures as shown in Table 7.2:

(a) manipulations not involving foreign DNA;
(b) manipulations involving foreign DNA .

7.2.2.1   Genome manipulations not involving Foreign DNA or Bases: Conventional Mutation

Nature of conventional mutation
The properties of any microorganism depend on the sequence of the four nucleic acid bases on its genome: adenine (A), thymine (T), cytosine (C), and guanine (G). The

Table 7.2   Methods of manipulating the genetic apparatus of industrial organisms

A. *Methods not involving foreign DNA*
     1. Conventional mutation

B. *Methods involving DNA foreign to the organism (i.e. recombination)*
     2. Transduction
     3. Conjugation
     4. Transformation
     5. Heterokaryosis
     6. Protoplast fusion
     7. Genetic engineering
     8. Metabolic engineering
     9. Site-directed mutation

arrangement of these DNA bases dictates the distribution of genes and hence the nature of proteins synthesized. A mutation can therefore be described as a change in the sequence of the bases in DNA (or RNA, in RNA viruses). It is clear that since it is the sequence of these bases which is responsible for the type of proteins (and hence enzymes) synthesized, any change in the sequence will lead ultimately to a change in the properties of the organism.

Mutations occur spontaneously at a low rate in a population of microorganisms. It is this low rate of mutations which is partly responsible for the variation found in natural populations. An increased rate can however be induced by mutagens, (or mutagenic agents) which can either be physical or chemical.

7.2.2.1.1   Physical agents

(i) ionizing radiations
(ii) ultraviolet light

(i) *Ionizing radiations*: X-rays, gamma rays, alpha-particles and fast neutrons are ionizing radiations and have all been successfully used to induce mutation. X-rays are produced by commercially available machines as well as van de Graaf generators. Gamma rays are emitted by the decay of radioactive materials such as Cobalt[60]. Fast neutrons are produced by a cyclotron or an atomic pile. Ionizing radiations are so called because they knock off the outer electrons in the atoms of biological materials (including DNA) thereby causing ionization in the molecules of DNA. As a result, highly reactive radicals are produced and these cause changes in the DNA. Some authors do not advise the use of ionizing radiations unless all other methods fail. This is party because the equipment is expensive and hence not always readily available, but also because ionizing radiations are apt to cause breakage in chromosomes.

(ii) *Ultraviolet light*: The mutagenic range of ultraviolet light lies between wave length 200 and 300 nm. 'Low pressure' UV lamps used for mutagenesis emit most of their rays in the 254 nm region. The suspension of cells or spores to be mutagenized is placed in a Petri dish 2-3 cm below a 15 watt lamp and stirred either by a rocking mechanism or by a magnetic stirrer. The organisms are exposed for varying periods lasting from about 300 seconds to about 20 minutes depending on the sensitivity of the organisms. Since UV damage can be repaired by exposure to light in a process known as photo-reactivation all manipulations should be conducted under a special light source such as 25 watt yellow or red bulbs. A proportion of the organisms ranging from about 60–99.9% should be killed by the radiation. The preference of workers as to the amount of kill varies, but the higher the kill the more the likelihood of producing desirable mutants. Furthermore, the higher the kill, the less likely it is that the killing is due to overheating consequent on having the organism too close to the lamp. The initial concentration of the organisms should also be in the order of 10^7 per ml.

The main effect of ultraviolet light on DNA is the formation of covalent bonds between adjacent pyrimidine (thymine and cytosine) bases. Thymine is mainly affected, and hence the major effect of UV light is thymine dimerization, although it can also cause thymine-cytosin and cytosin-cytosin dimers. Dimerization causes a distortion of the DNA double strand and the ultimate effect is to inhibit transcription and finally the organism dies (Fig. 7.1).

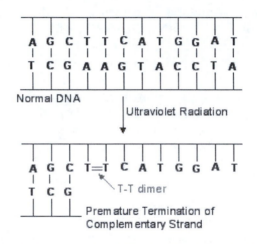

Fig. 7.1   Schematic Representation of Thymine Dimerization by UV Light on DNA

7.2.2.1.2   Chemical mutagens

These may be divided into three groups:

  (i) Those that act on DNA of resting or non-dividing organisms;
 (ii) DNA analogues which may be incorporated into DNA during replication;
(iii) Those that cause frame-shift mutations.

(i) *Chemicals acting on resting DNA*

Some chemical mutagens, such as nitrous acid and nitrosoguanidine work by causing chemical modifications of purine and pyrimidine bases that alter their hydrogen-bonding properties. For example, nitrous acid converts cytosine to uracil which then forms hydrogen bonds with adenine rather than guanine. These chemicals act on the non-dividing cell and include nitrous acid, alkylating agents and nitrosoguanidine (NTG) (also known as MNNG).

  (a) *Nitrous acid*: This acid is rather harmless and the mutation can be easily performed by adding 0.1 to 0.2 M of sodium nitrate to a suspension of the cells in an acid medium for various times. The acid is neutralized after suitable intervals by the addition of appropriate amounts of sodium hydroxide. The cells are plated out subsequently.

(b) *Alkylating agents*: These are compounds with one or more alkyl groups which can be transferred to DNA or other molecules. Many of them are known but the following have been routinely used as mutagens: EMS (ethyl methane sulphonate), EES (ethyl ethane sulphonate) and DES (Diethyl sulphonate). They are liquids and easy to handle. Cells are treated in solutions of about 1% concentration and allowed to react from ¼ hour to ½ hour and thereafter are plated out. Experimentation has to be done to decide the amount of kill that will provide a suitable amount of mutation. While some are carcinostatic (i.e., stop cancers), some are carcinogenic and must be handled carefully.

(c) *NTG* – nitrosoguanidine: also known as *M-methyl-N-nitro-M-guanidine* - MNNG: it is one of the most potent mutagens known and must therefore should be handled with care. Amounts ranging from 0.1 to 3.0 mg/ml have been used but for most mutations the lower quantity is used. It is reported to induce mutation in closely linked genes. It is widely used in industrial microbiology.

(d) *Nitrogen mustards*: The most commonly used of this group of compounds is methyl-bis (Beta-chlorethyl) amine also referred to as 'HN$_2$'. Nitrogen mustards were used for chemical warfare in World War I. Other members of the group are 'HN,' 'HN$_1$', or 'HN$_3$' from the wartime code name for mustard gas, H. The number after the H denotes the number of 2-chloroethyl groups which have replaced the methyl groups in trimethylamine. A spore or cell suspension is made in HN$_2$ (methyl-bis [Beta-chloroethyl amine]) and after exposure to various concentrations for about 30 minutes each, the reaction is ended by a decontaminating solution containing 0.7% NaHCO$_3$ and 0.6% glycine. The solution is then plated out for survivors. Between 0.05 and 0.1% HN$_2$ solutions in 2% sodium bicarbonate solutions have been found satisfactory for *Streptomyces*. Sometimes the exposure time may be extended

(ii) *Base analogues*

These are compounds which because they are similar to base nucleotides in composition may be incorporated into a dividing DNA in place of the natural base. However, this incorporation takes place only in special conditions. The best examples include 2-amino purine, a compound that resembles adenine, and 5-bromouracil (5BU), a compound that resembles thymine. The base analogs, however, do not have the hydrogen-bonding properties of the natural base. Base analogues are not useful as routine mutagens because suitable conditions for their use may be difficult to achieve. For example, with BU, incorporation occurs only when the organisms is starved of thymine.

(iii) *Frameshift mutagens (also known as intercalating agents)*

Frameshift or intercalating agents are planar three-ringed molecules that are about the same size as a nucleotide base pair. During DNA replication, these compounds can insert or intercalate between adjacent base pairs thus pushing the nucleotides far enough apart that an extra nucleotide is often added to the growing chain during DNA replication. A mutation of this sort changes all the amino acids downstream and is very likely to create a nonfunctional product since it may differ greatly from the normal

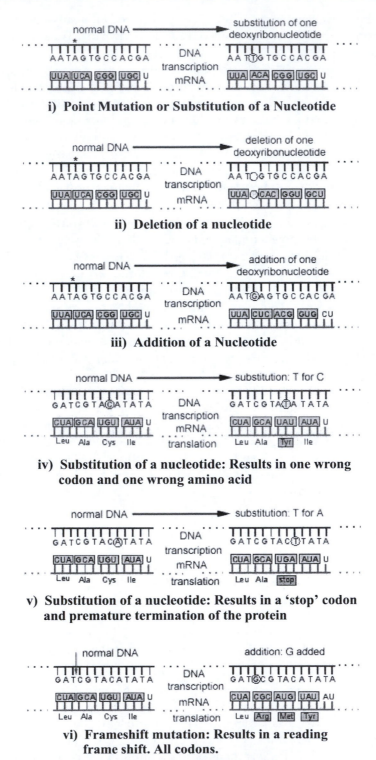

Fig. 7.2 Different Types of Mutation

protein. Furthermore, reading frames (i.e., the DNA base sequences) other than the correct one often contain stop codons which will truncate the mutant protein prematurely.

Acridines are among the best known of these mutagens, which cause a displacement or shift in the sequence of the bases. Although strongly mutagenic for some bacteriophages, acridines have not been found useful for bacteria. However, certain compounds, ICR (Institute for Cancer Research), (eg, ICR191) compounds in which an acridine nucleus is linked to an alkylating side chain, induce mutations in bacteria.

Acridine, $C_{13}H_9N$, is an organic compound consisting of three fused benzene rings (Fig. 7.3). Acridine is colorless and was first isolated from crude coal tar. It is a raw material for the production of dyes. Acridines and their derivatives are DNA and RNA binding compounds due to their intercalation abilities. Acridine Orange (3,6-dimethylaminoacridine) is a nucleic acid selective metachromatic stain useful for cell cycle determination. Another example is ethidium bromide, which is also used as a DNA dye.

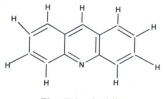

Fig. 7.3   Acridine

7.2.2.1.3   Choice of mutagen

Mutagenic agents are numerous but not necessarily equally effective in all organisms. Should one agent fail to produce mutations then another should be tried. Other factors besides effectiveness to be borne in mind are (a) the safety of the mutagen: many mutagens are carcinogens, (b) simplicity of technique, and (c) ready availability of the necessary equipment and chemicals.

Among physical agents, UV is to be preferred since it does not require much equipment, and is relatively effective and has been widely used in industry. Chemical methods other than NTG are probably best used in combination with UV. The disadvantage of UV is that it is absorbed by glass; it is also not effective in opaque or colored organisms.

7.2.2.1.4   The practical isolation of mutants

There are three stages before a mutant can come into use: the organisms must be exposed to a suitable mutagen under suitable conditions; the treated cells must be exposed to conditions which ideally select for the mutant; and finally, the mutant must then be tested for productivity.

(i) *Exposing organisms to the mutagen*: The organism undergoing mutation should be in the haploid stage during the exposure. Bacterial cells are haploid; in fungi and actinomycetes the haploid stage is found in the spores. However, in non-sporing strains of these organisms hyphae, preferable the tips, may be used. The use of haploid is essential because many mutant genes are recessive in comparison to the parent or wild-type gene.

(ii) *Selection for mutants*: Following exposure to the mutagen the cells should be suitably diluted and plated out to yield 50 – 100 colonies per plate. The selection of mutants is greatly facilitated by relying on the morphology of the mutants or on some selectivity in-built into the medium on which the treated cells or spores are plated.

When morphological mutants are selected, it is in the hope that the desired mutation is pleotropic (i.e., a mutation in which change in one property is linked with a mutation in another character). The classic example of a pleotropic mutation is to be seen in the development of penicillin-yielding strains of *Penicillium chrysogenum*. It was found in the early days of the development work on penicillin production that after irradiation, strains of *Penicillium chrysogenum* with smaller colonies and which also sporulated poorly were better producers of penicillin. Similar increases of metabolite production associated with a morphological change have been observed in organisms producing other antibiotics: cycloheximide, nystatin, and tetracyclines. In citric acid production it was observed that mutants with color in the conidia produced more of the acid; in some bacteria strains overproducing nucleic acid had a different morphological characteristic from those which did not.

In-built selectivity of the medium for mutants over the parent cells may be achieved by manipulating the medium. If, for example, it is desired to select for mutants able to stand a higher concentration of alcohol, an antibiotic, or some other chemical substance, then the desired level of the material is added to the medium on which the organisms are plated. Only mutants able to survive the higher concentration will develop. Toxic analogues may also be incorporated. Mutants resisting the analogues develop and may, for reasons discussed in Chapter 6, be higher yielding than the parent.

(iii) *Screening*: Screening must be carefully carried out with statistically organized experimentation to enable one to accept with confidence any apparent improvement in a producing organism. Shake cultures are preferred and about 6 of these of 500 ml capacity should be used. Accurate methods of identifying the desired product among a possible multitude of others should be worked out. It may also be better in industrial practice where time is important to carry out as soon as possible a series of mutations using ultraviolet, and a combination of ultraviolet and chemicals and then to test all the mutants.

Isolation of auxotrophic mutants

Auxotrophic mutants are those which lack the enzymes to manufacture certain required nutrients; consequently, such nutrients must therefore be added to the growth medium. In contrast the wild-type or prototrophic organisms possess all the enzymes needed to synthesize all growth requirements. As auxotrophic mutants are often used in industrial microbiology, e.g., for the production of amino acids, nucleotides, etc., their production will be described briefly below.

A procedure for producing auxotrophic mutants is illustrated in Fig. 7.4. The organism (prototroph) is transferred from a slant to a broth of the minimal medium (mm) which is the basic medium that will support the growth of the prototroph but not that of the auxotroph. The auxotroph will only grow on the complete medium, i.e., the minimal

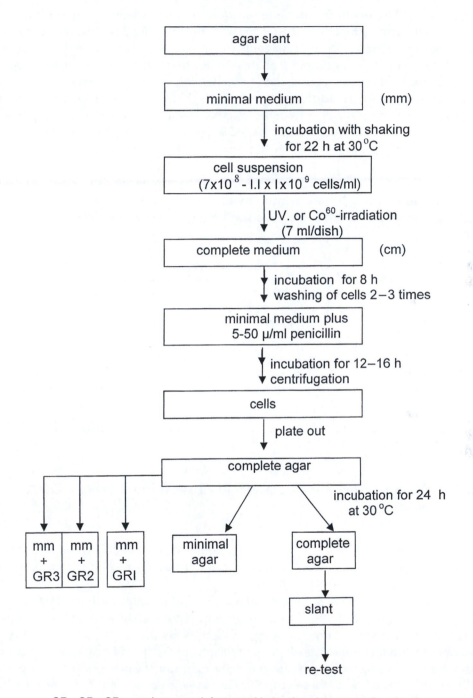

GR$_1$, GR$_2$, GR$_3$ = various growth factors added to the minimum medium (mm)

Fig. 7.4   Procedure for Isolating Auxotrophic Mutants

medium plus the growth factor, amino-acid or vitamin which the auxotroph cannot synthesize. The prototroph is shaken in the minimal broth for 22–24 hours, at the end of which period it is subjected to mutagenic treatment. The mutagenized cells are now grown on the complete medium for about 8 hours after which they are washed several times. The washed cells are then shaken again in minimal medium to which penicillin is added. The reason for the addition of penicillin is that the antibiotic kills only dividing cells; as only prototrophs will grow in the minimal medium these are killed off leaving the auxotrophs. The cells are washed and plated out on the complete agar medium.

In order to determine the growth factor or compound which the auxotroph cannot manufacture, an agar culture is replica-plated on to each of several plates which contain the minimal medium and various growth factors either single or mixed. The composition of the medium on which the auxotroph will grow indicates the metabolite it cannot synthesize; for example when the auxotroph requires lysine it is designated a 'lysine-less' mutant (Table 7.3).

Table 7.3 Growth of various mutants, produced after treatment of a wild-type organism

S/N	Complete medium (cm)	Minimal medium (mm)	mm + lysine	Growth on mm + biotin	mm + valine	Remarks
1	+	-	-	+	-	Biotin-less mutant
2	+	-	+	-	-	Lysine-less mutant
3	+	-	-	-	+	Valine-less mutant
4	+	-	-	+	+	Biotin-and valine-less
5	+	+	+	+	+	Parent Prototype

Key: + = growth
    – = no growth

7.2.2.2 Strain Improvement Methods Involving Foreign DNA or Bases

7.2.2.2.1 Transduction

Transduction is the transfer of bacterial DNA from one bacterial cell to another by means of a bacteriophage. In this process a phage attaches to, and lyses, the cell wall of its host. It then injects its DNA (or RNA) into the host.

Once inside the cell the viral genome may become attached to the host DNA or remain unattached forming a plasmid. Such a phage, which does not lyse the cell, is a temperate phage and the situation is known as lysogeny. Sometimes the viral genome may direct the host DNA to produce hundreds of copies of the phage. At the end of this manufacture the host is lysed releasing the viral particles into the medium; the new phages carry portions of the host DNA. If one of these viral particles now invades another bacterium, but is lysogenic in the new host, the new host will acquire some nucleic acid, and hence, some properties from the previous bacterial host. This process of the acquisition of new DNA from another bacterium through a phage is transduction.

Transduction is two broad types: *general transduction* and *specialized transduction*. In general transduction, host DNA from any part of the host's genetic apparatus is integrated into the virus DNA; in specialized transduction, which occurs only in some temperate phages, DNA from a specific region of the host DNA is integrated into the viral DNA and replaces some of the virus' genes.

It is now possible by methods which will be discussed later under the section on genetic engineering to excise genes responsible for producing certain enzymes and attach them on the special mutant viral particles, which do not cause the lysis of their hosts. Several hundreds of virus particles carrying the attached gene may therefore be present in one single bacterial cell following viral replication in it. The result is that the enzyme specified by the attached gene may be produced up to 1,000-fold. Gene amplification by phage is much higher than that obtained by plasmids (see below). The method is a well-established research tool in bacteria including actinomycetes but prospects for its use in fungi appear limited.

7.2.2.2.2   Transformation

Transformation is a change in genetic property of a bacterium which is brought about when foreign DNA is absorbed by, and integrates with the genome of, the donor cell. Cells in which transformation can occur are 'competent' cells. In some cases competence is artificially induced by treatment with a calcium salt. The transforming DNA must have a certain minimum length before it can be transformed. It is cut by enzymes, endonucleases, produced by the host before it is absorbed.

Reports of transformation in *Streptomyces* spp have been made. Transformation has been used to introduce streptomycin production into *Streptomyces olivaceus* with DNA from *Streptomycin grisesus*. Oxytetracycline producing ability was transformed into irradiated wild-type *S. rimosus*, using DNA from a wild-type strain. The technique has also been used to transform the production of the antifungal antibiotic thiolutin from *S. pimpirin* to a chlortetracycline producing *S. aureofaciens* which subsequently produced both antibiotics. An inactive strain of *Bacillus* was transformed to one producing the antibiotic bacitracin with the same method. The method has also been used to increase the level of protease and amylase production in *Bacillus* spp. The method therefore has good industrial potential.

7.2.2.2.3   Conjugation

Conjugation involves cell to cell contact or through sex pili (*singular*, pilus) and the transfer of plasmids. Conjugation involves a donor cell which contains a particular type of conjugative plasmid, and a recipient cell which does not. The donor strain's plasmid must possess a sex factor as a prerequisite for conjugation; only donor cells produce pili. The sex factor may on occasion transfer part of the hosts' DNA. Mycelial 'conjugation' takes place among actinomycetes with DNA transfer as in the case of eubacteria. Among sex plasmids of actinomycetes, perhaps the two best known are plasmids SCP1 and SCP2. Plasmids play an important role in the formation of some industrial products, including many antibiotics. Plasmids will be discussed in more detail later in this chapter.

7.2.2.2.4   Parasexual recombination

Parasexuality is a rare form of sexual reproduction which occurs in some fungi. In parasexual recombination of nuclei in hyphae from different strains fuse, resulting in the formation of new genes. Parasexuality is important in those fungi such as *Penicillium chrysogenum* and *Aspergiluss niger* in which no sexual cycles have been observed. It has been used to select organisms with higher yields of various industrial product such as phenoxy methyl penicillin, citric acid, and gluconic acid. Parasexuality has not become widely successful in industry because the diploid strains are unstable and tend to revert to their lower-yielding wild-type parents. More importantly is that the diploids are not always as high yielding as the parents.

7.2.2.2.5   Protoplast fusion

Protoplasts are formed from bacteria, fungi, yeasts and actinomycetes when dividing cells are caused to lose their cell walls. Protoplasts may be produced in bacteria with the enzyme lysozyme, an enzyme found in tears and saliva, and capable of breaking the β-1-4 bonds linking the building blocks of the bacterial cell wall. Protoplast fusion enables recombination in strains without efficient means of conjugation such as actinomycetes. It has also been used previously to produce plant recombinants. The technique involves the formation of stable protoplasts, fusion of protoplasts and subsequent regeneration of viable cells from the protoplasts. Fusion from mixed populations of protoplasts is greatly enhanced by the use of polyethylene glycol (PEG). Protoplast fusion has been successfully done with *Bacillus subtilis* and *B. megaterium* and among several species of *Streptomyces (S. coeli-color, S. acrimycini, S. olividans, S. pravulies)* has been done between the fungi *Geotrichum* and *Aspergillus*. The method has great industrial potential and experimentally has been used to achieve higher yields of antibiotics through fusion with protoplasts from different fungi.

7.2.2.2.6   Site-directed mutation

The outcome of conventional mutation which we have discussed so far, is random, the result being totally unpredictable. Recombinant DNA technology and the use of synthetic DNA now make it possible to have mutations at specific sites on the genome of the organism in a technique known as **Site-Directed Mutagenesis**. The mutation is caused by *in vitro* change directed at a specific site in a DNA molecule. The most common method involves use of a chemically synthesized oligonucleotide mutant which can hybridize with the DNA target molecule; the resulting mismatch-carrying DNA duplex may then be transfected into a bacterial cell line and the mutant strands recovered. The DNA of the specific gene to be mutated is isolated, and the sequence of bases in the gene determined (Chapter 3). Certain pre-determined bases are replaced and the 'new' gene is reinserted into the organism. Site-directed mutagenesis creates specific, well-defined mutations (i.e., specific changes in the protein product). It has helped to raise the industrial production of enzymes, as well as to produce specific enzymes.

7.2.2.2.7   Metabolic engineering

Metabolic engineering is the science which enables the rational designing or redesigning of metabolic pathways of an organism through the manipulation of the genes so as to maximize the production of biotechnological goods. In metabolic engineering, existing pathways are modified, or entirely new ones introduced through the manipulation of the genes so as to improve the yields of the microbial product, eliminate or reduce undesirable side products or shift to the production of an entirely new product. It is a modern evolution of an existing procedure which as described earlier in Chapter 6, is used to induce over production of products by blocking some pathways so as to shunt productivity through another. In the older procedure the pathways are shut off by producing mutants in which the pathways are lacking using the various mutation methods described earlier. In metabolic engineering the desired genes are isolated, modified and reintroduced into the organism. Metabolic engineering is the logical end of site-directed mutagenesis. It has been used to overproduce the amino acid isoluecine in *Corynebacterium glutamicum,* and ethanol by *E. coli* and has been employed to introduce the gene for utilizing lactose into *Corynebacterium glutamicum* thus making it possible for the organism to utilize whey which is plentiful and cheap. Through metabolic engineering the gene for the utization of xylose was introduced into *Klebsiella* sp making it possible for the bacterium to utilize the wood sugar.

It is equally applicable to primary and secondary metabolites alike. Among primary metabolites the alcohol producing *adhB* gene from the high alcohol yielding bacterium, *Zymomonas mobilis* was introduced into *E. coli* and *Klebsiella oxytoca,* enabling these organisms to produce alcohol from a wide range of sugars, hexose and pentose. Other primary metabolites which have been produced in other organisms by introducing genes from extraneous sources are carotenoids, the intermediates in the manufacture of vitamin A in the animal body, and 1,3 propanediol (1,3 PD) an intermediate in the synthesis of polyesters. 1,3 PD is currently derived from petroleum and is expensive to produce. 1,3 PD has been produced by *E. coli* carrying genes from *Klebsiella pneumoniae* able to anaerobically produce the diol.

Among secondary metabolites, increase in the production of existing antibiotics, and the production of new antibiotics and anti-tumor agents have been enabled by metabolic engineering. The transfer of genes from *Streptomyces erythreus* to *Strep lividans* facilitated the production of erythromycin in the latter organism. In the field of anti-tumor drugs, epirubicin has less cardiotoxicity than others such as the more frequently prescribed doxorubicin. The chemical production of epirubicin is complicated and requires seven steps. However using a metabolic engineering method in which the erythromycin biosynthetic gene was introduced into *Strep peucetius* it has been possible to produce it directly by fermentation.

7.2.2.2.8   Genetic engineering

Genetic engineering, also known as recombinant DNA technology, molecular cloning or gene cloning. has been defined as the formation of new combinations of heritable material by the insertion of nucleic acid molecules produced by whatever means outside the cell, into any virus, bacterial plasmid or other vector system so as to allow their incorporation into host organisms in which they do not naturally occur but in which they are capable of continued propagation

The DNA to be inserted into the host bacterium may come from a eucaryotic cell, a prokaryotic cell or may even be synthesized chemically. The vector-foreign DNA complex which is introduced into the host DNA is sometimes known as a DNA chimera after the Chimera of classical Greek mythology which had the head of lion, the body of a goat and the tail of a snake.

A species has been described as a group of organisms which can mate and produce fertile offspring. A dog cannot mate with a cat; even if they did the offspring would not be fertile. A horse and the donkey are not the same species. Although they can mate, the offspring the mule, is not fertile. Genetic engineering has enabled the crossing of the species barrier, in that DNA from one organism can now be introduced into another where such exchange would not be possible under natural conditions. With this technology engineered cells are now capable of producing metabolic products vastly different from those of the unaltered natural recipient.

Procedure for the Transfer of the Gene in Recombinant DNA Technology (Genetic Engineering)

In broad items the following are the steps involved in *in vitro* recombination or genetic engineering. The bulk of the work done so far has been with *E. coli* as the recipient organism

1. Dissecting a specific portion from the DNA of the donor organism.
2. Attachment of the spliced DNA piece to a replicating piece of DNA (or vector), which can be from either a bacteriophage or a plasmid.
3. Transfer of the vector along with the attached DNA (i.e., the DNA chimera) into the host cell.
4. Isolation (or recognition) of cells successfully receiving and maintaining the vector and its attached DNA.

7.2.2.2.8.1   Dissection of a portion of the DNA of the donor organism
The donor DNA may come from a plant, an animal, a microorganisms or may even be synthesized in the laboratory.

The dissection of DNA at specific sites is done by enzymes obtained from various bacteria and known as *restriction endonucleases*. They will be discussed briefly below.

(i) Nature and Types of restriction endonucleases

Restriction endonucleases are nucleic acid-splitting enzymes and are termed 'restriction' because they help a host cell destroy or restrict foreign DNA which enter the cell. The host protects its DNA from its own restriction endonucleases by the introduction of methyl groups at recognition sites where the cleavage of the DNA occurs. The host DNA so protected is said to be 'modified.' For every restriction enzyme there is a modification one hence the enzymes exist as restriction-modification complexes. Their

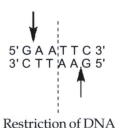

Restriction of DNA

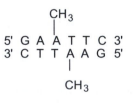

Modification of DNA

discovery was an important landmark in molecular biology. Daniel Nathans and Hamilton Smith received the 1978 Nobel Prize in Physiology and Medicine for their isolation of restriction endonucleases, which are able to cut DNA at specific sites.

Conventionally restriction enzymes are denoted as the single stranded DNA; the position of the restriction is written / while the position of the modification is written as an asterisk *. Thus the representation for the above enzyme would 5'G/AA*TTC3'.

There are four different types of restriction endonucleases: Types I, II, III and IV (Type IV is designated Type II S by some authors), but only Type II is used extensively in gene manipulations. In Types I and III, one enzyme is involved for recognition of specific DNA sequences for cleavage and methylation, but the cutting positions are at variable distances from these sites (sometimes up to 1000 base pairs (bps)) away from these sites. Type IV cuts only methylated DNA. As most molecular biology work is done with Type II endonucleases and only they will be discussed.

Type II endonucleases have the following advantages over the others. Firstly in Type II systems, restriction and modification are brought about by different enzymes and hence it is possible to cut DNA in the absence of modification (note that in Types I and III a single enzyme is involved); secondly, Type II enzymes are easier to use because they do not require enzyme cofactors. Finally as will be seen below they recognize a defined symmetrical sequence and cut within this sequence.

Type II restriction endonucleases recognize and cut DNA within particular sequences of 4 to 8 nucleotides in an axis of symmetry in such a way that the sequences of the top strand when read backwards are exactly like the bottom on the other side of the axis thus:

$$5'\text{-A T G} \mid \text{C A T-}3'$$
$$3'\text{-T A C} \mid \text{G T A-}5'$$
Axis of symmetry

Such sequences are referred to as palindromes. Type II restriction endonucleases were discovered in *Haemophilus influenzae* in 1970. About 3,000 of theses enzymes have now been discovered and they cut in about 200 patterns; many of them are available commercially.

(ii) Nomenclature of restriction endonucleases

The nomenclature of restriction endonucleases is based on the proposals of Smith and Nathans and the currently adopted procedure is as follows:

(a) The species name of the host organisms is identified by the first letter of the genus name and the first two letters of the species name to form a three-letter abbreviation written in italics. For example, *E. coli* is Eco and *Haemophilus inflenzae*, Hin.

(b) Strain or type identification is supposed to be written as a subscript. Thus, *E. coli* strain K, Eco_K. In practice it is all written in one line Ecok.

(c) Where a particular host has several different restriction and modification systems, these are identified by Roman numerals. Thus, those from *H. influenzae* strain Rd. would be *Hin*d I, *Hin*d II, *Hin*d III, in the order of their discovery.

(d) Restriction enzymes have the general name endonuclease R and in addition carry the system name, thus endonuclease *R. Hin*d I. Modification enzymes are named methylase M; thus the modification enzyme from *H. influenzae Rd.* is named methylase M. *Hin*d I. Where the context makes it clear that restriction enzymes are being discussed, 'endonuclease R' is left out leaving *Hin*d I as in the example quoted above.

(iii) Cutting DNA by Type II endonucleases

Type II endonucleases recognize and break DNA within particular sequences of four, five, six, or seven nucleotides (Table 7.4A, B) which have a symmetry along a central axis.

Table 7.4 Restriction Endonucleases

A: Patterns of Endonucleases Cutting);

1 **5' overhangs:** The enzyme cuts asymmetrically within the recognition site such that a short single-stranded segment extends from the 5' ends as wth BamHI.

2 **3' overhangs:** Again, we see asymmetrical cutting within the recognition site, but the result is a single-stranded overhang from the two 3' ends as KpnI does.

3 **Blunts:** Enzymes that cut at precisely opposite sites in the two strands of DNA generate blunt ends without overhangs. SmaI cuts in this way.

B: Some Restriction Endoncleases and their Recognition Sequences

ORGANISM	ENZYME	RECOGNITION SEQUENCES
Bacillus subtilis	*BsuRI*	CC /ĊC
Brevibacterium albdium	*BalI*	TGG / ĊCA
Escherichia coli	*EcoRI*	G / ÅATTC
Haemophilus influenzae	*HindII*	GTPy / PuÅC
Nocardiaotitidis-caviarum	*NotI*	GC / GGĊCGC

The same restriction endonuclease is used to cut the foreign DNA to be inserted into a vector, as well as the vector itself, in order to open it up.

Restriction enzymes cut DNA between deoxyribose and phosphate groups, leaving a phosphate at 5' end and an OH group at the 3'. The restriction enzymes used in genetic engineering cut within their recognition sites and generate one of three types of ends (see Table 74.4A):

a) Single-stranded, "sticky" or cohesive ends as cut by *Bam* H1 (1, 5' overhangs).

b) Single-stranded, "sticky" or cohesive ends as cut by *Kpn* 1 (2, 3' overhangs).

c) Double-stranded, "blunt" ends as cut by *Sma* 1. (3, Blunts)

The single-stranded sticky or cohesive ends of DNA ends (Table 7.4A and Fig. 7.5) will join (anneal) with any DNA with sticky ends, having complimentary bases no matter the origin of the DNA, provided that both DNA samples have been cut with the same restriction enzyme.

Some restriction endonucleases and their recognition sequences are given in Table 7.4B.

7.2.2.2.8.2 The attachment of the spliced piece of DNA to a vector

(i) *Joining DNA molecules*: Three methods are used for the *in vivo* 'tying' of DNA molecules. The first method uses an enzyme DNA ligase to tie sticky ends produced by restriction endonucleases; the second is the use of another DNA ligase produced by *E. coli* infected by T4 bacteriophages to link blunt ended DNA fragments. The third method uses an enzyme terminal deoxynucleotidyl – transferase isolated from calf thymus to introduce single-stranded complimentary tails to two different DNA populations after which they anneal when mixed. Only the first, method, i.e., the use of DNA ligase, will be discussed, because this has been used extensively.

(ii) *The use of DNA ligase to join foreign DNA to the vector*: High concentrations of the DNA of the previously circular vector (usually a plasmid) and of the foreign DNA to be cloned onto the vector, are mixed. Both DNA types have sticky ends having been treated with the same restriction endonuclease: in the case of the foreign DNA to cut it from its source and in the case of the vector, to open it up. Complimentary sticky ends from the foreign DNA and the vector anneal leaving however gaps created by the absence of a few base pairs in opposite strands (Fig. 7.5).

The enzyme DNA ligase can repair these gaps to create an intact duplex. DNA ligase is produced by *E. coli* and phage T4. The ligase from T4 can, however, join blunt-ended DNA whereas that from *E. coli* cannot. The vector-foreign DNA chimaera is then introduced into the bacterial cell by transformation.

To prevent recircularization of the linearized vector, it may be treated with alkaline phosphotase. When it is so treated circularization can only occur when a foreign DNA is introduced. A gap is left at each joint. These gaps are closed after transformation by the hosts' repair system.

(iii) *Vectors used in recombinant DNA work*: Two broad groups of cloning vehicles have been used, namely plasmids and lamda phages. Both have replication systems that are independent of that of the host cell.

7.2.2.2.8.3 Plasmids

Plasmids are circular DNA molecules with molecular weights ranging from a few million to a few hundred million Daltons. Plasmids appear to be associated with virtually all known bacterial genera. They replicate within the cell. Some of the larger plasmids, known as *conjugative plasmids*, carry a set of genes which promote their own transfer in a sexual process known as conjugation which has already been discussed. Smaller plasmids are usually non-conjugative but their transfer can usually be promoted by the presence of a conjugative plasmid in the same cell.

Besides genes for sexual transfer, plasmids usually carry genes for antibiotic or heavy metal resistance. They often also carry genes for the production of toxins, bacteriocins,

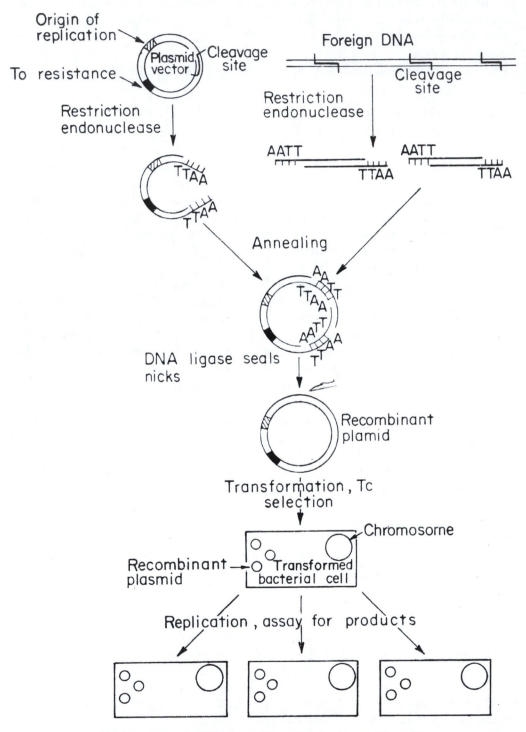

Fig. 7.5   Generalized Diagram of Procedure for Genetic Engineering

antibiotics, and unusual metabolites. In some cases they may carry genes for unusual capabilities such as the breakdown of complex organic compounds. Plasmids are, however, not essential for the cell's survival.

Two important features of plasmids to be used in genetic experiments may be compared by examining two plasmids. Plasmid psC101 has only two to five copies per cell and replicates with its host DNA. It is said to be under 'stringent' control. However, another plasmid pCol E 1 is found in about 25–30 copies per cell. It has a 'relaxed control' independent of the host and replicates without reference to the host DNA. When the host cell is starved of amino-acids or its protein synthesis is inhibited in some other manner, such as with the use of chloramphenicol, the Col E 1 plasmid continues to replicate for several hours until there are 1,000 to 3,000 copies per cell. Due to this high level of gene dosage (also referred to as gene amplification), products synthesized because of the presence of these plasmids are produced in extremely high amounts, a property of immense importance in biotechnology and industrial microbiology. Generally conjugative plasmids are large, exhibit stringent control of DNA replication, and are present in low copy numbers; on the other hand, non-conjugative plasmids are small, show relaxed DNA replication, and are present in high numbers. Many other plasmid vectors exist, some constructed in the laboratory (Table 7.5).

(i) *Ideal properties in a plasmid used as a vector*

A plasmid to be used in genetic engineering should ideally have the following properties:

(a) the plasmid should be as small as possible so the unwanted genes are not transmitted, as well as to facilitate handling;

(b) it should have an *origin of replication*, the site where DNA replication initiates;

(c) it should have a relaxed mode of replication;

(d) it should have sites for several restriction enzymes;

(e) it should carry, preferably, two marker genes. Marker genes are those which express characteristics by which the plasmid can be identified. Such characteristics include resistance to one or more antibiotics. A marker of great importance is the ability to satisfy auxotrophy, i.e., the ability to produce an amino acid or other nutritional component which the host's chromosome is incapable of producing.

Table 7.5   Some commonly used plasmid cloning vehicles

Plasmid	Molecular weight (x 10^{-6})	Marker*	Single restriction sites
pSC101	5.8	Tc	BamHI, EcoRI, HindIII, Hpal, SalI, Smal
Col E1	4.2	Colimm	EcoRI
pMB9	3.6	Tc^r, Colimm	BamHI, EcoRI, HindIII, Hpal, SalI Smal
pBR313	5.8	Tc^r, Ap^r Colimm	BamHI, EcorI, HindIII, Hpal, SalI, Smal
pBR322	2.6		BamHI, EcoRI, HindIII, pstI, SalI

*Tc^r: tetracycline resistance. Ap^r: ampicillin resistance
    colimm: colicin immunity

(f) the nucleotide sequence of the plasmid should be known;

(g) for safety reasons the plasmid should not be able to replicate at mammalian body temperatures so that should it enter the human body and be able to produce deleterious substances, it should fail to replicate;

(h) for safety reasons also, it should not be highly transmissible by conjugation if it controls the production of any material harmful to the mammalian body;

(i) the plasmid as a cloning vehicle should have a site for inducing transcription across the inserted fragment. The plasmid-initiated transcription should be controlled by the host (by induction or repression). Uncontrolled transcription could be harmful to the host.

Table 7.5 shows some commonly used plasmid cloning vehicles. They carry various markers based on tetracycline or ampicillin resistance or immunity against colicin attack. The marker may be carried either on the plasmid or on the inserted DNA. If neither of them carries a marker then DNA carrying a marker can be grafted on to either the vector or the insert.

(ii) *Plasmids currently in use for cloning*

In the early years of genetic engineering, naturally occurring plasmids such as Col E1 and pSC 101 were used as cloning vectors. They were small and had single sites for the common endonucleases. However they lacked markers which would help select transformed organisms. New plasmids were therefore developed. The best and most commonly used is pBR322 developed by Francisco Bolivar. (In naming plasmids p is used to show it is a plasmid; p is followed by the initials of the worker who isolated or developed the plasmid; numbers are used to denote the particular strain). Plasmid pBR322 has all the properties expected in a plasmid vector: low molecular weight, two markers, (resistance to ampicillin, Ap^R and tetracycline, Tc^R) an origin of replication, and several single-cut replication sites. (see map of pBR322 in Fig 7.6). Modifications of the original pBR322 have been made to suit special purposes, and consequently many variants exist in the pBR322 family. A widely used variant of pBR322 is pAT153, which some consider a better vector than its parent because it is present in more copies per cell than pBR322 Another series of popular vectors is the pUC family of vectors (Fig. 7.7). It has several unique restriction sites in a short stretch of DNA, which is an advantage in some kinds of work.

7.2.2.2.8.4   Phages

Two types of phages have been developed for cloning, λ lamda, and M13. Most of the phages used for cloning are derivatives of the lamda phage of *E. coli* because so much is already known about this phage. Derivatives are used because the wild-type phage is not suitable as a vector as it has several targets of sites for most of the most commonly used endonucleases. The chromosome of phage must be folded and encapsulated into the head of the virus in order to provide a mature virion. The amount of DNA that can enter the head is limited, and hence the available DNA in a phage is also limited. Therefore unwanted phage DNA must be removed as well as all but one of restriction targets for the chosen enzyme.

The DNA of phage lambda when it is isolated from the phage particle is linear and double-stranded. At each end of the chain are single-stranded portions which are

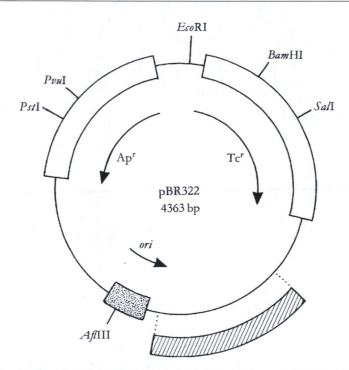

Fig. 7.6 Genetic Map of Plasmid pBR322 Showing Unique Recognition Endonuclease Sites and Genes for Tetracycline and Ampicillin Resistance

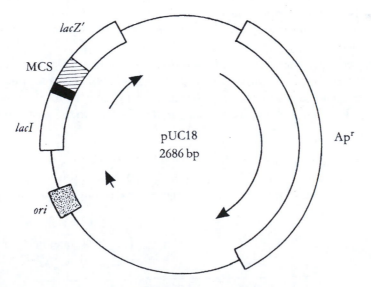

Fig. 7.7 Genetic Map of pUC18

complimentary to each other, much like the 'sticky ends' produced from DNA cutting by restriction endonucleases (Fig. 7.9). These lamba DNA pieces are able to circularize and replicate independently within the host.

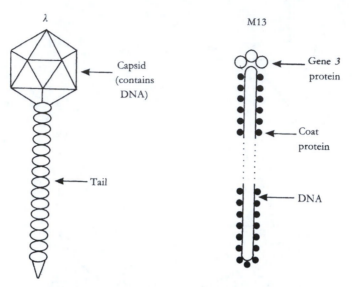

Fig. 7.8 Structure of λ and M13 Bacteriophages

The middle portion of the linear double-stranded phage DNA is non-essential for phage growth and it is here that the foreign DNA is introduced. The more distal positions carry genes which code for essential components such as the head, tail of the bacteriophage and the host lysis (Fig. 7.9)

(i) *Transfection*: The linear chimera can be introduced by transformation. (When virus DNA is transformed the process is known as transfection.) However, much of introduced chimeras are restricted in comparison to when pure phage DNA is transfected.

(ii) *Packaging the chimeras into virus heads*: The recombinant DNA or chimera may be packaged into a virus head and a tail attached by *in vitro* means. The procedure for this packaging is outside the scope of this book but may be found elsewhere. Once packaged, the synthetic virus can then inject its DNA into the host in the usual way.

7.2.2.2.8.5 Cosmids

Cosmids are plasmids constructed from phage DNA by circularization at the 'sticky,' single stranded ends or *cos sites*. Foreign DNA is attached to the cosmid which is then packaged into a phage. When the cosmid is injected it circularizes like other virus DNA

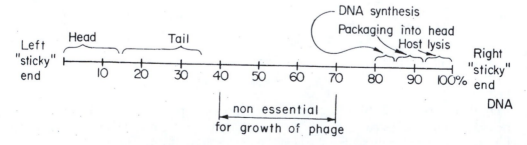

Fig. 7.9 Map of Chromosome of Lamba (λ) Phage

but it does not behave as a phage, rather it replicates like a plasmid. Drug resistance markers carried on it help to identify it.

A number of commercially available vectors are based on phages and some are shown in Table 7.6

Table 7.6 Phage-based vectors

Vector	Features	Applications
pBR322	$Ap^r Tc^r$	General cloning and sub-
	Single cloning sites	cloning in E. coli
pAT 153	$Ap^r Tc^r$	General cloning and sub-
	Single cloning sites	cloning in E. coli
pGEM™-32	Ap^r	General cloning and
	MCS	in vitro transcription in E.
	SP6/T7 promoters	coli and mammalian cells
	lacZ α-peptide	
pCI™	Ap^r	Expression of genes in
	MCS	mammalian cells
	T7 promoter	
	CMV enhance/promoter	
pCMV-Script™	Neo^r	Expression of genes in
	Large MCS	mammalian cells
	CMV enhancer/promoter	Cloning of PCR products

7.2.2.2.8.6 Transfer of the vector along with the attached DNA into the host all

The vector once spliced with the endonuclease cannot reform into a circular structure unless a suitable fragment of the foreign DNA with a complimentary 'sticky end' fits in. Foreign DNA digests produced by physical inactivation may also be used. If a large enough amount of foreign DNA digest is used the probability is that a piece with the appropriate complementary end will fit in. The new hybrid DNA is introduced into the host cell by transformation. Transformation is facilitated by treating the host cells in calcium salts, after washing them in magnesium salts.

7.2.2.2.8.7 Recognizing the transformed cell

The introduction of the new property into the host may be detected by growing the cells in a medium containing antibiotics whose resistance is specified in the introduced foreign DNA. Growth should occur if the resistance gene was transferred. If genes for the synthesis of some products were introduced via the chimera, the transformed bacteria should grow on the selective medium. The products are then examined for the synthesis of the new compound. For the introduced gene to lead to protein synthesis a suitable promoter must be present; it may be introduced from other organisms if an appropriate one is not indigenous.

7.2.2.2.8.8 Gene transfer into organisms other than *E. coli*, including plants and animals

The methods discussed above are those developed primarily for *E. coli*. The discussion below will look at the introduction of DNA into bacteria other than *E. coli* as well into other organisms, including plants and animals. Some of the methods to be discussed are also used on *E. coli*.

(i) *Delivery into bacteria other than E. coli*

(a) **Electroporation**

In the process of electroporation, cells into which DNA is to be introduced (i.e., cells to be **transfected**) are exposed to high-voltage electric pulses. This creates temporary holes in the cell membrane through which DNA can pass. Electroporation can be used for transfecting cells of *Bacilli* spp and actinomycetes, especially when protoplasts are produced from the cells. Electroporation is the short form for electric field-mediated membrane permeabilization. It is still used for *E. coli*, especially when chimera longer than about 100 kilobases (100 kb) are to be used. In general electroporation can be used for transfecting bacteria and archae, after the appropriate electric voltage and other parameters have been worked out. As will be seen below, it is also used for transfecting plant cells.

(b) **Conjugation**

In some bacteria where other means of introducing DNA appear difficult or have failed, the natural means of transferring DNA by plasmid mediated conjugation has been exploited. A conjugative plasmid which is carrying the insert and which has the genes for its own transfer is used. However where this is not possible, a conjugative plasmid with its own transfer gene may first be introduced, followed by the non-conjugative (i.e., does not promote the transfer of DNA through pili) plasmid carrying the DNA insert. This has been used in some strains of *Pseudomonas*.

(c) **Use of Liposomes**

When the DNA to be introduced is first entrapped in phospholipid droplets known as liposomes, it enhances the entry of the DNA into protoplasts of Gram-positive *Bacillus* and actinomycetes. Liposomes have also been used for delivering DNA into animal cells. Liposomes are microscopic, fluid-filled vesicles whose walls are made of layers of *phospholipids* identical to the phospholipids that make up cell membranes. The outer layer of the vesicle is hydrophobic, while the inner layer is hydrophilic; this enables the liposome to carry water soluble materials within it. They can be designed so that they have cationic, anionic or neutral charges at the hydrophobic end depending on the purpose for which they are meant. They are used for introducing DNA into animal, plant or bacterial cells. When used for introducing DNA into plant cells, such cells must have their cell walls removed, yielding protoplasts; with bacteria, the cell walls must also be removed to yield sphaeroplasts. Liposomes have been used experimentally to carry normal genes into a cell to replace defective, disease-causing genes in gene therapy. Liposomes are used to deliver certain vaccines, enzymes, or drugs (e.g., insulin and some cancer drugs) to the animal body. Liposomes are sometimes used in cosmetics because of their moisturizing qualities.

(ii) *Delivery of DNA into Plant cells*

Plants are peculiar in that most single plant cells can be caused to develop into the entire plants. Successfully transfecting (i.e., introducing foreign DNA into) a plant cell will result in having the foreign DNA as part of the genetic apparatus of the transfected plant.

The introduction of foreign DNA into plants is done for the improvement of the agricultural, ornamental, nutritional, or horticultural value of the plant. It is also done to convert the plants into 'living fermentors' which with the appropriate genes can manufacture cheaply, some industrially important materials, which it may not even be possible to produce by chemical means. Several methods are available for the delivery of DNA into plants

(a) Plant Transfection with Ti plasmid of Agrobacterium tumefaciens

The Gram-negative soil bacterium, *Agrobacterium tumefaciens*, is the causative agent of crown gall, a disease which produces tumors in plants (mainly dicots) on entering through a wounded plant cell.

The pathogenic properties of the bacterium are due to the Ti (tumor inducing) plasmid which it carries. Part of the Ti plasmid, known as the T (transfer) DNA, is transferred to the plant cell and is integrated into the genome of the host under the direction of the virulence gene (Fig. 7.10). It is within the TDNA that foreign DNA can be introduced. A section of the TDNA codes for the production of auxins and cytokins which lead to the formation of galls or tumors. Another section codes for the production of conjugates of amino acids and sugars known as opines and which are metabolized by *Agrobacterium tumefaciens* residing in the tumor. The oncogenic (tumor-causing) and the opine-producing portions of the TDNA of wild-type *Agrobacterium tumefaciens* are removed, when it is to be used for cloning. Furthermore, marker genes (e.g., for kamycin resistance) are introduced into the plasmids so that transformed plants can be identified. Because the marker genes are of bacterical origin an origin of replication from *E. coli* is also introduced. The TDNA is defined by the left and right borders. The sequences of the right border are essential for the TDNA transfer and integration into the host plant. The transfected plant cells therefore result in normal plants. Ti plasmids in which the oncogenic section has been removed is said to be 'disarmed'. Such disarmed plasmids lack the sequences necessary to produce the phytohormones which give rise to diseased conditions, gall or tumor. Other properties such as the transfer of DNA are still active and the regeneration of healthy plants can still occur. The *Agrobacterium tumefaciens* Ti plasmid has been successfully and widely used in cloning in plants. However it has been more successful in dicots than in monocots.

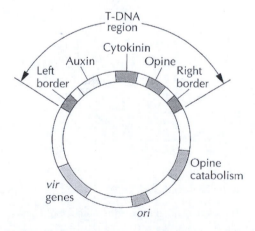

Fig. 7.10 Map of *Agrobacterium tumefaciens* Ti Plasmid

(b) **Use of Viruses**

The cauliflower mosaic virus (caMV) has been used as a powerful vector for introducing DNA into plants. It is a double-stranded DNA virus with 8025 base pairs.

Certain portions of the virus are dispensable and foreign genes can be replace them. However, it has a limited host range; furthermore foreign sequences are often unstable in the caMV genome.

(c) **Electroporation**

Electroporation is widely used for transfecting plant cells. When plants are to be transfected, protoplasts or whole plant cells placed in contact with exogenous DNA in foil-lined cuvettes and exposed to high electrical current. The cells become permeable and take up exogenous DNA, some of which integrate with the plant genome. It has been successfully used in a wide variety of species using equipment which is relatively inexpensive.

(d) **Biollistic or Microprojectile methods**

This is one of the commonest methods used for transfecting plants. In this method a so-called gene gun or particle gun is used to shoot tiny pellets of tungsten or gold coated with the foreign DNA in question into the leaves or stem of the plant to be transformed. It is a widely used and highly successful method of transfecting plants using plant protoplasts, plant cell suspensions, callus cultures, even chloroplasts and mitochondria, and indeed any form of plant preparation capable of regeneration in dicots, monocots,

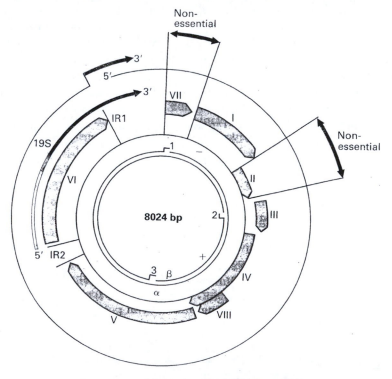

The eight shaded boxes are the coding regions

Fig. 7.11   Genetic Map of the Cauliflower Mosaic Virus

and conifers. Success occurs more with linear DNA than with circular; furthermore very large DNA inserts tend to be broken during the projection. When inside the cell, some of the introduced DNA get integrated with the plant DNA.

(e) **Microinjection**

This method involves immobilizing the cells and injecting DNA into protoplasts, walled cells, or embryos. It is done with a fine needle under the microscope. The technique needs a lot of skill. Some authors do not think it has much future because only one cell can be injected at a time.

(iii) *Delivery of DNA into Animal cells*

Genetic engineering in plants differs in at least two respects from that in animals. Firstly while plant cells are mostly totipotent (i.e., most plant cells are able to give rise to a new plant), animal cells cannot give rise to whole animals once differentiated into specialized cells. In animals the cells that become reproductive cells separate early from those that are ordinary body (somatic) cells. Somatic cells do not give rise to new animals To create transgenic animals the foreign DNA must be introduced into cells while they are still totipotent and differentiation has not occurred. Generally this involves introducing the DNA into stem cells (yet undifferentiated cells), an egg, the fertilized egg, (oocyte or zygote) or early embryo.

Some of the methods discussed above for introducing foreign DNA into bacteria and plants are also applicable to animal cells: electroporation, biollistic methods and microinjection have all been successfully used in animals. In addition the liposome (phospholipid) delivery seen in bacteria is also used in animal cells.

Genes are introduced into animal cells as well as *in vivo* by transduction via viruses in gene therapy. Four groups of viral vectors are used for gene therapy in humans: adenoviruses, baculoviruses, herpesvirus vectors, and retroviruses.

Changing the genetic make-up of animals, in large domesticated mammals such as cows, pigs and sheep, allows a number of commercial applications. These applications include the production of animals which express large quantities of exogenous proteins in an easily harvested form (e.g., expression into the milk), the production of animals which are resistant to infection by specific microorganisms and the production of animals having enhanced growth rates or reproductive performance.

Most of the work on transgenic animals has been done with mice on account of their small size and low cost of housing in comparison to that for larger vertebrates, their short generation time, and their fairly well defined genetics. Foreign DNA is introduced in mice in one of the following ways: DNA microinjection, embryonic stem cell-mediated gene transfer and retrovirus-mediated gene transfer, sperm-mediated transfer, transfer into unfertilized ova.

(a) DNA microinjection

This method involves the direct microinjection of a chosen gene construct (a single gene or a combination of genes) from another member of the same species or from a different species, into the pronucleus of a fertilized ovum. The introduced DNA may lead to the over- or under-expression of certain genes or to the expression of genes entirely new to the animal species. The insertion of DNA is, however, a random process, and there is a high probability that the introduced gene will not insert itself into a site on the host DNA that will permit its expression. The manipulated fertilized ovum is transferred into the oviduct of a recipient female, or foster mother that has been induced to act as a recipient by mating with a vasectomized male. Such males cannot inject sperms into the female because the tubes carrying the sperms, the vas deferens, have been cut. The major advantage of this method is its applicability to a wide variety of species.

(b) Embryonic stem cell-mediated gene transfer

This method involves prior insertion of the DNA sequence by homologous recombination into an *in vitro* culture of embryonic stem (ES) cells. Stem cells are undifferentiated cells that have the potential to differentiate into any type of cell (somatic and germ cells) and therefore to give rise to a complete organism. These cells are then incorporated into an embryo at the blastocyst stage of development. The result is a chimeric animal. ES cell-mediated gene transfer is the method of choice for gene inactivation, the so-called knock-out method. This technique is of particular importance for the study of the genetic control of developmental processes. This technique works particularly well in mice. It has the advantage of allowing precise targeting of defined mutations.

(c) Retrovirus-mediated gene transfer

To increase the probability of expression, gene transfer is mediated by means of a carrier or vector, generally a virus or a plasmid. Retroviruses are commonly used as vectors to transfer genetic material into the cell, taking advantage of their ability to infect host cells in this way. Offspring derived from this method are chimeric, i.e., not all cells carry the retrovirus. Transmission of the transgene is possible only if the retrovirus integrates into some of the germ cells.

(d) Sperm-mediated Gene Transfer

Sperms may be coated with the target DNA or attached to the sperm through a linker protein, and introduced through surgical oviduct insemination. It has been successfully used in pigs.

With the above techniques the success rate in terms of live birth of animals containing the transgene is extremely low. If there is birth, the result is a first generation (F1) of animals that need to be tested for the expression of the transgene. The F1 generation may result in chimeras. When the transgene has integrated into the germ cells, the germ line chimeras are then inbred for 10 to 20 generations until homozygous transgenic animals are obtained and the transgene is present in every cell. At this stage embryos carrying the transgene can be frozen and stored for subsequent implantation.

7.2.2.2.8.9   Application of genetic engineering in industrial microbiology and biotechnology in general

The unparalleled ability of DNA to replicate and reproduce itself is truly remarkable. What this means is that, put crudely, DNA of a given sequence coding for the production of a polypeptide or protein in organism A will lead to the production of the same polypeptide or protein if the same sequence is put into organism B. This is the basic assumption underlying the numerous advances in our manipulation of the biotic world for the benefit of humans. This section looks only at some of the numerous positive changes recombinant DNA technology has contributed to spreading a better quality of life to millions of people around the world through improvements in agriculture, health care delivery and industrial productivity.

(i) Production of Industrial Enzymes

Genetically engineered bulk enzymes are used mostly in the food industry (baking, starch manufacture, fruit juices), the animal feed industry, in textile manufacture, and in detergents. A leading manufacturer of these enzymes among world manufacturer is Novo Enzymes of Denmark.

The advantages of using engineered enzymes are as follows:

(a) such enzymes have a higher specificity and purity;

(b) it is possible to obtain enzymes which would otherwise not be available due to economical, occupational health or environmental reasons;

(c) on account of the higher production efficiency there is an additional environmental benefit through reducing energy consumption and waste from the production plants;

(d) for enzymes used in the food industry particular benefits are for example a better use of raw materials (juice industry), better shelf life of the final food and thereby less wastage of food (baking industry) and a reduced use of chemicals in the production process (starch industry);

(e) for enzymes used in the animal feed industry particular benefits include a significant reduction in the amount of phosphorus released to the environment from farming.

Two enzymes will be discussed briefly: chymosin (rennets) and bovine somatotropin (BST).

Chymosin is also known as rennets or chymase and is used in the manufacture of cheese. It used to be produced from rennets of farm animals, namely calves. Later it was produced from fungi, *Rhizomucor* spp. Over 90% of the chymosin used today is produced by *E. coli*, and the fungi, *Kluyveromyces lactis* and *Aspergillus niger*. Genetically engineered

Table 7.7 Some genetically engineered industrial enzymes (selected from brochure by Novo Industries of Denmark)

Type of enzyme	Main application
Alpha-amylase/Bacillolysin/Xylanase	Brewing industry, starch industry, baking industry
Amyloglucosidase	Alcohol industry, fruit processing
Cellulase	Detergent industry; textiles
Decarboxylase	Brewing industry
Glucoamylase	Alcohol industry, starch industry
Glucose oxidase	Baking industry
Lipase	Oils and fats industry; baking industry; dairy industry; leather industry
Lipase	Pasta/noodles
Maltogenic amylase	Starch industry, baking industry
Pectate lyase	Textile industry, fruit processing
Pectinesterase	Fruit processing
Phytase	Animal feed industry
Protease	Meat industry; detergent industry
Pullulanase/Amyloglucosidase	Starch industry, fruit processing

chymosin is preferred by manufacturers because while it behaves in exactly the same way as calf chymosin, it is purer than calf chymosin and is more predictable. Furthermore, it is preferred by vegetarians and some religious organizations.

Bovine Somatotropin (BST) is a growth hormone produced by the pituitary glands of cattle and it helps adult cows produce milk. It is produced by genetic engineering in *E. coli* using a plasmid vector. Supplementing dairy cows with bovine somatotropin safely enhances milk production and serves as an important tool to help dairy producers improve the efficiency of their operations. The use of supplemental BST allows dairy farmers to produce more milk with fewer cows, thereby providing them with additional economic security. It is marketed by Monsanto as Posilac.

(ii) Enhancing the activities of Industrial Enzymes

Through protein engineering it has been possible to enhance the properties of proteins to make them more stable to denaturation, more active in their biocatalytic ability and even to design new properties in existing enzymes. The properties of proteins are due to their conformation which is a result of their amino acid sequence. Certain amino acids in a protein play important parts in determining the stability of the protein to high temperatures, specificity and stability to acidity. In protein engineering changes are caused to occur in the protein by changes in the nucleotide sequence; a change of even a single nucleotide could lead to a drastic change in a protein. Many industrial processes are carried out at elevated temperatures, which can unfold the proteins and cause them to denature. The addition of disulphide bonds helps to stabilize them. Disulphide bonds are usually added by engineering cysteine in positions where it is desired to have the disulphide bonds. The addition of disulphide bonds not only increases stability towards elevated temperatures, but in some instances also increases stability towards organic solvents and extremes of pH. An example of the increase of stability to elevated temperatures due to the addition of disulphide bonds is seen in xylanase.

Xylanase is produced from *Bacillus circulans*. During paper manufacture, wood pulp is treated with chemicals to remove hemicelluloses. This treatment however leads to the release of undesirable toxic effluents. It is possible to use xylanase to breakdown the hemicellulose. However, at the time when bleaching is done, the pulp is highly acidic as a result of the acid used to digest the wood chips to produce wood pulp. The acid is neutralized with alkali, but the temperature is still high and would denature native xylanase. *In silico* (i.e. computer) modeling showed the sites where disulphide bonds can be added without affecting the enzyme's activity. The introduction of the disulphide bonds did increase the thermostability of the enzyme, making it possible to keep 85% of its activity after 2 hours at 60°C whereas the native enzyme lost its activity after about 30 minutes at the same temperature.

Another way in which enzyme activity can be enhanced by protein engineering is to actually increase the activity of the enzyme. This can be done only with an enzyme whose conformation, including the active sites, is thoroughly understood. Using *in silico* modeling, it is possible to predict the effect of changing amino acids at the active site of an enzyme. This has been done with the enzyme tRNA synthase from *Bacillus stearothermophilus*.

Various other properties have been engineered into proteins including a modification of the metal co-factor and even a change in the specificity of enzymes.

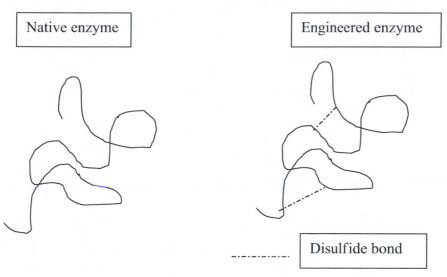

Fig. 7.12   Stabilizing Enzymes through the Introduction of Disulfide Bonds

(iii) Engineered Products or Activities Used for the Enhancement of Human Health

Engineered health care products and activities can be divided into: a) those used to replace or supplement proteins produced by the human body in insufficient quantities or not produced at all; b) those involving the replacement of a defective gene; c). those that are used to treat disease, d) those that are used for prophylaxis or prevention of disease, i.e., vaccines, or e) those that are used for the diagnosis of disease (Table 7.8). Only insulin and edible vaccines will be discussed.

Table 7.8   Some genetically engineered health related products

Product	Application
Hormones	
Insulin	Treatment of diabetes
Human growth hormone (somatotropin)	Treatment of dwarfism
Follicle stimulating hormone	Treatment of some disorders of the reproductive system
Immune System Participants	
Tumor necrosis factor	Anti-tumor agent
Interleukin 2	Treatment of some cancers
Lysozyme	Anti-inflammatory agent
A-Interferon	Antiviral
Blood components	
Erythropoeitin	Treatment of anemia and cancers
Tissue plasminogen activate	Dissolves blood clots
Factor VIII	Treating hemophilia
Enzymes	
Human DNase I	Treatment of cystic fibrosis

(1) *Insulin*

Insulin is a hormone produced by the pancreas; hormones are small proteins. Insulin is used to treat diabetes of which there are there three types, only two of which are relevant to this discussion.

Type 1 diabetes (previously known as insulin-dependent diabetes) is an auto-immune disease where the body's immune system destroys the insulin-producing beta cells in the pancreas. This type of diabetes, also known as juvenile-onset diabetes, accounts for 10-15% of all people with the disease. It can appear at any age, although common under 40, and is triggered by environmental factors such as viruses, diet or chemicals in people genetically predisposed. To live, people with type 1 diabetes must inject themselves with insulin several times a day and follow a careful diet and exercise plan.

Type 2 diabetes (previously known as non-insulin dependent diabetes) is the most common form of diabetes, affecting 85-90% of all people with the disease. This type of diabetes, also known as late-onset diabetes, is characterized by relative insulin deficiency. The disease is strongly genetic in origin but lifestyle factors such as excess weight, inactivity, high blood pressure and poor diet are major risk factors for its development. Symptoms may not show for many years and, by the time they appear, significant problems may have developed. People with type 2 diabetes are twice as likely to suffer cardiovascular disease. Type 2 diabetes may be treated by dietary changes, exercise and/or medications. Insulin injections may later be required.

The third type affects pregnant women, is less common, and will not be discussed.

Genetically engineered insulin was the first major product of biotechnology. As insulin from some animals is similar to human insulin, beginning from the 1920s, insulin isolated from the pancreas of farm animals, mainly pigs and cows, was used to treat diabetes. There were several problems with this product. First it takes several months for animals to mature and be ready to be slaughtered for their pancreas. This made animal-based insulin expensive since it was difficult to meet the demand. Furthermore such animal insulin caused immune reactions in some patients and a few became intolerant or resistant to animal insulin. For a more effective solution the then new technology of recombinant DNA was resorted to. In 1978, in the laboratory of Herbert Boyer at the University of California at San Francisco, a synthetic version of the human insulin gene was constructed and inserted *E. coli*. In 1982 Eli Lilly Corporation was granted approval for its genetically engineered insulin. Insulin is a small protein, and today's insulin is produced with a synthesized gene, which is expressed in a yeast.

Insulin consists of two amino acid chains: the **A peptide chain** which is acidic and with 21 amino acids and the **B peptide chain** which is basic and has 30 amino acids. When synthesized the A and B chains are further linked by a 30 amino acid **C peptide chain** to produce a structure known as pro-insulin. Pro-insulin is cleaved enzymatically to yield insulin. (Fig. 7.13).

(2) Edible vaccines

An innovative new approach to vaccine production is the surface expression of the antigen of a bacterium in a plant. Most current immunization is done by injection (parenteral delivery) and rarely results in specific protective immune responses at the mucosal surfaces of the respiratory, gastrointestinal and genito-urinary tracts. Mucosal

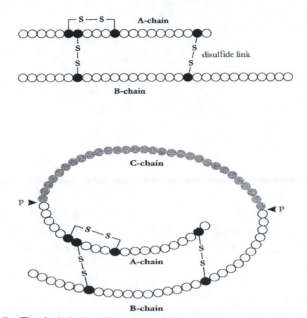

Top: Structure of insulin. The A chain has 21 amino acids (represented by circles) and the B chain has 30 amino acids. The A and B chains are linked by disulfide bond between cysteine residues (filled circles)

Bottom: Synthesis of Insulin: Proinsulin, the Precursor of Insulin
When synthesized proinsulin consists of an 81-amino acid polypeptide. The C chain is then cleaved off by a protease (P) to yield insulin.

Fig. 7.13   Structure and Synthesis of Insulin

immune responses represent a first line of defense against most pathogens. In contrast, mucosally targeted vaccines achieve stimulation of both the systemic as well as the mucosal immune networks. In addition, mucosal vaccines delivered orally increase safety and compliance by eliminating the need for needles. In addition many vaccines depend on the need of maintenance of a 'cold chain' (refrigeration) for delivery. Many developing countries, lack the resources to maintain the chain giving rise to many cases of vaccine failure, along with the fact they lack the resources and technology for fermentation industries. Both of these factors create constraints in vaccine use in the developing world, where these vaccines are needed the most. Combining a cost-effective production system with a safe and efficacious delivery system, plant edible vaccines, provide a compelling new opportunity.

Plant-based oral vaccines are cheap, safe and efficient. From the point of the little child receiving the vaccine, a smile might be elicited when a 'banana' vaccine is eaten rather the sharp cry of the pain of a needle! Vaccines have been produced in several plants, including a vaccine against dental carries caused by *Streptococcus mutans*, which was produced in the tobacco. Two other interesting examples are the expression of the rabies external antigen in tomatoes and the hepatitis virus antigen in lettuce.

7.2.2.2.8.10   Genetically engineered plants

Plants have been engineered for the introduction of many new desirable properties. Collectively these attainments represent a major triumph of biotechnology, enabling us to

Table 7.9 Vaccines produced in plants

Vaccine against	Plant
Cholera	Potato
Foot and mouth disease	Arabidopsis
Herpes virus	Tobacco
Norwark (diarrhea) virus	Potato
Rabies	Tobacco

achieve in a few years what would take traditional plant breeding decades to attain, if at all. Some genetic engineering achievements would be impossible with traditional plant breeding methods since in the latter, the introduction of new genetic properties occurs only through the exchange of sexual materials (in the pollen grains) of the same species. In genetic engineering the natural species barrier is not recognized since the DNA sequence introduced into a plant can come from another plant of a different species or even from a non-plant source such as a bacterium, and indeed may even be synthesized. The introduction of some genetically engineered foods has met with public resistance, although many have been shown to be safe. What is required is continued public education about their safety before their introduction, and constant sensitivity to public opinion thereafter.

The ensuing discussion will be under two headings, a) improving field, production or agronomic traits and b) modification of consumer products.

Improving Field or Production Characteristics

Numerous improvements have been made in agricultural crops by introducing into them genes coding for the desired properties, but only plant engineering for herbicide resistance, resistance to viral diseases, resistance to insect pests, and resistance to salt stress will be discussed.

(i) Engineering Plants for Herbicide Resistance

An estimated US $10 billion is spent annually on weed killers. In spite of this about 10% of world crop production is lost to weeds. Herbicides (weed killers) target processes that are essential and unique to plants. These processes are however important to plants and weeds alike, and getting methods that are selective for either is difficult. One method that is used is to engineer crops so they become resistant to the weed killer. Plants can become resistant to herbicides in one of the four following ways:

(a) overproduction of the herbicide sensitive target, so that some is still left for the proper cell function despite presence of the herbicide in the cell;

(b) reduction of the ability of the herbicide-sensitive target protein to bind to the herbicide;

(c) engineering into plants the ability to metabolically inhibit the herbicide;

(d) inhibition of the uptake of the herbicide.

It is important to have some idea of the modes of action of herbicides so as to understand how the genetic engineering is to proceed. The most common modes of action are given below and examples of herbicides in each group are given in parenthesis:

- **Auxin mimics** (2,4-D, clopyralid, picloram, and triclopyr), which mimic the plant growth hormone auxin causing uncontrolled and disorganized growth in susceptible plant species;
- **Mitosis inhibitors** (fosamine), which prevent re-budding in spring and new growth in summer (also known as dormancy enforcers);
- **Photosynthesis inhibitors** (hexazinone, bromoxynil), which block specific reactions in photosynthesis leading to cell breakdown;
- **Amino acid synthesis inhibitors** (glyphosate, imazapyr, and imazapic), which prevent the synthesis of amino acids required for construction of proteins;
- **Lipid biosynthesis inhibitors** (fluazifop-p-butyl and sethoxydim), that prevent the synthesis of lipids required for growth and maintenance of cell membranes.

Engineering plants for resistance to glyphosate will be discussed. Glyphosate (N-phosphonomethylglycine) is a non-selective, broad spectrum herbicide that is systematically translocated to the meristems of growing plants. It causes shikimate accumulation through inhibition of the chloroplast localized EPSP synthase (5-*enol*pyruvylshikimate-3-phosphate synthase; EPSPs) [EC 2.5.1.19]. Resistance to the herbicide glyphosate has been developed for soybean. Glyphosate is said to be environmentally-friendly as it does not accumulate in the environment because it is easily broken by soil bacteria. Glyphosate kills plants by preventing the synthesis of certain amino acids produced by plants but not animals. It acts by inhibiting the enzyme 5-enolpyruvyshikimate-3-phosphate synthase (EPSPS), an enzyme in the shikimate pathway (Fig 7.14) and plays an important part in the synthesis of aromatic amino acids in plants and bacteria. An EPSPS gene isolated from a glyphosate-resistant *E. coli* was linked to a plant promoter and termination transcription sequence and cloned into plant cells. Tobacco, tomato, potato, petunia, and cotton have been transfected with glyphosate

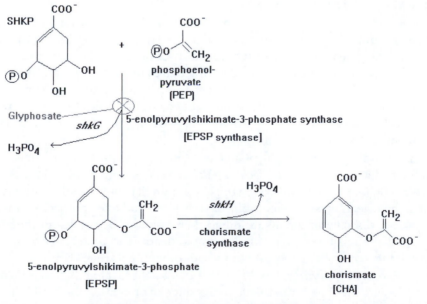

Fig. 7.14 Mode of Action of Action Glyphosphate

resistance in this way. They produce large amounts of EPSPS, enough to leave some for the plants cells to utilize for their metabolism after neutralization of a portion by glyphosate.

An example of a case where the plant is engineered to inactivate the herbicide before it can act is bromoxynil, a photosynthesis inhibitor. Plants were made resistant to this herbicide by engineering into them the gene for nitrilase obtained from the bacterium, *Klebsiella ozaenae*. Nitrilase inactivates bromoxynil before it can act.

(ii) Engineering Plants for Pathogenic Microbe Resistance

The majority of microbes attacking plants are fungi, but some bacterial diseases of plants do exist. Plants are conventionally sprayed with chemicals to eliminate fungal pathogens. Such chemicals sometimes are not always easily biodegradable, and they may also find their way into food. A genetic approach which bypasses this problem is to engineer into plants anti-fungal proteins such as the gene coding for chitinase, an enzyme which hydrolyzes chitin, a polymer of the amino sugar N-acetyl glucosamine. Chitinase gene from bean has been cloned into tobacco where chitinase stopped the attack by the fungus, *Rhizoctonia solani*. Chitinase is one of the 'pathogen-related proteins' (PRs) synthesized by plants; they also synthesize ant-fungal peptides known as defensins. Genes coding for these are sought from source of high productivity and cloned into plant to protect them.

Plant resistance to bacterial disease has also been genetically engineered. For example, the α-thionin gene from barley has been shown to confer resistance to a bacterial pathogen, *Pseudomonas syringae* in transgenic tobacco.

With regard to engineering plants against viruses, when the viral coat of a plant virus is engineered into a plant, that plant usually becomes resistant to the virus from which the coat comes. Often the plant is also resistant against other unrelated viruses.

(iii) Engineering Plants for Insect Resistance

Insect pests are devastating to crops, about US $5 billion are currently being used to control them annually with chemicals. The advantages of using biological means of controlling insect pest have been highlighted in Chapter 17. The methods described relate to the use of biological insecticides which are sprayed on plants. Such sprayed insecticides have the disadvantage that thet are inactivated by ultraviolet rays from the sun or may be washed away by the rain. Genetic engineering of crops for resistance against insect pests has the advantage that the active constituents are protected from the environment and remain within the plant.

The major strategy of producing plants resistant to insect pests is to engineer the gene for producing the toxic crystals of *Bacillus thuringiensis* (Bt) into plants. These crystals are produced in Bt but in no other *Bacillus* sp. They are small proteins and are highly specific against given insects. In such susceptible insects they bind to receptors in the gut lining of the insects, dissolve in the alkali milieu therein and create holes in the gut lining through which gut contents leak out, leading to death. The gene for Bt toxin has been engineered into cotton, tomatoes and numerous other plants (Fig. 7.15).

Alternative strategies which have been inspired by the fact that Bt toxins do not affect some insects, is to engineer into plants two groups of enzymes which inhibit digestive enzymes in the insect gut: amylase inhibitors and protease inhibitors. In effect the insect starves to death.

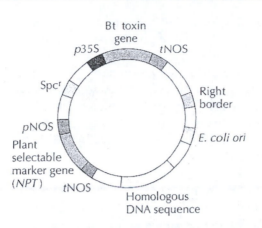

Fig. 7.15 Cloning Vector Carrying a B. *thuringiensis* (Bt) Insecticidal Toxin Gene

Another strategy for developing insect resistance in plants is to engineer into the plant the gene for cholesterol oxidase, which is present in many bacteria. Cholesterol oxidase catalyzes 3-hydroxysteroids to ketosteroids and hydrogen peroxide (Chapter 26). Small amounts of this enzyme are very lethal to the larvae of boll weevil which attacks cotton. It is possible that the cholesterol oxidase acts by disrupting the insect larva's alimentary canal epithelium leading to its death.

(iv) Genetically Engineering Plants to Survive Water and Salt Stress

Many parts of the world have desert or near desert conditions where water is in short supply. Added to this is the fact that salt used for treating ice in the winter finds its way into agricultural land. These factors create conditions which bring plants into conditions of water (drought) and salt stress. To survive under these conditions, many plants synthesize compounds known as osmoprotectants. They help the plant increase its water uptake as well as retain the water absorbed. Osmoprotectants include sugars, alcohols and quartenary ammonium compounds. The quartenary ammonium compound, betaine, is a powerful osmoprotectant and the gene encoding it

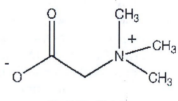

Fig 7.16 Betaine

obtained from *E. coli* has enabled plants into which it was cloned survive drought better than un-engineered plants.

Modification of Plant Consumer Products

This section looks at how genetic engineering has been used to modify the plant food which comes to the consumer as opposed to the previous section which dealt with the concerns of the farmer or the producer.

(i) Maintenance of Hardness and Delayed Ripeness in Fruits

During post-harvest transportation of fruits to supermarkets these fruits sometimes ripen and become soft due to the natural processes which go on within the fruit. These natural processes include the production of polygalacturonase (PG) (which hydrolyzes pectin) and cellulases by the fruit. In tomatoes the softening of the fruit is inhibited by engineering an anti-sense PG producing gene into the plant, enabling the fruit to ripen on the plant before harvesting instead of harvesting them while still green. Such tomatoes have a longer shelf life while retaining the taste of regular tomatoes. The genetically engineered tomato known as Flavr Savr was approved by the FDA in 1994 as safe for human use. In anti-sense technology, a gene sequence is inserted in the opposite direction, so that during transcription, mRNA complimentary to the normal RNA is produced. The anti-sense mRNA therefore binds to the normal inhibiting translation. The net result is that the gene is shut off and in the particular case of PG the fruit-softening enzyme is reduced to about 1% of the normal, thereby inhibiting softening of the fruit and possible microbial attack thereafter.

In climacteric fruits (i.e., fruits that are picked before they are ripe) such as tomatoes, avocados, and bananas, the initiation of ripening is associated with a burst in ethylene biosynthesis. After harvesting unripe fruits such as bananas may be treated with ethylene to induce simultaneous ripening. Ethylene has been described as a gaseous plant hormone: extraneous ethylene and ethylene generated by the plant equally induce ripening. Ethylene is a gaseous effector with a very simple structure. In higher plants, ethylene is produced from L-methionine (Fig 7.17). A major step is the production of the non-protein amino acid 1-aminocyclopropane-1-carboxylic acid (ACC), catalysed by the enzyme ACC synthase. It has numerous functions in higher plants. It stimulates the following activities: the release of dormancy, leaf and shoot abscission, leaf and flower senescence, flower opening and fruit ripeneing.

Two biotechnological strategies have been pursued to control ethylene action on fruit ripening. One approach taken in tomato was designed to inhibit biosynthesis of ethylene within the plant by the use of antisense expression of ACC synthase. In a second approach, a mutated ethylene receptor from *Arabidopsis* was introduced into tomato and petunia. This resulted in delayed fruit ripening.

(ii) Engineering Sweetness into Foods

The taste of fresh tomatoes and lettuce is well known in sandwiches. Some enjoy these items with greater relish with the addition of sweet tasting tomato ketchup. Sweet taste has been engineered into tomatoes and lettuce by cloning into them the synthesized gene coding for monellin. Monellin is a protein which is 3,000 times sweeter than sucrose by weight; it is naturally obtained from the red berries of the West African plant, *Dioscoreophyllum comminsii* Diels, and has been expressed in yeast. A major attraction of sweeting tomatoes and lettuce with this protein is that it is 'weight-friendly'. Several sweet proteins which might be similarly engineered into foods are shown in Table 7.11.

(iii) Modification of Starch for Industrial Purposes

Starch consists of amylose in which the glucose molecules are configured in a straight chain in the α-1-4 linkage, and the branched chain amylopectin which has α-1,4 and α-1,6 linkages (Chapter 4). Starches from different plants have different percentages of amylase and amylopectin, but generally in the order of 30% amylase to 70 to 80% amylopectin. Starch is used for making several industrial products such as glue, gelling agent or thickener. For some purposes it may be desirable to have starch that has a preponderance of amylase. When that is the case, antisense technology has been used to block the formation of the amylopectin component of starch, giving rise to a product with only about 20%.

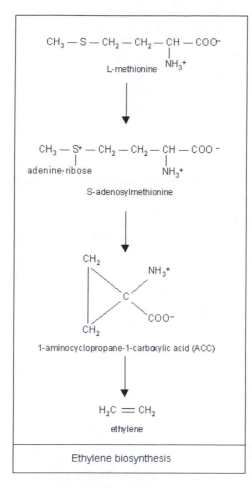

Ethylene biosynthesis

Fig. 7.17   Synthesis of Ethylene

Table 7.10   Some sweet tasting proteins produced by plants

S/No	Name	Plant	Sweetness ratio over sucrose (w/w)
1	Thaumatin	*Thaumatococcus danielli* Benth	3,000
2	Monellin	*Dioscoreophyllum cumminsii* Diels	3,000
3	Brazzein	Pentadiplandra brazzeana	2,000
4	Curculin	Curculingo latifolia	550

One further modification is the engineering into a starch source the enzymes needed to convert starch to high fructose syrup. In the production of high fructose syrup, the starch is first converted to glucose by α-amylase and thereafter the resulting glucose is converted to high fructose syrup by glucose isomerase. Both operations are normally done sequentially. However, both enzymes have been linked together and engineered into potato and the potato starch converted into fructose in one operation with consequent saving in costs.

Table 7.11 Modification of Canola oil for different purposes

Seed product	Commercial use(s)
40% Stearic	Margarine, cocoa butter
40% Lauric	Detergents
60% Lauric	Detergents
80% Oleic	Food, lubricants, inks
Petroselinic	Polymers, detergents
"Jojoba" wax	Cosmetics, lubricants
40% Myristate	Detergents, soaps, personal care items
90% Erucic	Polymers, cosmetics, inks, pharmaceuticals
Ricinoleic	Lubricants, plasticizers, cosmetics, pharmaceuticals

(iv) Modifying Flower Pigmentation and Delaying Wilting and Abscision in Flowers

The flower business is of the order of many billions of dollars annually. Most of the market centers around four flowers: roses, carnations, tulips, and chrysanthemums. Hundreds of different flowers differing in shape, size, color, fragrance, and structure have become available through tradional plant breeding. But the usual shortcomings have also affected this industry: the slow pace of the plant breeding, the uncertainty of the the results of the efforts and the limitation imposed by the paucity of the genes available in traditional plant breeding.

Genetic engineering has now been introduced and has helped to extend the range of the variety of flowers. A group of flavonoids, anthocyanins (Chapter 22) are commonest pigments in flowers. Anthocyanins are glucosides of phenolic compounds produced in plants, some being colorless, while many are responsible for the colors in plants. The aglycone (non-sugar) protions of anthocyanins are derived from the amino acid phenylalanine. The color which they bear is determined by the chemical nature of the side chain substituent. By blocking some of the genes in the pathway of anthocyanin synthesis using anti-sense technology or introducing toally new genes it is possible to create flowers with new colors (Fig. 7.18).

It is also known that flower wilting and abscission are controlled by ethylene in the same way as it does with fruits. When a mutated ethylene receptor from *Arabidopsis* was introduced into petunia, it led to delayed petal fading, and in delayed flower abscission.

(v) Modification of Nutritional Capabilities of Crops

Genetic engineering has enabled the introduction of new nutritional capabilities in crops, in a much shorter time and in a range of qualities impossible with traditional breeding. Unlike genetic engineering which can cross the species barrier, plant breeding deals with the collection of genes within the species. The amino acid content of foods, the lipid composition, the amylose/amylopectin ratio of starch, the vitamin contents and even the mineral contents of foods have all been modified by genetic engineering.

(a) Engineering Vitamin A into Rice

'Golden rice' has been prepared by engineering it beta-carotene, a substance which the body can convert to Vitamin A to combat vitamin A deficiency (VAD), a condition which afflicts millions of people in developing countries, especially children and pregnant women. Severe Vitamin A deficiency (VAD) can cause partial or total blindness; less severe deficiencies weaken the immune system, increasing the risk of infections such as

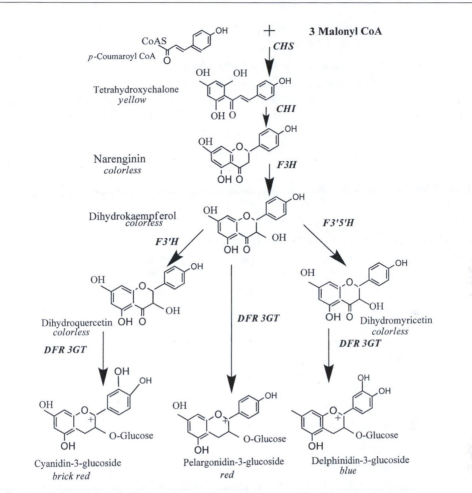

Most flowers derive their color from anthocyanins which are synthesized from the amino acid phenyl alanine. The color of the flower depends on the possession by the plant of genes which can code for the the enzymes whose reactions result in the various colors in flowers. In the figure above the plant must possess the gene coding for the enzyme CHI (chalcone synthase) which produces 4,2',4',6'-Tetrahydroxychalcone from the two intermediates indicated and gives rise to yellow flowers. Lower down the chain the critical enzymes are DFR (Dihydroflavonol 4-reductase) and 3GT (UDP-glucose: flavonoid 3-O-glucosytransferase. These two enzymes DFR and 3GT will convert intermediates to compounds which will give rise brick red, red, or yellow flowers. By manipulating the pathway through introducing various genes, flowers of different colors can be produced at will (see text).

Fig. 7.18   Synthesis of Anthocyanins in Flowers

measles and malaria. Women with VAD are more likely to die during or after childbirth. Each year, it is estimated that VAD causes blindness in 350,000 preschool age children, and it is implicated in over one million deaths. Golden rice was created by transforming rice with three beta-carotene biosynthesis genes: *psy* (phytoene synthase) and *lyc* (lycopene cyclase) both from daffodil (*Narcissus pseudonarcissus*), and *crt1* from the soil bacterium *Erwinia uredovora*. The *psy*, *lyc*, and *crt1* genes were transformed into the nuclear genome and placed under the control of an endosperm specific promoter, so that they are only expressed in the endosperm. The plant endogenous enzymes process the lycopene to beta-carotene in the endosperm, giving the rice the distinctive yellow color which gave it the name 'golden'.

(b) Engineering Amino Acids into Legumes and Cereals

The seed storage compounds in cereals such as corn usually contain proteins deficient in the essential amino acids lysine and methionine. Storage proteins found in many legumes are sometimes deficient in these two essential amino acids, or cysteine. Corn and many grain legumes are used as animal feeds, and feeds made from them have to be supplemented with the deficient amino acids. Legumes such as lupine have been engineered to express sunflower seed albumin which is unusually rich in the sulfur-containing amino acids methionine and cysteine. High-lysine corn is currently available, but engineering lysine, methionine and or cysteine into corn is almost certainly a matter of time

(c) Modifying Fats and Oils for Various Purposes

Plant oils are derived from soybean, oil palm, sunflower, and rapeseed (canola) and to a lesser extent from the endosperm of corn. Most of the oils is used for margarine manufacture, as fats for baking, for salads and for frying. The extent to which an oil is liquid at room temperature depends on the degree of unsaturation, i.e., the number of double bonds it has. For industrial purposes oils are also used in cosmetics, in detergents, soaps, confectionaries, and as drying agents in paints and inks. Each use to which the oils are put requires a different property. For example oils which contain conjugated double bonds (in contrast to those which contain double bonds separated by methyl groups– CH_2 (Fig. 7.19) require less oxgene for polymerization and hence dry more quickly in paints and inks. Genetic engineering has been used to modify oils for various uses. Thus canola oil from rapeseed has been genetically modified for use in various products.

$$CH_2{=}CH{-}CH{=}CH{-}CH_3 \qquad\qquad CH_2{=}CH{-}CH_2{-}CH{=}CH_2$$

Fig. 7.19 Cojugated Double Bonds

Soyoil has been genetically modified by DuPont to make it more suitable as an edible oil and also for certain industrial uses including the manufacture of inks, paints, varnishes, resins, plastics, and biodiesel.

Soybean oil is a complex mixture of five fatty acids (palmitic, stearic, oleic, linoleic, and linolenic acids) that have vastly differing melting points, oxidative stabilities, and chemical functionalities. The most notable example, developed by researchers at DuPont, is the transgenic production of soybean seeds with oleic acid content of approximately 80% of the total oil. Conventional soybean oil, by comparison, contains oleic acid at levels of 25% of the total oil. The high oleic acid trait was obtained by down regulating the expression of *FAD2* genes that encode the enzyme, which converts the monounsaturated oleic acid to the polyunsaturated linoleic acid. High-oleic oils with elevated oleic acid content are generally considered to be healthier oils than conventional soybean oil, which is an omega-6 or linoleic acid-rich oil. From an industrial perspective, the high content of oleic acid and low content of polyunsaturated fatty acids result in an oil that has high oxidative stability. In addition, soybean oil is naturally rich in the vitamin E antioxidant gamma-tocopherol, which also contributes to the oxidative stability of high oleic acid soybean oil. High oxidative stability is a critical property for lubricants.

Genetic engineering can also be used to produce soybean oil with high levels of linolenic acid, a polyunsaturated fatty acid with low oxidative stability. Soybean seeds with linolenic acid content in excess of 50% of the total oil have been generated by increasing the expression of the *FAD3* gene, which encodes the enzyme that converts linoleic acid to linolenic acid The linolenic acid content of conventional soybean oil, in contrast, is approximately 10% of the total oil. The low oxidative stability associated with high linolenic acid oil is a desirable property for drying oils that are used in coating applications, such as paints, inks, and varnishes.

Significant progress has been made in the development of chemical methods for enhancing the functionality of soybean oil for the production of polyols from soybean oil which may eventually lead to a number of industrial applications, including the production of polyurethanes.

7.2.2.2.8.1   Transgenic animals and plants as biological fermentors (or Bioreactors)

Transgenic animals and plants have been used to produce high-quality pharmaceutical substances or diagnostics. The procedure is known as 'pharming' from a parody of the word pharmaceutical; it is also known as 'molecular farming' or 'gene pharming' and the transgenic plants or animals used are sometimes referred to as animal or plant 'bioreactors' or 'fermentors'. Therapeutically active proteins already on the market are usually produced in bacteria, fungi, or animal cell cultures. However microorganisms usually produce comparatively simple proteins; furthermore microorganisms are not always able to correctly assemble and fold complex proteins. If the protein structure is very complicated, such microorganisms may produce defective clumps.

In pharming, transgenic animals are mostly used to make human proteins that have medicinal value. The protein encoded by the transgene is secreted into the animal's milk, eggs or blood or even urine, and then collected and purified. Livestock such as cattle, sheep, goats, chickens, rabbits and pigs have already been modified in this way to produce several useful proteins and drugs. Some human proteins that are used as drugs require biological modifications that only the cells of mammals, such as cows, goats and sheep, can provide. For these drugs, production in transgenic animals is a good option. Using farm animals for drug production has many advantages: they are reproducible, have flexible production, and are easily maintained. Since the mammary gland and milk are not part of the main life support systems of the animal, there is not much risk of harm to the animal making the transgenic protein. To ensure that the protein coded in the transgene is secreted in the milk, the transgene is attached to a promoter which is only active in the mammary gland. Although the transgene is present in every cell of the animal, it is only active where the milk is made. Some examples of the drugs currently being tested for production in animals are antithrombin III and tissue plasminogen activator used to treat blood clots, erythropoietin for anemia, blood clotting factors VIII and IX for hemophilia, and alpha-1-antitrypsin for emphysema and cystic fibrosis.

A good example of the need for processing a protein in an animal is seen in the silk of the golden spider, *Nephila clavipes*. The dragline form of spider silk is regarded as the strongest material known; it is five times stronger than steel. People have actually tried starting 'spider farms' to harvest silk, but the spiders are too aggressive and territorial to live close together. They also like to eat each other. Though the genes for dragline silk were isolated several years ago, attempts to produce it in bacterial and mammalian cell culture have failed. When the genes were put into a goat and expressed in the mammary glands, however, the animal produced silk proteins in its milk that could be spun into a fine thread with all the properties of spider-made silk. This can be used to make lighter, stronger bulletproof vests, thinner thread for surgery and stitches or indestructible clothes.

The advantage of using biological fermentors have been put as follows: lower drug prices for consumers, production of drugs unavailable any other way, new value-added products for farmers, and inexpensive vaccines for the developing world.

(a) Lower drug costs

Expected savings on infrastructure and production costs lead companies producing 'pharm' and industrial crops to predict drug prices 10 to 100 times lower than current prices. The cost of treating a patient with Fabry's disease, currently as much as US $400,000 a year, for example, is predicted to drop to approximately US $40,000 annually. Similarly, it is claimed that the leaves from only 26 tobacco plants could make enough glucocerebrosidase, currently one of the most expensive drugs in the world, to treat a patient with Gaucher's disease for a whole year. Regarding plants, the biggest factor in reducing costs is the high yields of recombinant proteins attainable in transgenic plants. Production costs for corn systems are estimated at between US $10 and US $100 per gram for proteins that currently cost as much as US $1,000 per gram. Dollar figures based on large-scale tobacco production vary widely from less than US $10 per gram to US $1000 per gram. If realized, these projections would represent substantial savings over current costs.

(b) Faster, more flexible manufacturing

Abundantly available commodity like corn and the environment as a production inputs could cut not only production costs but also capital investment in, and the time it takes to increase, manufacturing infrastructure. Very rapid scale-up (or scale-down) of a production pipeline in response to the market or other factors and new drugs could, theoretically, become available sooner.

(c) Drugs unavailable any other way

Cheap production also means that drugs that could not be produced cheaply enough at high volume through conventional methods might become economically viable using genetically engineered crops. Monoclonal antibodies ('plantibodies') fall into this category. One company's idea for such a product is a monoclonal antibody against bacteria responsible for tooth decay, which could be used as a dental prophylactic. A topical therapeutic for herpes, as well as antibodies for the treatment of many other diseases, are also under development.

(d) New value-added agricultural products

These crops producing pharmaceutical products could be a boon to farmers, as they could be economically viable alternatives to commodity production of corn or tobacco.

Special Advantages of Plants as Biological Fermentors

Apart from the advantages accruing in the use of plants and animals as bioreactors, a sector of the biotechnology industry views plants (in comparison with animals) as preferable for protein production for the following reasons.

Table 7.12   Human proteins synthesized in animals

Protein	Use	Animal(s)
α-1-anti-protease inhibitor	α-1-antirypsin deficiency	Goat
α-1-antirypsin	Anti-inflammatory	Goat, sheep
Antithrombin III	Sepsis and disseminated intravascular coagulation (DIC)	Goat
Collagen		Cow
Factor VII and IX	Burns, bone fracture, incontinence	Sheep, pig
Fibrinogen	Hemophilia	Pig, sheep
Human fertility hormones	"Fibrin glue," burns, surgery localized chemotherapeutic drug deliver	Goat, cow
Human hemoglobin	Infertility, contraceptive vaccines	Pig
Human serum albumin	Blood replacement for transfusion	Goat, cow
Lactoferrin	Burns, shock, trauma, surgery	Cow
LAtPA	Bacterial gastrointestinal infection	Goat
Monoclonal antibodies	Venous stasis ulcers	Goat
Tissue plasminogen activator	Colon cancer	Goat
	Heart attacks, deep vein thrombosis, pulomary embolism	

Table 7.13   Some therapeutic agents produced in transgenic plants

Protein	Plant(s)	Application
Human protein C	Tobacco	Anticoagulant
Human hirudin variant 2	Tobacco, canola, Ethiopian mustard	Anticoagulant
Human gramulocyte-macrophage colony-stimulating factor	Tobacco	Neutropenia
Human erythropoientin	Tobacco	Anemia
Human enkephalins	Thale cress, canola	Antihyperanalgesic by opiate activity
Human epidermal growth factor	Tobacco	Wound repair/control of cell proliferation
Human α-interferon	Rice, turnip	Hepatitis C and B
Human serum albumin	Potato, tobacco	Liver cirrhosis
Human hemoglobin	Tobacco	Blood substitute
Human homotrimetic collagen I	Tobacco	Collagen synthesis
Human α-1-antitrypsin	Rice	Cystic fibrosis, liver disease, hemorrhage
Human growth hormone	Tobacco	Dwarfism, wound healing
Human aprotinin	Corn	Trypsin inhibitor for transplantation surgery
Angiotension-1-converting enzyme	Tobacco, tomato	Hypertension
α-Tricosanthin	Tobacco	HIV therapy
Glucocerebrosidase	Tobacco	Gaucher disease

(a) Plants are less controversial than animal production systems

Plants do not normally transmit animal pathogens (that can cause diseases like mad cow). Products obtained by 'pharming' plants are promoted as safer than animal-sourced proteins. It must, however, be borne in mind that plants as bioreactors may also present their own risks of product contamination from mycotoxins, pesticides, herbicides or endogenous plant secondary metabolites such as nicotine and glycoalkaloids.

Using plants as bioreactors also avoids any animal welfare and certain ethical concerns associated with cloning animals and using them as bioreactors

(b) Inexpensive, easily delivered vaccines

Food plants engineered to contain pieces of disease agents can function as orally administered vaccines, avoiding the need for injection and syringes. Currently, tomatoes and other vegetables are under development for that purpose. Although the vaccines would still have to be standardized for dose and delivered to patients one at a time, it is hoped that the lower production costs and the convenience of avoiding refrigeration would make the products attractive to the developing world.

SUGGESTED READINGS

Bleecker A.B., Kende, H. 2000. Ethylene: a gaseous signal molecule in plants. Annual Review of Cell Devision and Biology 16, 1-18.

Deacon, J. 2006. *Fungal Biology*. 4th Blackwell. Malden USA.

Gelvin, S.B. 2003. Agrobacterium-Mediated Plant Transformation: the Biology behind the "Gene-Jockeying" Tool Microbiology and Molecular Biology Reviews, 67, 16-37.

Glick, B.R., Pasternak, J.J. 2003. Molecular Biotechnology: Principles and Applications of Recombinant DNA. ASM Press Washington DC, USA.

Glover, S.W., Hopwood, D.A. 1981 eds. Genetics as a Tool in Microbiology Cambridge University Press, Cambridge, UK.

Keiji, K.Y., Miura, Y., Sone, H., Kobayashi, K., Hiroshi Iijima, H., Parek, S. 2004. Strain Improvement: High-level expression of a sweet protein, monellin, in the food yeast *Candida utilis*.

Koffas, M., Roberge, C., Lee, K., Stephanopoulos, G. 1999. Metabolic Engineering. Annual Reviews of Biomedical Engineering. 1, 535- 557.

Kondo, K., Miura, Y., Sone, H., Kobayashi, K., Iijima, H. 1997. High-level expression of a sweet protein, monellin, in the food yeast *Candida utilis*. Nature Biotechnology, 15, 453-457.

Labeda, D.P., Shearer, M.C. 1990. Isolation of Actinomycetes for Biotechnological Aplications. In: Isolation of Biotechnological Organisms from Nature. D.P. Labeda. (ed) McGraw-Hill, New York, USA. pp. 1-20.

Murooka, Y., and Imanaka, T. (eds) 1994. Recombinant Microbes for Industrial and Agricultural Applications. Marcel Dekker, New York, USA.

Parek, S. 2004. Strain Improvement. In: the motherland. The Desk Encyclopedia of Microbiology. M Schaechter (ed.) Elsevier Amsterdam: pp. 960–973.

Steele, D.B., Stowers, M.D. 1991. Techniques for the Selection of Industrially Important Microorganisms. Annual Review of Microbiology, 45, 89–106.

Velandez, W., Lubon, H., Drohan, W. 1997. Transgenic Livestock as Drug Factories. Scientific American, 1: 55.

Wei, L.N. 1997. Transgenic Animals as New Approaches in Pharmacological Studies. Annual Reviews of Pharmacology and Toxicology 37, 119–141.

CHAPTER 8

The Preservation of the Gene Pool in Industrial Organisms: Culture Collections

8.1 THE PLACE OF CULTURE COLLECTIONS IN INDUSTRIAL MICROBIOLOGY AND BIOTECHNOLOGY

The central importance of a microorganism in an industrial microbiological establishment may sometimes be taken for granted. While a raw material may be fairly easily substituted, the use of an organism different from one already in existence may involve extensive experimentation and modification of established processes if the usual products are to be obtained. It is therefore important that organisms whose genetic potentials remain unchanged be constantly available. In other words, the gene pool of organisms with desirable properties must be preserved and be constantly available.

The gene pool is the group of genes which collectively define a species and create the distinctions which exist between one species and another. Thus, while genes which give rise to the variations among humans exist in the gene pool of humans, such that they are short or tall or fat or thin humans, humans are clearly distinct from other animals such as cats, which have a completely different gene pool. It should also be mentioned that even within the gene pool, there are groups of genes which define *strains* within the species. In industrial microbiology, the *strain* is often more valuable than the species as the ability to produce the unique characteristics of a product resides in the strain.

Industrial microbiological establishments usually keep a collection of the microorganisms which possess the gene pools for producing the goods manufactured by the establishment. This stock of organisms is known as a *culture collection* and ensures a regular supply of organisms to be used in the manufacturing process. Organisms in a culture collection are maintained in a low metabolic state in which replication of the cells is kept to a minimum or even entirely restricted. Industrially important microorganisms are often mutants, and the condition of low metabolism in which they are kept, limits their tendency to revert to their low-yielding ancestors.

In some circumstances organisms are maintained for comparatively short periods of days in an active state in which they are immediately ready for use in fermentations; such organisms are called *working stock*. In many breweries, for example, the producing yeasts are reused sometimes for up to eight runs or more before being discarded. In the interval between inoculations such yeasts are regarded by some workers as working stocks. It must be borne in mind that working stocks stand the chance of contamination and/or mutation, two serious problems inherent in industrial fermentations.

8.2 TYPES OF CULTURE COLLECTIONS

Culture collections in general, are an important part of the science of microbiology but, as will be shown below, they are specially important in industrial microbiology. Culture collections maintained by industrial establishments are usually specialized and store mainly those used in that particular organization.

There are various kinds of culture collections. Some national culture collections handle a wide variety of organisms, of whatever kind. The best known in this category is the American Type Culture Collection (ATCC). Other collections are specialized and may handle only pathogenic microorganisms, such as the National Collection of Type Cultures (NCTC) in Colindale, London, UK or industrial microorganisms, such as National Collection of Industrial Bacteria (NCIB) in Aberdeen, Scotland. Still others almost exclusively handle one type of organism such as Center vor Braunsveitzer (CBS) in Holland, which handles fungi exclusively. Many universities all over the world have culture collections which reflect their range of microbiological interests.

Culture Collections around the world are linked by the World Federation of Culture Collections (WFCC). The WFCC is an affiliate of the International Union of Microbiological Societies (IUMS) the organization which links national microbiological societies world wide. The WFCC is concerned with the collection, authentication, maintenance, and distribution of cultures of microorganisms and cultured cells. Its aim is to promote and support the establishment of culture collections and related services, to provide liaison and set up an information network between the collections and their users, and work to ensure the long term perpetuation of important collections.The WFCC pioneered the development of an international database on culture resources worldwide. The result is the **WFCC World Data Center for Microorganisms (WDCM).**

Culture Collections are organized on regional and international basis for the exchange of cultures and ideas and include the Asian Network on Microbial Research (ANMR), BCCCM (Belgium Co-ordinated Collections of Microorganisms), ECCO (European Culture Collection Organization), JFCC (Japanese Federation of Culture Collections), MICRO-NET (Microbial Information Network of China), MSDN (Microbial Strain Data Network, UK), UKNCC (United Kingdom National Culture Collection), USFCC (United States Federation of Culture Collections, USA). The WFCC maintains a World Data Center for Microorganisms (WDCM) at the National Institute of Genetics (NIG) in Japan, and has records on about 500 culture collections from 60 countries. A list of culture collections around the world will be found in the Kirsop and Doyle, 1991.

Culture collections may be specialized and in-house such as those in industrial establishments. Others are public and have the function of acquiring, identifying, preserving and distributing microorganisms and for a fee will supply cultures for in teaching, research or to industry. Such culture collections receive cultures from all over the world and thus serve the overall purpose of maintaining worldwide microbiological biodiversity.

In addition to making available organisms for industrial use, the major culture collections serve the important function of acting as depositories for microorganisms mentioned in the patenting of microbiological processes.

8.3   HANDLING CULTURE COLLECTIONS

Cultures are expensive to purchase. They are usually, however, supplied at a discount when used for reaching. Universities can however build their own cultures collections by preserving cultures arising from their research.

An industrial process may be initiated with organisms obtained through the Patent Office in connection with a patent. Often only one vial of such an organism is usually available. Once growth has been obtained from that vial the organism should be multiplied and stored in one or more of the several manners described below for the preservation of primary stock organisms in a Culture Collection. No matter what the source of a valuable organism, it is important that several replicates are stored immediately for fear of contamination while tests are carried out to ascertain its potential for fulfilling the expected activity. If the tests show that the expected antibiotic or other desired metabolite is being produced in the expected quantity then stored organisms are retained. The stocks of those organisms which proved negative at first sampling should not be discarded in a hurry because further examination may show that poor productivity was due to factors extrinsic to the organism such as an inadequate medium. In order to identify the organisms they must be properly labeled and accurate records kept of the handling of the organism. Date of transfer, the medium and the temperature of growth, etc., must all be carefully recorded to afford a means of assessing the effect of the preservation method.

8.4   METHODS OF PRESERVING MICROORGANISMS

Several methods have been devised for preserving microbial cultures. None of them can be said to apply exclusively to industrial microorganisms. Furthermore, no one method is suitable for preserving all organisms. The method most suited to any particular organism must therefore be determined by experimentation unless the information is already available.

Methods employed in the preservation of microorganisms all involve some limitation on the rate of metabolism of the organism. A low rate of spontaneous mutation exists during the growth of microorganisms, about once in every 10^9 division. Lowering the metabolic rate of the organism will further reduce the chances of occurrence of mutations. Preservation methods will be discussed under the following three headings, although it should be understood that in practice the methods combine one or more of the following three principles. The principles involved in preserving microorganisms are:

(a) reduction in the temperature of growth of the organism;

(b) dehydration or desiccation of the medium of growth;

(c) limitation of nutrients available to the organism.

All three principles lead to a reduction in the organisms metabolism.

8.4.1   Microbial Preservation Methods Based on the Reduction of the Temperature of Growth

8.4.1.1   Preservation on agar with ordinary refrigeration (4 – 10°C)

Organisms growing on suitable agar at normal growth temperatures attain the stationary phase and begin to die because of the release of toxic materials and the exhaustion of the nutrients. Agar-grown organisms are therefore **refrigerated** as soon as adequate growth is attained as to preserve them.

a) Aerobic organisms

Agar slants: Aerobic organisms may be grown on **agar slants** and refrigerated at 4 – 10°C as soon as they have shown growth.

Petri dishes: Aerobic organisms may also be stored on Petri dishes. The plates may be sealed with electrical tapes to prevent the plates from drying out on account of evaporation. Electrical tapes of different colors may be used to identify special attributes or groups among the cultures.

b) Anaerobic organisms: Anaerobic organisms may be stored on *agar stabs* which are then sealed with sterile molten petroleum jelly.

Storage using the above agar methods has advantages and disadvantages.

The **advantage** is that agar storage methods are inexpensive because they do not require any specialized equipment.

The **disadvantage**s are

(a) The organisms must be sub-cultured at intervals which have to be worked for each organism, medium used, laboratory practice, etc. This is because the temperature of the refrigeration is not low enough to limit growth completely.

(b) Consequent on regular sub-culturing is the possibility that contaminations and or mutations may occur.

(c) The third disadvantage is that Petri dishes occupies a lot of space in the refrigerator when compared with agar slants. But even agar slants are too bulky in comparison with the small vials in which lyophilized (freeze-dried) cultures are stored. Since plates occupy a lot of space, test tubes are usually preferred for storage in refrigerators.

(d) The process of sub-culturing is tedious apart from the possibility of contamination and mutation.

(e) When petroleum jelly is used as a seal, the arrangement can be messy.

Oil overlay

With the method of oil overlay whose function is to limit oxygen diffusion many bacteria, especially anaerobes and facultatives, and fungi survive for up to three years, and most of them for at least one year.

Medium for storing organisms on agar

The nature of the agar medium on which organisms are stored is of importance. A medium prepared from natural components rather than a chemically defined material is preferable, since a defined medium may, because it lacks some components present in the natural components, select for organisms specifically capable of growing on it. A stock culture medium should also not be unduly rich in carbohydrates such as glucose which will lead to early production of acid and hence possible early microbial death. Where glucose is used, such as for lactic bacteria, the medium should be buffered with calcium carbonate.

Popularity of agar storage methods

In spite of its shortcomings storage on agar is very popular and is the most widely used after lyophilization.

8.4.1.2 Preservation in Deep Freezers at about -20°C, or between -60°C and -80°C

The regular home freezer attains a temperature of about -20 °C.

Laboratory deep freezers used for molecular biology work range in temperature between -60 °C and -80 °C. It is possible to store microorganisms in either type of deep freezers in the form of agar plugs or on sterile glass beads coated with the organism to be stored.

Preservation on glass beads

The bacteria to be preserved are placed in broth containing cryoprotective compounds such as glycerol, raffinose, lactose, or trehalose. Sterile glass beads are placed in the glass vials containing the bacterial cultures. The vials are gently shaken before being put in the deep freezers.

To initiate a culture a glass bead is picked up with a pair of sterile forceps and dropped into warm broth. Growth develops from the organisms coating the bead. The growth is introduced onto an agar plate containing the appropriate medium and checked for purity before use.

Storage of agar cores with microbial growth

Bacteria, but especially moulds, yeasts, and actinomycetes may be stored as agar plugs made from plates of the confluent growths bacteria or of hyphe of filamentous organisms. It consists of placing agar plugs of confluent growth of bacteria and yeasts and hyphe of moulds or actinomycete in glass vials containing a suitable cryoprotectant and freezing the vials in deep freezers as above. To initiate growth a plug is placed in warm broth and plated out.

Freezing is rapidly gaining acceptance for preserving organisms because of its dual use for working and primary stock maintenance as well as its storage effectiveness for up to three years. It is useful for a wide range of organisms, and survival rates have been shown to be as good as freeze-drying in many organisms.

Advantages of the above freezing methods

   (a) the methods are simple to use and require a minimum of equipment;

(b) they save space as many hundreds of cultures can be stored in a small space;

(c) beads thaw rapidly and hence the method saves time,

(d) differently bead colors can represent different bacteria and so recognizing them is easy;

(e) the methods can be adapted for both aerobic and anaerobic organisms;

(f) the methods are suitable for situations or countries where power outages occur, as the freezer can remain cold for some time during power failures.

8.4.1.3 Storage in low temperature liquid or vapor phase nitrogen (-156°C to -196°C)

The liquid or vapor phase of nitrogen at -156°C to -196°C is widely used for preserving microorganisms and cultured cells. Fungi, bacteriophages, viruses, algae, protozoa, bacteria, yeasts, animal and plant cells, and tissue cultures have all been successfully preserved in it. It is a major method for organisms which will not survive freeze-drying. The period of survival and the number of surviving organisms are higher for most organisms than when freeze drying is used. In many laboratories it is the choice method for storing very valuable organisms. Some organisms are prone to losing numbers with this method, but the loss is reduced with the use of cryoprotectants. Some of the most commonly used cryoprotectants are (vol/vol) 10-20% glycerol and 5-10% dimethyl sulfoxide (DMSO) in broth culture of the organism in vials which are then frozen in liquid nitrogen. Vials for storing organisms in low temperature nitrogen may be made of glass or fashioned from ordinary polypropylene (plastic) drinking straws. Straws (4 mm diameter) are usually cut into pieces 40 mm long and made into ampoules by sealing the ends with heat.

Freezing at –156°C to -196°C has the following disadvantages:

(a) As liquid nitrogen evaporates, it has to be replenished regularly; if not replenished the cultures may be lost.

(b) A risk of explosion exists when cultures are frozen in liquid nitrogen in improperly sealed glass vials which permit entry of liquid nitrogen into the vials. Such vials may explode when warmed to thaw them. Discarding poorly sealed glass vials removes such risks; vapor phase storage removes such dangers.

(c) Although it is not labor intensive the equipment is expensive.

(d) Finally it is not a convenient method for transporting organisms.

8.4.2 Microbial Preservation Methods Based on Dehydration

Just as reduction in temperature limits the metabolism of the organism, dehydration removes water a necessity for the metabolism of the organism. Several methods may used to achieve desiccation as a basis for preserving microorganisms.

8.4.2.1 Drying on sterile silica gel

Many organisms including actinomycetes and fungi are dried by this method. Screw-cap tubes half-filled silica gel are sterilized in an oven. On cooling a skim-milk suspension of spores and the cells of the fungus or actinomycetes is placed over the silica gel and cooled. They are dried at 25°C, cooled and stored in closed containers containing desiccants.

8.4.2.2   Preservation on sterile filter paper

Spore-forming microorganisms such as fungi, actinomycetes, or *Bacillus* spp may be preserved on sterile filter paper by placing drops of broth containing the spores on sterile filter paper in a Petri dish and drying in a low temperature oven or in a dessicator. Alternatively, sterile filter paper may be soaked in the broth culture of the organism to be dried, placed in a tube, which is then evacuated and sealed. After drying the filter paper may be placed in sterile screw caps bottles and stored either at room temperature or in the refrigerator.

8.4.2.3   Preservation in sterile dry soil

The most commonly used form of storage in a dry state is the use of *dry sterile soil*. In this method dry soil is sterilized by autoclaving. It is then inoculated with a broth or agar culture of the organism. The soil is protected from contamination and allowed to dry over a period of time. Subsequently it may be refrigerated. The method has been widely and successfully used to store sporulating organisms especially clostridia and fungi; it has also been used for bacilli and *Azotobacter* sp. Some non-sporulating bacteria which do not survive well under Lyophilization, may be stored in soil.

8.4.2.4   Freeze-drying (drying with freezing), lyophilization

Freeze-drying or lyophilization is widely employed and a lot has been written about it. The principle of the method is that the organism is first frozen. Subsequently, water is removed by direct vaporization of the ice with the introduction of a vacuum. As the suspension is not in the liquid state, distortion of shape and consequent cell damage is minimized. At the end of the drying the ampoule containing the organism may be stored under refrigeration although survival for many years has also been obtained by storage at room temperature. The initial freezing (before the drying) may be achieved in a number of ways including the use of freezing mixtures of CO_2 and alcohol, salt and ice, or in a chamber of a freeze-drying machine in which the evaporation of water vapor from the material causes enough cooling to freeze the material. A desiccant, usually phosphorous pentoxide, is used to absorb water vapor during the freezing.

The suspending medium must be carefully chosen, because of differences in the cryoprotection properties of different substrates. Horse blood is usually used; others which have been successfully used are inositol, various disaccharides, and polyalcohols. Unless the information already exists the best suspending medium can only be decided by experimentation. The ampoule is usually evacuated after freeze-drying. It may however be filled with nitrogen; CO_2 or argon but the survival of organisms with them is lower than in vacuum, or with nitrogen.

Lyophilization is preferred for the preservation of most organisms because of its success with a large number of organisms, the relatively inexpensive equipment, the scant demand on space made by ampoules, but above all, the longevity (up to 10 years or more in some organisms) of most organisms stored by lyophilization.

8.4.2.5   L-drying (liquid drying, drying without refrigeration)

This is considered a modification of drying methods, since unlike freeze-drying, the organisms are not frozen, but dried from the liquid state. It has been used to preserve non-spore formers sensitive to freeze-drying, such as *Cytophaga, Spirillum and Vibrio*. Liquid drying has been effectively used to preserve organisms such as anaerobes that are damaged by freezing.

Small vials made of glass are filled with a mixture of skim milk, medical grade activated charcoal and myo-inositol , autoclaved and thereafter frozen at about -40°C for a few hours. The vials are then freeze-dried and this leads to a disc of freeze-dried carrier material in the vials. The broth of the organism to be dried is placed on the disc and the material is subjected to a vacuum in the liquid unfrozen state at 20°C.

8.4.3   Microbial Preservation Methods Based on the Reduction of Nutrients

8.4.3.1   Storage in distilled water

Many organisms die in distilled water because of water absorption by osmosis. However some have been known to survive for long periods in sterile distilled water. Usually such storage is accompanied by refrigeration; some organisms are however, harmed by refrigeration. Among organisms which have been stored for long periods with this method are *Pseudomonas solanaceanum, Saccharomyces cerevisiae*, and *Sarcina lutea*. The attractiveness of this method is its simplicity and inexpensiveness; since so few organisms seem to be storable in this manner, it should not, for fear of losing the organism, be adopted as the sole method for storing a newly acquired or isolated organism until it has been shown to be suitable.

8.4.4   The Need for Experimentation to Determine the Most Appropriate Method of Preserving an Organism

No one method can be said to suitable for the preservation of all and every organism. The appropriate method must be determined for each organism unless prior literature information exists. Even then such information must be used with caution, because a minor change in the medium composition may affect the outcome of the effort. The criterion to be used for determining the success of a method may not always necessarily be growth.

The preservation method must retain the characteristics which are desirable in the organism and this is crucial for industrial microorganisms. For example, the characteristic brick-red color of *Sarcina lutea* was lost in some preservation methods, while the production of rennet by *Rhizomucor* sp and of antibiotics by some actinomycetes were respectively affected by the method used for their preservation.

SUGGESTED READINGS

Anony. 1980. National Work Conference on Microbial Collections of Major Importance to Agriculture. American Phytopathological Society St Paul, Minnesota, USA.

Calam, C.T. 1980. The long-term storage of microbial cultures in industrial practice. The Stability of Industrial Organisms. B.E. Kirsop, (ed) In: Commonwealth Mycological Institute, Kew, England.

Demain, A.L., Solomon, N.A. (eds) 1985. Biology of Industrial Microorganisms. The Benjamin/Cummings Publishing Co., California, USA.

Kirsop, B.E., Doyle, A. (eds) 1991. Maintenance of Microorganisms and Cultured Cells. Academic Press London and San Diego.

Kurtzman, C.P. 1992. Culture Collections: Methods and Distribution In: Encyclopedia of Microbiology J, Lederberg, (ed) Vol 1 Academic Press, San Diego, USA. pp. 621–625.

Lamana, C. 1976. The Role of Culture Collections in the Era of Molecular Biology. Rita Colwell, (ed) In: The Role of Culture Collections in the Era of Molecular Biology. American Society for Microbiology Washington, DC, USA.

Monaghan, R.L., Gagliardi, M.M., Streicher, S.L. 1999. Culture Collections and Inoculum Development. In: Manual of Industrial Microbiology and Biotechnology. A.L. Demain, J.E. Davies, ASM Press, Washington, DC, USA, pp. 29-48; 2nd Ed.

Newman, Y.M., Ring, S.G., Colago, C. 1993. In: Biotechnology and Genetic Engineering Reviews M. P. Tombs, (ed). Vol 11. Intercept Press, Andover USA, pp. 263–294.

Stevenson, R.E., Hatt, H. 1992. Culture Collections, Functions In: Encyclopedia of Microbiology J, Lederberg, (ed), Vol 1 Academic Press, San Diego, USA, pp. 615-625.

Basic Operations in Industrial Fermentations

CHAPTER 9

Fermentors and Fermentor Operation

9.1 DEFINITION OF A FERMENTOR

A fermentor (or fermenter) is a vessel for the growth of microorganisms which, while not permitting contamination, enables the provision of conditions necessary for the maximal production of the desired products. In other words, the fermentor ideally should make it possible to provide the organism growing within it with optimal pH, temperature, oxygen, and other environmental conditions. In the chemical industry, vessels in which reactions take place are called reactors. Fermentors are therefore also known as bioreactors.

Fermentors may be **liquid**, also known as **submerged** or **solid state**, also known as **surface**. Most fermentors used in industry are of the submerged type, because the submerged fermentor saves space and is more amenable to engineering control and design. The discussions in most of the chapter will be therefore be on submerged fermentors; solid state fermentors will be discussed at the end of the chapter.

Depending on the purpose, a fermentor can be as small as 1 liter or up to about 20 liters in laboratory-scale fermentors and range from 100,000 liters to 500,000 liters (approximately 25,000 – 125,000 gallons) for factory or production fermentors. Between these extremes are found pilot fermentors which will be discussed later in this chapter. It should be noted that while fermentor size is measured by the total volume, only about 75% of the volume is usually utilized for actual fermentation, the rest being left for foam and exhaust gases. Several types of fermentors are known and they may be grouped in several ways: shape or configuration, whether aerated or anaerobic and whether they are batch or continuous. The most commonly used type of fermentor is the Aerated Stirred Tank Batch Fermentor. So widely used is this type that unless specifically qualified, the word fermentor usually refers to the Aerated Stirred Tank Batch Fermentor. This type will be discussed early in the chapter. Other types will be discussed later. Major differences between this and other fermentor types in configuration and operation will also be discussed.

The construction and design of a fermentor are the province of the engineer and only enough as will help the biotechnologist or microbiologist understand and utilize it efficiently will be discussed.

9.2 THE AERATED STIRRED TANK BATCH FERMENTOR

A *typical* fermentor of this type (Fig. 9.1) is an upright closed cylindrical tank fitted with four or more baffles attached to the side of the wall, a water jacket or coil for heating and/ or cooling, a device for forcible aeration (known as sparger), a mechanical agitator usually carrying a pair or more impellers, means of introducing organisms and nutrients and of taking samples, and outlets for exhaust gases. Modern fermentors are highly automated and usually have means of continuously monitoring, controlling or recording pH, oxidation-reduction potential, dissolved oxygen, effluent O_2 and CO_2, and chemical components of the fermentation broth (or fermentation *beer* as the broth is called before it is extracted). Nevertheless the fermentor need not have all these gadgets and many automated activities can also be prosecuted manually.

It is important that the type of fermentation required be clearly understood when a fermentor is being planned; a fermentor is expensive and once installed it may be unnecessarily expensive to drastically remodify it. Furthermore, because of its expense, a

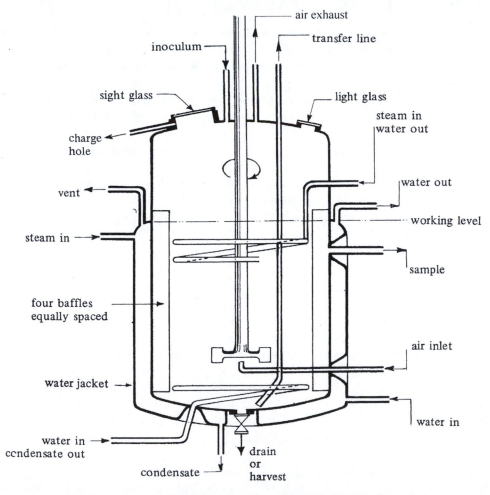

Fig. 9.1 Structure of a Typical Fermentor (Stirred Tank Batch Bioreactor)

fermentor will be expected to enable the organization to recover the outlay made on it by being in use over a reasonably long period. It may therefore be wise for small establishments to set up general-purpose fermentors such as has been described above, with provision for, if not for actual installation of, as many components as are likely to be needed in the future.

9.2.1   Construction Materials for Fermentors

A simple batch fermentor may consist of no more than an open tank made of wood, concrete or carbon-steel if contamination is not a serious problem and provided that no need exits for strict pH and temperature control, or that the temperature is controlled in the building. Thus, many breweries, particularly those making top-fermented beers for many years had open fermentors. Although it is not the practice, feed yeasts for the consumption of farm animals may also be grown in open fermentors. Serious contamination is restricted because of the acidity of the medium usually used. However, for fermentations with strict sterility requirements and closely controlled environmental needs, such as in the antibiotic industry, a material which can withstand regular steam sterilization is necessary. Furthermore, the hydrostatic pressure of a large volume of liquid can be enormous. Stainless steel is therefore normally used for pilot and production fermentors. Laboratory scale fermentors are usually made of Pyrex glass to enable autoclaving.

Where a highly corrosive material is fermented, e.g. citric acid, the fermentor should definitely be made of stainless steel. It is inevitable that small quantities of the material of which the fermentor is made will dissolve in the medium. Some materials, e.g., iron may inhibit the productivity of organisms in certain fermentations. It is for this reason that carbon-steel fermentors are often lined with glass, or 'plastic' materials e.g. a phenolic-epoxy coating. The material used for lining depends on the expected abrasion on the wall of the fermentor by medium constituents. Glass lining is employed only for small fermentors because of the high cost and the possibility of breakage.

In order to avoid contamination, fermentor vessels of all types should be of welded construction throughout. The welds should be free of pinpoints where organisms can develop in small bits of old media, and shielded from sterilization. The joint inlets and outlets of the fermentor should be designed so as to provide smooth surfaces and eliminate pockets difficult to sterilize. If gaskets are used at joints these should be non-porous.

9.2.2   Aeration and Agitation in a Fermentor

Oxygen is essential for growth and yield of metabolites in aerobic organisms. In those fermentations where aerobic organisms are used, the supply of oxygen is therefore critical. For the oxygen to be absorbed by microorganisms it must be dissolved in aqueous solution along with the nutrients. Unfortunately not only does air ordinarily contain only 20% of oxygen, but oxygen is also highly insoluble in water. At 20°C for example, water holds only about nine parts per million of oxygen. Furthermore, the higher the temperature the less oxygen (and other gases) water can hold. For some highly aerobic fermentations such as the growth of yeast or production of citric acid, oxygen is so critical that even if the broth were entirely saturated with oxygen it would contain only a 15 second supply for the organisms. In other fermentations, the aeration requirement need not be as intense but must be presented to the organisms at a controlled level. The foregoing would have shown that oxygen control in industrial fermentations is as important as pH, temperature and other environmental controls.

The air used in most fermentation is sterile and produced as discussed in Chapter 11. However, in some fermentations where sterility is not necessary such as in yeast fermentation, the air is merely scrubbed by passing it through glycerol. The air used in fermentation, whether, sterile or not, is forced under pressure into the bottom of the fermentor just below the lowest impeller the air enters through a sparger which is a pipe with fine holes. The smaller the holes the finer the bubbles and the more effective the supply of oxygen to the microorganisms. However, if the holes are too small, then a greater pressure will be required to force the air through, with consequent higher consumption of energy and therefore of costs. A balance must be struck between wide holes which may become plugged and holes small enough to release fine bubbles. Plugging by hyphae of filamentous fungi or by other particles in the medium may occur. Usually holes of about 0.25-3.0 cm in diameter meet this compromise. Since the size of the holes is fixed, the amount of oxygen fed into the medium (usually measured in feet/sec) can be controlled by altering the pressure of the incoming air.

For many fermentations especially where filamentous fungi and actinomycetes are involved, or the broth is viscous, it is necessary to agitate the medium with the aid of impellers. In large-scale operations, where aeration is maintained by agitator-created swarms of tiny air bubbles floating through the medium, the cost is very high and for this reason careful aeration is done based on mathematical calculations conducted by chemical engineers.

Agitators with their attached impellers serve a number of ends. They help to distribute the incoming air as fine bubbles, mix organisms uniformly, create local turbulence, as well as ensure a uniform temperature. The optimal number and arrangement of impellers have to be worked out by engineers using information from pilot plant experiments. The viscosity of the broth affects the effectiveness of the impellers. Since the viscosity of the broth may alter as fermentation proceeds, a satisfactory compromise of size, shape, and number of impellers must be worked out. In unbaffled fermentors a vortex or inverted pyramid of liquid forms and liquid is thrown up on the side of the fermentor. The result is that heavier particles sediment and thorough agitation is not achieved. The insertion of baffles helps eliminate the formation of a vortex and interferes with the upward throw of liquid against the side of the fermentor. A similar effect can be observed by stirring a cup of coffee or water rapidly with the handle of a spoon and inserting the handle of the spoon thereafter along the side of the cup. If four spoon ends were stuck simultaneously in the (storm in a) tea cup (!) the effective mixing of the liquid can be easily visualized. The use of baffles thus ensures not only a more thorough mixing of the nutrient and air but also the breakup of the air bubbles. In order to understand the importance of fine bubbles, it is important to appreciate the several barriers through which oxygen must theoretically pass before reaching the organism in the two film gas model which is commonly used (Fig. 9.2).

These barriers are indicated in Fig. 9.2 and include the following:

(i)   Gas-film resistance between gas and interface;

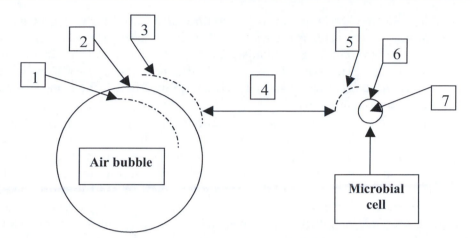

Fig. 9.2   The Various Barriers through which Gas Passes to Reach the Microorganism in a Liquid

 (ii) Gas-liquid interfacial resistance;
 (iii) Liquid-film resistance between interface and the bulk of the liquid;
 (iv) Liquid-path resistance characterized by oxygen gradient in bulk;
 (v) Liquid-film resistance around cell or cell-clump;
 (vi) Inter-cellular or intra-clump resistance;
 (vii) Resistance to reaction ('absorption') of oxygen with the cell respiratory enzymes.

In this model, in the transfer of oxygen from a gas bubble to a liquid, a stagnant gas film and a stagnant liquid film exist on both sides of a gas-liquid inter-phase. The resistance of these films to the transfer of solutes depends to a large extent on the degree of agitation. In the case of oxygen the only significant resistance is that of the liquid film which is broken by agitation. On the other hand, cell-liquid resistance, becomes important when there is clumping of organisms.

In terms of the above theory, the function of agitation of the fermentation may be taken as follows:

 (i) Gas dispersion or the creation of a large air-liquid interfacial area;
 (ii) Reduction in the thickness, and hence to resistance to oxygen diffusion of the liquid film which surrounds each bubble;
 (iii) Bulk mixture of the culture;
 (iv) Control of clump size.

It is clear from the figure that the finer the bubbles, the greater will be the total surface area of oxygen presented to the organism by a given volume of air. Provisions for agitation and aeration are thus very important components of an Aerated Stirred Tank Fermentor. In large-scale operations, where aeration is achieved by swarms of tiny air bubbles floating through thousand of liters of medium, the cost of aeration and agitation could be high, hence aeration and agitation have been, and are still, the subject of intense study by chemical engineers. From such studies the size and shape of the impellers in comparison to the rest of the fermentor (i.e., tank geometry), the airflow, the power requirement, etc., are calculated.

The exhaust air from the fermentor is passed through a filter which is sterilized with steam from time to time. This is especially necessary if the organism being grown is pathogenic (e.g., for vaccines). The exhaust pipe is positioned away from the incoming sterile air to avoid any chance of contamination. Furthermore, the agitation shaft which is impelled by a motor is fitted with a special seal at the point where it enters the fermentor in order to avoid contamination.

Sterile air is needed in some aerobic fermentations and it is produced in several ways including irradiation, electrostatic absorption of particles, the use of heat resulting from the compression of the gas. But the most commonly used method is the passage of the air through filters either made of materials such as cellulose nitrate, or more commonly of cotton and sometimes other materials. Sterility will be discussed in more detail in Chapter 11. Besides supplying oxygen to the organisms, the provision of air under pressure helps remove inhibitory volatile metabolites, and contributes to the reduction of contaminants by providing a positive air pressure.

9.2.3 Temperature Control in a Fermentor

Many fermentation processes release heat, which must be removed so as to maintain the optimum temperature for the productivity of the organism. In small laboratory fermentors temperature control may be achieved by immersing the tank in a water bath; in medium-sized ones control may be achieved by a jacket of cold water circulating outside the tank or merely by bathing the unjacketed cylinder with water. In large fermentors temperature is maintained by circulating refrigerated water in pipes within the fermentor and sometimes outside it as well. A heating coil is also provided to raise the temperature when necessary.

The area required for the transfer of heat may be determined theoretically on the basis of the expected heat release from the fermentation, the energy input from the agitator, the work done by the air stream, and the amount of heat involved if the broth were sterilized *in situ* in the fermentor. Heat losses to be taken account of include that lost by the effluent air and to the cooling water.

9.2.4 Foam Production and Control

Foams are dispersions of gas in liquid. In fermentation they usually occur as a result of agitation and aeration. In a few industrial processes, e.g. in the beer industry (where foam head retention is a desirable quality), or in the manufacture of foam latex, foam is a welcome property. However, in most industrial fermentations, foam has undesirable microbiological, economic and chemical engineering consequences, as follows:

(i) The need to accommodate foams means that a substantial head space is left in industrial fermentations. By reducing foaming it has been possible to increase the total fermentation by 30-45%.

(ii) If the fermentation medium is such that it encourages rapid foaming, then the maximum aeration and agitation possible cannot be introduced because of excessive foaming. The effect of this is that the oxygen transfer rate is reduced.

(iii) If the foam escapes, then contamination may be introduced when foam bubbles coalesce and fall back into the medium after wetting the filters and other non-sterile portions of the fermentor.

(iv) Organic nutrients or inorganic ions with complex organic compounds may be removed from the medium by foam floatation, a phenomenon well known in beer fermentation, when proteins, hop-resins, dextrin's, etc., concentrate in the foam layer. A loss of nutrient from fermentations in this way could lead to reduced yield.

(v) It can be seen that the fermentation product may also be removed should it be amenable to foam floatation. Such a loss has actually been observed in a laboratory experiment with the antibiotic, monamycin.

(vi) Loss of microorganisms could also easily occur by floatation thereby leading also to reduced yields.

The above has dealt with surface foams which occur on the top of the medium. The more stable surface foams are the most troublesome. The unstable ones breakdown in about 20 seconds and cause no further havoc. In contrast with surface foams are the so-called fluid foams which occur within the broth. These are common in highly viscous mycelial fermentations and in unbaffled vortex fermentors.

9.2.4.1 Foaming patterns

In order to understand methods of dealing with foams, it is important to discuss some factors leading to their formation and their behavior during the progress of fermentation. Fermentation media are usually made up of complex materials whose compositions are not always precisely known. Of the compounds which give rise to foams, proteins produce the most stable foams. A medium consisting of only inorganic compounds will not foam unless suitable metabolites are produced by the organisms.

It is sometimes possible to reduce foaming by altering the medium composition of the fermentation. Thus, it was possible to use a larger broth volume by reducing foam from a yeast fermentation following the absorption of caramel and organic acids with bentonite from a sample of molasses. Furthermore, the concentration of nutrients, the pH, the method of preparing the medium components e.g. sterilization time, etc., can all affect foam formation and stability

The pattern of peak foam formation and disappearance during the course of fermentation depends on the composition of the medium and the nature and the activity of microorganisms taking part in the fermentation. Four or five foaming patterns have been recognized (Fig. 9.3).

In the first type (designated 1 in Fig. 9.3) the foam remains constant throughout the fermentation. This is not common in media made of complex materials and is more frequent in defined media consisting mainly of inorganic components. In the second type the foam falls from a fairly high level to a low but constant level, following the utilization of foam stabilizers in the nutrients by micro-organisms. In this type the microorganisms themselves produce neither foam stabilizers nor defoamers. In the third type foam life-time falls at first, but then rises. Under this condition the foam stabilizers in the original medium are metabolized but the organism also produces foam-stabilizing metabolites. In the fourth type the medium initially contains only a low amount of foam stabilizers. These increase as autolysis of the mycelium sets in. If these are later metabolized the foaming may once more drop resulting in a fifth pattern. In practice combinations of all or some of these may occur simultanously.

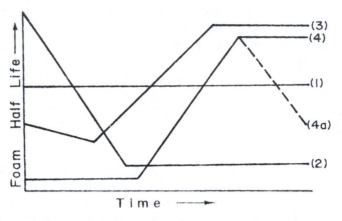

Fig. 9.3   Foaming Patterns in Industrial Fermentations

9.2.4.2   Foam control

Foams in industrial fermentations are controlled either by chemical or mechanical means. *Chemicals controlling foams* have been classified into **antifoams**, which are added in the medium to prevent foam formation, and **defoamers** which are added to knock down foams once these are formed. Some may not see much in the distinction and in this discussion the term antifoam will refer to both.

Foams are formed via froths which are temporary dispersals of gas bubbles in a liquid of no foam formation ability. Bubbles in a froth coalesce as they rise to the surface. In a foam however, they do not coalesce. Rather, the liquid film between two bubbles thins to a lamella. Materials which yield foam forming aqueous solutions such as proteins, peptides, synthetic detergents, soaps, and natural products such as saponin, lower the surface tension of the solution and permit foam formation. An analogy which seems to explain this is to imagine a fermentation liquid as being covered by various sheets of rubber (or other elastic materials) of varying thickness representing surface tensions. A thick sheet in this analogy represents high surface tension and thin sheets, low surface tension. The thinner the sheet the easier it is to blow balloons from it. Solutes which lower the surface tension of water are surfactants (although this name applies to a particular group of chemicals). In general, surfactants have a positive hydrophobic or water repellent end and a negative, hydrophilic on water-absorbing end.

The foam-forming properties of a surfactant may be seen as resulting from the repulsion of positive charges surrounding the bubbles. Some commercial surfactants can lower the surface tension of water from 92 dyne cm^{-1} (7.2×10^{-2}Nm^{-1}) to about 27 dyne cm^{-1} (27×10^{-2}Nm^{-1}). The positively absorbed surfactant layer confers on the liquid a phenomenon known as film elasticity which prevents local thinning during bubble formation in the same manner as a rubber sheet stretches and holds together.

Basically, antifoams enter the lamella between the bubbles by spreading over or mixing with the positively absorbed surfactants monolayer and thus destroying the film elasticity. The result is that the film collapses. Ideally, therefore, the antifoam should be miscible with the foaming liquid. Antifoams used in industrial fermentation should ideally have the following properties. They should:

(i) be non-toxic to microorganisms and higher animals, especially if the fermentation product is for internal use.

(ii) have no effect on taste and odor; a change in the usual organoleptic properties of the finished goods due to the antifoam or other components of the medium may result in consumer rejection of the goods. It is significant that a silicone antifoam has been used to limit foaming during wort boiling in beer manufacture. The silicone was subsequently removed and had no effect on the quality, including the foam head retention, of the beer.

(iii) be autoclavable.

(iv) not be metabolized by the microorganisms; sometimes as when natural oils are used, the antifoam may be utilizable in which case they must be replaced regularly.

(v) not impair oxygen transfer.

(vi) be active in small concentrations, cheap, and persistent.

Some chemical antifoams are discussed in Table 9.1. Many antifoams work best if dispersed in suitable carriers. Thus for Alkaterge C (Trade name) paraffin oil was found to be the best carrier.

Table 9.1 Some antifoams which have been used in industry

Category	Example	Chemical nature	Remarks
Natrual oils and fats	Peanut oil, soybean oil	Esters of glycerol and long chain mono-basic acids	Not very efficient. Used as carriers for other antifoams; may be metabolized.
Alcohols	Sorbitan alcohol	Mainly alcohols with 8-12 carbon atoms	Not very efficient; may be toxic or may be metabolized.
Sorbitan derivatives	Sorbitan monolaureate (Span 20-Atlas)	Derivatives of sorbitol produced by reacting it with H_2SO_4 or ethylene	Span 20 active in extremely small amounts.
Polyethers	P400, P1200, P2000 (Dow Chemical Co.)	Polymers of ethylene oxide & propylene oxide	Active, but varies with fermentation.
Silicones	Antifoam A (Dow Corning Ltd.)	Polymers of polydimethyl-siloxane fluids	Very active; inert, highly dispersable, low toxity; expensive.

Antifoams may be added manually when foam is observed. This entails a close watch and may be expensive. Automatic antifoam additions are now very common and depend on a probe which is activated when foams rise and make contact with the probe. One of the earliest is the wick defoamer (Fig. 9.4) in which the foam drew some antifoam on making contact with a wick. Modern methods are electrically activated systems. Other systems which have been used include antifoam introduction via the sparging air, or continous drip-feeding.

Mechnanical defoamers of various designs have been described. In general they act by physically dispersing the foams by rapidly breaking them up.

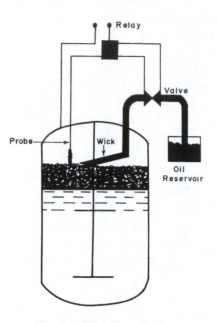

Fig. 9.4   Wick Type Anti-foam

9.2.5   Process Control in a Fermentor

The course of a fermentation may be followed by monitoring various operational parameters within the fermentor e.g. pH, air input, effluent gases, temperature; factors such as cell yield, or the output of metabolites may also be followed. The degree of accuracy of the monitoring depends of course on the instruments being used for the purpose. The purpose of this section is to discuss the principles involved in the operation of some of the various instruments used. Lists of manufacturers will be found in various publications, and on the Internet. It is not considered important to lay emphasis on manufacturers and their equipment as these are subject to changes dictated by the market.

9.2.5.1   pH measurement and control

The importance of the control of pH in microbial growth is well known. In some industrial fermentations, good yield depends on accurate control (and hence accurate measurement) of the pH of the fermentation broth. Sometimes the control of pH is achieved by natural buffers present in the medium; phosphates and calcium carbonate may also be used for this purpose. The buffering effect of these compounds is however usually temporary. The broth must therefore be sampled and the pH adjusted as desired with either acid or base. This method is laborious and may not accurately reflect the continuous change taking place in the pH of the broth. Sterilizable pH probes have become available and these are inserted in the fermentor or in a suitable projection therefrom in which the broth bathes the electrode. With these electrodes it is now possible to use an arrangement which will monitor pH changes and automatically induce the introduction into the medium of either acid or alkali. In many fermentations acidity rather alkalinity is the situation to be combated. Such acidity usually arises from microbial activity. It is therefore usual to arrange for the introduction of anhydrous ammonia as acidity increases.

9.2.5.2 Carbon dioxide measurement

Water and carbon dioxide are two of the most common end-products of aerobic fermentations. The measurement of CO_2 therefore helps determine the course of the fermentation as well as the carbon balance. At least three principles are employed in current equipment for CO_2 determination. The first method, which is the most widely used, depends on the ability of CO_2 to absorb infrared rays. A sensitive sensor translates this absorption to a gauge or record, from which it can be read off. In another principle, the effluent gas emerging from the broth is bubbled through a dilute solution of NaOH containing phenol red. The change in color of the phenol red is reflected in a photocell and the amount of CO_2 may be calculated from a standard curve. The third method depends on the thermal conductivities of the various gases in a mixture.

9.2.5.3 Oxygen determination and control

A number of methods are available for determining the oxygen concentration in a fermentation broth.

Of the *chemical* methods, the best known is that of Winkler which is routinely used to determine the biochemical oxygen demand (B.O.D) of water (Chapter 29). This method relies on the back-titration, using iodine and starch, of unoxidized manganous salt added to the liquid to be analyzed. Interfering substances are usually present in fermentation broths. Furthermore, the method is cumbersome. Modern sensing methods are not, however, based on this method. They rather sample the dissolved oxygen (DO) in the medium. Modern dissolved oxygen probes are autoclavable and are based on one or the other of two principles: the polarographic or the galvanic method.

In the polarographic method, a negative electric current 0.6-0.8 in voltage is passed through an electrode immersed in an electrolyte made of neutral potassium chloride. This negative electrode (cathode) is made of a noble metal such as platinum or gold. The anode is calomel or Ag/Ag CI. Under this condition the dissolved oxygen is reduced at the surface of the cathode according to the following reactions:

Cathode:	$O_2 + 2H_2O + 2e \rightarrow H_2O_2 + 2OH-$
	$H_2O_2 + 2e- \rightarrow 2OH-$
Anode:	$Ag + Cl- \rightarrow Ag\ Cl + e-$
Overall:	$4\ Ag + O_2 + 2H_2 + 4\ Cl \rightarrow 4\ Ag\ Cl + 4OH-$

The current which is measured after it has passed through the electrolyte is proportional to the dissolved oxygen reacting at the cathode. A plastic membrane permeable to gases but not ions separates the cathode, anode, and the electrolyte from the liquid to be studied. The dissolved oxygen diffuses through the membrane and its reaction at the cathode is measured at the current meter (Fig. 9.5). The electrolyte soon becomes depleted by the constant replacement of Cl by OH⁻ (see equations above) and the electrolyte has to be replaced.

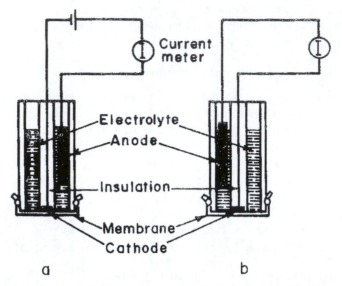

Fig. 9.5   Structure of Oxygen Electrodes: (A) Polarographic (B) Galvanic

In the galvanic method, no external source of electricity is applied. Instead the electricity generated between a base metal anode (zinc, lead, or cadmium) and noble metal cathode (silver or gold) is sufficient to cause the reduction of oxygen at the cathode. The reactions are thus:

$$\text{Cathode:} \qquad O_2 + 2H_2 + 4e \rightarrow 4\,OH$$
$$\text{Anode:} \qquad Pb \rightarrow Pb^{2+} + 2e$$
$$\text{Overall:} \qquad O_2 + 2Pb + 2H_2O \rightarrow 2Pb(OH)_2$$

The principle remains the same otherwise. The electric current generated in the system is proportional to the quantity of oxygen reacting at the cathode. The electrolyte does not however participate but the anode surface is gradually oxidized.

9.2.5.4   Pressure

It is important to know the pressure of gases in order to ensure that a positive pressure is maintained. A positive pressure helps eliminate contamination and contributes to the maintenance of proper aeration. Pressure may be determined with aid of a manometer.

9.2.5.5   Computer control

The fermentation industry, especially the antibiotic manufacturing aspect, usually compares its operations with those in the chemical industry. Leaders in the fermentation industry usually point to the fact that the fermentation industry in the early 1970s lagged behind chemical industries in applying computers in regulating and managing fermentations. The situation today is different and fermentation procedures are now highly automated. Automation is an engineering problem and the expected advantages of computerization have been given as follows:

(i) It should reduce labor by eliminating manual intervention.

(ii) The use of a computer should render an operator's work easier and reduce human error; it should, however, be possible to make changes while fermentation is on.

(iii) Automatic recording of all aspects of the fermentation is possible with a computer and is useful in meeting any regulatory requirements as well as in improving fermentation operations.

(iv) Experimentation should be easier as it should be much easier to study the effect of altering any variables such as dissolved oxygen, temperature, pH, air flow, nutrient addition, etc.

(v) Quality control should be easier to carry out.

(vi) In the event of power failure, and other emergencies, the system should be able to shut up itself and restart and gradually build up to the original level of activity.

Commercial sensors are available for a wide variety of parameters in a fermentor. This includes various ions, redox potential, cell mass measurement, carbohydrate measurements to name a few. The computerization of these parameters makes it fairly easy to monitor the operations in modern fermentation operations.

9.3 ANAEROBIC BATCH FERMENTORS

Some processes do not require the high levels of oxygen needed in aerobic fermentation; indeed some, such as clostridial fermentations do not require oxygen at all. These are collectively referred to as 'anaerobic' fermentations although in strict terms some may be micro-aerophilic. Anaerobic fermentors, whether strict or micro-aerophilic (i.e., requiring small amounts of oxygen) are not commonly used in industry. When they do (i.e., require oxygen), they are essentially the same as the descriptions given above in the typical (aerobic) fermentor. They, however, differ in the construction and operation as given below.

(i) Vigorous aeration through air sparging is absent, as oxygen is not required.

(ii) Agitation when done is aimed only at achieving an even distribution of organisms, nutrients and temperature, but not for aeration. In some cases agitation may be essential only initially; the evolution of CO_2 and H_2 in anaerobic fermentors may stir the medium.

(iii) The medium is introduced into the fermentor while hot to prevent the absorption of gases; and usually it is also introduced at the bottom of the fermentor.

(iv) The fermentor itself is filled as much as possible, in order to avoid an airspace which would introduce oxygen.

(v) If strict anaerobiosis is desirable, then an inert gas such as nitrogen may be blown through the fermentation, at least initially, to remove oxygen.

(vi) Some low redox compounds, such as cysteine, may be introduced into the medium.

The same typical fermentor already described may be used for both aerobic and anaerobic fermentations. It is especially important that it be possible for aerobic or anaerobic fermentations to be carried in the same vessel as some fermentatons such as alcohol manufacture require an earlier aerobic stage in which cells are produced in large numbers and a later stage in which alcohol is produced anaerobically. But even the strictly anaerobic fermentations can be carried out in the stirred tank batch fermentor already described above.

Two strictly anaerobic fermentaiton processes include acetone-butanol fermentation (*Clostridium acetobutylicum*) and anti-tetanus toxoid production (*Clostridium tetani*). An example of a micro-aerophilic fermentation which requires only a small amount of oxygen is lactic acid production, while one which has a primary aerobic and a secondary anaerobic system is alcohol production. Other examples are dextran production and the production of 2-3 butylene-glycol.

9.4 FERMENTOR CONFIGURATIONS

Based on the nomenclature of the chemical engineering industry fermentors have been grouped into four:

   (i) **Batch fermentors** (Stirred Tank Batch Fermentors): (designated BF in Fig. 9.6) The major features of this type of fermentor have been described already in Section 9.2 and Fig. 9.1. The other three are continuous fermentors and these are described below:

  (ii) **Continuous stirred tank fermentors**: (CSTF in Fig. 9.6) The tank used in this system is essentially similar to that of the batch fermentor. It differs only in so far as there is provision for the inlet of medium and the outlet of broth. The system has been described under continuous cultivation.

 (iii) **Tubular fermentors**: (TF in Fig. 9.6) The tubular fermentor was originally so named because it resembled a tube. In general tubular fermentors are continuous unstirred fermentors in which the reactants move in a general direction. Reactants enter at one end and leave from the other and no attempt is made to mix them. Due to the absence of mixing, there is a gradual fall in the substrate concentration between the entry point and the outlet while there is an increase in the product in the same direction.

 (iv) **The fluidized bed fermentor**: This is essentially similar to the tubular fermentor. In both the continuous stirred fermentor and the tubular fermentor there is a real danger of the organisms being washed out (Fig. 9.10). The fluidized bed reactor is an answer to this problem because it is intermediate in nature between the stirred tank and the tubular fermentor. The microorganisms which are in a fluidized bed fermentor are kept in suspension by a medium flow rate whose force just balances the gravitational force. If the flow were lower, the bed would remain 'fixed' and if the flow rate was at a force higher than the weight of the cells then 'elutriation' would occur with the particles being washed away from the tube. The tower fermentor for the brewing of beer and production of vinegar (Chapter 14) is an example of a fluidized bed fermentor.

9.4.1 Continuous Fermentations

Continuous fermentations are those in which nutrients are continuously added, and products are also continuously removed. Continuous fermentations contrast with batch fermentations in which the products are harvested, the fermentor cleaned up and recharged for another round of fermentation. In the chemical industry continuous processing has replaced many batch processes. This is because for products for which there is a high and constant demand continuous processing offers several advantages.

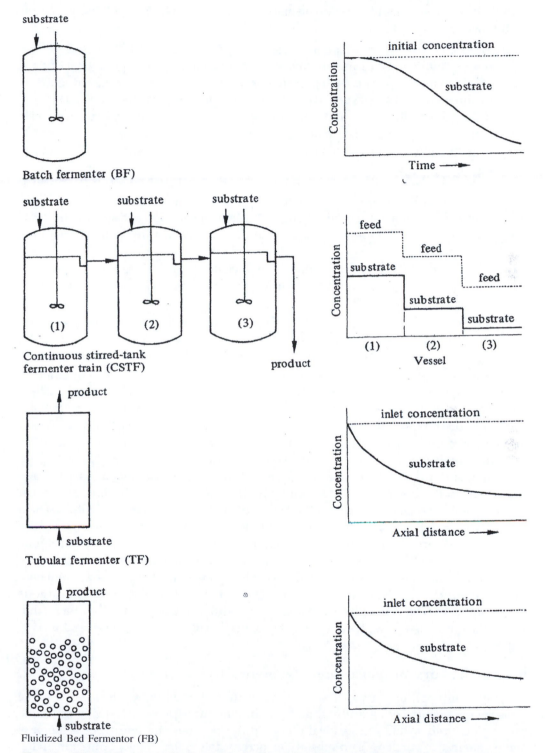

Fig. 9.6   Different Fermentor Configurations (Left) and Graphs Depicting Substrate Usage in the Various Configurations (Left) (see also Fig. 9.7)

These advantages when adapted to the microbiological industries, potentially include the following:

(i) More intensive use of the equipment, especially the fermentor, and therefore greater return on the initial capital outlay made in installing them. A great deal of time involved in the cycle of batch production is not employed in direct production of the final goods. Part of such 'dead' time is used in emptying the batch fermentor during harvest, for cleaning, sterilizing, cooling and recharging with fresh medium in between each batch. Furthermore, much of the period of a batch fermentation is required for a lag period when the organisms are merely growing and not yet producing (where the product is a metabolite), or the maximum population has not been attained (where the product is the cell itself). In a continuous fermentation, as soon as the steady state has been attained and provided no contaminations occur, and other production activities permit the plant to run for a reasonably long time, the 'dead' time required for all the above is eliminated.

(ii) Allied to the above are savings in labor which do not have to repeatedly perform the various operations linked with the 'unproductive' portions of batch fermentation.

(iii) Continuous processes are more easily automated. This helps eliminate human error and thus ensures greater uniformity in the quality of the products. Automation also further saves labor costs additional to those mentioned in (ii).

Despite the possible advantages of continuous fermentation, the fermentation industry has not in general adopted it. The areas where it has been employed include beer brewing, food and feed yeast production, vinegar manufacture, and sewage treatment.

The reasons for the slow adoption of continuous fermentation since interest developed in it several decades ago, are to be found in technical and economic factors. One of the early deterrents was the fact that many early continuous fermentations became easily contaminated. It is easy to see that while slow growing contaminants might not have developed to the point where they can be noticed in the 4, 5, or 10 days of a batch fermentation they can pose a serious threat to production, in a continuous culture which goes on for up to three, six, or nine months. If the contaminant is fast growing then the danger while serious in a batch fermentation is infinitely more so in a continuous fermentation. Another problem was that mutants better adapted to the environment of the continuous fermentor are easily selected. Where they perform better than the parent type the difference was hardly noticed, except perhaps that a particular continuous fermentation was inbued with an apparently inexplicable efficiency. On the other hand, where the mutants were less productive, the reputation of continuous fermentation was not helped.

9.4.1.1 Theory of continuous fermentation

In a batch culture four or five phases of growth are well recognized: the lag phase, the phase of exponential or logarithmic growth, the stationary phase, the death or decline phase. Some others add the survival phase. In the lag phase individual cells increase somewhat in size but there is no substantial increase in the size of the population. In the exponential phase, the population doubles at a constant rate, in an environment in which the various nutritional requirements are present in excess. As the population increases, various nutrients are used up and inhibitory materials, including acids, are produced; in other words the environment changes. The change in the environment soon leads to the death of some organisms. In the stationary phase the rate of growth of the organisms is the same as the rate of death. The net result is a constant population. In the death phase, the rate of death exceeds the growth rate and the population declines at an exponential rate.

If however during the exponential phase of growth, a constant volume is maintained by ensuring an arrangement for a rate of broth outflow which equals the rate of inflow of fresh medium, then the microbial density (i.e., cells per unit volume) remains constant. This is the principle of one method of the continuous culture in the laboratory, namely, the *turbidostat*.

As discussed above, the stationary phase sets in partly because of the exhaustion of various *nutrients* and partly because of the introduction of an unfavorable environment produced by *metabolites* such as acid. Either of these two groups of factors can be used to maintain the culture at a constant density. Usually nutrients are used and their use for this purpose will be discussed.

In a batch culture the various nutrients required by an organism are usually initially present in excess. If all but one of the nutrients are present in adequate amount, then the rate of growth of the organisms will depend on the proportion of the limiting nutrient that is added. Thus if 100 grams per liter of the limiting nutrients are required for maximum growth but only 90 grams per liter are added, then the rate of growth will be 90% of the maximum. It is then possible to control the growth at any given rate but which rate is less than the maximum possible, by letting in fresh nutrient at the same rate as broth is released and also supplying one of the nutrients at a level slightly less than the maximum. This principle is employed in the *chemostat* method of continuous growth.

In both the chemostat and the turbidostat the rate of nutrient inflow and broth outflow must relate to the generation time or growth rate of the organism. If the rate of nutrient addition is too high, then sufficient time is denied to the organism to develop an adequate population. The organisms are then washed out in the outflow. If on the other hand the rate of nutrient addition is too low, a stationary phase may set in and the population may begin to decline.

The above is a simple non-mathematical description of the two basic procedures which have been employed in the laboratory study and industrial application of continuous individual cultivation. More detailed studies are widely available in texts on microbial physiology.

To summarize, in the **turbidostat** a device exists for ensuring that a constant volume of a microbial culture is maintained at constant density or turbidity. All the nutrients are present in excess and the density or turbidity is monitored by a photo-cell which translates any change to a mechanism which automatically reduces or increases the rate of medium inlet and broth output, as necessary.

In the **chemostat** method a constant population is maintained in a constant volume by the use of sub-maximal amounts of nutrient(s).

In the laboratory and in practice the chemostat is far more widely used than the turbidostat, probably because of the slightly more complex set up of the turbidostat which follows from the need for constant density monitoring of the broth.

9.4.1.2 Classification of continuous microbial cultivation

It is important to understand the physiology of the production of the fermentation product in order to enable the designing of an efficient continuous fermentation set-up. The classification given below enables such a selection (Fig. 9.7).

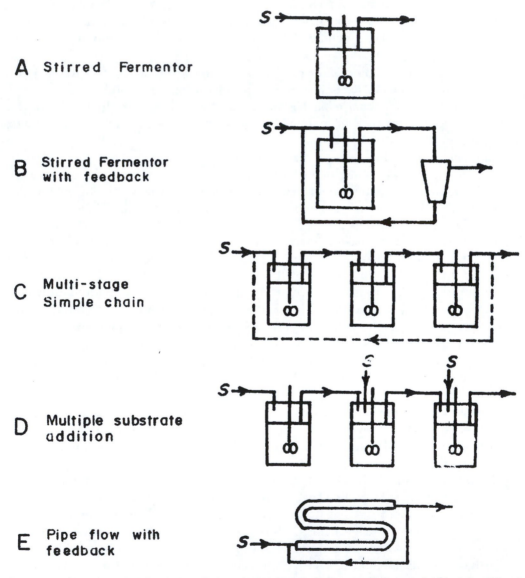

A Stirred Fermentor

B Stirred Fermentor with feedback

C Multi-stage Simple chain

D Multiple substrate addition

E Pipe flow with feedback

Fig. 9.7 Various Types of Continuous Culture Arrangements (S Denotes Substrate Addition) A = Stirred Batch Fermentor. B-E, Continuous Fermentors

9.4.1.2.1 Single-state continuous fermentations

There are fermentations in which the entire operation is carried out in one vessel, the nutrient being added simultaneously with broth outflow. This system is suited for growth related fermentations such as yeast, alcohol, or organic acid production.

9.4.1.2.2 Multiple-stage continuous fermentation

This consists of a battery of fermentation tanks. The medium is led into the first and the outflow into the second, third, or fourth as the case may be. This is most frequently used for the fermentation involving metabolites. The first tank may be used for the growth phase and subsequent tanks for production, depending on the various requirements identified for maximal productivity.

9.4.1.2.3 Recycled single or multiple stage continuous fermentation

The out flowing broth may be freed of the organisms by centrifugation and the supernatant returned to the system. This system is particularly useful where the substance is difficult to degrade or not easily miscible with water such as in hydrocarbons. Recycling can be applied in a single stage fermentor. In a multiple stage fermentor, recycling may involve all or some of the fermentation vessels in the series depending on the need.

9.4.1.2.4 Semi-continuous fermentations

In semi-continuous fermentations, simultaneous nutrient addition and outflow withdrawal are carried out intermittently, rather than continuously. There are two types of semi-continuous fermentation, namely;

(i) 'cyclic-continuous'; (ii) 'cell reuse'.

In *Cyclic-continuous*, a single vessel is usually employed, although a series of vessels may be used. Fermentation proceeds to completion or near completion and a volume of the fermentation broth is removed. Fresh medium of a volume equivalent to that withdrawn is introduced into the vessel. As the size of the fresh medium is reduced, the time taken to complete the fermentation cycle is reduced until eventually the intermittent feeding becomes continuous. This system has been said to ensure a compromise, between the desirable and undesirable features of batch and continuous fermentation; productivity has however been shown theoretically and experimentally to be lower than in continuous fermentation.

In *cell reuse*, cells are centrifuged from the fermentation broth and used to reinoculate fresh medium. It is continuous only in the sense that cells are reused; in essence it is a batch fermentation.

9.4.1.3 Applications of continuous cultivation

The literature is full of various areas of potential application of continuous fermentation, experimented upon either in the laboratory or in pilot plants. These include single cell protein production, organic solvents such as ethanol, acetone, butanol, isopropanol, acetic acid from traditional raw materials such as sugar, starch, and molasses. Cellulose is also being considered as a substance for these and the continuous culture of cellulose digesting enzymes from *Trichoderma* is an important step. In agriculture, continuous cheese making, continuous yoghurt starter production and continuous use of lactore in whey are being vigorously pursued. Medical and veterinary applications include the continuous production of vaccines, and cell cultivation.

Continuous waste digestion for sewage chemical wastes outside the activated sludge exist as also do the continuous brewing of beer, the continuous production of wine and the continuous manufacture of yeasts, vinegar and alcohol.

9.5   FED-BATCH CULTIVATION

Fed-batch cultivation is a modification of batch cultivation in which the nutrient is added intermittently to a batch culture. It was developed out of cultivation of yeasts on malt, where it was noticed that too high a malt concentration lead to excessively high yeast growth leading to anaerobic conditions and the production of ethanol instead of yeast cells.

After its successful introduction in yeast cultivation, the original method or modifications of it have been used to achieve higher yields or more efficient media utilization in the production of various antibiotics, amino acids, vitamins, glycerol, acetone, butanol, and lactic acid. Some of the modifications include continuous (rather than intermittent) addition of single or multiple media components, withdrawal of a portion of the broth from the growth vessel and immediate dilution of the residue with fresh medium and the use of diffusion capsules. The latter are cylindrical capsules to one end of which a semi-permeable membrane is fixed. The nutrient diffuses slowly out through the membrane into the medium.

9.6   DESIGN OF NEW FERMENTORS ON THE BASIS OF PHYSIOLOGY OF THE ORGANISMS: AIR LIFT FERMENTORS

The Stirred Tank Batch Fermentor already described is the most widely used type of fermentor. Increasing knowledge of the physiology of industrial microorganisms and better instrumentation have provided the bases for more efficient manipulation of the organisms in the existing batch fermentors:

(i)   More sophisticated instrumentation is now used to monitor such fermentor parameters as dissolved oxygen and carbon dioxide, redox potential, and control of the fermentation leading to higher yields.

(ii)   Different levels of pH, temperature, and phosphate concentration are sometimes needed during the trophophase and the idiophase for the production of secondary metabolites. These differences have been exploited in some fermentations for higher yields.

(iii)   By careful monitoring using automated sensing devices, it is now possible to add just enough of the nutrients required by a growing culture so that feedback inhibition is avoided.

The above are a few examples to show that the existing fermentors can be better utilized when greater knowledge of microbial physiology is harnessed for that purpose. Despite these improvements, needs have arisen for drastic change from the typical stirred tank batch fermentor, and these needs would appear not to be fully met by automation of batch fermentors. Some of the needs call for the design of new fermentors based on the following:

(i) The diversification of fermentation products and new attendant problems. Examples are the production of single cell protein by continuous fermentation; production of microbial polysaccharides; fermentor cultivation of animal and plant cells; the growing re-emergence of anaerobic fermentations such as for ethanol.

(ii) The unusual properties of the substrates or products involved in this diversification such as insolubility in water (for example, of petroleum fractions, agricultural wastes, or hydrogen gas) or high viscosity (for example, microbial capsules).

(iii) Greater knowledge or awareness of the physiology of the organisms during their growth in a fermentor especially:

   (a) The need for high amounts of dissolved oxygen.
   (b) Adequate mixing of fermentation broths
   (c) The problem of clumping or aggregation especially in filamentous organisms such as actinomycetes.
   (d) The need to avoid feedback inhibition by the removal of inhibitory products.

To solve some of these problems most of the newly designed fermentors have moved away from the structure of the commonly used Stirred Tank Batch Fermentor. They in fact lack stirrers; instead they are of the recycle, loop, or airlift type in which stirring is replaced by pumping of air. Some of the problems these fermentors or arrangements are designed to solve are given below.

(i) Need for high amounts of dissolved oxygen: Many industrial fermentations require large amounts of oxygen, and yields are severely limited when the gas is in short supply. To solve this problem especially in regard to the utilization of novel carbon sources, from hydrocarbons, the airlift fermentor was designed (Fig. 9.8). In this fermentor high levels of dissolved oxygen are achieved by using the air pressure to lift the broth. According to

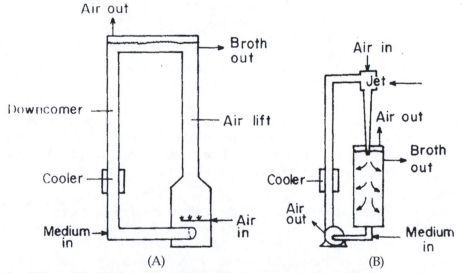

In the airlift type (A), air is forced through a sparger; in the plunging jet type (B) air is forced into the broth in a jet. There are no moving parts in loop fermenters.

Fig. 9.8 Loop Fermentors

some authors the airlift fermentor is a modification of the batch fermentor differing in the absence of stirring. It is in fact one of several types of loop fermentors.

(ii) Mixing of the broth: Poor mixing reduces yields in yeasts grown on alkanes. Aggregates consisting of alkane droplets and yeast cells float to the top of the broth in poorly mixed fermentations. The nutrients cannot therefore get to the yeasts which become starved as a consequence. The problem was solved using a completely filled circulating fermentor which operates on the same principle as a shake flask (Fig. 9.9).

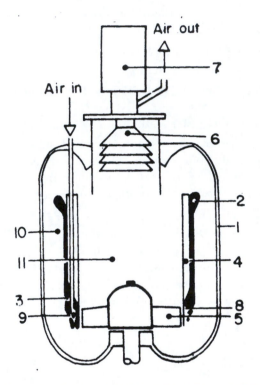

Fig. 9.9 Circulating Fermentor: 1, Vessel; 2&3, Draught Type; 4, Baffle; 5, Stirrer; 6 & 7, Foam Breaker; 8 & 9, Air Sparger; 10, Outer Section; 11, Inner Section.

(iii) Aggregate of cells: In filamentous cells, e.g., actinomycetes and fungi, the cells tend to aggregate and only those at the periphery of the clump grow. A steep gradient concentration of the product therefore exists from the outside to the inside. The avoidance of clumps and the production of loosely organized cells are achieved in the airlift fermentor.

(iv) Removal of inhibitory products: In the high concentration of components of a fermentation broth, feedback inhibition easily limits production. One manner of dealing with the problem is to subject the broth to dialysis. This can be achieved by constantly circulating the broth in an external membrane in contact with water. Volatile end products may be removed as they are formed by applying reduced pressure.

9.7 MICROBIAL EXPERIMENTATION IN THE FERMENTATION INDUSTRY: THE PLACE OF THE PILOT PLANT

When the microorganism used in a fermentation is new, experimentation must be carried out to determine conditions for its maximum productivity. It is usual to initiate the studies in a series of conical flasks of increasing size and to progress through a 10-20 liter fermentor to a pilot plant (100-500 liter) and finally to a production plant (10,000-200,000 liters). The processes involved in the increasing scale of operation culminating in the production plant are known as *scaling up*.

On the other hand, in a well-established fermentation procedure, any change to be introduced must be experimented on and tested out in a pilot plant whose function is to simulate the conditions and structures of the production plant. This procedure is often referred to as *scaling down*. The processes of scaling up and scaling down are essentially in the domain of the chemical engineer who depends on data supplied by the microbiologist.

Information gathered at the shake flask stage is used to predict requirements in the *pilot plant* which itself serves a similar purpose for the production plant. The optimum requirements of medium composition, aeration, temperature, redox potential, pH, foaming, etc., are determined and extrapolated for the next higher scale. The pilot fermentor is also used for training new recruits in the fermentation industry; it may also be used for continuous fermentation where a large enough number of them exist.

One approach which helps facilitate translation of information from the pilot plant to production is to reproduce the production plant as a geometrical replica of the pilot plant. Baffles, agitators, etc., are increased exactly according to a predetermined scale. This, however, does not entirely solve the problem because the mere increase in volume immediately poses its own problems. If the same level of productivity as encountered in the pilot study is to be maintained, then agitation and aeration may be applied at a level higher than that expected in a proportional increase in the production fermentor.

9.8 INOCULUM PREPARATION

The conditions needed for the development of industrial fermentations often differ from those in the production plant. This is because except in a few examples where the cells themselves are the required product, e.g., in single cell protein, or in yeast manufacture, most fermentation products are metabolites. Cells to be used must be actively growing, young and vigorous and must therefore be in the phase of logarithmic growth. Since organisms used in most fermentations are aerobes, the inocula will usually be vigorously aerated in order to encourage maximum cell development, although they may need less aeration in subsequent incubation. The chemical composition of the medium may differ in the inoculum and production stages. The inoculum usually forms 5-20% of the final size of the fermentation. By having an inoculum of this size the actual production time is considerably shortened.

The initial source of the inoculum is usually a single lyophilized tube. If the content of such a tube were introduced directly into a 100,000 liter pilot fermentor, the likelihood is that it would take an intolerably long time to achieve a production population, during which period the chance of contamination is created. For these reasons inocula are prepared in several stages of increasing volume. At each step, the growth is checked for the absence of contamination by plating. When the lyophilized vial is initially plated out and shown to be pure, the *entire* plate instead of a single colony is scraped off and transferred to the shake flask so as to avoid picking mutants (Fig. 1.2, Chapter 1)

9.9 SURFACE OR SOLID STATE FERMENTORS

In solid state fermentors rice bran or some such solid is used. Molasses may be added and a nutrient solution of ammonium and phosphate may be introduced. It is used mainly in Japan for enzyme production, and has been used for citric acid production. Fungal bioinsecticides are also cultivated as surface cultures. Certain mushrooms are also grown in tray fermentors.

In the surface fermentor shown in Fig. 9.10, a series of shallow trays no more than about 7 cm in depth is used, the solid medium not being more than about 5 cm so that air can penetrate into the solid medium. Humid air is blown into the chamber containing the trays. The incoming air and the out going air may be filtered especially when fungi are used to save the dissemination of the spores in the atmosphere. In some fermentations some form of temperature control is imposed through blowing cold air into the fermentor and also by cooling the room where the fermentor is located.

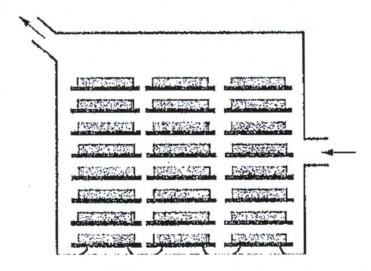

Fig. 9.10 Diagram of a Solid-state (Surface)Tray Fermentor Humid Cooled, Sometimes Filtered, Air is let into the Fermentor; the Exhaust Air is also Filtered (see text)

SUGGESTED READINGS

Ahuja, S. 2000. Handbook of Bioseparations. Vol 2 Academic Press. San Diego, USA.

Dobie, M., Kruthiventi, A.K., Gaikar, V.G. 2004. Biotransformations and Bioprocesses. Marcel Dekker, New York, USA.

Endo, I., Nagamune, T., Katoh, S., Yonemoto (eds) 1999. Bioseparation Engineering. Elsevier Amsterdam the Netherlands.

Flickinger, M.C., Drew, S.W. (eds) 1999. Encyclopedia of Bioprocess Technology - Fermentation, Biocatalysis, and Bioseparation, Vol 1-5. John Wiley, New York, USA.

Garcia, A.A., Bonem, M.R., Ramirez-Vick, J., Saddaka, M., Vuppu, A. 1999. Bioseparation Process Science. Blackwell Science Massachussets USA.

Harrison, R.G., Todd, P., Rudge, S.R., Petrides, D.P. 2003. Bioseparation Science and Engineering. Oxford University Press, New York, USA.

Kalyanpur, M. 2000. Downstream Processing in Biotechnology In: Downstream Processing of Proteins: Methods and Protocols. M Desai, (ed) Humana. Totowa, NJ, USA pp. 1–10.

Naglak, T.J., Hettwer, D.J., Wang, H.Y. 1990. Chemical permeabilization of cells for intracellular product release In: Separation Processes In Biotechnology, Marcel Dekker, New York, USA.

CHAPTER 10

Extraction of Fermentation Products

Judging from the extent of discussion on the fermentor and its accessories one might be led to feel that they consume nearly all of the capital investment in a fermentation industry. This is however not so: not only is the investment in recovery equipment high, but isolation costs represent a good proportion (sometimes up to 60%) of the cost of the final product. In one antibiotic factory, recovery equipment cost four times more than the fermentor. The necessity of having a well-planned and reliable recovery process and an efficient recovery plant is therefore of utmost importance. In this discussion only broad outlines of the principles of extraction will be given, more detailed consideration being given when each product is discussed.

The central problem in the extraction of fermentation products from the fermentation 'beer' or broth is that the required product usually (but not always) forms a small proportion of a complex heterogeneous mixture of cell debris, other metabolic product, and unused portions of the medium. The following are the factors borne in mind in deciding the extraction method to be used:

(i)   the value of the final product;
(ii) the degree of purity required to make the final product acceptable, bearing in mind its revenue-yielding potential;
(iii) the chemical and physical properties of the product;
(iv) the location of the product in the mixture i.e. whether it is free within the medium or is cell-bound;
(v)   the location and properties of the impurities; and finally;
(vi) the cost-effectiveness or the economic attractiveness of the available alternate isolation procedures.

The various steps followed in the extraction of fermentation products together with the approximate level of purification obtained in each stage are given in Table 10.1.

The procedure followed within each stage depends of course on the material being extracted, and are discussed hereunder. The product sought could be the cells themselves such as in yeast manufacture, or lodged in the cells (such as in streptomycin or some enzymes) or free in the medium as with penicillin.

10.1 SOLIDS (INSOLUBLES) REMOVAL

In general the initial step separates solids from the liquid fraction thereby facilitating further extractive steps, such as sorption, solvent extraction which would be wasteful or near impossible if the cells were not separated. When the required product is solid or is lodged in the insoluble portion liquid removal helps concentrate the solids.

Table 10.1 Conventional steps* followed in the purification of products in the soluble portion of 'beer'

Step Process	Hypothetical degree of purity (%)
1a. Removal of insolubles	0.1-1.0 (if product solubles)
Filtration	90-99 if product is cell
Centrifugation	such as yeasts
Decantation	
1b. Disruption of cells	
2 Primary foam isolation of the product	1-10
Sorption physical and/or ion exchange	
Solvent extraction	
Precipitation	
Ultracentrifugation	
3 Purification	50-80
Fractional precipitation	
Chromatography (adsorption, partition, ion exchange, affinity)	
Chemical derivatization	
Decolorization	
4. Final product isolation	90-100
Crystallization	
Drying	
Solvent removal	

*Some modern extraction methods combine steps 1 and 2

In a few cases no separation takes place such as in the acetone butanol fermentation, where the entire beer is used. In most cases, however, the separation methods used are filtration, centrifugation, decantation, and foam fractionation. Where the required fraction is in the cells then much of the impurities are removed with the filtrate after the cells have been isolated. The various methods used in solids removal are discussed below.

10.1.1 Filtration

The rotary vacuum filter: One of the most commonly used filters in industry is the *rotary vacuum filter* which is available in several forms. Essentially the filter consists of a hollow rotating cylinder divided into four partitions and covered with a metal or cloth gauze. A vacuum is applied in the cylinder and as it rotates the vacuum sucks liquid materials from the shallow trough in which the rotating cylinder is immersed. For thick slurries which are difficult to filter (e.g. aminoglycoside broths) a thin layer of filter aid (e.g. Kiesselghur) is first allowed to be absorbed on the cylinder. Later the filter cylinder with its thin coating of the filter aid is allowed to rotate in the trough in which the broth is now placed. The rotating cylinder, the vacuum still on, is washed with a sprinkle of water; a knife whose edge is positioned just short of the layer of filter aid scrapes off the solids picked up from the broth.

When it is used for easily filtered broth such as in penicillin broth no filter aid is used. Instead an arrangement of strings coupled with a release of the vacuum in the segment of the cylinder helps release the material picked up from the broth.

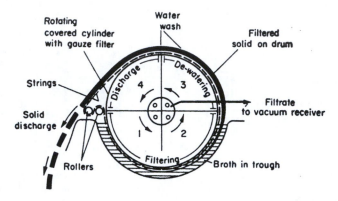

Fig. 10.1 Transverse Section of the Rotary Vacuum Filter Illustrating its Operation

Ring and wire type filters: These filters consist of a coating of diatomaceous earth on a wire-mesh supported by a frame of metal rods. The liquid to be filtered is introduced under a pressure of 75 p.s.i rather than under a vacuum as in the rotary vacuum filter. They are used when the load is light such as for polishing beer or fruit juices. They can be cleaned by back flushing with water.

10.1.2 Centrifugation

Centrifugation is not widely used for the primary separation of solids from broth in fermentation beer because of the thickness of these slurries and the fact that many industries have operated successfully with filters. Only in a few cases will a centrifuge de-water a broth to anywhere near the extend a filter would. In the enzyme isolation industry, however, centrifugation is preferred to filtration, probably because unwanted cell debris are quite efficiently removed by this method. A large number of centrifuges are available in the market and a new fermentation industry or a change in the production method of old processes may require the use of centrifuges for primary separation.

10.1.3 Coagulation and Flocculation

Coagulation is the cohesion of dispersed colloids into small flocs; in flocculation these flocs aggregate to form larger masses. The first is induced by electrolytes and the latter by polyelectrolytes, high molecular weight, water soluble compounds that can be obtained in ionic, anionic, or cationic forms. Bacteria and proteins being negatively charged colloids are easily flocculated by electrolytes or polyelectrolytes. Sometimes clay, or activated charcoal may be used. The net effect of the flocculation is that colloid removal facilitates filtration. It may even be possible to merely decant the supernatant once large enough flocs remove the solid portion of the 'beer' of them which to use and low much to use among the various flocculants must be worked out by experimentation. Since flocculation depends on cell wall characteristics, the agents must meet the following requirements especially if the cells, and not the liquid, are the required products. The flocculants should have the following properties.

  (i)   They must react rapidly with the cells.
 (ii)   They must be non-toxic.
(iii)   They should not alter the chemical constituents of the cell.
(iv)   They should have a minimum cohesive power in order to allow for effective subsequent water removal by filtration.
 (v)   Neither high acidity nor high alkalinity should result from their addition.
(vi)   They should be effective in small amounts and be low in cost.
(vii) They should preferably be washable for reuse.

10.1.4   Foam Fractionation

Foam formation has been described in Chapter 9. The principle of foam fractionation is that in a liquid foam system the chemical composition of a given substance in the bulk liquid is usually different from the chemical composition of some substance in the foam. Foam is formed by sparging the bulk liquid containing the substance to be fractionated with an inert gas. The gas is fed at the bottom (Fig. 10.2) of a tower and the foam created overflows at the top carrying with it the solutes to be fractionated. Surfactants or (surface

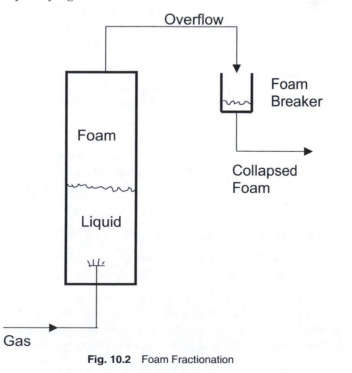

Fig. 10.2   Foam Fractionation

active substances that reduce surface tension e.g. teepol) may be added in liquids that do not foam. This method has been used to collect a wide range of microorganisms and although mainly experimental it may be used on a large scale in industry.

10.1.5 Whole-broth Treatment

As had been indicated earlier, in some fermentations such as the acetonebutanol fermentation, the whole unseparated broth is stripped of its content of the required product. In the antibiotic industry a similar situation was achieved before it became possible to directly absorb the antibiotics streptomycin (using cationic-exchange resin) and novobiocin (on an anionic resin.) The antibiotics are eluted from the resins and then crystallized. This process saves the capital and recurrent expense of the initial separation of solids from the broth.

10.2 PRIMARY PRODUCT ISOLATION

After separation of the broth into soluble and insoluble fractions, the next process depends on the location of desired product as follows: the cells themselves as in yeasts may be desired product; they are dried or refrigerated and the liquid discarded. Further treatment such as drying is discussed later in the chapter.

The required product may be bound to the mycelia or to bacterial cells as in the case of bound enzymes or antibiotics. The cells then have to be disrupted with any of the several ways available – heat, mechanical disruption, etc. The cell debris are now removed by centrifugation, filtration or any of the other methods for removing solids, described above.

Where the material is extracellularly available or if it has been obtained by leaching with or without cell disruption then it is treated by one of the following methods: liquid extraction, dissociation extraction, sorption, or precipitation.

10.2.1 Cell Disruption

A lot of biological molecules are inside the cell, and they must be released from it. This is achieved by cell disruption (lysis). Cell disruption is a sensitive process because of the cell wall's resistance to the high osmotic pressure inside them. Furthermore, difficulties arise from a non-controlled cell disruption, that results from an unhindered release of all intracellular products (proteins nucleic acids, cell debris) as well as the requirements for cell disruption without the desired product's denaturation. There are mechanical and non-mechanical cell disruption methods.

10.2.1.1 Mechanical methods

When the target material is intracellular, the means microorganisms are disrupted mainly by mechanical disruption of the cells. Equipment for cell disruption includes:

i) *Homogenizers.* These pump slurries through restricted orifice or valves at very high pressure (up to 1500 bar) followed by an instant expansion through a special exiting nozzle. The sudden pressure drop upon discharge, causes an explosion of the cell. The method is applied mainly for the release of intracellular molecules.

ii) *Ball Mills.* In a ball mill, cells are agitated in suspension with small abrasive particles. Cells break because of shear forces, grinding between beads, and collisions with beads. The beads disrupt the cells to release biomolecules.

iii) *Ultrasonic disruption.* This method of cell lysis is achieved with high frequency sound that is produced electronically and transported through a metallic tip to an appropriately concentrated cellular suspension. It is expensive and is used mainly in laboratories.

10.2.1.2   Non-mechanical methods

Cells can be caused to disrupt by permeabilization thorough a number of ways:

(i) Chemical Permeabilization. Many chemical methods have been employed in order to extract intra cellular components from microorganisms by permeabilizing (i.e., making them permeable) the outer-wall barriers. It can be achieved with organic solvents that act by the creation of canals through the cell membrane: toluene, ether, phenylethyl alcohol DMSO, benzene, methanol, chloroform. Chemical permeabilization can also be achieved with antibiotics, thionins, surfactants (Triton, Brij, Duponal), chaotropic agents, and chelates. A very important chemical is EDTA (chelating agent) which is widely used for permeabilization of Gram negative microorganisms. Its effectiveness is a result of its ability to bond the divalent cations of $Ca++$, $Mg++$. These cation stabilize the structure of outer membranes by bonding the lipopolysaccharides to each other. The removal of these cations EDTA, increases the permeability areas of the outer walls.

(ii) Mechanical Permeabilization. One method of mechanical permeabilization is osmotic shock. While cells exposed to slowly varying extracellular osmotic pressure are usually able to adapt to such changes, cells exposed to rapid changes in external osmolarity, can be mechanically injured. This procedure is typically conducted by first allowing the cells to equilibrate internal and external osmotic pressure in a high sucrose medium, and then rapidly diluting away the sucrose. The resulting immediate overpressure of the cytosol is assumed to damage the cell membrane. Enzymes released by this method are believed to be periplasmic, or at least located near the surface of the cell.

(iii) Enzymatic Permeabilization. Enzymes can also be employed to permeabilize cells, but this method is often limited to releasing periplasmic or surface enzymes. In these procedures, they often use EDTA in order to destabilize the outer membrane of Gram negative cells, making the peptidoclycan layer accessible to the enzyme used. Enzymes used for enzymatic permeabilization are: beta(1-6) and beta(1-3) glycanases, proteases, and mannase.

10.2.2   Liquid Extraction

Also known as solvent extraction, or liquid-liquid extraction this procedure is widely used in industry. It is used to transfer a solute from one solvent into another in which it is more soluble. It also can be used to separate soluble solids from the mixture with insoluble material by treatment with a solvent.

The law on which liquid-liquid extraction is based states that when an organic solute is exposed to a two-phase immiscible liquid system the ratio of the solute concentration in the two phases is constant for a given temperature. This ratio, K, is the partition or distribution coefficient, given as:

$$K = C_1/C_2$$

where C_1 and C_2 are the concentrations in phase 1 and phase 2 respectively. The equation is effective, however, for dilute solutions and breaks down for very concentrated solutions. The selectivity of a solvent solution is indicated by the ratio of the distribution coefficients of the components in question. Which solvents are actually used will depend on a number of factors including the distribution properties of the solute in them, volatility, ease of recovery and cost.

In this method the broth to be extracted is shaken with a hydrophobic solvent (i.e., one that will not mix with water), allowed to settle and the solvent which should contain more of the material to be extracted is removed. This may be done in a small laboratory scale in separating funnels or in a stirred tank in industry.

The separation may be done in a stirred tank in one of several ways (Fig. 10.3): (1) batch wise in a single tank and the solvent with its solute drained; or (2) continuous with a mixing and a setting tank. More efficient extractions are achieved with continuous addition of solvent in (3) a cross-current arrangement in which successive solvent extracts will be progressively more dilute or in a (4) counter-current fashion in which efficient extraction is achieved with less solvent usage.

The counter-current multi-stage system is most commonly used and a wide variety of equipment incorporating this system exist. In many applications a vertical column may be used with the heavier liquid introduced from the top and the lighter from the bottom. Mixing of the liquid may occur via a stirring shaft or by the turbulence created by a series of plates placed in the column. A series of such columns may be set up with provision for automatic transfer of liquid from one column to the next. A set-up similar to this has been used to separate radio-active materials. This principle has also found use in penicillin separation, but because penicillin would be destroyed in the acidified broth by prolonged contact and also because such prolonged contact enable protein present in the medium to stabilize, separation is done quickly.

10.2.3 Dissociation Extraction

Dissociation extraction is a special case of liquid-liquid extraction. Many fermentation products are either weak bases or acids. When solvent extraction is employed the pH is so selected that the material to be isolated is unionized since the ionized form is soluble in the aqueous phase and the unionized form is soluble in the solvent phase. Weak bases are therefore extracted under high pH conditions and weak acids under low pH conditions. The result is a rapid and complete extraction of the solute and materials similar to it.

10.2.4 Ion-exchange Adsorption

Ion exchange adsorption is one of several adoption methods which include chromatography, and charcoal adsorption. These will be discussed later.

Ionic filtrates of fermentation broths can be purified and concentrated using ion exchange resins packed in columns. An ion exchange resin is a polymer (normally

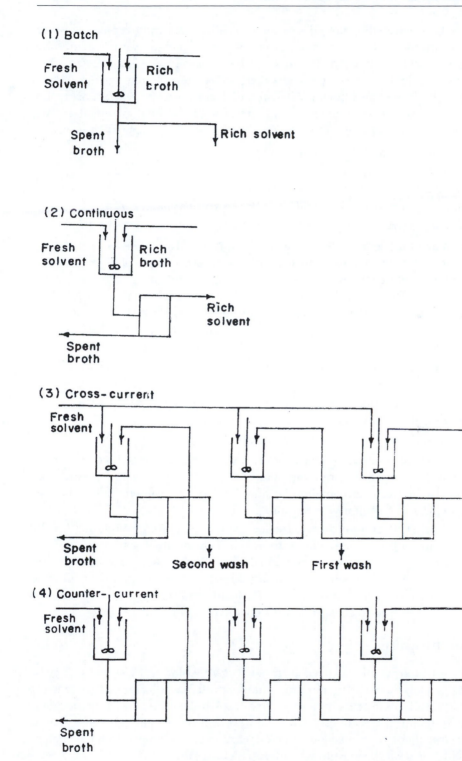

Fig. 10.3   Schematic Representations of Various Methods of Solvent Extraction

polystyrene) with electrically charged sites at which one ion may replace another. Synthetic ion exchange resins are usually cast as porous beads with considerable external and pore surface where ions can attach. Whenever there is a great surface area, adsorption plays a role. If a substance is adsorbed to an ion exchange resin, no ion is liberated. Testing for ions in the effluent will distinguish between removal by adsorption and removal by ion exchange. While there are numerous functional groups that have charges, only a few are commonly used for manmade ion exchange resins. These are:

- -COOH which is weakly ionized to $-COO^-$
- $-SO_3H$ which is strongly ionized to $-SO_3^-$
- $-NH_2$ that weakly attracts protons to form NH_3^+
- -secondary and tertiary amines that also attract protons weakly
- $-NR_3^+$ that has a strong, permanent charge (R stands for some organic group)

These groups are sufficient to allow selection of a resin with either weak or strong positive or negative charge. The resins are usually branched polymers of high molecular weight-containing easily exchanged ions which are in equilibrium with ions in the surrounding solution. The resins are however usually used in neutral salt forms: cation exchangers in the sodium form and anion exchangers in the chloride form. The resins lose the labile ions and in exchange bind suitable materials in the liquid percolating down the column. The efficiency of the exchange depends on the following factors:

(i)   The capacity of the resin for the ion to be adsorbed, usually expressed in milli-equivalents.

(ii) The size of the resin spheres: the smaller, the more the exchange.

(iii) The flow rate; the slower, the greater the adsorption.

(iv) Temperature: the higher, the more rapid the exchange.

The choice of the resin depends on the chemical and physical properties of both resin and product as well as on the contaminating materials. $CaCO_3$, for example, is often left out of media for streptomycin fermentation, because Ca^{++} ions are preferentially adsorbed onto the resin in place of streptomycin cations.

As indicated earlier, streptomycin is extracted over a resin (a carboxylic acid resin) with prior separation of the soluble from the insolubles. The broth is passed successively through two resin columns which have previously been flushed with NaOH to convert them to the sodium phase. The resin absorbs a large amount of the streptomycin which is eluted with HCI converting the streptomycin to chloride and the resin to the hydrogen form. In this way the streptomycin is both purified and concentrated.

10.2.5   Precipitation

The insolubility of many salts is used in the selective isolation of some industrial products. It is particularly useful in the elimination of proteinaceous impurities or in the isolation of enzymes. Salts are precipitated by one of several methods: adding inorganic salts and (or) reducing the solubility with the addition of organic solvents such as alcohol in the case of enzymes. Lactate and oxalate salts of erythromycin have been so isolated and citric acid has been isolated with its calcium salt.

10.3   PURIFICATION

The methods described earlier isolate mixtures of materials similar in chemical and physical properties to the required product. The methods used in this section are finer and further eliminate the impurities thus leaving the desired product purer.

10.3.1   Chromatography

In chromatography, the components of a mixture of solutes migrate at different rates on a solid because of varying solubilities of the solutes in a particular solvent. The mixture of solutes is introduced (usually as a solution) at one end of the solid phase and the solvent (i.e., the solution which separates the mixture) flows through this initial point of the mixture application. Fermentation products are separated by any of the following chromatographic methods, where the separation of the solids occur for the reasons given in each of the following.

(i) *Adsorption chromatography*: (e.g., paper chromatography) variations in the weak (Van der Wall) forces binding solutes to the solid phase;

(ii) *Partition chromatography*: A mobile solvent is passed through a column containing an immobilized liquid phase; the solvent and immobilized liquid phase are immiscible. Separation occurs by the different distribution or partition coefficients of the solutes between the mobile and immobilized liquid phases.

(iii) *Ion exchange chromatography*: The difference in the strength of the chemical bonding between the various solutes and the resin constitutes the basis for this method.

(iv) *Gel Filtration*: This depends on the ability of molecules of different sizes and shapes to permeate the matrix of a gel swollen in the desired solvent. The gel can be considered as containing two types of solvent; that within the gel particle and that outside it. Large particles which cannot penetrate the gel appear in the column effluent after a volume equivalent to the solvent outside the gel has emerged from the column. Small molecules which permeate the matrix appear in the effluent after a volume equivalent to the total liquid volume within the matrix has emerged.

10.3.2   Carbon Decolorization

Some solids are able to adsorb and concentrate certain substances on their surfaces when in contact with a liquid solution (or gaseous mixture). These include activated charcoal, oxides of silicon, aluminum, and titanium and various types of absorbent clays. Absorbents have been used for the adsorption of antibiotics from broths, removal of colored impurities from a solution of an antibiotic, sugar or even from gasoline. In the fermentation industry activated charcoal has been most widely used because of its extensive pores which confer on it a large surface. Furthermore, the pores are large enough to allow the passage of the solvent.

Activated carbon, powdered or granular, is used to remove color. Thus penicillin solution is usually treated with activated carbon before the crystallization of the amino salt. A single-stage batch-wide system of mixing the solution with carbon followed by filtration may be used. Multi-stage counter-current decolorization is far more efficient per unit of carbon than batch. Before using an adsorbent it is important to determine experimentally the most efficient depth of the absorption zone which will thoroughly remove all color.

10.3.3   Crystallization

Crystallization is the final purification method for those materials which can stand heat. The solution is concentrated by heating and evaporation at atmospheric pressure to produce a super saturated solution. Many fermentation products will not however stand heat and the initial water removal is made by heating at reduced pressure or by lowering the temperature to form crystals which can be centrifuged off leaving a concentrated liquor. It yields compounds which are highly potent more stable and free from colored impurities. To obtain crystals, first a super saturated solution is produced; secondly, minute nuclei or seeds are formed and thirdly, the molecules of the solute build on the nuclei. Crystalline particles from a previous preparation may be deliberately introduced to produce the nuclei. In procaine penicillin production, fine crystals are used to induce crystallization whereas in dehydrostreptomycin sulfate, addition of methanol brings about crystallization.

10.4   PRODUCT ISOLATION

The final isolation of the product is done in one of the two following ways:

   (i) processing of crystalline products.
   (ii) drying of products direct from solution.

10.4.1   Crystalline Processing

Crystalline products are free-filtering and non-compressible and therefore may be filtered on thick beds under high pressure. This is usually done on a centrifugal machine capable of developing very high (about 1,000 fold) gravitational force. The crystals are washed to remove adhering mother liquor. After washing they are dried by spinning for further drying or solvent removal.

10.4.2   Drying

Drying consists of liquid removal (either organic solvent or water) from wet crystals such as was described above from a solution, or from solids or cells isolated from the very earliest operation. Several methods of drying exist and the one adopted will depend on such factors as the physical nature of the finished product, its heat sensitivity, the form acceptable to the consumer, and the competitiveness of the various methods in relation to the cost of the finished product. Drying can be considered under two heads: (i) liquid-phase moisture removal, and (ii) solid-phase moisture removal.

10.4.2.1   Liquid-phase moisture removal

Liquid-phase moisture removal involves drying by heat. When drying is done by heating, the processes may be broken down to the supply of heat to the material and the removal of the resulting water vapor. The simplest method is by direct heating in which heated atmospheric air both heats the material and removes the water vapor. In others, the heating is done at reduced pressure to facilitate evaluation of the water vapor. Under such conditions, indirect heating from a heated surface, radiation (e.g., infra-red) or both is used to supplement the heat introduced by reduced vapor pressure.

The actual method of heating is done in a number of different mechanical contraptions which will be mentioned briefly below.

(i) *Tray Driers*: The most commonly used in some fermentation industries is the vacuum tray drier. It is versatile and consists simply of heated shelves in a single cabinet which can be vacuum evacuated. In some, the trays may have provision for vibration or shaking to hasten evaporation. As it can be evacuated, heating at fairly low temperature is possible and hence it is useful for heat-labile materials.

(ii) *Drum dryers*: In this method the broth or slurry is applied to the periphery of a revolving heated drum. The drum may be single or in pairs. High temperature is applied though for a short time on the material to be dried and some destruction may occur. One arrangement of drum driers is illustrated in Fig. 10.4.

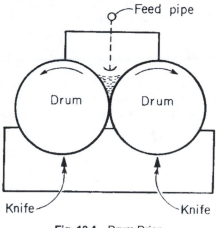

Fig. 10.4   Drum Drier

(iii) *Spray drying*: This method is used extensively in the food and fermentation industries for drying heat-sensitive materials such as drugs, plasma and milk. The conventional spray consists of an arrangement for introducing a fine spray of the liquid to be dried against a counter-current of hot air. As the material is exposed to high temperature for only a short while, a matter of a few seconds, very little damage usually occurs. Furthermore, it is convenient because of its continuous nature.

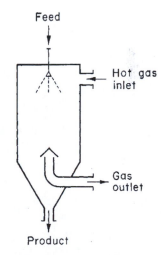

Fig. 10.5 Conventional Spray Drying

Sometimes the material is introduced simultaneously with air (Fig. 10.5).

10.4.2.2   Solid-phase moisture removal (freeze-drying)

The equipment used in freeze-drying is essentially the same as in the vacuum drier described earlier. The main difference is that the material is first frozen. In this frozen state, the water evaporates straight from the material. It is useful for heat-labile materials such as enzymes, bacteria, and antibiotics.

SUGGESTED READINGS

Ahuja, S. 2000. Handbook of Bioseparations. Vol 2 Academic Press. San Diego, USA.

Dobie, M., Kruthiventi, A.K., Gaikar, V.G. 2004. Biotransformations and Bioprocesses. Marcel Dekker, New York, USA.

Endo, I., Nagamune, T., Katoh, S., Yonemoto (eds) 1999. Bioseparation Engineering. Elsevier Amsterdam, the Netherlands.

Garcia, A.A., Bonem, M.R., Ramirez-Vick, J., Saddaka, M., Vuppu, A. 1999. Bioseparation Process Science. Blackwell Science, Massachussets, USA.

Harrison, R.G., Todd, P., Rudge, S.R., Petrides, D.P. 2003. Bioseparation Science and Engineering. Oxford University Press, New York, USA.

Kalyanpur, M. 2000. Downstream Processing in Biotechnology In: Downstream Processing of Proteins: Methods and Protocols. M. Desai, (ed) Humana. Totowa, NJ: USA. pp. 1–10.

Naglak, T.J., Hettwer, D.J., Wang, H.Y, 1990. Chemical Permeabilization of cells for intracellular product release. In: Separation Processes In Biotechnology, Marcel Dekker, New York, USA.

CHAPTER 11

Sterility in Industrial Microbiology

In the microbiology laboratory, sterility is a most important consideration and ways of achieving it form the earliest portions of the training of a microbiologist. In the fermentation industry contamination by unwanted organisms could pose serious problems because of the vastly increased scale of the operation in comparison with laboratory work. If *Pediococus streptococcus damnosus* which causes sourness in beer were to contaminate the fermentation tanks of a brewery then hundreds of thousand of liters of beer may have to be discarded, with consequent loss in revenue to the brewery. The situation would be similar if a penicillianase-producing *Bacillus* sp were to contaminate a penicillin fermentation, or lytic phages an acetone-butanol mash.

11.1 THE BASIS OF LOSS BY CONTAMINANTS

Contaminations in industrial microbiology as seen above could lead to huge financial losses to a fermentation firm. Losses due to contaminations may be explained in one or more of the following ways:

(i) The contaminant may utilize the components of the fermentation broth to produce unwanted end-products and therefore reduce yield. When slime-forming *Leuconostoc mesenteroides* invades a sugar factory, it utilizes sucrose to form the polysaccharide in its capsule which forms the slime. Similarly, in the beer industry when lactic acid bacteria contaminate the fermentating wort, they utilize sugars present therein to produce unwanted lactic acid which renders the beer sour.

(ii) The contaminant may alter the environmental conditions such as the pH or oxidation-reduction potential of the fermentation and render it unsuitable for maximum production of the required product. Thus, if *E. coli* which grows much more rapidly than the highly aerobic *Streptomyces griseus* should contaminate a streptomycin fermentation it may use up a large proportion of the oxygen thereby reducing the yield of the antibiotic, because less than optimal amounts of oxygen are available to the actinomycete.

(iii) Contamination by lytic organisms such as bacteriophages or *Bdellovibrio* could lead to the entire destruction of the producing organism.

(iv) Finally, it is conceivable that contaminants could even, if they did not reduce yield in a product, produce by-products not removable in the extraction process already established in the factory. The result could be losses in manpower time needed to devise means of dealing with the product.

Although contaminants are generally undesirable, not all fermentation need to be carried out under strict asepsis, depending on the selling price of the end-product. Thus while the high cost of antibiotics justifies strict sterility during production, such sterility is not called for in such bulk products as yeasts or industrial alcohol.

11.2 METHODS OF ACHIEVING STERILITY

The various methods for achieving sterility are well-known and include physical and chemical methods.

11.2.1 Physical Methods

11.2.1.1 Asepsis

Asepsis involves general cleanliness and is a procedure routinely observed in many microbiological, pharmaceutical and food industries. In such organizations, laboratory coats, face masks, gloves, and other protective clothing are often worn to prevent the transfer of organisms from the individual to the product. Hands are regularly washed; pipes, utensils, fermentation vats, and floors are washed with water and disinfectants. In some industries such as those concerned with *parenteral* (injection) material, or with vaccines, even the incoming air must be sterile. The maintenance of asepsis does not sterilize but it helps reduce the load of microorganisms and hence lessens the stringency of the sterility measures employed. It also helps to remove foci of microbial growth such as particles of food, or media which could be sources of future contaminations.

11.2.1.2 Filtration

Filtration is used in industry and in the laboratory to free fluids (i.e., gases and liquids) of dust and other particles and microorganisms. If properly used, it is highly effective and also relatively inexpensive. Large volumes of sterile air and other gases are sometimes required for 'sterile' areas where in the pharmaceutical industries, injections and vaccines are handled, and for aeration in most fermentations.

Two types of air filters are available, the so-called absolute filters which are usually made of ceramic and are so called because their pores are not large enough to admit a microorganism and hence, they should theoretically be highly efficient. Their disadvantage is that they are suitable for only small volumes of the gas being sterilized. The second group, fibrous filters, is made of fibers of wool, cotton, glass or mineral slag, whose diameters are in the order of 0.5-15 μ. Fibrous filters are not absolute; nevertheless they are quite effective and hold back organisms of the diameter of about 1.0μ or even viruses. The factors which contribute to their removal of microorganisms include direct interception by the fibers, settlement by gravity electrostatic attraction between fiber and particles, Brownian movement and convection (Fig. 11.1).

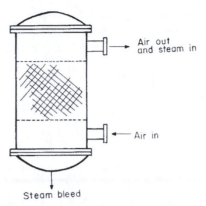

Air out
and steam in

Air in

Steam bleed

Note the fibers placed in the central portion of a steel casing.

Fig. 11.1 Fibrous Filter

Prefilters usually consist of discs of mats of asbestos of the type used in Seitz filters. They, however, let in fine fibers, which are undesirable in injectable materials. The fine fibers are removed in the final filter. Prefilters also absorb large amounts of the liquid, although such absorbed liquid can be re-extracted by flushing the filter at the end of filtration with sterile nitrogen. The filters may also be made of compressed paper pulp; filter paper coated with Kieselghur may be placed between the filter pads.

Final filters which may be of unglazed porcelain are usually made in the form of cylimerial candles over which the liquid to be filtered flows. The filtrate is drained from the inside of the cylinder. This type of arrangement increases the surface area available for filtration. The candles may be sterilized by autoclaving. Sintered glass is usually made in form of discs and, like porcelain, they are fragile. Membrane ('Millipore') filters of cellulose acetate may also be used as final filters. They can be autoclaved.

Sterilizing filters should have pores with maximum diameters of 0.2 μ. They should be themselves sterilized before being used. Membrane filters can be sterilized by chemical sterilants (such as ethylene oxide, hydrogen peroxide in vapor form, propylene oxide, formaldehyde, and glutaraldehyde), radiant energy sterilization (such as c-irradiation) or steam sterilization. The most common method of sterilization is steam sterilization.

Steam sterilization of a membrane filter can be accomplished either by an autoclave or by *in situ* steam sterilization.

11.2.1.3 Heat

Heat may be applied dry or moist:

Dry heat: Not only is dry heat used to sterilize glassware on a small scale in industry associated laboratories, more importantly it is used on a large scale in industry for sterilizing some types of air filters. Principally, however, it is used for sterilizing air by compression. When air is compression the temperature rises in accordance with the gas law, $PV = RT$ where P is the pressure, V the volume, R the gas constant and T temperature. If P and V are increased, T, the temperature would rise as shown in Table 11.1. However, compression is expensive. Furthermore, heated air must be at a high temperature (at a

Table 11.1   Temperature of air after compression

Final pressure (p.s.i.g.)	*Temp. (^{o}C)*
20	78
40	117
60	140
80	169
100	189
150	229
200	261

much higher pressure than that at which it will be used) and for fairly long holding periods. Although not a very practical method, compression could reduce the microbial population of air.

Other methods which have been explored include direct or indirect heating with the gases and also with electrical heating. In each case while the procedure was effective, it was too expensive.

Moist heat: Moist heat can be employed in industry to kill microorganisms during boiling, tyndallization, and autoclaving.

Tyndallization consists of boiling the material for one half hour on three consecutive days. Vegetative cells are killed on the first day's boiling. Spores are not but they germinate. During the second day's boiling, the vegetative spores resulting from the spores not killed on the first day, are killed. Any spore still surviving after the second day will be killed during boiling on the third day as the spores would have germinated. After the third day's boiling the medium is expected to be sterile. It is a method which can be used for sterilizing heat-labile media where filtration is not possible for whatever reason, including that the medium is too viscous for filtration.

Pasteurization is very widely used in the food industry. It is used for treating beer and wine. It consists of exposing the food or material to a temperature for a sufficiently long period to destroy pathogenic or spoilage organisms. Pasteurization can either be batch or continuous. The low temperature long time (LTLT) technique usually involves heating at about 60°C for one half hour and is used in batch pasteurization whereas the high temperature short time (HTST) of flash method involves heating at about 70°C for about 15 seconds. The flash method is employed in continuous pasteurizing.

When batch pasteurization is used on a large scale the final temperature of pasteurization is attained by gradual increases. Similarly, the temperature is lowered gradually to cool it; for 600 ml bottles in many breweries batch pasteurization time is a total of about 90 minutes divided equally between raising the temperature, holding at the pasteurization temperature, and cooling. This prolonged time during which the material is exposed to high temperature and which may give rise to a 'burnt' odor is the major deficiency of batch in comparison with continuous pasteurization.

Steam under pressure: Steam is useful as a sterilizer for the following reasons:

(i)   It has a high heat content and hence a high sterilizing ability per unit weight or volume; this heat is rapidly released.

(ii) Steam releases its heat at a readily controlled and constant temperature.

(iii) It can be fairly easily produced and distributed.

(iv) No obnoxious waste products result from its use and it is clean, odorless and tasteless.

Its disadvantages are that it is not suitable for sterilizing anhydrous soils, greases, powders, and its effectiveness, as will be seen later, may be limited in the presence of air.

Steam is widely used for the sterilization of equipment in the laboratory as well as in industry. Pipes, fermentors and media are all sterilized with the steam. Steam used for this purpose is under pressure because the higher the pressure the higher the temperature. The relationship between steam temperature and pressure will be discussed further later in this section.

There are three 'types' of steam.

Wet steam is steam in which sufficient heat is lacking to keep all the steam in the gaseous vapor phase. The effect of this is that some liquid water is present in the steam.

In '*saturated*' (or sometimes wrongly called dry saturated) steam, all the steam is in the vapor phase; its heat content is such that there is an equilibrium between it and water at any temperature and pressure. Saturated steam is water vapor in the condition in which it is generated from the water with which it is in contact. Saturated steam cannot undergo a reduction in temperature without a lowering of its pressure, nor can the temperature be increased without expanding the pressure. When steam is saturated therefore, it can be described either by its pressure or its temperature, with which the two characteristics are linked. Wet steam has far less heat than saturated steam per unit weight of steam. Furthermore, wet steam introduces a lot more water than necessary in the material being sterilized; for example media in fermentors may become diluted. One major reason for the occurrence of wet steam is the use of long poorly insulated pipes.

In *superheated* steam no liquid water is present, and the temperature is higher than that of saturated steam at the same pressure. Superheated steam is produced by, for example, passing it over heated surfaces or coils. For the purposes of sterilization, saturated steam is the most dependable, efficient and effective of the three types of steam. Superheated steam behaves more like a gas than vapor and takes up water avidly. Although it has a higher temperature than saturated steam at the same pressure, it does not sterilize to the extent of saturated steam. This is because it lacks moisture which enables heat to kill micro-organisms at considerably lower temperatures than dry air. Superheated steam, like dry air, would require that the organisms be exposed for periods as long as glassware is exposed in a dry air oven. For transportation over long distances steam is transported in the superheated form in pipes in order to reduce heat losses; it is returned to saturated steam at the end of the transportation and at the point of use by the introduction of water.

The temperature of steam sterilization is 121°C for media both in industry and in the laboratory, although other time-temperature combinations are equally satisfactory (Table 11.2). When industrial media are sterilized by heat, steam is forced into the medium which is gently agitated; heating is supplemented when necessary by passing steam through coils running along the fermentor wall. The dilution resulting from steam injection is calculated from the quantity of steam introduced. In some instances the medium may be autoclaved in a much larger version of a laboratory autoclave known as a retort.

Table 11.2   Minimum time/temperature relationship arrangements

Time (min)	Temp ^0C
30	116
18	118
12	121
8	121
2	132

The major difference between sterilization of media in industry and in the laboratory is the much greater scale of the former. Due to the greater scale it takes a much longer time to attain the sterilizing temperature and to cool down than would be the case in the laboratory. In the laboratory a liter of medium would probably require ten minutes to attain the sterilizing temperature. It would remain there for 15 minutes and cool down gradually over another 10-15 minutes, making a total of 40-45 minutes. With a 10,000 liter medium the equivalent periods may well take several hours for each of the three periods.

11.2.1.4   Radiations

The electromagnetic spectrum is given in Fig. 11.2. The shorter the wavelength the more powerful the radiation. Thus on the electromagnetic spectrum the most powerful wavelengths are those of gamma rays, while the least powerful are radio waves. The radiations used for sterilizing ultra violet light, x-rays and gamma rays.

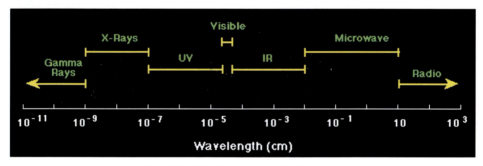

Fig. 11.2   The Electromagnetic Spectrum

Ionizing radiations: These are extremely high frequency electromagnetic waves (X-rays and gamma rays), which have enough photon energy to produce **ionization** (create positive and negative electrically charged atoms or parts of molecules) by knocking off the electrons on the outer orbits of atoms of the materials through which they pass. The atoms knocked out are accepted by other atoms. The atoms losing the electrons and those accepting them become ionized on account of the electron changes. It is this ability of x-rays and gamma rays to create ions that has earned them the name ionizing radiations. Gamma rays are generated from x-ray machines such as those used in hospitals to take x-ray pictures. Gamma rays are also produced by the spontaneous decay of radioactive metals such as cobalt 60 (Co^{60}). Ionizing radiations can be used to sterilize plastic syringes, rubber gloves, and other materials which are liable to damage by heat or chemicals.

Ultraviolet light: Visible light falls between wavelengths of 400 and 700 nm. Ultraviolet light (UV) ranges from 100 to 400 nm. Not all uv is germicidal. The 'germicidal range' is approximately 200 – 300 nm, with a peak germicidal effectiveness at 254 nm. The process of the killing of microbes by UV involves absorption of a UV photon by DNA chains. This causes a disruption in the DNA chain by causing adjacent thymine bases to dimerize or become linked. The organism's metabolism is disrupted and it may eventually die.

Unfortunately, ultraviolet light does not penetrate, and acts mainly on the surface. Therefore its use would be limited to laboratory work such as sterilizing the laboratory air, for creating mutations in culture improvement. In industry it is used for sterilizing the air in fermentation halls and other such large open spaces.

11.2.2   Chemical Methods

These can be divided into two groups: chemosterilants (which kill both vegetative cells as well as spores of bacteria, fungi, viruses, and protozoa) and disinfectants which may no-kill spores, or even some vegetative cells, but at least kill unwanted (pathogenic or spoilage) organisms.

11.2.2.1   Chemosterilants

For a chemical to be useful as a sterilant it should have the following properties:

   (i)   It should be effective at low concentrations.
  (ii)   The components of the medium should not be affected, when used for media.
 (iii)   Any breakdown products resulting from its use should be easily removed or be innocuous.
 (iv)   It should be effective under ambient conditions.
  (v)   It should act rapidly, be inexpensive and be readily available.
 (vi)   It should be non-flammable, non-explosive, and non-toxic.

The discussion on chemosterilants will focus on gaseous sterilants because they have special advantages when parts of the materials to be sterilized are difficult to reach or when they are of heat-labile.

11.2.2.2   Gaseous Sterilants

(i) *Ethylene oxide*: Ethylene oxide $CH^2 - CH^2$ has become accepted as a gaseous sterilant and a lot of information about it has accumulated. It reacts with water, alcohol, ammonia, amines, organic acids and mineral acids. Above $10.7^\circ C$ it is gaseous. It is very penetrating and is widely used in the food and pharmaceutical industries where it is capable of killing all forms of microorganisms. Bacterial spores are however 3-10 more resistant than vegetative cells.

Spores of some bacteria e.g. the thermophilic *Bacillus stearothermophilus* are in fact less resistant than vegetative cells of some bacteria e.g. *Staphylocous aureus*, *Micrococcus radiodurans*, and *Streptococcus faecalis*.

Relative humidity is very important in deciding the bacterial activity of ethylene oxide; it is most effective in the range of 28-33% relative humidity. At humidities higher than 33% it is converted to ethylene glycol which has a weaker anti-bacterial activity. For effectiveness ethylene oxide requires a much longer time of exposure than steam sterilization.

It is widely used in the pharmaceutical industry for sterilizing rubber and plastic bottles, vials, catheters and sometimes, sutures, syringes and needles and some antibiotics and microbiological media. Residual ethylene oxide must however be removed by allowing it to evaporate and this takes some time.

One of the main disadvantages of the sterilant is that the liquid (which form it assumes below 107°C is highly inflammable; the gas also forms explosive mixtures with air from 3 to 80 by volume. For this reason it is mixed with inert gases such as CO_2 often in a ratio of 10% ethylene oxide and 90% CO_2. The explosive nature of ethylene oxide is made even worse by the fact that the pure ethylene gas has an unpleasant odor. For use it is introduced into large containers constructed like autoclaves.

(ii) *Propylene oxide*: This is only about half as active as ethylene oxide. It is liquid at room temperature. It hydrolyzes less slowly than ethylene oxide in the presence of moisture. It is used for room fumigation, and for food because some countries discourage the use of ethylene oxide for this purpose. Propylene oxide has been used in industry for sterilizing culture media, powdered and flaked foods, barley seeds and dried fruits. For these dried foods an exposure of 1,000-2,000 mg/liter of the sterilant for 2-4 hours resulted in 90–99% kill of various microorganisms, including bacteria and fungi. Like ethylene oxide it is an alkylating agent and should be handled carefully since it is a potential carcinogen.

(iii) *β-propiolactone*: β-propiolactone is a heterocyclic colorless pungent liquid. It is highly active as an anti-bacteria agent, but it has a low penetrative power. Its probable carcinogenicity has lowered its general use, although it has been used to fumigate houses. It is used in the pharmaceutical industry to sterilize plasma and vaccines; when it was used to sterilize bacterial medium all the spores introduced were killed. Subsequently, *E. coli* grew indicating that no residual toxicity resulted. Indeed β-propioplactone breaks down to the non-toxic and less carcinogenic β-hydroxypropionic acid. Under maximum operating conditions (temperature, humidity, etc.) it has been claimed that β-propiolactone in the vapor phase is 25 times more effective than formaldehyde, 4000 times more than ethylene oxide and 50,000 more active than methyl bromide. The relative humidity for maximum activity is 75%.

(iv) *Formaldehyde*: Formaldehyde is a gas which is highly soluble in water. Like other gaeous sterilants relative humidity is important, but it is most active between 60-90% humidity. It does not penetrate deeply and it should be used at 22°C or above to be effective. An exposure of at least 12 hours is necessary. Formaldehyde oxidizes to formic acid and this breakdown product could be corrosive to metals. It is used in the pharmaceutical industries where it is used to preserve pathological specimens of animals used for tests.

(v) *Methylbromide*: Methyl bromide is widely used for fumigation and disinfection in cereal mills, warehouses, granaries, seed houses, and food processing plants. As it is highly toxic ethylene oxide is sometimes preferred to it. Furthermore, it has been reported to be only about one tenth as effective as ethylene oxide.

(vi) *Sulfur dioxide*: This is a colorless pungent gas. Due to its corrosiveness it is of limited use, but it is used in the food industries; in wineries, it is used to partially 'sterilize' the grape must before fermentation, to destroy wild yeasts and other unwanted organisms.

11.2.2.3 Other sterilants

(i) *Chorine*: is widely used in industry as solutions of hypochloride. It is used for washing pipes in breweries and other establishments and in the dairy industry for sterilizing utensils.

(ii) *Phenol*: Phenol and phenol-derivatives are widely used as disinfectants. Other compounds which could find use in some aspects of industry include ozone, hydrogen peroxide, and quaternary ammonium compounds.

11.3 ASPECTS OF STERILIZATION IN INDUSTRY

In the foregoing, principles of dealing with unwanted organisms have been stressed; where it was possible some aspects of practice were discussed. In this section the practical methods of dealing with contaminations and the potential for contaminations to occur in industry will be discussed.

11.3.1 The Sterilization of the Fermentor and its Accessories

The fermentor itself, unless sterilized, is a source of contamination. Of the various methods discussed above, steam is the most practical for fermentor sterilization. Steam is used to sterilize the medium in situ in the fermentor but sometimes the medium may be sterilized separately in a retort or autoclave and subsequently transferred aseptically to a fermentor. In order to avoid microbial growth within the fermentor when not in use, crevices and rough edges are avoided in the construction of fermentors, because these provide pockets of media in which undesirable microorganisms can grow. These crevices and rough edges may also protect any such organisms from the lethal effects of sterilization. For the reasons discussed earlier, saturated steam should be used and should remain in contact with all parts of the fermentor for at least half an hour. Pipes which lead into the fermentor should be steam-sealed using saturated steam. The various probes used for monitoring fermentor activities, namely probes for dissolved oxygen, CO_2, pH, foam, etc., should also be sterilized.

11.3.2 Media Sterilization

The following should be borne in mind when sterilizing industrial media with steam:

  (i) Breakdown products may result from heating and may render the medium less available to the microorganisms; some of the breakdown products may even be toxic;

 (ii) pH usually falls with sterilization and the usual laboratory practice of making the pH slightly higher than the expected final pH should be followed;

(iii) Most media would have been sterilized if heat was available to all parts at a temperature of 120-125°C for 15-20 minutes. Oils (sometimes used as anti-foams) are generally more difficult to sterilize. If immiscible with water they may need to be sterilized separately at a much higher temperature than the above and/or for a longer period.

(iv) The order and number of the addition of the various components of the medium could be important. Thus, when powders such as corn starch are to be added it is advisable to dissolve them separately and to add the slurry into the fermentor with vigorous stirring; otherwise lumps could form. Such lumps may not only protect some organisms, but may even render the powdered material unavailable as nourishment for the target organisms. Some commercial autoclaves therefore have an arrangement for stirring the medium to break up clumps of medium as well as distribute the heat.

Sterilization of heat labile medium: Thermolabile media may be sterilized by tyndalization. For this procedure the temperature of the medium is raised to boiling on three consecutive days. The theory behind tydalization is that while boiling destroys the vegetative cells, the bacterial spores survive. After the first day's boiling the vegetative cells are killed and the spores germinate. On the second day's boiling the vegetative cells resulting from the germinated spores surviving the first day's boiling, are killed. In the unlikely event that any spores still survive – after two days of boiling–they will germinate and the resulting vegetative cells will be killed with the third day's boiling. With the third day's boiling the medium in all likelihood will be sterile.

Chemical sterilization of the medium may be done with β-propiolactone. Filtration may also be used. Filtration is especially useful in the pharmaceutical industry where in addition to sterilization it also removes pyrogens (fever-producing agents resulting from walls of Gram-negative bacteria), when filtration is combined with charcoal adsorption.

Batch vs. continuous Sterilization: The various advantages of continuous over batch fermentation (Section 7.4) can be extended, with appropriate modifications, to sterilization. Exposure to sterilization temperature and cooling thereafter are achieved in continuous sterilization in much shorter periods than with batch sterilization (Fig. 11.3). The two methods generally used for continuous sterilization are shown in Fig. 11.4

11.4 VIRUSES (PHAGES) IN INDUSTRIAL MICROBIOLOGY

Viruses are non-cellular entities which consist basically of protein and either DNA or RNA and replicate only within specific living cells. They have no cellular metabolism of their own and their genomes direct the genetic apparatuses of their hosts once they are within them. Viruses are important in the industrial microbiology for at least two reasons:

(i) Those that are pathogenic to man and animals are used to make vaccines against disease caused by the viruses.

(ii) Viruses can cause economic losses by destroying microorganisms used in a fermentations.

It is this second aspect which will be considered in this section. We will therefore look at those viruses which attack organisms of industrial importance, namely bacteria (including actinomycetes) and fungi. Such viruses are known as bacterophages, actinophages, or mycophages depending on whether they attack bacteria, actinomycetes, or fungi. Most of the discussion will center around baceriophages since more information exist on them than on the other two groups.

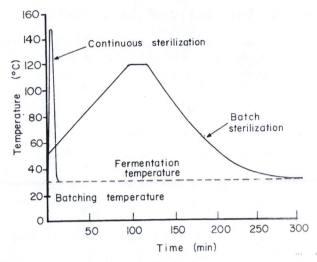

Fig. 11.3   Temperature-time Relationships in Continuous and Batch Sterilization

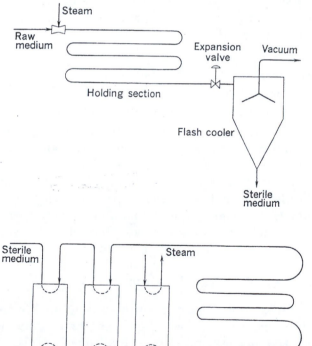

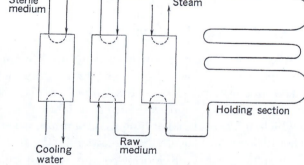

Top: Direct injection of steam into medium
Bottom: Medium sterilization via heated plates, steam is not injected directly into the medium
Note that in both methods the medium is held in a holding section for a period of time

Fig. 11.4   Methods of Continuous Sterilization

11.4.1 Morphological Grouping of Bacteriophages

Bacteriophages can be divided into six broad morphological groups (Table 11.3). Most phages attacking industrial organisms are to be found in groups A, B, C. Groups D, E, and F attack industrial organisms less frequently if at all.

11.4.2 Lysis of Hosts by Phages

The growth cycle of a phage has three steps: adsorption onto the host cell, multiplication within the cell, and liberation of the prepotency phages by the lysis of the host cell. Phages may be classified as virulent or temperate according to how they react to their host. In temperate phages, the phage genome (known as a prophage) integrates with the genetic apparatus of the host, replicates with it and can be lysogenic and are known as lysogons.

Temperate phages may become virulent and lyse their hosts, either spontaneously or after induction by various agents e.g. mitomyin C or UV light. They may also mutate to lytic phages, complete the virus growth cycle and lyse their hosts. It is for this reason that lysogenic phages should be avoided in industrial microorganisms.

The nature of the disturbance caused by phages in an industrial fermentation depends on a number of factors.

(i)   The kinds of phages.
(ii) The time and period of phage infect.
(iii) The medium composition.
(iv) The general physical and chemical conditions of the fermentor.

The manifestation of infection is variable and the same phage does not always cause the same symptom. In general the symptom of infection could be a slowing down of the process resulting in poor yield, this being the case when the infection is light. When it is heavy, the cells may be completely lysed. The length of time taken before a phage manifests itself is variable depending on its latent period (i.e., the period before the cell is lysed) and the number of phages released. In a continuously operated fermentation, phages may take up to three months to manifest themselves. In general, however, the period is much shorter, being noticeable in a matter of days.

11.4.3 Prevention of Phage Contamination

Phages are as ubiquitous as microorganisms in general and are present in the air, water, soil, etc. The first cardinal rule in avoiding phage contamination therefore is routine general cleanliness and asepsis. Pipes, fermentors, utensils, and media, should all be well sterilized. The culture should be protected from aerial phage contamination, an insidious situation, which unlike bacterial or fungal contamination cannot be observed on agar. Air filters should be replaced or sterilized regularly.

Aerosol sterilization of the factory with chlorine compounds, and other disinfectants, as well as UV irradiation of fermentation halls should be done routinely.

Table 11.3   Morphology, nucleic acid composition and examples of the six groups of phages (II)

Group	I	IIA	IIB	III	IVA	IVB	V	VI
General morphology								
Structure Characteristics	Tail contractile	Tail long non contractile	Tail short non contractile	Tailless lipid envelope	Tailless large capsomeres	Tailless double capsid	Tailless smal capsomees	Headless filamentous
Nucleic acid type	2 DNA	2 DNA	2 DNA	2 DNA	1 DNA	1 DNA	1 RNA	1 DNA
Example	E coli T4	E coli λ	E coli T3	Pseudo monas PM2	E coli φ X174	Bdello vibrio HDC-1	E coli f2	E coli fd

11.4.4 Use of Phage Resistant Mutants

Phages may of course be introduced as direct contaminants or be lysogenic in the organism being used in the fermentation. Mutants as productive as the original parent but resistant to various contaminating phages should be developed. Such mutants should have no tendency to revert to the phage-sensitive type. Freedom from particular phages can be checked by treating the organisms with antisera against phages normally or likely to attach to the surfaces of the organisms. Lysogenic bacteria which are resistant to some phages even when high yielding should be avoided.

It must be remembered that phage-resistant mutants may become infected by new phages to which the organisms have no resistance.

11.4.5 Inhibition of Phage Multiplication with Chemicals

Specific chemicals selectively active on phages and which spare bacteria may be used in the fermentation medium.

(i) To prevent infection by phages requiring divalent cations (Mg^{2+}; Ca^{2+}) for adsorption to host cell or for DNA injection into the host cell, chelating agents have been used. These sequester the cations from the medium and hence the phage cannot adsorb onto its host. Examples of the chelating agents are 0.2-0.3% tripolyphosphate and 0.1-0.2% citrate.

(ii) Non-ionic detergents e.g. tween 20, tween 60, polyethylene glycol monoester also inhibit the adsorption of some phages or the multiplication of the phages in the cell. The above two agents usually have no effect on the growth of many industrial organisms.

(iii) The addition of Fe^{2+} suppresses cell lysis by phages.

(iv) Certain antibiotics may be added to prevent growth of phages, but only the selective ones should be used. Chloramphenicol has been used. It has no direct action on the phage, but it inhibits protein replication in phage-infected cells, probably due to selective absorption of the antibiotic by phage-infected cells.

11.4.6 Use of Adequate Media Conditions and Other Practices

Fermentation conditions and practices which adversely affect phage should be selected. Media unfavorable to phages (high pH, low Ca2+, citrates, salts with cations reacting with –SH groups) should be developed. Pasteurization of the final beer and high temperature of incubation consistent with production should be used; both of these adversely affect phage development.

SUGGESTED READINGS

Flickinger, M.C., Drew, S.W. (eds) 1999. Encyclopedia of Bioprocess Technology - Fermentation, Biocatalysis, and Bioseparation, Vol 1-5. John Wiley, New York, USA.

Soares, C. 2002. Process Engineering Equipment Handbook Publisher. McGraw-Hill.

Zeng, A. 1999. Continuous culture. In: Manual of Industrial Microbiology and Biotechnology. A.L. Demain, J.E. Davies (eds) 2nd Ed. ASM Press. Washington, DC, USA, pp. 151–164.

Vogel, H.C., Tadaro, C.L. 1997. Fermentation and Biochemical Engineering Handbook - Principles, Process Design, and Equipment. (2nd Ed) Noyes.

Alcohol-based
Fermentation Industries

CHAPTER 12

Production of Beer

12.1 BARLEY BEERS

The word beer derives from the Latin word *bibere* meaning to drink. The process of producing beer is known as brewing. Beer brewing from barley was practiced by the ancient Egyptians as far back as 4,000 years ago, but investigations suggest Egyptians learnt the art from the peoples of the Tigris and Euphrates where man's civilization is said to have originated. The use of hops is however much more recent and can be traced back to a few hundred years ago.

12.1.1 Types of Barley Beers

Barley beers can be divided into two broad groups: top-fermented beers and bottom-fermented beers. This distinction is based on whether the yeast remains at the top of brew (top-fermented beers) or sediments to the bottom (bottom-fermented beers) at the end of the fermentation.

12.1.1.1 Bottom-fermented beers

Bottom-fermented beers are also known as lager beers because they were stored or 'lagered' (from German *lagern* = to store) in cold cellars after fermentation for clarification and maturation. Yeasts used in bottom-fermented beers are strains of *Saccharomyces uvarum* (formerly *Saccharomyces carlsbergensis*). Several types of lager beers are known. They are Pilsener, Dortumund and Munich, and named after Pilsen (former Czechoslovakia) Dortmund and Munich (Germany), the cities where they originated. Most of the lager (70%-80%) beers drunk in the world is of the Pilsener type.

Bottom-fermentation was a closely guarded secret in the Bavarian region of Germany, of which Munich is the capital. Legend has it that 1842 a monk passed the technique and the yeasts to Pilsen. Three years later they found their way to Copenhagen, Denmark. Shortly after, German immigrants transported bottom brewing to the US.

(i) *Pilsener beer*: This is a pale beer with a medium hop taste. Its alcohol content is 3.0-3.8% by weight. Classically it is lagered for two to three months, but modern breweries have substantially reduced the lagering time, which has been cut down to about two weeks in many breweries around the world. The water for Pilsener brew is soft, containing comparatively little calcium and magnesium ions.

(ii) *Dortmund beer*: This is a pale beer, but it contains less hops (and therefore is less bitter) than Pilsener. However it has more body (i.e., it is thicker) and aroma. The alcohol content is also 3.0-3.8%, and is classically lagered for slightly longer: 3-4 months. The brewing water is hard, containing large amounts of carbonates, sulphates and chlorides.

(iii) *Munich:* This is a dark, aromatic and full-bodied beer with a slightly sweet taste, because it is only slightly hopped. The alcohol content could be quite high, varying from 2 to 5% alcohol. The brewing water is high in carbonates but low in other ions.

(iv) *Weiss:* Weiss beer of Germany made from wheat and *steam* beer of California, USA are both bottom fermented beers which are characterized by being highly effervescent.

12.1.1.2   Top-fermented beers

Top fermented beers are brewed with strains of *Saccharomyces cerevisiae*.

(i) *Ale*: Whereas lager beer can be said to be of German or continental European origin, ale (Pale ale) is England's own beer. Unless the term 'lager?' is specifically used, beer always used to refer to ale in England. Lager is now becoming known in the UK especially since the UK joined European Economic Community. English ale is a pale, highly hopped beer with an alcohol content of 4.0 to 5.0% (w/v) and sometimes as high as 8.0% Hops are added during and sometimes after fermentation. It is therefore very bitter and has a sharp acid taste and an aroma of wine because of its high ester content. *Mild ale* is sweeter because it is less strongly hopped than the standard Pale ale. In Burton-on-Trent where the best ales are made, the water is rich in gypsum (calcium sulfate). When ale is produced in places with less suitable water, such water may be 'burtonized'? by the addition of calcium sulfate.

(ii) *Porter*: This is a dark-brown, heavy bodied, strongly foaming beer produced from dark malts. It contains less hops than ale and consequently is sweeter. It has an alcohol content of about 5.0%.

(iii) *Stout*: Stout is a very dark heavily bodied and highly hopped beer with a strong malt aroma. It is produced from dark or caramelized malt; sometimes caramel may be added. It has a comparatively high alcohol content, 5.0-6.5% (w/v) and is classically stored for up to six months, fermentation sometimes proceeding in the bottle. Some stouts are sweet, being less hopped than usual.

12.1.2   Raw Materials for Brewing

The raw materials used in brewing are: barley, malt, adjuncts, yeasts, hops, and water.

12.1.2.1   Barley malt

As a brewing cereal, barley has the following advantages. Its husks are thick, difficult to crush and adhere to the kernel. This makes malting as well as filtration after mashing, much easier than with other cereals, such as wheat. The second advantage is that the thick husk is a protection against fungal attack during storage. Thirdly, the gelatinization temperature (i.e., the temperature at which the starch is converted into a water soluble gel) is 52-59°C much lower than the optimum temperature of alpha-amylase (70°C) as well as of beta-amylase (65°C) of barley malt. The effect of this is that it is possible to bring the starch into solution and to hydrolyze it in one operation. Finally, the barley grain even before malting contains very high amounts of beta-amylase unlike wheat, rice and sorghum. (Alpha-amylase is produced only in the germinated seed).

Two distinct barley types are known. One with six rows of fertile kernel (*Hordeum vulgare*) and the other with two rows of fertile kernels (*Hordeum distichon*). These differ in many other properties and as a result there are thousands of varieties. The six-row variety is used extensively in the United States, whereas the two-row variety is used in Europe as well as in parts of the US. The six-row varieties are richer in protein and enzyme content than the two-row varieties. This high enzymic content is one of the reasons why adjuncts are so widely used in breweries in the United States. Adjuncts dilute out the proteins i.e. increase the carbohydrate/protein ratio. If an all-malt beer were brewed from malts as rich in protein as the six-row varieties, this protein would find its way into the beer and give rise to hazes. The process of malting, during which enzymes (amylases and proteases) are produced by the germinating seedling will be discussed later.

10.1.1.2 Adjuncts

Adjuncts are starchy materials which were originally introduced because the six-row barley varieties grown in the United States produced a malt that had more diastatic power (i.e. amylases) than was required to hydrolyze the starch in the malt. The term has since come to include materials other than would be hydrolyzed by amylase. For example the term now includes sugars (e.g. sucrose) added to increase the alcoholic content of the beer. Starchy adjuncts, which usually contain little protein contribute, after their hydrolysis, to fermentable sugars which in turn increase the alcoholic content of the beverage.

Adjuncts thus help bring down the cost of brewing because they are much cheaper than malt. They do not play much part in imparting aroma, color, or taste. Starch sources such as sorghum, maize, rice, unmalted barley, cassava, potatoes can or have been used, depending on the price. Corn grits (defatted and ground), corn syrup, and rice are most widely used in the United States.

When corn is used as an adjunct it is so milled as to remove as much as possible of the germ and the husk which contain most of the oil of maize, which could form 7% of the maize grain. The oil may become rancid in the beer aid thus adversely affecting the flavor of the beverage if it were not removed. The de-fatted ground maize is known as corn grits. Corn syrups produced by enzymic or acid hydrolysis, are also used in brewing. Since adjuncts contain little nitrogen, all the needs for the growth of the yeast must come from the malt. The malt/adjunct ratio hardly exceeds 60/40. Soy bean powder (preferably defatted) may be added to brews to help nourish the yeast. It is rich nitrogen and in B vitamins.

12.1.2.3 Hops

Hops are the dried cone-shaped female flower of hop-plant *Humulus lupulus* (synomyn: *H. americanus, H. heomexicams, H. cordifolius*). It is a temperate climate crop and grows wild in northern parts of Europe, Asia and North America. It is botanically related to the genus

Cannabis, whose only representative is *Cannabis sativa* (Indian hemp, marijuana, or hashish). Nowadays hop extracts are becoming favored in place of the dried hops. The importance of hops in brewing lies in its resins which provide the precursors of the bitter principles in beer and the essential (volatile) oils which provide the hop aroma. Both the resin and the essential oils are lodged in lupulin glands borne on the flower.

In the original Pilsener beer the amount of hops added is about 4 g/liter, but smaller amounts varying 0.4-4.0 g/liter are used elsewhere. The addition of hops has several effects:

(a) Originally it was to replace the flat taste of unhopped beer with the characteristic bitterness and pleasant aroma of hops.

(b) Hops have some anti-microbial effects especially against beer sarcina (*Pediococus damnosus*) and other beer spoiling bacteria.

(c) Due to the colloidal nature of the bitter substances they contribute to the body, colloidal stability and foam head retention of beer.

(d) The tannins in the hops help precipitate proteins during the boiling of the wort; these proteins if not removed cause a haze (chill haze) in the beer at low temperature. This is further discussed under beer defects later in this chapter.

12.1.2.4 Water

The mineral and ionic content and the pH of the water have profound effects on the type of beer produced. Some ions are undesirable in brewing water: nitrates slow down fermentation, while iron destroys the colloidal stability of the beer. In general calcium ions lead to a better flavor than magnesium and sodium ions. The pH of the water and that of malt extract produced with it control the various enzyme systems in malt, the degree of extraction of soluble materials from the malt, the solution of tannins and other coloring components, isomerization rate of hop humulone and the stability of the beer itself and the foam on it. Calcium and bicarbonate ions are most important because of their effect on pH. Water is so important that the natural water available in great brewing centers of the world lent special character to beers peculiar to these centers. Water with a large calcium and bicarbonate ions content as is the case with Munich, Copenhagen, Dublin, and Burton-on-Trent are suited to the production of the darker, sweeter beers. The reason for this is not clear but carbonates in particular tend to increase the pH, a condition which appears to enhance the extraction of dark colored components of the malt. Water low in minerals such as that of Pilsen (Table 12.1) is suitable for the production of a pale, light colored beer, such as Pilsen has made famous.

Water of a composition ideal for brewing may not always be naturally available. If the production is of a pale beer without too heavy a taste of hops, and the water is rich in carbonates then it is treated in one of the following ways:

(a) The water may be 'burtonized' by the addition of calcium sulfate (gypsum). Addition of gypsum neutralizes the alkalinity of the carbonates in an equation which probably runs thus:

$$2Ca\,(HCO_3)_2 + 2Ca(SO_4) \rightarrow 2Ca^{++} + 2H_2SO_4 + 4CO_2$$

(b) An acid may be added: lactic acid, phosphoric, sulfuric or hydrochloric. CO_2 is released, but there is an undesirable chance that the resulting salt may remain. The CO_2 released is removed by gas stripping.

Table 12.1   Mineral content of water in some cities with breweries

Place	Total Solids	Ca^{2+}	Mg^{2+}	SO_4^{2-}	NO_3^-	Cl^-	HCO_3^-
Miwaukee	148	34	11	20	0.8	6.6	
New York	28	6	1	8	0.5	0.5	11
St. Louis	201	22	12	77	4	10	65
Pilsen	63	9	3	3		5	37
Munich	270	71	19	18		2	283
Dublin	3	100	4	45		16	266
Copenhangen	480	114	16	62		60	347
Burton-on-trent	1,206	268	62	638	31	36	287

(Mineral content in ppm)

(c) The water may be decarbonated by boiling or by the addition of lime calcium hydroxide.

(d) The water may be improved by ion exchange, which may if it is so desired remove all the ions.

One or more of the above methods may be used simultaneously.

12.1.2.5   Brewer's yeasts

Yeasts in general will produce alcohol from sugars under anaerobic conditions, but not all yeasts are necessarily suitable for brewing. Brewing yeasts are able, besides producing alcohol, to produce from wort sugars and proteins a balanced proportion of esters, acids, higher alcohols, and ketones which contribute to the peculiar flavor of beer. A number of characteristics distinguish the two types of brewers' yeasts (i.e. the top and the bottom-fermenting yeasts).

(a) Under the microscope *Sacch. uvarum (Sacch. carlsbergensis)* usually occurs singly or in pairs. *Sacch. cerevisiae* usually forms chains and occasionally cross-chains as well. These characteristics must however be taken together with other more diagnostic (particularly the biochemical) tests given below.

(b) *Sacch. cerevisiae* sporulates more readily than does *Sacch. uvarum.*

(c) Perhaps the most diagnostic distinction between them is that *Sacch. uvarum* is able to ferment the trisaccharide, raffinose, made up of galactose, glucose, and fructose. *Sacch. cereisiae* is capable of fermenting only the fructose moiety; in other words, it lacks the enzyme system needed to ferment melibiose which is formed from galactose and glucose.

(d) *Sacch. cerevisiae* strains have a stronger respiratory system than *Sacch uvarum* and this is reflected in the different cytochrome spectra of the two groups.

(e) Bottom-fermenters are able to flocculate and sink to the bottom of the brew, a characteristic lacking in most strains of *Sacch. cerevisiae.* Bottom ferments are classified into rapid settling or slow-settling (powdery); settling characteristics affect the rate of production, some secondary yeast metabolites, and hence beer quality.

Yeasts are reused after fermentation for a number of times which depend on the practice of the particular brewery. Mutation and contamination are two hazards in this practice, but they are inherent in all inocula.

12.1.3   Brewery Processes

The processes involved in the conversion of barley malt to beer may be divided into the following:

1. Malting
2. Cleaning and milling of the malt
3. Mashing
4. Mash operation
5. Wort boiling treatment
6. Fermentation
7. Storage or lagering
8. Packaging

Of the above processes, malting is specialized and is not carried out in the brew house. Rather, breweries purchase their malt from specialized malsters (or malt producers). The description to be given will in general relate to lager beers and where the processes differ from those of ales this will be pointed out.

12.1.3.1   Malting

The purpose of malting is to develop amylases and proteases in the grain. These enzymes are produced by the germinated barley to enable it to break down the carbohydrates and proteins in the grain to nourish the germinated seedling before its photosynthetic systems are developed enough to support the plant. However, as soon as the enzymes are formed and before the young seedling has made any appreciable in-road into the nutrient reserve of the grain, the development of the seedling is halted by drying, but at temperatures which will not completely inactivate the enzymes in the grain. These enzymes are reactivated during mashing and used to hydrolyze starch and proteins and release nutrients for the nourishment of the yeasts.

Not all barley strains are suitable for brewing; some are better used for fodder. During malting, barley grains are cleaned; broken barley grains as well as foreign seeds, sand, bits of metal etc. are removed. The grains are then steeped in water at 10-15°C. The grain absorbs water and increases in volume ultimately by about 4%. Respiration of the embryo commences as soon as water is absorbed. Microorganisms grow in the steep and in order not to allow grain deterioration the steep water is changed approximately at 12-hourly intervals until the moisture content of the grain is about 45%. Steeping takes two to three days.

The grains are then drained of the moisture and may be transferred to a malting floor or a revolving drum to germinate. The heat generated by the sprouts further hastens germination. Sometimes moist warm air is blown through beds of germinating seedlings about 30 cm deep. Water may also be sprinkled on them. The plant hormone gibberellic acid is sometimes added to the grains to shorten germination time. The grain itself synthesizes gibberellic acid and it is this acid which triggers off the synthesis of various hydrolytic enzymes by the aleurone layer situated on the periphery of the grain. The enzymes so formed diffuse into the center of the grain where the endosperm is located.

In the endosperm, the starch granules are harbored within cells. These cell walls are made up of hemicellulose, which is broken down by hemicellulases before amylases can

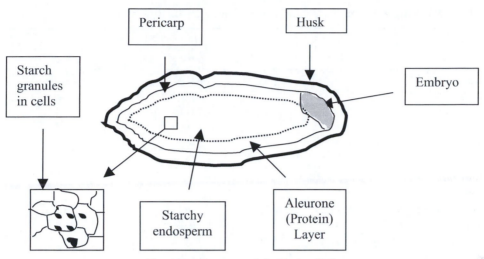

Fig. 12.1   Structure of the Barley Grain

attack the starch. Alpha-amylase (see discussion on mashing below) is also synthesized by the grain. Beta-amylase is already present and is not synthesized but is bound to proteins and is released by proteolytic enzymes.

'Modification' or production of enzymes is complete in four to five days of the growth of the seedling; the extent being tested roughly by the sweet taste developed in the grain and by the length of the young plumule. The various enzymes formed break down some quantities of their respective substrates but the major breakdown takes place during mashing.

Further reactions in the grain are halted by kilning, which consists of heating the 'green' malt in an oven, first with a relatively mild temperature until the moisture content is reduced from about 40% to about 6%. Subsequently the temperature of heating depends on the type of beer to be produced. For beer of the Pilsener type the malt is pale and has no pronounced aroma and kilning takes 20-24 hours at 80–90°C. For the darker Munich beers with a strong aroma drying takes up to 48 hours at 100 – 110°C. For the caramelized malts used for stout and other very dark beers, kilning temperature can be as high as 120°C. Such malts contain little enzymic activity.

At the end of malting, some changes occur in the gross composition of the barley grain as seen in Table 12.2. The rootlets are removed and used as cattle feed.

Weight loss known as malting loss occurs at each stage of malting and the accumulated loss may be as high 15%. The barley malt with its rich enzyme content resembles swollen grains of unthreshed rice and can be stored for considerable periods before being used.

10.1.3.2 Cleaning and milling of malt

The barley is transported to the top of the brewing tower. Subsequent processes in the brewery process occur at progressively lower floors. Lagering and bottling are usually done on the ground level floor. In this way gravity is used to transport the materials and the expense of pumping is eliminated. At the top of the brewing tower, the barley malt is

Table 12.2   Composition of barley grain before and after malting

Fraction	Proportion (% dry weight)	
	Barley	Malt
Starch	63-65	58-60
Sucrose	1-2	3-5
Reducing sugars	0.1-0.2	3-4
Other sugars	1	2
Soluble gums	1-1.5	2-4
Hemicelluloses	8-10	6-8
Cellulose	4-5	5
Lipids	2-3	2-3
Crude protein (N x 6.25)	8-11	8-11
Albumin	0.5	5
Globulin	3	-
Hordein-protein	3-4	2
Glutelin-protein	3-4	3-4
Amino acids and peptides	0.5	1-2
Nucleic acids	0.2-0.3	0.2-0.3
Minerals	2	3
Others	5-6	6-7

cleaned of dirt and passed over a magnet to remove pieces of metals, particularly iron. It is then milled.

The purpose of milling is to expose particles of the malt to the hydrolytic effects of malt enzymes during the mashing process. The finer the particles therefore the greater the extract from the malt. However, very fine particles hinder filtration and prolong it unduly. The brewer has therefore to find a compromise particle size which will give him maximum extraction, and yet permit reasonably rapid filtration rate. No matter what is chosen the crushing is so done as to preserve the husks which contribute to filtration, while reducing the endosperm to fine grits.

12.1.3.3 Mashing

Mashing is the central part of brewing. It determines the nature of the wort, hence the nature of the nutrients available to the yeasts and therefore the type of beer produced. The purpose of mashing is to extract as much as possible the soluble portion of the malt and to enzymatically hydrolyze insoluble portions of the malt and adjuncts. In the sense of the latter objective, mashing may be regarded as an extension of malting. In essence mashing consists of mixing the ground malt and adjuncts at temperatures optimal for amylases and proteases derived from the malt. The aqueous solution resulting from mashing is known as wort.

The two largest components in terms of dry weight of the grain are starch (55%) and protein (10-12%). The controlled breakdown of these two components has tremendous influence on beer character and will be considered below.

12.1.3.3.1   Starch breakdown during mashing

Starch forms about 55% of the dry weight of barley malt. Of the malt starch 20-25% is made up of amylose. The key enzymes in the break down of malt starch are the alpha and beta-amylases. The temperature of optimal activity and destruction of these enzymes as well as their optimum pH are given in Table 12.3 (Starch and its breakdown are also discussed in Chapter 4).

Table 12.3   Temperature optima of alpha- and beta-amylases

Enzyme	Optimum temperature	Temperature of destruction	Optimal pH
Alpha-amylase	70°C	80°C	5.8
Beta-amylase	60-65°C	75°C	5.4

12.1.3.3.2   Protein breakdown during mashing

The breakdown of the malt proteins, albumins, globulins, hordeins, and gluteins starts during malting and continues during mashing by proteases which breakdown proteins through peptones to polypeptides and polypeitidases which breakdown the polypetides to amino acids. Protein breakdown has no pronounced optimum temperature, but during mashing it occurs evenly up to 60°C, beyond which temperature proteases and polypeptidases are greatly retarded. Proteoloytic activity in wort is however dependent on pH and for this reason wort pH is maintained at 5.2-5.5 with lactic acid, mineral acids, or calcium sulphate.

12.1.3.3.3   General environmental conditions affecting mashing

The progress of mashing is affected by a combination of temperature, pH, time, and concentration of the wort. When the temperature is held at 60-65°C for long periods a wort rich in maltose occurs because beta amylase activity is at its optimum and this enzyme yields mainly maltose. On the other hand, when a higher temperature around 70°C is employed dextrins predominate. Dextrins contribute to the body of the beer but are not utilized by yeast. Mash exposed to too high a temperature will therefore be low in alcohol due to insufficient maltose production.

The pH optima for amylases and proteolytic enzymes have already been discussed. The optimum pH for beta-amylase activity is about the same as that of proteolysis and as can be seen in Table 12.3, a fortunate coincidence for the maximum production of maltose and the breakdown of protein.

The concentration of the mash is important. The thinner the mash the higher the extract (i.e., the materials dissolved from the malt) and the maltose content.

12.1.3.3.4   Mashing methods

There are three broad mashing methods:

(a) Decoction methods, where part of the mash is transferred from the mash tun to the mash kettle where it is boiled.

(b) Infusion methods, where the mash is never boiled, but the temperature is gradually raised.

(c) The double mash method in where the starchy adjuncts are boiled and added to the malt.

(i) **Decoction methods**: In these methods the mash is mixed at an initial temperature of 35-37°C and the temperature is raised in steps to about 75°C. About one-third of the initial mash is withdrawn, transferred to the mash kettle, and heated slowly to boil, and returned to the mash tun, the temperature of the mash becoming raised in the process. The enzymes in the heated portion become destroyed but the starch grains are cooked, gelatinized and exposed. Another portion may be removed, boiled and returned. In this way the process may be a one, two or three-mash process. In a three-mash process (Fig. 12.2) the initial temperature of 35-40°C favors proteolysis; the mash is held for about half hour at 50°C for full proteolysis, for about one hour at 60-65°C for saccharification and production of maltose, and at 70-75°C for two or three hours for dextrin production. The three-mash method is the oldest and best known and it was originated in Bavaria, West Germany. Figure 12.2 shows the temperature relations in a three-mash decoction. The decoction is used in continental Europe.

(ii) **Infusion method**: The infusion method is the one used in Britain and is typically used to produce top-fermenting beers. It is carried out in a mash tun, which resembles a lauter tub of lager beer, but it is deeper. The method involves grinding malt and a smaller amount of unmalted cereal, which may sometimes be precooked. The ground material, or grist, is mixed thoroughly with hot water (2:1 by weight) to produce a thick porridge-like mash and the temperature is carefully raised to about 65°C. It is then held at this temperature for a period varying from 30 minutes to several hours. On the average the holding is for 1-2 hours. The enzyme acts principally on the starch and its degradation products in both the malted and unmalted cereal, and only a little protein breakdown occurs. Further hot water at 75-78°C is sprayed on the mash to obtain as much extract as possible and to halt the enzyme action. It is believed by some authors that this method is not as efficient as the double mash or decoction method in extracting materials from the malt. No part of the mash is boiled from mashing-in to mashing-off. It is, however, more easily automated, but a malt in which the proteins are already well degraded must be used since the high temperature of mashing rapidly destroys the proteolytic enzymes.

(iii) **The double-mash** (also called the cooker method): This method was developed in the US because of its use of adjuncts. It has features in common with the infusion and the decoction method. Indeed some authors have described it as the downward infusion method whilst describing the infusion method mentioned above as an upward infusion method. In a typical US double mash method ground malt is mashed with water at a temperature of 35°C. It is then held for an hour during the 'protein rest' for proteolysis. Adjuncts are then cooked in an adjunct cooker for 60-90 minutes. Sometimes about 10% malt is added during the cooking. Hot cooked adjunct is then added to the mash of ground malt to raise the temperature to 65-68°C for starch hydrolysis and maintained at this level for about half hour. The temperature is then increased to 75°C-80°C after which the mashing is terminated. During starch hydrolysis completion of the process is tested with the iodine test.

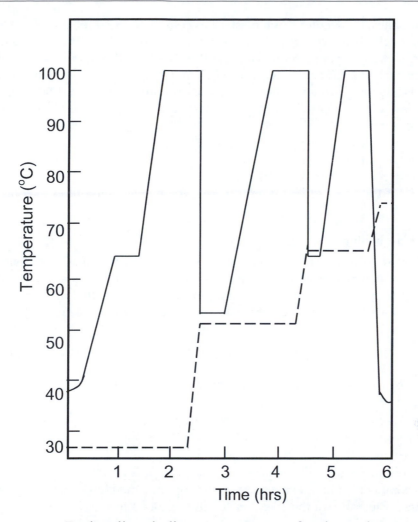

Broken lines indicate temperature of main mash
Unbroken lines indicate temperature of added portion of mash

Fig. 12.2   Three-stage Decoction Method

Various combinations of the above methods may be used, depending on the type of beer, the type of malt, and the nature of the adjunct.

12.1.3.3.5   Mash separation

At the end of mashing, husks and other insoluble materials are removed from the wort in two steps. First, the wort is separated from the solids. Second, the solids themselves are freed of any further extractable material by washing or sparging with hot water.

The conventional method of separating the husks and other solids from the mash is to strain the mash in a lauter (German for clarifying) tub which is a vessel with a perforated false-bottom about 10 mm above the real bottom on which the husks themselves form a bed through which the filtration takes place. In recent times in large breweries, especially in the United States, the Nooter strain master has come into use. Like the Lauter tub, filtration is through a bed formed by the husks, but instead of a false bottom, straining is through a series of triangular perforated pipes placed at different heights of the bed. The strain master itself is rectangular with a conical bottom whereas the Lauter tub is cylindrical. Its advantage among others is that it can handle larger quantities than the Lauter tub. Besides the Lauter tub and the strainmaster, cloth filters located in plate filters and screening centrifuges are also used.

The sparging (or washing with hot water) of the mash solids is done with water at about 80°C and is continued till the extraction is deemed complete. The material which is left after sparging is known as spent grain and is used as animal feed. Sometimes liquid is extracted from the spent grain by centrifuging, the extract being used to cook the adjuncts.

12.1.3.3.6 Wort boiling

The wort is boiled for 1-1½ hours in a brew kettle (or copper) which used to be made of copper (hence the name) but which, in many modern breweries, is now made of stainless steel. When corn syrup or sucrose is used as an adjunct it is added at the beginning of the boiling. Hops are also added, some before and some at the end of the boiling. The purpose of boiling is as follows.

(a) To *concentrate* the wort, which loses 5-8% of its volume by evaporation during the boiling;

(b) To *sterilize* the wort to reduce its microbial load before its introduction into the fermentor.

(c) To *inactivate* any enzymes so that no change occurs in the composition of the wort.

(d) To *extract* soluble materials from the hops, which not only aid in protein removal, but also in introducing the bitterness of hops.

(e) To precipitate protein, which forms large flocs because of heat denaturation and complexing with tannins extracted from the hops and malt husks. Unprecipitated proteins form hazes in the beer, but too little protein leads to poor foam head formation.

(f) To develop color in the beer; some of the color in beer comes from malting but the bulk develops during wort boiling. Color is formed by several chemical reactions including caramelization of sugars, oxidation of phenolic compounds, and reactions between amino acids and reducing sugars.

(g) Removal of volatile compounds: volatile compounds such as fatty acids which could lead to rancidity in the beer are removed.

During the boiling, agitation and circulation of the wort help increase the amount of precipitation and flock formation.

Pre-fermentation treatment of wort: The hot wort is not sent directly to the fermentation tanks. If dried hops are used then they are usually removed in a hop strainer. During boiling proteins and tannins are precipitated while the liquid is still warm. Some more precipitation takes place when it has cooled to about 50°C. The warm precipitate is known as "trub" and consists of 50-60% protein, 16-20% hop resins, 20-30% polyphenols and about 3% ash. Trub is removed either with a centrifuge, or a whirlpool separator which is now more common. In this equipment the wort which is fed into a flat centrifuge, is thrown at the side of the equipment and finds its way out through an outlet on the periphery. The heavier particles (the trub) are thrown to the center and withdrawn through a centrally located outlet. The separated wort is cooled in a heat exchanger. When the temperature has fallen to about 50°C further sludge known as 'cold break' begins to settle, but it cannot be separated in a centrifuge because it is too fine. In many breweries the wort is filtered at this stage with kieselghur, a white distomaceous earth.

The cooled wort is now ready for fermentation. It contains no enzymes but it is a rich medium for fermentation. It has therefore to be protected from contamination. During the transfer to the fermentor the wort is oxygenated at about 8 mg/liter of wort in order to provide the yeasts with the necessary oxygen for initial growth.

12.1.3.4   Fermentation

The cooled wort is pumped or allowed to flow by gravity into fermentation tanks and yeast is inoculated or 'pitched in' at a rate of $7\text{-}15 \times 10^6$ yeast cells/ml, usually collected from a previous brew.

12.1.3.4.1   Top fermentation

This is used in the UK for the production of stout and ale, using strains of *Saccharomyces cerevisiae*. Traditionally an open fermentor is used. Wort is introduced by a fish tail spray so that it becomes aerated to the tune of 5-10 ml/liter of oxygen for the initial growth of the yeasts. Yeast is pitched in at the rate of 0.15 to 0.30 kg/hl at a temperature of 15-16°C. The temperature is allowed to rise gradually to 20°C over a period of about three days. At this point it is cooled to a constant temperature. The entire primary fermentation takes about six days. Yeasts float to the top during this period, they are scooped off and used for future pitching. In the last three days the yeasts turn to a hard leathery layer, which is also skimmed off. Sometimes the wort is transferred to another vessel in the so-called dropping system after the first 24-36 hours. The transfer helps aerate the system and also enables the discarding of the cold-break sediments. Sometimes the aeration is also achieved by circulation with paddles and by the means of pumps. Nowadays cyclindrical vertical closed tanks are replacing the traditional open tanks. A typical top fermentation cycle is shown in Fig. 12.3.

12.1.3.4.2   Bottom fermentation

Wort is inoculated to the tune of $7\text{-}15 \times 10^6$ yeast cells per ml of wort. The yeasts then increase four to five times in number over three to four days. Yeast is pitched in at 6-10°C and is allowed to rise to 10-12°C, which takes some three to four days; it is cooled to about 5°C at the end of the fermentation. CO_2 is released and this creates a head called Krausen, which begins to collapse after four to five days as the yeasts begin to settle. The total fermentation period may last from 7-12 days (Fig. 12.4).

12.1.3.4.3   Formation of some beer components

During wort fermentation in both top and bottom fermentation anaerobic conditions predominate; the initial oxygen is only required for cell growth. Fermentable sugars are converted to alcohol, CO_2 and heat which must be removed by cooling. Dextrins and

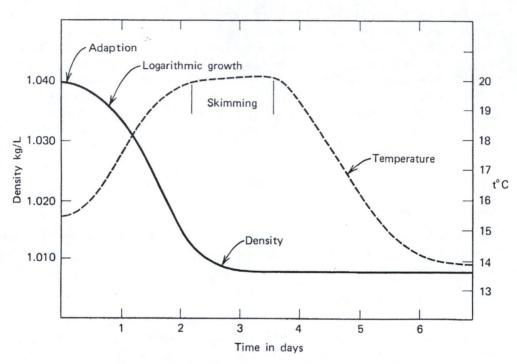

Fig. 12.3 Typical Fermentation with Top-fermenting Yeast

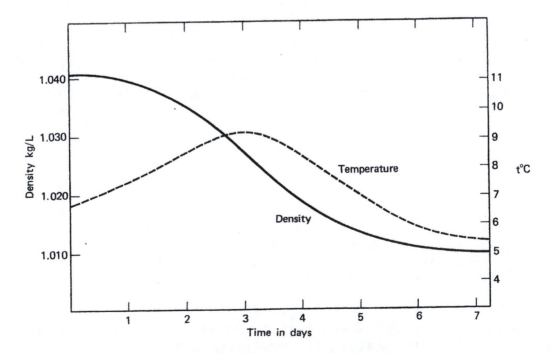

Fig. 12.4 Typical Fermentation with Bottom-fermenting Yeast

maltoteraose are not fermented. Higher alcohols (sometimes known as fusel oils) including propanol and isobutanol are generated from amino acids. Organic acids such as acetic, lactic, pyruvic, citric, and malic are also derived from carbohydrates via the tricarboxylie acid cycle.

12.1.3.4.4   Monitoring following fermentation progress

The progress of fermentation is followed by wort specific gravity. During fermentation the gravity of the wort gradually decreases because yeasts are using up the extract. However alcohol is also being formed. As alcohol has a lower gravity than wort the reading of the special hydrometer (known as a saccharometer) is even lower. The saccharometer reading does not therefore reflect the real extract, but an apparent extract, which is always lower than the real extract because of the presence of alcohol. In the UK and some other countries the extract is measured as the direct specific gravity as 60°F (15.5°C) x 1000. Hence, with wort with sp. Gr. of 1.053 the extract would be 1053°. Outside the UK extract is measured in °Balling or °Plato. Both systems measure the percentage of sucrose required to give solutions of the same specific gravity. The original tables were designed by Von Balling. Improvements and greater accuracy were made on Von Balling's tables first by Brix and later by Plato but the figures were not changed drastically. For this reason °Balling, °Brix, °Plato are the same except for the fifth and sixth decimal places (Table 12.4). °Brix is used in the sugar industry, whereas Balling (United States) and °Plato (continental Europe) are used in the brewing industry.

Table 12.4   Comparison between original gravity and percent extract

Original gravity	°B	°P
1.01968	4.925	5.00
1.02370	5.931	6.00
1.02774	6.920	7.00
1.03180	7.913	8.00
1.03591	8.917	9.00
1.04003	9.925	10.00
1.04419	10.921	11.00
1.04837	11.920	12.00
1.05260	12.928	13.00
1.05684	13.943	14.00

The apparent extract, real, extract, and alcohol content are related to each other as well as to the original extract, i.e., the solids in the original worts and may be read from tables. The degree of attenuation is the amount of extract fermented, measured as a percentage of the original or total extract, hence an apparent and a real degree of attenuation both exist.

12.1.3.5   Lagering (bottom-fermented beers) and treatment (top-fermented beers)

(a) **Lagering**: At the end of the primary fermentation above, the beer, known as 'green' beer, is harsh and bitter. It has a yeasty taste arising probably from higher alcohols and aldehydes.

The green beer is stored in closed vats at a low temperature (around 0°C), for periods which used to be as long as six months in some cases to mature and make it ready for drinking.

During lagering secondary fermentation occurs. Yeasts are sometimes added to induce this secondary fermentation, utilizing some sugars in the green beer. The secondary fermentation saturates the beer with CO_2, indeed the progress of secondary fermentation is followed by the rate of CO_2 escape from a safety valve. Sometimes actively fermenting wort or Kraeusen may be added. At other times CO_2 may be added artificially into the lagering beer. Materials which might undesirably affect flavor and which are present in green beer e.g. diacetyl, hydrogen sulfide, mercaptans and acetaldehyde are decreased by evaporation during secondary fermentation. An increase occurs in the desirable components of the beer such as esters. Any tannins, proteins, and hop resins still left are precipitated during the lagering period.

Lagering used to take up to nine months in some cases. The time is now considerably shorter and in some countries the turnover time from brewing, lagering, and consumption could be as short as three weeks. This reduction has been achieved by artificial carbonation and by the manipulation of the beer due to greater understanding of the lagering processes. Thus, in one method used to reduce lagering time, beer is stored at high temperature (14°C) to drive off volatile compounds e.g. H_2S, and acetaldehyde. The beer is then chilled at -2°C to remove chill haze materials, and thereafter it is carbonated. In this way lagering could be reduced from 2 months to 10 days.

Lagering gives the beer its final desirable organoleptic qualities, but it is hazy due to protein-tannin complexes and yeast cells. The beer is filtered through kieselghur or through membrane filters to remove these. Some properties of lager beer are given in Table 12.5.

Table 12.5 Some properties of lager beer

Property	Pilsener	United States lager beer	Danish Pilsener	English ale	English stout	Mounich Lowenbrau	Dortmund
Original extract content °pc	12.1	11.5-12.0	10.6	15.0	21.1	13.3	13.6
Real extract content °pc	5.3	5.5	3.1	5.0	8.7	6.4	5.5
Alcoholic content, wt %	3.5	3.4-3.8	3.9	5.2	6.7	3.6	4.2
Protein content, wt %	0.28-0.35	0.3	0.6	0.6	0.5	0.8	
CO_2 content %		0.53	0.5	0.4	0.41		0.42
Color, EBC	10	2.7	5			40	8
Air in bottle, mL		1.5	2	8	10		6
pH		4.2-4.50	4				
Real degree of attenuation, %		60-75	69	66	59	48	60

(b) **Beer treatment (for top-fermented beers)**: Top-fermented beers do not undergo the extensive lagering of bottom-fermented beers. They are treated in casks or bottles in various ways. In some processes the beer is transferred to casks at the end of fermentation with a load of 0.2-4.00 million yeast cells/ml. It is 'primed' to improve its taste and appearance by the addition of a small amount of sugar mixed with caramel. The yeasts grow in the sugar and carbonate the beer. Hops are also sometimes added at this stage. It is stored for seven days or less at about 15°C. After 'priming', the beer is 'fined' by the addition of isinglass. Isinglass, a gelatinous material from the swim bladder of fish, precipitates yeast cells, tannins and protein-tannin complexes. The beer is thereafter pasteurized and distributed.

12.1.3.6 Packaging

The beer is transferred to pressure tanks from where it is distributed to cans, bottles and other containers. The beer is not allowed to come in contact with oxygen during this operation; it is also not allowed to lose CO_2, or to become contaminated with micro-organisms. To achieve these objectives, the beer is added to the tanks under a CO_2, atmosphere, bottled under a counter pressure of CO_2, and all the equipment is cleaned and disinfected regularly.

Bottles are thoroughly washed with hot water and sodium hydroxide before being filled. The filled and crowned bottles are passed through a pasteurizer, set to heat the bottles at 60°C for half hour. The bottles take about half hour to attain the pasteurizing temperature, remain in the pasteurizer for half hour and take another half hour to cool down. This method of pasteurization sometimes causes hazes and some of the larger breweries now carry out bulk pasteurization and fill containers aseptically.

12.1.4 Beer Defects

The most important beer defect is the presence of haze or turbidity, which can be of biological or physico-chemical origin.

12.1.4.1 Biological turbidities

Biological turbidities are caused by spoilage organisms and arise because of poor brewery hygiene (i.e. poorly washed pipes) and poor pasteurization. Spoilage organisms in beer must be able to survive the following stringent conditions found in beer: low pH, the antiseptic substances in hops, pasteurization of beer, and anaerobic conditions. Yeasts and certain bacteria are responsible for biological spoilage because they can withstand these. Wild or unwanted yeasts which have been identified in beer spoilage are spread into many genera including *Kloeckera, Hansenula,* and *Brettanomyces,* but *Saccharomyces* spp appear to be commonest, particularly in top-fermented beers. These include *Sacch. cerevisiae* var. *turblidans,* and *Sacch. diastaticus.* The latter is important because of its ability to grow on dextrins in beer, thereby causing hazes and off flavors.

Among the bacteria, *Acetobacter,* and the lactic acid bacteria, *Lactobacillus* and *Streptococcus* are the most important. The latter are tolerant of low pH and hop antiseptics and are micro-aerophilic hence they grow well in beer. *Acetobacter* is an acetic acid bacterium and produces acetic acid from alcohol thereby giving rise to sourness in beer.

Lactobacillus pastorianus is the typical beer spoiling lactobacilli, in top-fermented beers, where it produces sourness and a silky type of turbidity. *Streptococcus damnosus* (*Pediococcus damnosus, Pediococcus cerevisiae*) is known as 'beer sarcina' and gives rise to 'sarcina sickness' or beer which is characterized by a honey-like odor.

12.1.4.2 Physico-chemical turbidities

Non-biological hazes developing beer may be due to one or more of the following:

  (i)   Hazes induced by metals.
  (ii) Protein-tannin hazes.
  (iii) Polysaccharide sediments.
  (iv) Oxalate hazes and sediments.

(i) **Hazes induced by metals**: Tin, iron, copper have all been identified as causing hazes in beer. An amount of only 0.1 ppm of tin will immediately produce haze in beer. It does not unlike other metals, acts as an oxidation catalyst, but precipitates haze precursors directly. It may occur in some canned beers.

Copper and iron act as catalysts in the oxidation of the polyphenolic moiety of the protein-haze precursors of beer. They appear to be derived from both malt and hop (from copper insecticides) and also from the brewing plant. It has been suggested that EDTA (ethylenediaminetetraacetic acid) be used to form chelates with copper and iron and thereby prevent their deleterious action.

(ii) **Protein-tannin hazes**: The polyphenols of beer have often been solely and incorrectly referred to as tannins. Tannins proper are used to convert hides to leather but beer polyphenols cannot be so used. Polyphenols are widely distributed in plants. Beer tannings or polyphenols (Fig. 12.5) are derived from hops and barley husks. They react with proteins to form complex molecules which become insoluble in the form of haze. Hazes contain polypeptides, polyphenols, carbohydrates and a small amount of minerals.

Beer hazes are divided into two: *Chill hazes* (0.1–2 nm diameter particles) form at O°C and re-dissolve at 20°C. *Permanent hazes* (1.0–10 nm) remains above 20°C.

(I)   (II)

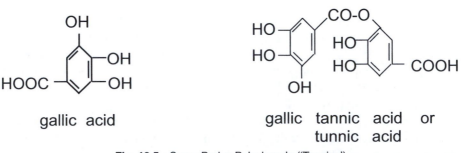

gallic acid   gallic tannic acid or tunnic acid

Fig. 12.5 Some Barley Polyphenols ('Tannins')

Protein-tannin hazes may be removed by:

(a) addition of papain which hydrolyzes the polypetides to low molecular weight components which cannot form hazes;

(b) adsorption of the polypeptides by silica gel and bentonite;

(c) precipitation of polypetides by tannic acid;

(d) adsorption of the polyphenols by polyamide resins e.g. Nylon 66.

(iii) **Polysaccharide sediments**: Freezing and thawing of beer may cause an unpredictable haze which can appear in the form of flakes. This haze differs from chill haze in being distinctly carbohydrate in nature. They were found in lager chilled to −10°C and consisted mainly of Beta glucans derived from malt.

(iv) **Oxalate sediments**: Oxalate sediments may appear after several week's storage in beers rich in oxalate as a result of a low calcium content.

(v) **Other beer defects: Wild or gushing** beer is a defect observed as a violent over-foaming when a bottle of beer is opened. The taste is unaffected. Gushing is due to the formation of micro-bubbles; excess pressure may force the micro-bubbles back into solution. Gushing beers have been identified with malt made from old barley and trial brews have shown them to be associated with the presence of mycelia of *Fusarium* during the steeping.

The *off-flavor* developed when beer is exposed to sunlight is due to the formation of mercaptans by photochemical reaction in the blue-green region (420-520 nm) of visible light.

12.1.5   Some Developments in Beer Brewing

The description made above is of conventional beer brewing. Some developments have taken place both in the manner of the production of beer as well as in the type of beer produced: This section will look briefly at some of these.

12.1.5.1   Continuous brewing

Although it is not yet widely used, continuous brewing is gaining gradual acceptance in many countries. In the current commercial continuous brewing systems, it is mainly fermentation that is continuous, secondary fermentation and lagering are usually batch.

Two systems of continuous fermentation are known: the open and the partially closed.

(i) **The open system of continuous fermentation**: In the open system wort is fed continuously into the fermentor, while beer flows out at the same rate. The yeast is allowed to attain its natural concentration or steady state. In the system described here wort is collected batch wise from the brew house and may be stored for up to 14 days at 2°C before use. The wort is sterilized in a heat-exchanger prior to oxygenation. It is then passed through the bottom into the first tank, which is continuously stirred and where aerobic growth occurs. It is later passed into a second tank where conditions are anaerobic; alcohol and CO_2, are formed in this tank. From there the beer with its suspended yeasts overflows into a third vessel for sedimentation. Finished beer is removed from the top and yeast cells from the bottom. The amount of yeast in the beer is just adequate for secondary fermentation. CO_2 is collected from the top. The yeast employed is a special one which apart from imparting the right flavor, must be able to remain in active fermenting condition in suspension in the anaerobic vessel and yet be able to flocculate rapidly once in the cooled sedimentation tank. It is possible theoretically to use one tank, or more than two tanks for sedimentation. Indeed in another system three tanks are used, but two afford flexibility of design and use (Fig. 12.6).

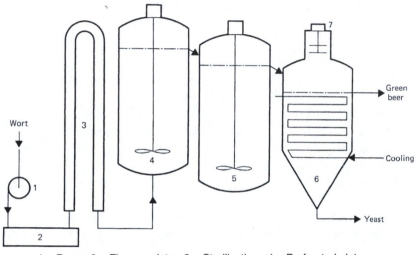

1 = Pump; 2 = Flow regulator; 3 = Sterilization; 4 = Perforated plates;
5 = Control of temperature; 6 = Yeast separator

Fig. 12.6   The Open System of Continuous Brewing

(ii) **Partially closed system of continuous fermentation**: In the closed system, yeast in held at a given concentration instead of allowing it to grow at its own steady state as in the open system. (The open system itself may indeed be modified to achieve a higher yeast concentration by recycling yeasts from the sedimentation tank into the first tank. The disadvantage of the modification is the possibility of contamination. Secondly, the returned yeasts are in a different physiological state of growth from those actively involved in fermentation, hence the wort and the beer quality may suffer).

In the closed system, typified by the tower fermentor (Fig. 12.6), sterilized wort is pumped into the base of the cylindrical tower with aeration, if necessary, and the beer is drawn off at the top at the same rate.

Yeasts attain a very high density, in excess of 350 gm/liter and wort becomes almost ready beer. The upper regions have a lower yeast concentration and serve partly as a final fermentation stage but especially as a means of separating the yeasts. Baffles enable the diversion of the rising CO_2 and beer from the beer outlet. Its over-riding advantage is that beer can be produced in four hours if the lower regions have the optimum yeast concentration of 350-400 gm/liter. Special yeasts able to maintain the high mass at the lower level and yet able to pass out of the fermentor if adequate amount must be used.

Although continuous brewing has not been generally adopted, its emergence forced brewers to make batch brewing more efficient and to find 'batch' answers to the advantages offered by continuous brewing.

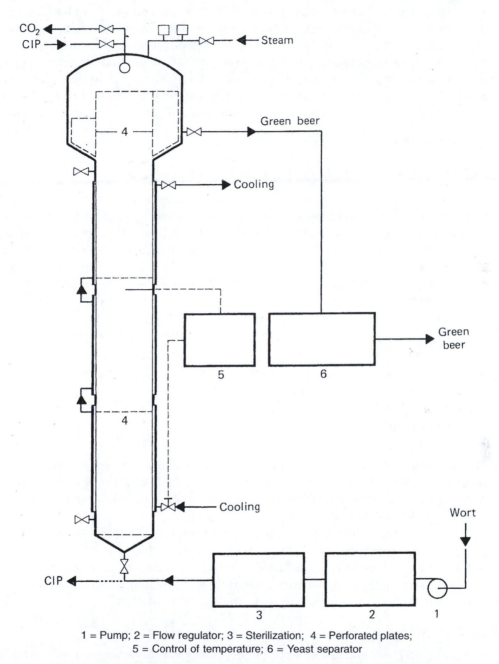

1 = Pump; 2 = Flow regulator; 3 = Sterilization; 4 = Perforated plates;
5 = Control of temperature; 6 = Yeast separator

Fig. 12.7 The Tower Fermentor (Closed) System of Continuous Brewing

12.1.5.2 Use of enzymes

A number of firms now market enzymes isolated from bacteria and fungi which can carry out the functions of malt. The advantage of the use of these enzymes is to greatly reduce costs since malting can be eliminated entirely. Despite the great potentials offered by this method, brewers are yet unwilling to accept it. The consequences of eliminating or reducing the need for malt from the barley farmer and the malting industry, two long-standing establishments, would pose great difficulties in adopting this method. When enzymes to become generally used, care must be taken to ensure that not only the major enzymes, amylases, and proteases, are included but that others such as Beta-glucanases which hydrolyze the gums of barley are also present in the enzyme mixture. It must also be certain that toxic microbial products are eliminated from the enzyme preparations.

12.2 SORGHUM BEERS

12.2.1 Kaffir Beer and Other Traditional Sorghum Beers

Barley is a temperate crop. In many parts of tropical Africa beer has been brewed for generations with locally available cereals. The commonest cereal used is *Sorghum bicolor* (= *Sorghum vulgare*) known in the United States as milo, in South Africa as kaffir corn and in some parts of West Africa as Guinea corn. The cereal which is indigenous to Africa is highly resistant to drought. Sorghum is often mixed with maize (*Zea mays*) or millets, (*Pennisetum* spp). In some cases such as in Central Africa e.g. Zimbabwe, maize may form the major cereal. Outside Africa sorghum is not used normally for brewing except in the United States where it is occasionally used as an adjunct.

The method for producing these sorghum beers of the African continent as well as their natures are remarkably similar. They

(i) are all pinkish in color; sour in taste; and of fairly heavy consistency imposed partly by starch particles, and also because they are

(ii) consumed without the removal of the organisms;

(iii) are not aged, or clarified, and

(iv) include a lactic fermentation.

The tropical beers are known by different names in different parts of the world: 'buru-kutu', 'otika', and 'pito' in Nigeria, , 'maujek' among the Nandi's in Kenya, 'mowa' in Malawi, 'kaffir beer' in South Africa, 'merisa' in Sudan, 'bouza' in Ethiopia and 'pombe' in many parts of East Africa.

It is only in South Africa that production has been undertaken in large breweries; elsewhere although considerable quantities are produced, this is done by small holders to satisfy small local clientele. In South Africa, in fact, it is reported that three or four times more kaffir beer is produced and drunk than is the case with barley beers. The processes of producing the beer include malting, mashing and fermentation.

12.2.1.1 Malting

For malting, sorghum grains are steeped in water for periods varying from 16-46 hours. They are then drained and allowed to germinate for five to seven days, water being sprinkled on the spread-out grains. At the end of this period, the grains are usually dried, often in the sun or in the South African system at 50°-60°C in driers. Kilning is however not done. In some parts, the dried malt may be stored and used over several months.

Contrary to opinions previously held by many, sorghum malt is rich in amylases, particularly α-amylase, although the ungerminated grain does not contain β-amylase as is the case with barley. Sorghum has not received much attention as a brewing material, except occasionally as an adjunct in the United States. However in recent times interest has grown in West Africa in its use for malting and it may be that strains which perform in malting as well as barley does may be found.

It has been suggested that the saccharification of sorghum starch is brought about partly by the fungi which grow on the grains during their germination as well as by the germinated sprout. This, however, has been disputed vehemently by some workers. The fungi so implicated are *Rhizopus oryzae, Botryodiplodia theobromae, Aspergillus flavus, Penicillium funiculosum*, and *P. citrinum*.

12.2.1.2 Mashing

The malt is ground coarsely and mixed in a rough 6:1 (v/v) proportion with water and boiled for about 2 hours. During the boiling starchy adjunct in the form of dried powder of plantains, cassava ('gari') or unmalted cereal may be added so that an approximate 1:2:6 proportion of the adjunct malt and water is attained. It is filtered and is then ready for fermentation. In South African kaffir beer breweries the adjunct consisting of boiled sorghum or maize grits is added after the initial souring of the mash.

12.2.1.3 Fermentation

Two fermentations take place during sorghum beer production: a lactic acid fermentation, and an alcoholic fermentation. In traditional fermentation, the dregs of a previous fermentation are inoculated into the boiled, filtered, and cooled wort. This inoculum consists of a mixture of yeasts, lactic acid and acetic acid bacteria. The first phase of the fermentation is brought about by lactic bacteria mainly *Lactobacillus mesenteroides*, and *Lactobacillus plantarum*.

In the sorghum beer breweries in South Africa, the temperature of the mash is held initially at 48-50°C to encourage the growth of thermophilic lactic acid bacteria which occur naturally on the grain, for 16-24 hours. The pH then drops to about 3-4. The sour malt is added to the previously cooked adjunct of unmalted sorghum or maize, and sometimes some more malt may be added. It is then cooled to 38°C and pitched with the top fermenting yeasts.

In the traditional method yeasts and lactic acid bacteria are present in the dregs. The yeasts which have been identified in Nigerian sorghum beer fermentation are: *Candida* spp, *Saccharomyces cervisiae*, and *Sacch. chevalieri*.

Fermentation is for about 48 hours during which lactic acid bacteria proliferate. Thereafter it is ready for distribution and consumption. No secondary fermentation of the kind seen in lager beer, lagering, or clarification is done. The live yeasts, and the lactic acid bacteria are consumed in much the same ways as they are done in palm wine. In some localities the fermentation lasts a little longer and the flavor is influenced by a slight vinegary taste introduced by the release of acetic acid by acetic acid bacteria.

Sorghum beers usually contain large amounts of solids (Table 12.6) mainly starch apart from the microorganisms. For this reason some authors have regarded them as much as foods as they are alcholic beverages an alcoholic beverage.

Table 12.6 Properties of South African sorghum beer

Properties	Small scale	Factory
pH	3.5	3.4
Alcohol	0.1	3.0
Solids (%w/v)		
Total	4.9	5.4
Insoluble	2.3	3.7
Nitrogen (%w/v)		
Total	0.084	0.093
Soluble	-	0.014

SUGGESTED READINGS

Amerine, M.A., Berg, H.W., Cruess, W.V. 1972. The Technology of Wine Making 3rd Ed. Avi Publications. West Port, USA, pp. 357-644.

Aniche, G.N. 1982. Brewing of Lager Beer from Sorghum. Ph.D *Thesis*, University of Nigeria, Nsukka, Nigeria.

Ault, R.G., Newton, R. 1971. In: Modern Brewing Technology, W.P.K. Findlay, (ed) Macmillan, London, UK. pp. 164-197.

Battcock, M., Azam-Ali, S. 2001. Fermented Fruits and Vegetables: A Global Perspective. FAO Agric. Services Bull. 134, Rome, Italy, pp. 96.

Doyle, M.P., Beuchat, L.R., Montville, T.J. 2004. Food Microbiology: Fundamentals and Frontiers, 2nd edit., ASM Press, Washington DC, USA.

Flickinger, M.C., Drew, S.W. (eds) 1999. Encyclopedia of Bioprocess Technology - Fermentation, Biocatalysis, and Bioseparation, Vol 1-5. John Wiley, New York, USA.

Gastineau, C.F., Darby, W.J., Turner, T.B. 1979. Fermented Food Beverages in Nutrition. Academic Press; New York, USA, pp. 133-186.

Gutcho, M.H. 1976. Alcoholic Beverages Processes. Noyes Data Corporation New Jersey and London. pp. 10-106.

Hammond, J.R.M., Bamforth, C.W. 1993. Progress in the development of new barley, hop, and yeast variants for malting and brewing. Biotechnology and Genetic Engineering Reviews 11, 147-169.

Hanum, H., Blumberg, S. 1976. Brandies and Liqeuers of the World. Doubleday Inc. New York USA.

Hough, J.S. 1985. The Biotechnology of Malting and Brewing. Cambridge University Press; Cambridge, UK. pp. 15-188.

Hough, J.S., Briggs, D.E., Stevens, R. 1971. Malting and Brewing Science, Chapman and Hall, London, UK.

Hoyrup, H.E. 1978. Beer Kirk-Othmer's Encyclopaedia of Chemical Technology, Wiley, New York, USA. Microbiology 24: 173-200.

Okafor, N. 1972. Palm-wine yeasts of Nigeria. Journal of the Science of Food and Agriculture 23, 1399-1407.

Okafor, N. 1978. Microbiology and biochemistry of oil palm wine. Advances in Applied.

Okafor, N. 1987. Industrial Microbiology. University of Ife Press Ile-Ife: Nigeria. pp. 201-210.

Okafor, N. 1990. Traditional alcoholic beverages of tropical Africa: strategies for scale-up. Process Biochemistry International 25, 213-220.

Okafor, N., Aniche, G.N. 1980. Lager beer from Nigerian Sorghum. Brewing and Distilling International. 10, 32-33, 35.

Packowski, G.W. 1978. Distilled Alcoholic Beverages, Encyclopadeia of Chemical Technology, 3rd Edit. Wiley, New York, USA. pp. 824-863.

Singleton, V.L., Butzke, C.E. 1998 Wine. Kirk-Othmer Encyclopedia of Chemical Technology. John Wiley & Sons, Inc. Article Online Posting Date: December 4, 2000.

Soares, C. 2002. Process Engineering Equipment Handbook Publisher. McGraw-Hill.

Taylor, J.R.N., Dewar, J. 2001. Developments in Sorghum Food Technologies. Advances in Food and Nutrition Research 43, 217-264.

Vogel, H.C., Tadaro, C.L. 1997. Fermentation and Biochemical Engineering Handbook - Principles, Process Design, and Equipment. 2nd Edit Noyes.

Zeng, A. 1999. Continuous culture. In: Manual of Industrial Microbiology and Biotechnology. A.L. Demain, J.E. Davies (eds) 2nd Edit. ASM Press. Washington, DC, USA. pp. 151-164.

CHAPTER 13

Production of Wines and Spirits

13.1 GRAPE WINES

Wine is by common usage defined as a product of the "normal alcoholic fermentation of the juice of sound ripe grapes". Nevertheless any fruit with a good proportion of sugar may be used for wine production. Thus, citrus, bananas, apples, pineapples, strawberries etc., may all be used to produce wine. Such wines are always qualified as fruit wines. If the term is not qualified then it is regarded as being derived from grapes, *Vitis vinifera*. The production of wine is simpler than that of beer in that no need exists for malting since sugars are already present in the fruit juice being used. This however exposes wine making to greater contamination hazards.

Wine is today principally produced in countries or regions with mild winters, cool summers, and an even distribution of rainfall throughout the year. In North America, the United States is the leading producer, most of the wine coming from the State of California and some from New York. In Europe the principal producers are Italy, Spain and France. In South America, Argentina, Chile, and Brazil are the major producers; and in Africa, they are Algeria, Morocco, and South Africa. Other producers are Turkey, Syria, Iran, and Australia.

13.1.1 Processes in Wine Making

13.1.1.1 Crushing of Grapes

Selected ripe grapes of 21° to 23° Balling (Chapter 12) are crushed to release the juice which is known as 'must', after the stalks which support the fruits have been removed. These stalks contain tannins which would give the wine a harsh taste if left in the must. The skin contains most of the materials which give wine its aroma and color. For the production of red wines the skins of black grapes are included, to impart the color. Grapes for sweet wines must have a sugar content of 24 to 28 Balling so that a residual sugar content is maintained after fermentation. The chief sugars in grapes are glocuse and fructose; in ripe fruits they occur in about the same proportion. Grape juice has an acidity of 0.60-0.65% and a pH of 3.0-4.0 due mainly to malic and tartaric acids with a little citric acid. During ripening both the levulose content and the tartaric acid contents rise.

Nitrogen is present in the form of amino acids, peptides, purines, small amounts of ammonium compounds and nitrates.

13.1.2 Fermentation

(i) **Yeast used**: The grapes themselves harbor a natural flora of microorganisms (the bloom) which in previous times brought about the fermentation and contributed to the special characters of various wines. Nowadays the must is partially 'sterilized' by the use of sulphur dioxide, a bisulphate or a metabisulphite which eliminates most micro-organisms in the must leaving wine yeasts. Yeasts are then inoculated into the must. The yeast which is used is *Saccaromyces cerevisiae* var, *ellipsoideus* (synonyms: *Sacch. cerevisiae, Sacch. ellipsoideus, Sacch, vini.)* Other yeasts which have been used for special wines are *Sacch. fermentati, Sacch. oyiformis* and *Sacch. bayanus.*

Wine yeasts have the following characteristics: (a) growth at the relatively high acidity (i.e., low pH) of grape juice; (b) resistance to high alcohol content (higher than 10%); (c) resistance to sulfite.

(ii) **Control of fermentation**

(a) **Temperature**: Heat is released during the fermentations. It has been calculated that on the basis of 24 Cal per 180 gm of sugar the temperature of a must containing 22% sugar would rise 52°F (11°C) if all the heat were stopped from escaping. If the initial temperature were 60°F (16°C) the temperature would be 100°F (38°C) and fermentation would halt while only 5% alcohol has been accumulated. For this reason the fermentation is cooled and the temperature is maintained at around 24°C with cooling coils mounted in the fermentor.

(b) **Yeast Nutrition**: Yeasts normally ferment the glucose preferentially although some yeasts e.g. *Sacch. elegans* prefer fructose. To produce sweet wine glucose-fermenting wine yeasts are used leaving the fructose which is much sweeter than glucose. Most nutrients including macro- and micro-nutrients are usually abundant in must; occasionally, however, nitrogeneous compounds are limiting. They are then made adequate with small amounts of $(NH_4)_2 SO4$ or $(NH_4)_2 HPO4$.

(c) **Oxygen**: As with beer, oxygen is required in the earlier stage of fermentation when yeast multiplication is occurring. In the second stage when alcohol is produced the growth is anaerobic and this forces the yeasts to utilize such intermediate products as acetaldehydes as hydrogen acceptors and hence encourage alcohol production.

(iii) **Flavor development**: Although some flavor materials come from the grape most of it come from yeast action. The flavor of wine has been elucidated with gas chromatography and has been shown to be due to alcohols, esters, fatty acids, and carbonyl compounds, the esters being the most important. Diacetyl, acetonin, fusel oils, volatile esters, and hydrogen sulfide have received special attention. Autolysates from yeasts also have a special influence on flavor.

13.1.3 Ageing and Storage

The fermentation is usually over in three to five days. At this time 'pomace' formed from grape skins (in red wines) will have risen to the top of the brew. As has been indicated earlier for white wine, the skin is not allowed in the fermentation. At the end of this fermentation the wine is allowed to flow through a perforated bottom if pomace had been allowed. When the pomace has been separated from wine and the fermentation is complete or stopped, the next stage is 'racking'. The wine is allowed to stand until a major portion of the yeast cells and other fine suspended materials have collected at the bottom of the container as sediment or 'lees'. It is then 'racked', during which process the clear wine is carefully pumped or siphoned off without disturbing the lees.

The wine is then transferred to wooden casks (100-1,000 gallons), barrels (about 50 gallons) or tanks (several thousand gallons). The wood allows the wine only slow access to oxygen. Water and ethanol evaporate slowly leading to air pockets which permit the growth of aerobic wine spoilers e.g. acetic acid bacteria and some yeasts. The casks are therefore regularly topped up to prevent the pockets. In modern tanks made of stainless steel the problem of air pockets is tackled by filling the airspace with an inert gas such as carbon dioxide or nitrogen.

During ageing desirable changes occur in the wine. These changes are due to a number of factors:

(a) Slow oxidation, since oxygen can only diffuse slowly through the wood. Small amounts of oxygen also enter during the filling up. Alcohols react with acids to form esters; tannins are oxidized.

(b) Wood extractives also affect ageing by affecting the flavor.

(c) In some wines microbial malo-lactic fermentation occurs. In this fermentation, malic acid is first converted to pyruvic acid and then to lactic acid. (Fig. 13.1)

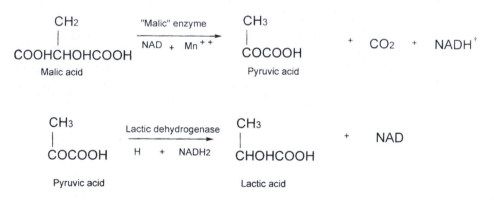
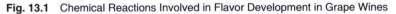

Fig. 13.1   Chemical Reactions Involved in Flavor Development in Grape Wines

The reaction is responsible for the rich flavor developed during the ageing of some wines e.g. Bordeaux. Cultures which have been implicated in this fermentation are *Lactobacillus* sp and *Leuconostoc* sp. A temperature of 11-16°C is best for ageing wines, High temperature probably functions by accelerating oxidation.

13.1.4   Clarification

The wine is allowed to age in a period ranging from two years to five years, depending on the type of wine. At the end of the period some will have cleared naturally. For others artificial clarification may be necessary. The addition of a fining agent is often practiced to help clarification. Fining agents react with the tannin, acid, protein or with some added substance to give heavy quick-settling coagulums. In the process of setting various suspended materials are adsorbed. The usual fining agents for wine are gelatin, casein, tannin, isinglass, egg albumin, and bentonite. In some countries the removal of metal ions is accomplished with potassium ferrocyanide known as 'blue fining'; it removes excess ions of copper, iron, manganese, and zinc from wines.

13.1.5 Packaging

Before packing in bottles the wine from various sources is sometimes blended and then pasteurized. In some wineries, the wine is not pasteurized, rather it is sterilized by filtration. In many countries the wine is packaged and distributed in casks.

13.1.6 Wine Defects

The most important cause of wine spoilage is microbial; less important defects are acidity and cloudiness. Factors which influence spoilage by bacteria and yeasts include the following (a) wine composition, specifically the sugar, alcohol, and sulfur dioxide content; (b) storage conditions e.g. high temperature and the amount of air space in the container; (c) the extent of the initial contamination by microorganism during the bottling process.

When proper hygiene is practiced bacterial spoilage is rare. When it does occur the microorganisms concerned are acetic acid bacteria which cause sourness in the wine. Lactic acid bacteria especially *Leuconostoc,* and sometimes *Lactobacillus* also spoil wines. Various spoilage yeasts may also grow in wine. The most prevalent is *Brettanomyces,* slow growing yeasts which grow in wine causing turbidities and off-flavors. Other wine spoilage yeasts are *Saccharomyces oviformis,* which may use up residual sugars in a sweet wine and *Saccharomyces bayanus* which may cause turbidity and sedimentation in dry wines with some residual sugar. *Pichia membanaefaciens* is an aerobic yeast which grows especially in young wines with sufficient oxygen.

Other defects of wine include cloudiness and acidity.

13.1.7 Wine Preservation

Wine is preserved either by chemicals or by some physical means. The chemicals which have been used include bisulphites, diethyl pyrocarbonate and sorbic acid. Physical means include pasteurization and sterile filtration. Pasteurization is avoided when possible because of its deleterious effect on wine flavor.

13.1.8 Classification of Wines

Grape wines may be classified in several ways. Some of the criteria include place of origin, color, alcohol content and sweetness. The one being adopted here is primarily used in the United States and is shown in Table 13.1. This system classifies wine into two groups: natural wines and fortified wines.

Table 13.1   Broad classification of grape wines

A. *Natural wines*: 9-14% alcohol; nature and keeping quality mostly dependent on 'complete' yeast fermentation and protection from air

    1. *Still wines* (known as 'Table' wines intended as part of meal); no carbon dioxide added.

        (a) *Dry table wines*: (no noticeable sweetness)

            (i) White; (ii) Rose (pink); (iii) Red

        (b) *Sweet table wines*

            (i) White (ii) Rose

        Further naming of above depends on grape type, or region of origin.

    2. *Sparkling wines* (appreciable CO_2 under pressure)

        (a) White (Champagne); (b) Rose (Pink Champagne); (c) Red (Sparkling burgundy; cold duck)

B. *Fortified (Dessert and appetizer) wines*: Contain 15 to 21% alcohol; nature and keeping quality depends heavily on addition of alcohol distilled from grape wine.

    1. *Sweet wines*

        (a) White (Muscatel, White port, angelica)

        (b) Pink (California tokay, tawny port)

        (c) Red (Port, black Muscat)

    2. *Sherries*: (White sweet or dry wines with oxidized flavors)

        (a) Aged types

        (b) Flor types

        (c) Baked types

    3. *Flavored specialty wines* (usually white Port base)

        (a) Vermouth (pale dry, French; Italian sweet types)

        (b) Proprietary brands

13.1.8.1   The natural wines

These result from complete natural fermentation. Further fermentation is prevented because the sugar is to a large extent exhausted. Spoilage organisms such as acetic acid bacteria do not grow if air is excluded. Owing to the natural limit of sugar in grapes, the alcohol content does not usually exceed 12%. They are sub-divided into still (without added CO_2) and sparkling (with added CO_2). Color and sweetness also subdivide the wines. The color, pink or red, is derived from the color of the grape; white wine comes from a grade whose skin is light-green, but whose juice is clear. White wine can also result from black (or deep-red) grapes if the skin is removed immediately after pressing before fermentation. Red wine results when the skin of black grapes are allowed in the fermenting must. Pink wine results when the skin is left just long enough for some material to extract some skin coloring.

Table wines: In general the natural wines are usually consumed at one sitting once they are opened. For this reason they are called 'table' wines and intended to be part of a meal.

They are usually served in generous amounts, partly because they contain less alcohol than the desserts and appetizers and partly because they do not have a high keeping quality once opened compared with appetizers and dessert wines.

Dessert and appetizer wines: The second broad group of wines are dessert or appetizer wines. As can be seen from their names they are served at the beginning (appetizers) or at the end (dessert) of meals. They contain extra alcohol from distilled wines, partly to make them more potent, but also to preserve them from yeast spoilage. These are divided into three categories: (a) Sweet e.g. port; (b) Sherries – sweet or dry, they originated from Portugal and are characterized by flavors induced by various degrees of oxidation; (c) Flavored wines e.g. vermouth; these are flavored with herbs and other components which are secrets of the producing firms (Table 13.1).

Sparkling wines (especially champagne), sherry, and flavored wines will be discussed briefly.

Sparkling Wines: Sparkling wines contain CO_2 under pressure before they are opened. They are called sparkling because the gentle release of carbon dioxide from the wine after the bottle is opened gives the wine a sparkle. The best known of the sparkling wines is produced in Champagne region in northern France which has given its name to the wine. Champagne is produced either in a bottle or in bulk.

(a) *Bottle Champagne*: Champagne is usually a clear sparkling wine made from white wine. Pink or red champagne made from wines of the same color are preferred by some. Champagne is produced by a second fermentation in the bottle of an already good wine. Not only does its production take a long time, it also requires a complicated method which is difficult to automate and hence must be handled manually. For these reasons the drink is expensive. The usual process is described below. The parlance of champagne manufacture is understandably French. *The method Champenoise* to be described is the one used for making the best sparkling wines. After the must is fully fermented, the wines to be used for champagne are racked, clarified, stabilized and fined. A blend is then made from several different wines to give the desired aroma. The blend is known as *cuvèe*. The *cuvèe* should have an alcohol content of 9.5 to 11.5%, adequate acidity (0.7 to 0.9% titrable as tartaric acid) and light straw or light yellow color. The SO_2 content should be low otherwise SO_2 odor would show when the wine is poured, or worse still, the yeasts might convert SO_2 to forms of hydrogen sulphide, which would give a rotten egg odor.

The cuvee is placed into thick walled bottles able to withstand the high pressure of CO_2 to be built up later in the bottle. For the secondary fermentation in the bottle, more yeast, more fermentable sugar, usually sucrose and nowadays sometimes a small amount (0.05 to 0.1%) of ammonium phosphate is added. The yeast usually used is *Saccharomyces bayanus*. It is selected because it meets the following requirements encountered in secondary fermentation for champagne production: it must grow at a fairly high alcohol content (10-12%), at high pressures (see below) and relatively low temperatures; the yeast must die or become inactive within a short time in order to prevent further fermentation after sugar (known as the *dosage*) is added once again before the final corking. Finally the yeast must be able to form a compact granulated sediment after fermentation.

The amount of sugar added depends on the expected CO_2 pressure of the sparkling wine. Sparkling wines have a pressure of about six atmospheres but not less than four. As a rough rule, 4 gm of sugar per liter will produce one atmosphere of CO_2 pressure. Therefore the sugar added is about 24 gm/liter. Account is taken of any sugar present in the wine. Although the bottles are thick, if the fermentation is too rapid, temperature too high or sugar too high the bottle may burst. Champagne bottles are therefore not reused since a scratched bottle may burst.

The bottle with its mixture of wine, sugar and yeast is placed horizontally and allowed to ferment at a temperature of about 15-16°C; this secondary fermentation takes two to three months. The secondarily fermented wine now known as *triage* is then stored still horizontally at a temperature of 10°C and remains undisturbed in that position for at least a year and sometimes up to five years. Much of the aroma of well-aged champagne appears to come from the reactions involving materials released from yeast autolysis.

The next stage is getting the sediment from the side to the neck of the bottle. Several methods are used by individual handlers. This transfer or *remuage* is achieved by first placing the bottle neck downwards at an angle in a rack with varying degrees of jolting. The bottles are next rotated clockwise and anti-clockwise on alternate days during which the bottle is gradually straightened to the perpendicular. The process may take anything from two or six weeks at the end of which the sediment finds it way to the neck of the bottle. To remove the sediment of yeast cells, the neck of the bottle is frozen to 1°C to 15°C; an ice plug which includes the sediment forms therein. The bottle is turned to an angle of about 45° and opened. The pressure in the bottle forces out the ice plug. During this process of *disgorging* some CO_2 is lost but sufficient is left to give the usual pop which launches a bride, a ship or a graduating student into the future! The lost wine is replaced from another bottle and the *dosage* is added. The *dosage* is a syrup consisting of about 60 gm of sugar in 100 ml of well-aged wine. All champagnes, even those labeled dry contain dosage; otherwise it would taste sour. Sweet champagne contains up to 10% sugar.

(b) ***Bulk production of Champagne:*** Champagne is sometimes produced in bulk in a large tank rather than in individual bottles. Bulk production is known as the Charmat process named after its inventor. When this is so produced it is usually declared on the label to save the more labor-intensive bottle-made versions from unfair competition. The tank which has a lining of a inert material such as glass holds 500 to 25,000 gallons and is built to withstand 10-20 atmospheres as a safety measure. Valves control the pressure and cooling coils the temperature. Since the tanks are aerated a rapid turnover is possible and 6-12 fermentations per year are made. Another reason for the rapid turnover is that a heavy yeast growth occurs which could lead to the production of off flavors especially H_2S if allowed prolonged contact with the wine.

Tank fermented champagne is usually given a cold-stabilizing treatment to remove excess tartarates. It is filtered still cold and under pressure to remove yeast cells. The wine is then filled into bottles with *dosage* of the right kind. It is usual to introduce some sulfur dioxide into it just in case some yeasts were not removed by filtration. Sulfur dioxide also helps to prevent darkening oxidation which may occur as the wine takes up oxygen during the transfer process. The sulphur dioxide odor is usually noticeable when Charmat-prepared champagne bottles are opened. Furthermore, they lack the aroma conferred on well-aged bottle brands by the autolysis of yeast. Bulk champagnes are amenable to bulk production because they are more difficult to ferment owing to their higher tannin contents; they may also require higher fermentation temperatures.

13.1.8.2   Fortified wines

The fortified wines can be divided into three main groups, which derive their names from the warmer more southern portion of the Iberian (Spain-Portugal) peninsula in Europe, and an island off that peninsula, where they were originally produced: sherry (Jerez de la Frontera area in Spain); port (Douro Valley in Portugal); Madeira (the Island of Madeira).

In other wines contact with oxygen no matter how small is undesirable. The fortified wines are however produced by the deliberate but controlled oxidation of wine. The oxidation is achieved by prolonged ageing in the pressure of air, by the growth of an aerobic yeast or by heating. The consequence of this oxidation is a product which has a dark, reddish-brown color with a characteristic flavor, whether the starting wine is white or red. For sherry therefore a white wine is used but for port or Madeira a red or a white wine may be used.

All the three fortified wines have a high alcohol content of ranging from 15-20% (v/v), derived from added alcohol hence their name. They are usually separated into two groups: (a) Vermouth (b) Other flavor wines (Special natural wines).

Table 13.2   Flow Sheet for the Production of Sherry, Port, and Madeira

	Sherry		*Port*	*Madeira*
1. Type of wine used	White		White/red	White/red
2. Grape sun dried to increase sugar content of juice	+	+	-	-
3. Skin of grapes left in contact with must during initial wine	_	_	+	-
4. fermentation	+	+	-	-
5. SO$_2$ added to juice	+	+	-	+
6. Juice fermented to dry wine Freshly fermented wine fortified to	15%	15%	17%	18%
7. alcohol content 15% given (% v/v)				
8. Matured in contact with air for flor film	+ _	— _	— -	- +
9. Maturation with heating Adjusted to given alcohol content				
10. About six months after fermentation (% v/v)	15% +	18% +	20% +	18% +
11. Matured in wood for several months Fortified wine	Fino, Oloros Amantillado		Port	Madeira

Key: + = Yes, — = No.

Vermouth may be of Italian (sweet) or French (dry) type. Vermouth comes from the German 'wermut' (wormwood, *Artemesia absinthium* a common herb) a frequent component of vermouth. The other flavored fortified wines, such as 'Campari', 'Dubonnet', 'Byrrh', like vermouth, contain 15-21% alcohol. In both cases the aromatic components of the herbs are extracted by immersing the herbs in wine or alcohol, or by distilling them. The nature and proportion of many of the components of flavored fortified wines are kept secret.

13.1.8.3 Fruit wines: cider and perry

Often fruits do not contain enough sugar to make a potent alcoholic beverage. Under such conditions, extra sugar in the form of sucrose is added to encourage fermentation. Fruit wines are popular in some countries where grapes cannot thrive.

Cider is derived from apples, (*Malus pumila*) and perry from pears or a mixture of pears and apples. They differ from other fruit wines in that their alcohol content is low (4-5% with a maximum of 7-8% v/v) because sugar is not usually added. The basic processes are similar to those of grape wine: pressing out the juice, fermenting, maturing, and bottling. Fruit wines have been made from cashew, pineapples, and other fruits.

13.2 PALM WINE

Palm wine is a general name for alcoholic beverages produced from the saps of palm trees. It differs from the grape wines in that it is opaque. It is drunk all over the tropical world in Africa, Asia, South America. Table 13.3 shows the palms from which palm wine is derived in the various areas.

Palm wine is usually a whitish and effervescent liquid both of which properties derive from the fact that the fermenting organisms are numerous and alive when the beverage is consumed. More information appears to exist on wine derived from the oil palm, *Elaeis guinieensis* than on any other and this will be discussed.

The sap of the palm is obtained from a variety of positions: the stem of the standing tree, the tip or trunk of the felled tree and the base of the immature male inflorescence. Which method is favored depends on the country concerned but most studies have centered on inflorescence wine. The sap produced by this method in *Elaeis guiniensis* contains about 12% sucrose, about 1% each of fructose, glucose, and raffinose, and small quantities of protein and some vitamins and is a clear, sweet, syrupy liquid.

To produce palm wine a succession of microorganisms occurs roughly: Gram-negative bacteria, lactic acid bacteria and yeasts and finally acetic acid bacteria. Yeasts in palm wine have been identified as coming from various genera (Table 13.4).

The organisms are not artificially inoculated and find their way into the wine from a variety of sources including the air, the tapping utensils including previous brews and the tree. The wine contains about 3% (v/v) alcohol and since the bacteria and yeasts are consumed live, it is a source of (single cell) protein and various vitamins.

The great problem with palm wine is that its shelf life is extremely short. It is best consumed within about 48 hours, but certainly not beyond about five days after tapping. For this reason various methods have been devised to preserve it. Pasteurization has met with some success, but methods which lower the microbial load of the wine by

Table 13.3   Palms from which palm wine is obtained

Name of palm	Location
Acromia vinifera Oerst	Nicargua, Panama, Costa Rica
Arenga pinnata (Wurmb.) Merr. (Syn. A. sacoharifera Labill.)	Far East
Attalea speciosa Mart.	Brazil, Guyana
Borassus aethipum Mart.	Tropical Africa
Broassus flabelifer Linn.	India, Cambodia, Java
Caryota urens Linn.	India
Cocos nucifera Linn.	India, Sri Lanka, Africa
Corypha umbraculifera L.	Sri Lanka
Elaesis guineensis Jacq.	Africa
Hyospathe elegans Mart.	Brazil, Guyana
Hyphaenae guineensis Thonn.	West Africa
Jubaea chilensis (Molina) Baillon	Chile
Mauritiella aculeate (H.B. and K.) Burret (Syn. Lepidococcus aculeatus H. Wendl and Drude)	Brazil, Venezuela
Marenia montana (Hump. and Bonpl.) Burret (Syn. Kunthia Montana Humb. and Bonpl.)	Brazil
Nypa fruticans Wurmb.	Sir Lanka, Bay of Bengal, The Phillippines, Carolines, Salmonon Islands
Orbignya cohune (Mart.) Dahlgreen ex Standley (Sun. Attalea cohune Mart.)	Hondruras, Mexico, Gautemala
Phoenix dactylifera Linn.	North Africa, Middle East
Phoenix reclinata Jacq. (Syn. Phoenix) Spinosa Schum. and Thonn.)	Central Africa
Phoenix sylvestris (L) Roxb.	India
Raphia hookeri Mann and Wendl.	Africa
Raphia sundanica A. Chev.	Africa
Raphia vinifera Beauv.	Africa
Scheelea princeps (Marts.) Karsten (Syn. Attalea princeps Mart.)	Brazil, Bolivia

centrifugation or filtration permit the use of milder pasteurization temperatures and lower quantities of chemicals (Table 13.5). Of the chemicals tried, sorbate and sulphite were found best. Fully fermented palm wine has 5% to 8% alcohol and is distilled for *kai-kai*, a gin with a distinct fruity flavor.

Other African alcoholic beverages

(i) Bouza is an alcoholic beverage produced in Egypt since the time of the Pharoahs. Formerly drunk by all classes, it is now drunk mainly among the lower income groups. Egyptian Bouza is prepared either from wheat or maize but the most popular is from wheat. To prepare bouza, coarsely ground wheat (about 20% of total) is placed in a large wooden basin and kneaded with water to form a dough which is cut into thick loaves and

Table 13.4 Yeasts identified in palm wines

Yeast	Type of Wine
Saccharomyces pastorianus	Oil palm
Saccharomyces ellipsoids	Oil palm,
Saccharomyces cereviside	Arenga palm
Saccharomyces cerevisiae	Oil palm,
Saccharomyces chevalieri	Oil palm,
Pichia sp.	Oil palm,
Schizosaccharomyces pombe	Oil palm,
Saccharomyces vafer	Oil palm,
	Palmyra palm
Endomycopsis sp.	Oil palm,
Saccharomyces markii	Oil palm,
Kloekera apculata	Oil palm,
Saccharomyces chevalieri	Palmyra palm
Saccharomyces rosei	Raphia
Candida spp.	Oil palm
Saccharomycoides luduigii	Oil palm
	Palmyra palm

Table 13.5 Counts of Bacteria on whole palm wine and on supernatant of centrifuged palm wine after various treatments

Treatment	Number	Percent Survival
Whole palm wine		
Untreated	$\times 10^1$	100.00
0.5% potassium sorbate	2.96×10^9	1.0571
0.10%	2.32×10^9	0.8286
0.15%	2.24×10^9	0.8000
0.15%	1.40×10^9	0.5000
0.15% sodium metaisulfie	1.20×10^8	0.4285
0.10%	4.80×10^5	0.00017
0.15%		
Supernatant after centrifuging		
Untreated	4.00×10^4	0.00001
Pasteurized at 60°C for ½ hour	2.00×10^2	0.0000007
Pasteurization plus 0.05% sorbate	2.00×10^2	0.00000007
Pasteurization plus 0.05% sodium metabisulfite	0.4×10^2	Virtually sterile

lightly baked. The rest of the wheat (about 80% of the total) is germinated for three to five days, sun dried and coarsely ground. The malted wheat and the crumbled bread are mixed in water in a wooden barrel. Bouza from a previous fermentation is added and the whole mixture is fermented for 24 hours at room temperature. The mash is sieved to remove large particles and the beverage is ready for drinking. The beverage has a pH of about 3.6 and an alcohol content of about 4%: but the pH drops while alcohol increases with further fermentation.

(ii) Talla (tella) is an Ethiopian small-producer beer with a smokey flavor derived from inverting the fermentation containers and talla collection pots over smouldering olive wood. Talla also acquires some smoke flavor from the toasted, milled and boiled cereal grains. During the toasting the grains are roasted until they begin to smoke slightly. In the production of talla, of which various types exist, powdered hop leaves and water are put in a fermentation vessel and allowed to stand for about three days. Ground barley or wheat malt and pieces of flat bread baked to burning on the outside. are added. On the fifth day hop stems are added in addition to cereal flour made by milling sorghum which is first boiled and then toasted. Water is added and the fermentation allowed for two days (i.e., to the seventh day). It is filtered and is ready to drink.

(iii) Busaa is an acidic alcoholic beverage drunk among the Luo. Abuluhya and Maragoli ethnic groups of Kenya. It is porridge-like and light-brown in color and is warmed to 35-40°C before being consumed. A stiff dough made from maize flour and water is incubated at room temperature for three to four days. The fermented dough is pounded and then heated on a metal plate till it turns brown.

Malt is made from millet, allowed to germinate for three to four days, sun dried, and ground to powder. A slurry is made in water of the millet malt powder and roasted maize flour dough and left to ferment for two to four days. It is filtered through cloth. The filtrate is busaa.

The organisms responsible for fermenting the uncooked maize dough include *Candida krusei. Saccharomyces cerevisiae. Lactobacillus helveticus, L. salivarus. L. brevis. L viridescens. L. plantarurn and Pediococcus damnosus*. The final product is the result of alcohol produced mainly by *S. cerevisiae:* the lactic acid in the beverage is produced by several lactobacilli:

(iv) Merissa is a sour Sudanese alcoholic (up to 6%) beverage made from sorghum. It has a pH of about four and a lactic acid content of about 2.5%. Sorghum grains are malted, dried, and ground into a coarse powder. Unmalted sorghum is milled into a fine powder. One third of this powder is mixed with a little water and allowed to ferment at room temperature for 24-36 hours. The resulting fermented sour dough is heated without further water addition until the product caramelizes to give rise to *soorji*. The cooled *soorji* is allowed to ferment for four in five hours in a mixture of malt, to which previous merissa is added as inoculum. The two are mixed together and portions of these are allowed to cool resulting in a product called 'futtura. Futtura is mixed from with malt flour (about 5%) and added to the bedoba from time to time. Fermentation lasts from 8 to 10 hours after which it is filtered to give rise to the drink, merissa.

(v) Tej is a mead (i.e., a wine made by fermenting honey) of Ethiopian origin. It is yellow, sweet, effervescent, and cloudy due to its yeast content. As it is expensive, it is beyond the reach of most Ethiopians and used only on special occasions. The wine may be flavored with spices or hops and also by passing smoke into the fermentation pot before it is used. To prepare Tej the honey is diluted with water by between 1:2 to 1:5 i.e., to a liquid of between 13 and 27% sugar since honey contains about 80% sugar. Yeasts of the genus *Saccharomyces* spontaneously ferment the brew in about five days to give a yellow wine of 7-14% alcohol (v/v).

(vi) Agadagidi wines are made from bananas and plantains and have the opaque, effervescent sweet-sour nature typical of African traditional alcoholic beverages. In Nigeria the best-known *agadagidi* is found in the cocoa-growing areas of south western Nigeria where plantains provide shade for the young cocoa trees. The ripe fruits are peeled and soaked in water where the sugars dissolving from the preparation permit the development of yeasts and lactic acid and giving rise to a typical opaque effervescent wine. The alcohol content is about 1%.

(vii) Mbege is consumed in Tanzania mainly by those living near Mount Kilimanjaro. It is not a wholly banana/plantain wine as is the case with Nigerian agadagidi. Rather it is produced from a mixture of malted millet and fermented banana juice. The juice is produced by boiling the ripe banana followed by decantation. The banana infusion is mixed with cooked and cooled millet malt and allowed to ferment for four to five days:

13.3 THE DISTILLED ALCOHOLIC (OR SPIRIT) BEVERAGES

The distilled alcoholic or spirit beverages are those potable products whose alcohol contents are increased by distillation. In the process of distillation volatile materials emanating directly from the fermented substrate or after microbial (especially yeast) metabolism introduce materials which have a great influence on the nature of beverage. The character of the beverage is also influenced by such post-distillation processes as ageing, blending, etc. The components of spirit beverages which confer specific aromas on them are known as congeners.

13.3.1 Measurement of the Alcoholic Strength of Distilled Beverages

(i) *Proof*: In English-speaking countries, such as the United States, Canada, the United Kingdom, and Australia, the alcoholic content of spirit beverages (and also of non-potable alcohol) is given by the term 'proof'. The reason for the term is historical. Before the advent of the use of the hydrometer in alcoholic measurements, an estimate of the alcoholic content of a spirit beverage was obtained by mixing it with gunpowder. If the gunpowder still ignited then it was satisfactory because it contained less than 50% (v/v) of alcohol. If it did not ignite it was 'under proof' because it contained less than 50% water. Proof has nowadays become more clearly defined, although slight differences occur among countries. In the United States proof spirit (i.e., 100 degrees proof, written 100°) shall be held to be that alcoholic liquor which contains one half its volume of alcohol of a specific gravity of 0.7939 at 15.6°C. In other words the proof is always exactly twice the alcoholic content by volume. Thus, 100 proof spirit contains 50% alcohol. In the British system proof spirit contains 57.1% by volume and 49.28% by weight of alcohol at 15.6°C.

A conversion factor of 1.142

$$\left(\frac{\text{Volume of alcohol of British Proof at } 15.6°\text{C}}{\text{Volume of alcohol of United States Proof at } 15.6°\text{C}} = \frac{57.1}{50} = 1.14 \right)$$

is applied to convert United States proof to a British proof.

It is customary to quantify large amounts of distilled alcoholic beverages or alcohol in *proof gallons* for tax and other purposes. This term simply specifies the amount of alcohol in a gallon of spirits. Thus a United States proof gallon contains 50% ethyl alcohol by volume; a gallon of liquor at 120° proof is 1.2 proof gallon and a gallon at 75° proof is 0.75 proof gallon. However, the ordinary United States gallon is 3.785 liters whereas the British (imperial) gallon is larger, 4.546 liters. The British proof gallon is multiplied by 1.37 to convert it to a United States gallon:

$$\frac{\text{Volume of British gallon}}{\text{Volume of United States gallon} \times \text{Conversion factor}} = \frac{4.546}{3.785} \times 1.142 = 1.375$$

The proof is read off a special hydrometer.

(ii) *Percentage by weight*: This is used in Germany for the ethanol content of a beverage or other liquid. The hydrometers are graduated at 15°C and the reference is with (Mendeleaf's) table of density. This results in a scale independent of the ambient temperature.

(iii) *Percentage by volume*: Many countries especially in Europe use the percentage by volume system. For most of them the hydrometer is calibrated at 15°C. France, Belgium, Spain, Sweden, Norway, Finland, Switzerland (which also used weight %), Brazil and Egypt, Russia use 20°C, Denmark and Italy 15.6°C whereas many South American countries use 12.5°C. In many of these countries the specific gravity of alcohol used as reference differ slightly.

13.3.2 General Principles in the Production of Spirit Beverages

In general the following steps are involved in the preparation of the above beverages. The details differ according to beverage.

(i) *Preparation of the medium*: In the grain beverages (whisky, vodka, gin) the grain starch is hydrolyzed to sugars with microbial enzymes or with the enzymes of barley malt. In all the others no hydrolysis is necessary as sugars are present in the fermenting substrate as in brandy (grape sugar) and rum (cane sugar).

(ii) *Propagation of yeast inoculum*: Large distilleries produce hundreds of liters of spirits daily for which fermentation broths many more times in volume are required. These broths are inoculated with up to 5% (v/v) of thick yeast broth. Although yeast is re-used there is still a need for regular inocula. In general the inocula are made of selected alcohol-tolerant yeast strains usually *Sacch. cerevisiae* grown aerobically with agitation and in a molasses base. Progressively larger volumes of culture may be developed before the desired volume is attained.

(iii) *Fermentation*: When the nitrogen content of the medium is insufficient nitrogen is added usually in the form of an ammonium salt. As in all alcohol fermentations the heat released must be reduced by cooling and temperatures are generally not permitted to exceed 35-37°C. The pH is usually in the range 4.5-4.7, when the buffering capacity of the medium is high. Higher pH values tend to lead to higher glycerol formation. When the buffering capacity is lower, the initial pH is 5.5 but this usually falls to about 3.5. During the fermentation contaminations can have serious effects on the process: sugars are used up leading to reduced yields; metabolic products from the contaminants may not only alter the flavor of the finished product, but metabolites such as acids affect the function of the yeast. The most important contaminants in distilling industries are lactic acid which affects the flavor of the product.

(iv) *Distillation*: Distillation is the separation of more volatile materials from less volatile ones by a process of vaporization and condensation. Three systems used in spirit distillation will be discussed.

   (a) **Rectifying Stills**: If the condensate is repeatedly distilled, the successive distillates will contain components which are more and more volatile. The process of repeated distillation is known as *rectification*. Rectification is done in columns, towers, or stills containing a series of plates at which contact occurs i.e., returned to the system. The alcohol-water mixture flows downwards and is stripped of alcohol by steam which is introduced from the bottom and flows upwards. Alcohol-rich distillate is withdrawn at the top of the column. Fusel oils higher alcohols separate out just above the point of entry of the mixture and are drawn off to another column. Volatile fractions are composed of esters and aldehydes. Whisky and brandy may be distilled successfully in a two-column still, but for high-strength distillates, at least three columns and possibly four or five may be required. The above description is of the modern still a modification of which is also used for producing industrial alcohol. Much older-type stills, such as are described below, continue to be used in some parts of the world.

   (b) **Pot Still**: These are traditional stills, usually made of copper. They are spherical at the lower-portion which is connected to a cooling coil. They are operated batchwise. The first portion, or 'heads' and the latter portion, 'tails', of the 'low wines' are usually discarded and only the middle portion is collected. Malt whisky, rum, and brandy are made in the post still. Its advantage is that most of the lesser aroma conferring compounds are collected in the beverage thereby conferring a rich aroma to it.

   (c) **Coffey (patent) Still**: The Coffey still was patented in 1830 and the various modifications since then have not added much to the original design of the still. Its main feature is that it has a rectifying column besides the wash column in which the beer is first distilled.

(v) *Maturation*: Some of the distilled alcoholic beverages are aged for some years, often prescribed by legislation.

(vi) *Blending*: Before packaging, samples of various batches of different types of a given beverage are blended together to develop a particular aroma.

13.3.3   The Spirit Beverages

The beverages to be discussed are whisky, brandy, rum, vodka, *kai-kai* (or akpeteshi), schnapps, and cordials.

13.3.3.1   Whisky

Whisky is the alcoholic beverage derived from the distillation of fermented cereal. Various types of whiskies are produced; they differ principally in the cereal used. Although many countries including Japan and Australia now produce whisky for export, the countries best associated with whisky are first and foremost, Scotland followed by Ireland, the USA and Canada.

In all whisky-producing countries the alcoholic content, the materials and the method of preparation are controlled by government regulations. Whiskies from various countries differ. In Scottish Malt Whisky the barley is malted just as in beer making, but during the kilning smoke from peat is allowed to permeate the green (fresh) malt, that the whisky made from the malt has a strong aroma of peat smoke, derived mainly from phenol. In the United States the principal types of whisky are rye and bourbon whiskies. Rye whisky is prepared from rye and rye malt, or rye and barley and barley malt. Bourbon whisky is prepared from preferably yellow maize, barley malt or wheat malt. A typical mash which will contain 51% corn, may have a composition of this type: 70% corn, 15% rye, and 15% barley malt.

The unmalted rye or corn is cooked to gelatinize it and hence to facilitate saccharification (or conversion to sugar) by the enzymes in the malt. The solids are not removed from the mash and the inoculating yeast sometimes contains *Lactobacillus*, whose lactic acid is said to improve the flavor of the whisky. Fermentation is usually in a two-column coffey-type still.

All whisky is matured in wooden casks for a number of years, usually three or more. They may then be blended with various types (usually controlled by law) before bottling.

13.3.3.2   Brandy

Brandy is a distillate of fermented fruit juice. Thus, brandy can be produced from any fruit-strawberries, paw-paw, or cashew. However, when it is unqualified, the word brandy refers to the distillate from fermented grape juice. It is subject to a distillation limitation of 170° proof (85%). The fermented liquor is double distilled, without previous storage, in pot stills. A minimum of two years maturation in oak casks is required for maturation.

Some of the best brandies come from the cognac region of France. Brandies produced in other parts of France are merely *eau de vie* (water of life) and are not called brandy. Certain parts of Europe (e.g., Spain, where brandy is distilled from Jerez sherry) and South America as well as the USA produce special brandies.

13.3.3.3   Rum

Rum is produced from cane or sugar by products especially molasses or cane juice. Rum production is associated with the Carribean especially Jamaica, Cuba, and Puerto Rico. It is also produced in the eastern USA. Rum with a heavy body is produced from molasses; while light rum is produced from cane syrup using continuous distillation.

During the fermentation the molasses is clarified to remove colloidal material which could block the still by the addition of sulphuric acid. The pH is adjusted to about 5.5 and a nitrogen source ammonium sulphate or urea may be added. For the heavier rums

Schizosaccharomyces pombe is used while *Saccharomyces cerevisiae* is used for the lighter types. For maturation rum is stored in oak casks for two to fifteen years.

13.3.3.4 Gin, Vodka, and Schnapps

These beverages differ from whisky, rum and brandy in the following ways:

(a) Brandy, rum and whisky are pale-yellow colored straw to deep brown by extractives from wooden casks in which they are aged and which have sometimes been used to store molasses or sherry. To obtain consistent color caramel is sometimes added. Gin, vodka, and schnapps are water clear.

(b) The flavor of brandy and whisky is due to congenerics present in the fermented mash or must. For gin, vodka, and schnapps the congenerics derived from fermentation are removed and flavoring is provided (except in vodka) with plant parts.

(c) The raw materials for their production is usually a cereal but potatoes or molasses may be used. For gin, maize is used, while for vodka rye is used. The cereals are gelatinized by cooking and mashed with malted barley. In recent times amylases produced by fungi or bacilli have been used since the flavor of malt is not necessary in the beverage. Congeners are removed by continuous distillation in multi-column stills.

In *gin* production, the grain-spirits (i.e., without the congeners) are distilled over juniper berries, *Juniperus communis,* dried angelica roots, *Angelica officinalis* and others including citrus peels, cinnamon, nutmeg, etc. Russian *vodka* is produced from rye spirit, which is passed over specially activated wood charcoal. In other countries it is sometimes produced from potatoes or molasses. *Schnapps* are gin flavored with herbs.

13.3.3.5 Cordials (Liqueurs)

Cordials are the American name for what are known as liqueurs in Europe. They are obtained by soaking herbs and other plants in grain spirits, brandy, or gin or by distilling these beverages over the plant parts mentioned above. The are usually very sweet, being required to contain 10% sugar. Some well-known brand names of cordials are Drambui, Crème de menthe, Triple Sac, Benedictine, and Anisete.

13.3.3.6 Kai-kai, Akpeteshi, or Ogogoro

Kai-kai is an alcoholic beverage widely drunk in West Africa. It is produced by distilling fermented palm-wine. It is the base for preparing some of the better known brands such as schnapps.

SUGGESTED READINGS

Amerine M.A., Berg H.W., Cruess, W.V. 1972. The Technology of Wine Making. 3rd Edit Avi Publications. West Port, USA. pp. 357–644.

Battcock, M., Azam-Ali, S. 2001. Fermented Fruits and Vegetables: A Global Perspective. FAO Agric. Services Bull. 134, Rome, Italy. pp. 96.

Doyle, M.P., Beuchat, L.R., Montville, T.J. 2004. Food Microbiology: Fundamentals and Frontiers, 2nd edit., ASM Press, Washington DC, USA.

CHAPTER 14

Production of Vinegar

Vinegar is a product resulting from the conversion of alcohol to acetic acid by acetic acid bacteria, *Acetobacter* spp. The name is derived from French (*Vin* = wine; *Aigre*-sour or sharp). With the ubiquity of acetic acid bacteria and the consequent ease with which wine is spoilt, vinegar must have been known to man for thousands of years since he apparently learnt to produce alcoholic beverages some 10,000 years ago. The Bible has many references to vinegar both in the Old and New Testaments, the best known of which, probably is: "It is consummated" which according to John (19, 29-30), was uttered by Christ before He bowed his head and died, after he had been offered vinegar while he was crucified on the cross. Vinegar may be regarded as wine spoilt by acetic acid bacteria, but for which other uses have been found.

Although acetic acid is the major component of vinegar, the material cannot be produced simply by dissolving acetic acid in water. When alcoholic fermentation occurs and later during acidifications many other compounds are produced, depending mostly on the nature of the material fermented and some of these find their way into vinegar. Furthermore, reactions also occur between these fermentation products. Ethyl acetate, for example, is formed from the reaction between acetic acid and ethanol. It is these other compounds which give the various vinegars their bouquets or organoleptic properties. The other compounds include non-volatile organic acids such as malic, citric, succinic and lactic acids; unfermented and unfermentable sugars; oxidized alcohol and acetaldelyde, acetoin, phosphate, chloride, and other ions.

14.1 USES

(i) *Ancient uses*: The ancient uses of vinegar which can be seen from various records include a wide variety of uses including use as a food condiment, treatment of wounds, and a wide variety of illnesses such as plague, ringworms, burns, lameness, variocose veins. It was also used as a general cleansing agent. Finally, it was used as a cosmetic aid.

(ii) **Modern uses**: Vinegar is used today mainly in the food industry as; (a) a food condiment, sprinkled on certain foods such as fish at the table; (b) for pickling and preserving meats and vegetables; vinegar is particularly useful in this respect as it can reduce the pH of food below that which even sporeformers may not survive; (c) It is an important component of sauces especially renowned French sauces many

Flickinger, M.C., Drew, S.W. (eds) 1999. Encyclopedia of Bioprocess Technology - Fermentation, Biocatalysis, and Bioseparation, Vol 1-5. John Wiley, New York, USA.

Gastineau, C.F., Darby, W.J., Turner T.B. 1979. Fermented Food Beverages in Nutrition. Academic Press; New York, USA. pp. 133-186.

Gutcho, M.H. 1976. Alcoholic Beverages Processes. Noyes Data Corporation, New Jersey and London, pp. 10 -106.

Hammond, J.R.M., Bamforth, C.W. 1993. Progress in the development of new barley, hop, and yeast variants for malting and brewing. Biotechnology and Genetic Engineering Reviews 11, 147-169.

Hanum, H., Blumberg, S. 1976. Brandies and Liqeuers of the World. Doubleday Inc. New York, USA.

Hough, J.S. 1985. The Biotechnology of Malting and Brewing. Cambridge University Press; Cambridge, UK. pp. 15-188.

Okafor, N. 1972. Palm-wine yeasts of Nigeria. Journal of the Science of Food and Agriculture 23, 1399-1407.

Okafor, N. 1978. Microbiology and biochemistry of oil palm wine. Advances in Applied Microbiology. 24: 173-200.

Okafor, N. 1987. Industrial Microbiology. University of Ife Press; Ile-Ife Nigeria. pp. 201-210.

Okafor, N. 1990. Traditional alcoholic beverages of tropical Africa: strategies for scale-up. Process Biochemistry International 25, 213-220.

Okafor, N., Aniche, G.N. 1980. Lager beer from Nigerian Sorghum. Brewing and Distilling International. 10, 32-33, 35.

Packowski, G.W. 1978. Distilled Alcoholic Beverages, Encyclopadeia of Chemical Technology, 3rd Edit. Wiley, New York, USA. pp. 824-863.

Singleton, V.L., Butzke, C.E. 1998. Wine. Kirk-Othmer Encyclopedia of Chemical Technology. John Wiley & Sons, Inc. Article Online Posting Date: December 4, 2000.

Soares, C. 2002. Process Engineering Equipment Handbook Publisher. McGraw-Hill.

Taylor, J.R.N., Dewar, J. 2001. Developments in Sorghum Food Technologies. Advances in Food and Nutrition Research 43, 217-264.

Vogel, H.C., Tadaro, C.L. 1997. Fermentation and Biochemical Engineering Handbook - Principles, Process Design, and Equipment. 2nd Edit Noyes.

Zeng, A. 1999. Continuous culture. In: Manual of Industrial Microbiology and Biotechnology. A.L. Demain, J.E. Davies (eds) 2nd Edit. ASM Press. Washington, DC, USA. pp. 151–164.

of which contain vinegar; (d) Nearly 70% of the vinegar produced today is supplied to various arms of the food industry where it finds use in the manufacture of sauces, salad dressings, mayonnaise, tomato productions, cheese dressings, mustard, and soft drinks. Most of the vinegar used in industry is the distilled or concentrated type (see below).

14.2   MEASUREMENT OF ACETIC ACID IN VINEGAR

Just as the alcoholic content of distilled alcoholic beverages is measured in proof, the acetic acid content is usually measured in 'grain'. Originally the strength of acetic acid was expressed in terms of the grains of sodium bicarbonate neutralized by one fluid ounce measure of vinegar and this was measured by the CO_2 evolved during the reaction of the two substances. The 'grain strength' is now measured differently and one-grain vinegar is now defined as that containing 0.1 gm of acetic acid in 100 ml at 20°C. Grain is derived by multiplying the acetic acid content (w/v) of a sample of vinegar by 10 or by moving the decimal point one place to the right. Thus, vinegar containing 8% acetic acid is 80 grain. Sometimes the percentage (w/v) is used.

14.3   TYPES OF VINEGAR

Vinegar is normally a product of two fermentations: alcoholic fermentation with a yeast and the production of acetic acid from the alcohol by acetic acid bacteria (Chapter 2). There is no distillation between the two fermentations, except in the production of spirit vinegar, which is described below. The vinegar may or may not be flavored. The substrate for the alcoholic fermentation for vinegar productions varies from one locality to the other. Thus, while wine vinegar made from grapes is common in continental Europe and other vine growing countries, malt vinegar is common in the United Kingdom; the United States on account of its great variety of climatic regions uses both malt and wine vinegars. Rice vinegar is common in the far Eastern countries of Japan and China and pineapple vinegar is used in Malaysia. In some tropical countries vinegar has been manufactured from palm wine derived from oil or raffia palm.

The composition and specifications of various types of vinegars are defined by regulations set up by the governments of different countries . In the United States, for example, vinegar should not contain less than 4.0% (w/v) acetic acid and not more than 0.5% ethanol (v/v). The various major vinegars are defined briefly

(i)   *Cider vinegar, apple vinegar*: Vinegar produced from fermented apple justice (US) and non-grape fruits.

(ii)   *Wine vinegar, grape vinegar*: Fermented grape juice malt.

(iii)   *Malt vinegar*: Produced from a fermented infusion of barley malt with or without adjuncts.

(iv)   *Sugar, glucose, dried fruits*: In the US vinegar from sugar syrup or molasses should be labeled sugar vinegar, while that from glucose (which should be dextrose-rotatory) and dried fruits should be labeled with 'glucose' or the particular fried fruit involved.

(v)   *Spirit vinegar*: Vinegar made from distilled alcohol. In the US synonyms for spirit vinegar are 'white distilled vinegar' and 'grain vinegar'. The alcohol used in the distillation is denatured for tax reasons with ethyl acetate. One gallon of ethyl acetate is usually added to 100 gal of 95% alcohol. The ethyl acetate is not deleterious and in any case is present in vinegar by the alcohol acetic acid reaction. It should be noted that in the Unites States the term 'distilled' refers to the ethanol used; in the United Kingdom, however, 'distilled' vinegar refers to a distillate of malt vinegar.

(vi) Some specialty vinegars: Specialty vinegars make up a category of vinegar products that are formulated or flavored to provide a special or unusual taste when added to foods.

Specialty vinegars are favorites in the gourmet market:

(a) Herbal vinegars: Wine or white distilled vinegars are sometimes flavored with the addition of herbs, spices or other seasonings. Popular flavorings are garlic, basil, and tarragon, but cinnamon, clove, and nutmeg flavored vinegars can be tasty and aromatic addition to dressings.

(b) Fruit vinegars: Fruit or fruit juice can also be infused with wine or white vinegar. Raspberry flavored vinegars, for example, create a sweetened vinegar with a sweet-sour taste.

(c) Balsamic vinegar: Traditional balsamic vinegar of Modena, Italy is made from white and sugary Trebbiano grapes grown on the hills around Modena. The grapes are harvested as late as possible to take advantage of the warmth of the weather. The traditional vinegar is made from the cooked grape 'must' (juice) matured by a long and slow process of natural fermentation, followed by progressive concentration by aging in a series of casks made from different types of wood and without the addition of any other spices or flavorings. The color is dark brown and the fragrance is distinct. Production of traditional balsamic vinegar is governed by the stringent standards imposed by the quasi-governmental Consortium of Producers of the Traditional Balsamic Vinegar of Modena.

(d) Raspberry red wine vinegar: Natural raspberry flavor is added to red wine vinegar, which is the aged and filtered product obtained from the acetous fermentation of select red wine. Raspberry red wine vinegar has a characteristic dark red color and a piquant, yet delicate raspberry flavor.

(e) Other specialty vinegars: Coconut and cane vinegars are common in India, the Phillipines and Indonesia with date vinegar being popular in the Middle East.

International definitions and standards are set by the joint efforts of the Food and Agriculture (FAO) as well as the World Health Organization (WHO) of the United Nations. Apart from these, various professional bodies such as the Vinegar Institute (a manufacturing association) also set standards.

14.4 ORGANISMS INVOLVED

Although vinegar had been known to man from time immemorial, like many other fermentative processes, the identity of the organism concerned is recent. Even then much more needs to be known about them, mainly because of the difficulty of cultivating them.

The bacteria converting alcohol to acetic acid under natural conditions are film-forming organisms on the surface of wine and beer. The film was known as 'mother of vinegar' before its bacteriological nature became known. The bacteria were first described as *Mycoderma* (viscous film) in 1822. Later other workers classified them in *M. vini* (forming film on wine) an *M. acetic* (forming film on beer). Pasteur confirmed that acetic acid is produced only in the presence of the bacteria, but he did not identify them. The genus name *Acetobacter* was put forward by Beijerinck in 1900. Suffice it to state that although *Acetobacter* spp are responsible for vinegar production, pure cultures are hardly used, except in submerged fermentation because of the difficulty of isolating and maintaining the organisms. The only member of the genus which is not useful, if not positively harmful in vinegar production is *Acetobacter xylinum* which tends to produce slime (Chapter 2). Recently a new species, *Acetobacter europaeus*, was described. Its distinguishing features are its strong tolerance of acetic acid of 4 to 8% in A–E agar, and its absolute requirement of acetic acid for growth.

Strains of acetic acid bacteria to be used in industrial production should a) tolerate high concentrations of acetic acid; b) require small amounts of nutrient; c) not over-oxidize the acetic acid formed; and d) be high yielding in terms of the acetic acid produced.

The biochemical processes are simple and are shown below:

$$CH_3CH_2OH + (O) \longrightarrow CH_3CHO + H_2O$$
Ethyl alcohol oxygen → Acetaldedyde Water

$$CH_3CHO + H_2O \longrightarrow CH_3CH + (OH)_2$$
→ Hydrated acetaldehyde

$$CH_3CH(OH)_2 + (O) \xrightarrow[\text{Dehydrogenase}]{\text{(Aldehyde)}} CH_3COOH + H_2O$$
Acetic acid

Theoretically, 1 gm of alcohol should yield 1.304 gm of acetic acid but this is hardly achieved and only in unusual cases is a yield of 1.1 attained. From the reactions one mole of ethanol will yield one more of acetic acid and more of water. It can be calculated that 1 gallon of 12% alcohol will yield 1 gal. of 12.4% acetic acid.

Over-oxidation can occur and it is undesirable. In over-oxidation acetic acid is converted to CO_2 and H_2O. It occurs when there is a lack or low level of alcohol. It occurs more frequently in submerged fermentations that in the trickle processes.

14.5 MANUFACTURE OF VINEGAR

The three methods used for the production of vinegar are the Orleans Method (also known as the slow method), the Trickling (or quick) Method and Submerged Fermentation. The last two are the most widely used in modern times.

14.5.1 The Orleans (or Slow) Method

The oldest method of vinegar production is the 'let alone' method in which wine left in open vats became converted to vinegar by acetic acid bacteria entering it from the atmosphere. Later the wine was put in casks and left in the open field in the 'fielding process'. A small amount of vinegar was introduced into a cask of wine to help initiate fermentation. The introduced vinegar not only lowered the pH to the disadvantage of many other organisms but also introduced an inoculum of acetic acid bacteria.

The casks were wooden and of approximately 200 liter capacity. It was never filled beyond about two-thirds of its capacity so that there was always a large amount of air available above the wine. A thick film of acetic acid bacteria formed on the wine and converted it in to vinegar in about five weeks. About 10-20% of the vinegar was drawn off at weekly intervals and replaced with new wine. The withdrawal and replenishment were done from the bottom of the cask so that the film would not be disturbed. Often a series of casks was present and the transfer was done from one cask to another.

Often due to its thickness and consequent weight and sometimes due to disturbance, the film sank. When this happened the whole process had to be restarted, since acetic acid bacteria are aerobic. Following Pasteur's (1868) suggestion, the film of bacteria often developed in wooden rafts is placed in the cask for this purpose. Later on the casks were stored, especially in the Orleans district of France, in a heated building or in an underground cellar to speed up the process. The process now derives its name from the district. The process had a number of disadvantages:

(a) It was slow in comparison with later methods, taking up to five weeks sometimes as against days, hence it is also known as the slow method.
(b) It was inefficient, yielding 75-85% of the theoretical amount.
(c) The 'mother of vinegar' usually gradually filled the cask and effectively killed the process.

Despite these disadvantages the product was of good quality and it continued to be used in many European countries long after the introduction of the Quick Process, described below. Modern vinegar production uses mainly the Trickling (quick) and submerged methods to be described below. There are fewer and fewer of the Orleans equipment in use today.

14.5.2 The Trickling Generators (Quick) Method

Credit for devising the fore-runner of the modern trickling generator is usually given to the Dutch Boerhaave who in 1732 devised the first trickling generator in which he used branches of vines, and grape stems as packing. Improvements were made by a number of other people from time to time. Later ventilation holes were drilled at the bottom of the generator and provided a mechanical means for the repeated distribution of the alcohol acetic acid mixture over the packing. The heat generated by the exothermic reaction in the generator caused a draft which provided oxygen for the aerobic conversion of alcohol to acetic acid. This latter model of the quick method (sometimes called the German method) enabled the production of vinegar in days instead of in weeks. It remained in vogue unmodified for just over a century when several modifications were introduced in the Frings method, including: (a) forced aeration, (b) temperature control, and (c) semi-continuous operation.

The modern vinegar generator consists of a tank constructed usually of wood preferably of cypress and occasionally of stainless steel. A false bottom supports the coils of birchwood shavings and separates them from the collection chamber which occupies about one fifth of the total capacity of the generator (Fig. 14.1). A pump circulates the alcohol-acetic acid mixture from the reservoir through a heat exchanger to the top of the generator where a spray mechanism distributes it over the packing in much the same way

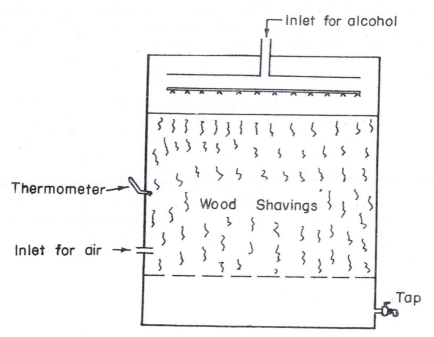

Fig. 14.1   Trickling Generator for Vinegar Manufacture

as a trickling filter functions in waste-water treatment (Chapter 29). Air is forced through the false bottom up through the set-up. The cooling water in the heat exchanger is used to regulate the temperature in the generator so that it is between 29°C and 35°C; this is determined with thermometers placed at different levels of the generator.

The top of the generator is covered but provision exists for exhaust air to be let out. Meters measure three parameters: (a) the circulation of the mash, (b) the flow of cooling water through the heat exchange, and (c) the amount of air delivered through the system. If the air flow rate is too high alcohol and vinegar are lost in effluent air.

Operation of the generator: The trickling or circulating Frings generator is reasonably efficient, achieving, when operating maximally, an efficiency of 91-92% and it is capable of producing 500–1000 gallons of 100-grain (i.e. 10%) vinegar every 24 hours. Although the wood shavings soften with age, well-maintained generators can proceed without much attention for twenty to thirty years. They are easy to maintain once airflow and recirculation rates as well as temperatures are maintained at the required level. The level of ethyl alcohol must be maintained so that it does not fall below 0.3-0.5% at any time. Complete exhaustion of the alcohol will lead to the death of the bacteria.

When wine and cider vinegar are made no nutrients need be added to the charge (i.e., the alcohol-containing material). However, when white vinegar (produced from synthetic alcohol is used) nutrients e.g. simple low concentration sugar-mineral salts solution sometimes containing a little yeast extraction may be added. Growth of the slime-forming *Acetobacter xylinum* is less with white vinegar (from pure alcohol) than with wine and cider vinegar. Generators for producing white vinegar therefore become blocked by slime much less quickly than those used for wine and cider vinegar, and can last far in excess of 20 to 30 years before the wood shavings are changed.

The finished acidity of the vinegar is about 12%; when it is higher, production drops off. In order not to exceed this level of acidity, when drawing off vinegar, the amount of alcohol in the replacement should be such that the total amount of alcohol is less than 5%.

14.5.3   Submerged Generators

With knowledge in submerged fermentation gained from the antibiotics and yeast industry it is not surprising that vinegar production was soon produced by the method. Several submerged growth vinegar generators have been described or are in operation. The common feature in all submerged vinegar production is that the aeration must be very vigorous as shortage of oxygen because of the highly acid conditions of submerged production, would result in the death of the bacteria within 30 seconds. Furthermore, because a lot of heat is released (over 30,000 calories are released per gallon of ethanol) an efficient cooling system must be provided. All submerged vinegar is turbid because of the high bacterial content and have to be filtered. Some submerged generators will be discussed below.

14.5.3.1   Frings acetator

First publicized in 1949, most of the world's vinegar is now produced with this fermentor. It consists of a stainless steel tank fitted with internal cooling coils and a high-speed agitator fitted through the bottom. Air is sucked in through an air-meter located at the top. It is then finely dispersed by the agitator and distributed throughout the liquid. Temperature is maintained at 30°C, although some strains can grow at a higher temperature. Foaming is interrupted with an automatic foam breaker. Essentially it is shaped like the typical aerated stirred tank fermentor described in Chapter 9).

It is operated batchwise and the cycle time for producing 12% vinegar is about 35 hours. Details of the parts of the Frings acetator are shown in Fig. 14.2. It is self-aspirating, no compressed air being needed. The hollow rotor is installed on the shaft of a motor mounted under the fermentor, connected to an air suction pipe and surrounded by a stator. It pumps liquid that enters the rotor from above outward through the channels of the stator that are formed by the wedges, thereby sucking air through the openings of the rotor and creating an air–liquid emulsion that is ejected outward at a given speed. This speed must be chosen adequately so that the turbulence of the stream causes a uniform distribution of the air over the whole cross section of the fermentor.

The Frings alkograph automatically monitors the alcohol content and signals the end of the batch when the alcohol content falls to 0.2% (v/v). At this stage about one third of the product is pumped out and fresh feed pumped to the original level. The aeration must continue throughout the period of the unloading and loading. The Frings Alkograph is an automatic instrument for measuring the amount of ethanol in the fermenting liquid. Small amounts of liquid flow through the analyzer continuously, at first through a heating vessel and then through three boiling vessels. The boiling temperature of the incoming liquid is measured in the first boiling vessel. While alcohol is distilled off continuously from the second and the third boiling vessel, the higher boiling point of the liquid from which ethanol has been removed is measured in the third boiling vessel. The difference in temperature corresponds to the ethanol content and is recorded

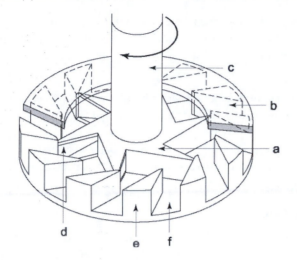

a = hollow rotor; b = stator;
c = air suction pipe; d = openings for air exit;
e = wedges to form the channels;
f = channels to form the beams of air–liquid emulsion

Fig. 14.2   The Frings Acetator

automatically. As the flow through the vessels takes some time, there is a delay of about 15 minutes between the beginning of the inflow of a sample and the appearance of the correct value on the recorder. However, because alcohol concentration is decreasing slowly during fermentation, this delay has no disadvantage on fermentation control. In more modern alkographs there is no time gap.

The more recent Frings acetators can be run on a semi-continuous basis. To carry out the single-stage semi-continuous process at a defined alcohol content, a contact in the Alkograph activates the vinegar discharge pump. As soon as a preset level has been reached, the mash pump starts adding fresh mash. This pump is controlled by the fermentation temperature in order to refill under constant temperature. The pump is stopped when the desired level is reached and an automatic cooling system is activated. A fermentation cycle takes 24 to 48 hours Since its first description, improvements and modifications have been made on the Frings acetator. One in recent operation is shown in Fig. 14.2 .

Advantages
   (a) The efficiency of the acetator is much higher than that of the trickling generator; the production rate of the acetator may be 10-fold higher than a trickling unit. Values of 94% and 85% of the theoretical have been recorded for both the acetator and the trickling filter.
   (b) The quality is more uniform and the inexplicable variability in quality noted for the trickling generator is absent.
   (c) A much smaller space is occupied (about one-sixth) in comparison with the trickling generator.

(d) It is easy and cheap to change from one type of vinegar to another.

(e) Continuous production and automation can take place more easily with Frings acetator than with trickling.

Disadvantages

(a) A risk exists of complete stoppage following death of bacteria from power failure even for a short time. Automatic stand-by generators have helped to solve this problem.

(b) It has a high rate of power consumption. Some authors have however argued that in fact in terms of power consumed per gallon of acetic produced the acetator is less power consuming.

14.5.3.2 The cavitators

The cavitator was originally designed to treat sewage: it was then modified for vinegar production. In many ways it resembled the acetator. However, the agitator was fixed to the top and finely dispersed air bubbles are introduced into the liquid. It operated on a continuous basis and was quite successful in producing cider and other vinegars as long as the grain strength was low. It was not successful with high grain vinegar and the manufacture of the 'cavitator' was discontinued in 1969. It is mentioned here only for its historical interest, although some are still being used in Japan and the US.

14.5.3.3 The tower fermentor

The tubular (tower) fermentor developed in the UK has been used on a commercial scale for the production of beer, vinegar, and citric acid. The fermentor is two feet in diameter, about 20 feet tall in the tubular section with an expansion chamber of about four feet in diameter and six feet high. It has a working volume of 3,000 liters and aeration is achieved by a stainless steel perforated plate covering the cross section of the tower and holding up the liquid. The charging wort is fed at the bottom. The vinegar overflows in a quiet Y-shaped area free of rising gas. The unit can produce up to 1 million gallons (450,000 liters) of 5% acetic acid per annum. The *Acetobacter* sp requires a month to adapt to the new system. The system can be batch, semi or fully-continuous without noticeable differences in the quality of the product.

14.6 PROCESSING OF VINEGAR

(a) *Clarification and bottling*: Irrespective of the method of manufacture, vinegar for retailing is clarified by careful filtration using a filter aid such as diatomaceous earth. Vinegar from trickling generators are however less turbid than those from submerged fermentations because a high proportion of the bacterial population responsible for the acetification is held back on the shavings. After clarification it is pasteurized at 60-65°C for 30 minutes.

(b) *Concentration of vinegar:* Vinegar can be concentrated by freezing; thereafter the resulting slurry is centrifuged to separate the ice and produce the concentrate. With this method 200° grain (i.e., 20% w/v) acetic acid can be produced. Concentration is necessitated by two considerations. One is the consequent reduction in transportation costs. The other is the need to prevent loss of activity of the vinegar when cucumbers were picked in it after first being soaked in brine.

SUGGESTED READINGS

Asai, T. 1968. Acetic Acid Bacteria University of Tokyo Press/University Park Press, Baltimore.

Ebner, H., Sellmer-Wilsberg, S. 1997. Vinegar, Acetic Acid Production. Kirk Othmer's Encyclopedia of Science and Technology. John Wiley, New York, USA.

Flickinger, M.C., Drew, S.W. (eds) 1999. Encyclopedia of Bioprocess Technology - Fermentation, Biocatalysis, and Bioseparation, Vol 1-5. John Wiley, New York, USA.

Greenshilds, R.N. 1978. In: Primary Products of Metabolism. A.H. Rose, (ed). Academic Press New York, USA. pp. 121-186.

Wagner, F.S. 2002. Acetic Acid. Kirk-Othmer Encyclopedia of Chemical Technology John Wiley & Sons, Inc. Article Online Posting Date: July 19, 2002.

Webb, A.D. 1997. Vinegar. Kirk-Othmer Encyclopedia of Chemical Technology John Wiley & Sons, Inc. All rights reserved. Article Online Posting Date: December 4, 2000.

Section E

Use of Whole Cells for Food Related Purposes

CHAPTER 15

Single Cell Protein (SCP)

The term 'Single Cell Protein' (SCP) is a euphemism for protein derived from micro-organisms. It was coined by Professor Wilson of the Massachusetts Institute of Technology to replace the less inviting 'microbial' or 'bacterial' protein or 'petroprotein' (for cells grown specifically on petroleum). The term has since become widely accepted. In the 1950s and 1960s concern grew about the 'food gap' between the industrialized and the less industrialized parts of the world, especially as there was rapid and continuing population growth in the latter. As a result of this concern, alternate and unconventional sources of food were sought. It was recognized that protein malnutrition is usually far more severe than that of other foods. The hope was that microorganisms would help meet this world protein deficiency. It was not thought that SCP would replace the need to increase proteins from plants such as oil beans or from animals such as fish. However, the limitations of conventional sources of proteins were recognized. These include: (a) possible crop failure due to unfavorable climatic conditions in the case of plants; (b) the need to allow a time lapse for the replenishment of stock in the case of fish; (c) the limited land available for farming in the case of plant production. On the other hand the production of SCP has a number of attractive features: (a) it was not subject to the vicissitudes of the weather and can be produced every minute of the year. (b) Microorganisms have a much more rapid growth than plants or animals. Thus a bullock weighing 10 hundred weight would synthesize less than 1 lb (or $\dfrac{1}{10,000}$ of its weight) of protein a day, 10 hundred weight of yeasts would produce over 50 tons (or over 100 times) of their own weight of protein a day. Furthermore, (c) waste products can be turned into food in the production of SCP.

SCP is itself not entirely lacking in disadvantages. One of the most obvious is that many developing countries, where protein malnutrition actually exists, lack the expertise and/or the financial resources to develop the highly capital intensive fermentation industries involved. But this short-coming can be bridged by the use of improvised fermentors and recovery methods which do not require sophisticated equipments. Other criticisms of SCP are that microorganisms contain high levels of RNA and that its consumption could lead to uric acid accumulation, kidney stone formation and gout. These are discussed later.

As had been stated earlier microorganisms began receiving attention as food on a worldwide basis in the late 1950s and early 1960s. Nevertheless they have for centuries been consumed in large amounts, either wholly or as part of a meal or alcoholic beverage by man, although he did not always recognize this. In many tropical countries, palm wine and sorghum beers which have high suspensions of bacteria and yeast have been consumed for centuries. Fermented milks and yoghurts which have been consumed from ancient times right up to the present day contain large amounts of bacteria and yeasts (10^{12}-10^{14} /ml). In Chad (Central Africa) the blue-green alga, *Spirulina* has been eaten for centuries as did also the Aztecs of South America.

The organized and deliberate cultivation of micro-organisms for food, however, is relatively recent. During the First World War (1914-1918) baker's yeasts, *Saccharomyces cerevisiae*, were grown on a molasses-ammonium medium. Development continued in between the wars and in the second world war (1939-1945), *Geotrichum lactis, (Odium lactis), Endomyces vernalis,* and *Candida utilis* were grown for food.

15.1 SUBSTRATES FOR SINGLE CELL PROTEIN PRODUCTION

A wide variety of substrates have been used for SCP production and include hydrocarbons, alcohols, and wastes from various sources.

15.1.1 Hydrocarbons

Patterns in the utilization of hydrocarbons by microorganisms have been summarized in the so-called Zobell's rules and shown in a modified form below:

a. Aliphatic hydrocarbons are assimilated by strains of yeasts in many genera. Other classes of hydrocarbons, including aromatics may be oxidized but are not usually efficiently assimilated, if at all.

b. n-Alkanes of chain length shorter than n-nonane are not usually assimilated, but may be oxidized. Yield factors increase but the rate of oxidation decreases with increasing chain length from n-nonane.

c. Unsaturated compounds are degraded less readily than saturated ones.

d. Branched chain compounds are degraded less readily than straight chain chemical compounds.

15.1.1.1 Gaseous hydrocarbons

Among the gaseous hydrocarbons, methane has been most widely studied as a source of SCP. Others which have been studied include propane and butane. Methane is the predominant gas in natural gas, (Table 15.1) whether such natural gas is associated with oil wells ('casinghead gas') or not. Natural gas is plentiful over the world and when present, is cheap. Indeed in many oil fields, it is flared. Perhaps its greatest advantage is the absence of residual hydrocarbon in the single cell protein produced from it, unlike the case with liquid hydrocarbons. One of its major disadvantages is that it is highly inflammable.

Table 15.1   Composition of natural gas

Gas	%
Methane	82-90
Ethane	4-8
Propane	2-3
Others	
iso-butane, n-Butane, iso-pentane,	
n-Potone, Heptanes plus CO_2,	
Nitrogen	Less than 1%

Methane is the most widely studied gaseous hydrocarbon for SCP production. Single cultures in methane are usually very slow growing. Single cell protein prodduction from methane has used continuous cultures and a mixed population of microorganisms. The advantages of a mixed methane are higher growth rate, higher yield coefficient, greater stability resistance to contaminations and a reduction in foam production. It has been suggested that the various members of a four-organism mixture had the following functions (Fig. 15.1): the unnamed methane bacterium utilizes methane slowly alone and produces methanol. *Hyphomicrobium* utilizes the methanol, whereas the other members, *Flavobacterium* and *Acinetobacter* (which do not grow on methane) remove waste products. The result is a fast growing mixture.

15.1.1.2   Liquid hydrocarbons

The major source of liquid hydrocarbons is crude petroleum. These hydrocarbons were first studied as a source of microbial vitamins and lipids. Later these studies were extended in the late 1940s to the feeding of whole paraffin-grown bacterial and yeast cells to rats. The first important move to grow cells on paraffin on a commercial scale was for 'dewaxing' i.e., removal of higher n-alkanes from crude petroleum fractions (see below). With the concern for world shortage of food, protein production soon became the goal.

Crude petroleum is highly variable in composition, differing from one part of the world to the other. However, most crude petroleum oils are made up of 90-95% hydrocarbons, which are most often saturated. During petroleum refining, the crude oil is first distilled at atmospheric pressure in a process known as 'topping'. The products of this primary distillation for various temperatures during topping and use of these fractions are shown in Table 15.2. The components left after topping contain large quantities of normal alkanes with carbon atoms longer than C_8. Such higher alkanes are crystalline solids at room temperature. It was this removal (or dewaxing of solid n-paraffins or waxes which first attracted the use of microorganisms. (After topping further distillation is done under vacuum). The petroleum hydrocarbons which have been used to grow SCP are diesel oil, gas oil, fuel oil, n-alkanes (C_{10}-C_{30} and C_{14}-C_{18}, C_{11}-C_{18}, C_{10}-C_{18}) n-hexadecane, n-dodecane.

British Petroleum (BP) pioneered the use of petroleum fractions in SCP production and by 1973 had the largest number of patents in the field. Soon after, many other oil companies and governments all over the world set-up research and pilot plants. Plans to build production plants were made by some.

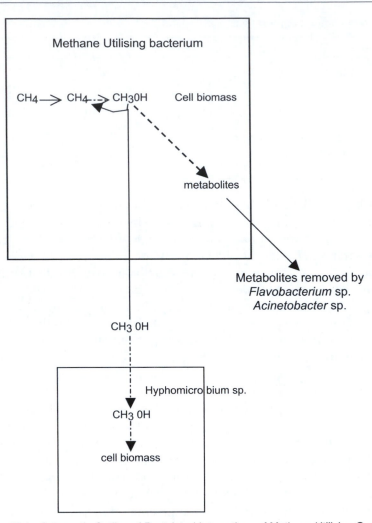

Fig. 15.1 Schematic Outline of Postulated Interactions of Methane Utilizing Organisms

Table 15.2 Products of the primary distillation of crude petroleum

Primary Fraction	Cut Point	Final Product and Boiling Range
Light gasoline	100°C	Gasoline 20-15°C
Medium naphtha	150°C	
Heavy naphtha	200°C	Kerosene 120–200°C
Light gas oil	300°C	Gas oil 200-350°C
Heavy gas oil	360°C	Blends of the appropriate
Residue		Primary fractions

In BP's plant in Scotland a material from the distillation column is passed through a molecular sieve so that only the part readily assimilated by micro-organism i.e. n-parraffins (specifically 97.5-99% $nC_{10} - C_{33}$ boiling range 30-33°C) is allowed into the fermentor under aseptic conditions. In the other plant in Lavera in the South of France, which was not aseptic gasoil was used untreated (i.e., not passed through a molecular sieve). About 10% of the gas oil was used and the remaining 90% returned to the refinery. Solvent extraction was used to remove the last traces of oil from the yeast creams (*Candida tropicalis*) including the 0.5 ton yeast lipids. By 1963, five tons dry weight of yeast was produced by this method per day. There was little difference between the two in terms of yeast composition. In terms of economics, marginal advantage accrued to the Lavera (dewaxing) process.

Since the oil boycott of 1973-1974 crude oil prices have risen sharply and the initial attraction to the use of crude oil as a substrate for SCP has been eroded. Consequently it is doubtful that the greatly raised expectations of SCP from petroleum is likely to be achieved. Indeed many of the plans announced by many oil companies for production stage fermentors were soon abandoned.

15.1.2   Alcohols

While work on SCP production from n-paraffin and gas oil was in progress, alternatives to petroleum based substrates were sought. Methanol and ethanol are such alternatives.

15.1.2.1   Methanol

Methanol is produced by the oxidation of paraffins in the gas or liquid phase or by the catalytic reduction by hydrogen of CO and CO_2, either singly or mixed. The catalysts are mixed zinc and chromium oxides. The source of the feed gas is natural gas, fermentation or fuel gas.

Methanol is suitable as a substrate for SCP for the following reasons: (a) it is highly soluble in water and this avoids the three-phase (water-paraffin-cell) transfer problems inherent in the use of paraffins; (b) the explosion hazard of methanol is minimized in comparison with methane-oxygen mixtures; (c) it is readily available in a wide range of hydrocarbon sources ranging from methane to naphtha; (d) it can be readily purified in a process which avoids the carry over of the most toxic polycyclic aromatic compounds; (e) it requires less oxygen than methane for metabolism by micro-organisms and hence a lower cooling load; (f) it is not utilized by many organisms.

The use of methanol as a SCP substrate has received attention by oil companies in Italy, West Germany, Norway, Sweden, Israel, the United Kingdom, and the United States. One of the most advanced in all these countries is the project of the Imperial Chemical Industries (ICI) which using the bacterium, *Methylophilus methylotropha* was due to start the annual production of several tons of proprietary 'Pruteen' in Billingham, the UK, using the loop fermentor, ('pressure cycle fermentor').

Over 20 species from the genera *Hansenula*, (*Hansenula polymorpha Pichia*, *Torulopsis* and *Candida* have been shown to grow on methanol.

15.1.2.2   Ethanol

Ethanol may, of course, be produced by the fermentative activity of yeasts. In the synthetic process however, it is produced by the hydration of ethylene which itself is obtained during petroleum refining from coke oven gas, the vapor-phase cracking of petroleum or the propane-butane cut of stabilizer gas. Ethylene is absorbed by concentrated H_2SO_4 to form ethyl hydrogen sulfate. The dilution of the acid with water causes hydrolysis to ethyl alcohol and H_2SO_4 The alcohol is then distilled off.

Although ethanol can be utilized ordinarily by many bacteria and yeasts, as a substrate for SCP, it is largely used by yeasts. Ethanol has the following advantages:

(a) Since it is already consumed in alcoholic beverages it is not quite as suspect a substrate for SCP as are gas oil and n-paraffins. (b) It is like methanol, highly miscible with water and hence more easily available than the three-phase paraffin system. (c) Ethanol in contrast with methane can be more safely stored and transported (d) As, unlike methanol, it is non-toxic it can be more easily handled. (e) Ethanol is partially oxidized. For these reasons, the fermentation of ethanol for SCP production requires comparatively less oxygen and hence releases considerably less heat than if it were unsaturated.

The major disadvantage in using ethanol for SCP production is that it is expensive, even when produced by the catalytic method described above. Despite this advantage yeast produced from ethanol is being produced and marketed as a flavor enhancer in baked foods, pizzas, sauces, etc., in the United States by the Amoco Oil Company in a plant which will ultimately produce 15 million lbs per annum. The yeast used is *Candida utilis*.

In Japan the Mitsubishi Oil Company has developed strains of *Candida acidothermophilum* which grow at a lower pH value and higher temperature than *Candida utilis*. These properties should help reduce costs by minimizing the need for asepticity and cooling, as also is the use of unpurified ethanol derived from the process described above. The pilot plant production is 100 tons per annum. In Spain *Hansenula anomala* is used. Ethanol-based SCP is also produced in Czechoslovakia and the USSR. In Switzerland a joint project between Nestles (the food Company) and Exxon (the US Oil Company) utilizes a bacterium *Acinetobacter caloaceticum* rather than a yeast. Unlike many other plants it is directed primarily towards human consumption hence reduction in the nucleic acid content is important.

15.1.3 Waste Products

In recent times petroleum prices have continued to soar; it is therefore unlikely that petroleum-based substrates such as synthetically produced methanol and ethanol, gas oil, etc., will be much less used in the future. Indeed many projects designed to operate on them are already being shelved. It is not however the end of the SCP story, because attention is being turned more and more to substrates derived from plants which are renewable during photosynthesis. Usually however these are obtained as waste products from various sources.

A large number of reports of SCP production from waste material lies scattered in the literature. They may be discussed under the following general headings:

(i) *Plant/wood wastes*: These are cellulose containing materials. The major difficulty with them is that cellulose is crystalline and highly resistant to fermentation without prior treatment. When lignin is present as is usually the case the resistance is even greater as it protects cellulose from direct attack. This is why wastes from manufacturing processes, such as sulfite pulping which must necessarily break down lignin, yield wastes which are comparatively easy to handle. Methods which convert lignocellulosic materials to fermentable sugars were discussed in Chapter 4.

Plant wastes containing cellulose include corn cobs, plant stems, leaves, stalks, husks, etc. For them to be used for SCP production, they usually have to be treated in some form such as ball-milling, acid, alkali, sodium chlorate or liquid ammonia treatment. The material may then be digested by a chemical means or by the use of microorganisms. Cellulosic agricultural wastes are available in large amounts all over the world; they are usually of little economic value, and are non-toxic. However, they are usually widely scattered and any process which aims at utilizing them must take into account the cost of collecting and storing them, as well as the fact that they vary widely in their content of cellulose and other materials. It is ironic that the tropical countries of the world which may be expected to have large amounts of plant wastes and which are also the areas most critically hit by protein shortage usually do not have the manpower, finance to purchase, or expertise to run, these fermentation equipments. It is encouraging that some studies aimed specifically at developing countries exist. For example the high points of the procedure being pursued by Tate and Lyle Ltd, the British sugar manufacturing Company, is the use of labor-intensive methods employing fermentor and other equipments fashioned from relatively cheap materials. Many developing countries in Africa/Asia and South American can indeed adopt these methods and produce SCP locally from agricultural wastes.

(ii) *Starch-wastes*: Starch-containing wastes from rice, potatoes, or cassava manufacturing industry are relatively easy to utilize in SCP production in comparison with cellulosic agricultural wastes. Starch hydrolysis is relatively easily achieved with either whole microbial cells or enzyme. A very interesting procedure is the Symba Process developed by the Swedish Sugar Corporation. In this process two yeasts are used symbiotically: *Endomycopsis fibuligera* hydrolyses starch to the sugars glucose and maltose with alpha and beta amylases. *Candida utilis* then utilizes these sugars for growth.

(iii) *Dairy wastes*: Whey is a by-product of the diary industry resulting from the removal of proteins (and fat) in cheese manufacture. It is a liquid rich in lactose which can be obtained in concentrated forms from cheese manufacturers and can then be suitably diluted to give the desired lactose concentration. *Saccharomyces fragilis* is grown in it for a high-quality edible food yeast. The process can be adjusted to produce either SCP or alcohol. Due to the cost of aeration, the authors recommend the concomitant manufacture of SCP and alcohol under anaerobic conditions. In the closed-loop continuous system described by the authors no effluent results.

(iv) *Wastes from chemical industries*: Various substrates from chemical industries can be utilized for SCP production, provided they contain sufficient amounts of utilizable carbon sources. Thus, *C. lipolutica* or *Trichosporon cutaneum* can be used for SCP production in oxanone water, a waste mixture of organic acids from the copralactam used for the manufacture of nylon.

(v) *Miscellaneous substrates*: Molasses the by-product of the sugar industry is a well-known raw material for microbial industries (Chapter 4). Its use for food yeast production, a form of SCP, will be described in the next chapter.

A wide variety of substrates may be, and have been, used for SCP production. These include coffee wastes, coconut wastes, palm-oil wastes, citrus waste, etc. In the study of any hitherto unexplored waste source, what is required is to determine what pretreatment, if any, is required and search for the appropriate organism which will grow in the hydrolysate.

15.2 MICROORGANISMS USED IN SCP PRODUCTION

A list of selected organisms currently used in SCP production is given in Table 15.3. Obviously that list is not exhaustive; new organisms may be discovered or new strains may be developed from existing strains.

The properties required in industrial organisms in general have been described in Chapter 1. In addition organisms to be used in SCP production should have the following properties:

(a) *Absence of pathogenicity and toxicity*: It is obvious that the large-scale cultivation of organisms which are pathogenic to animals or plants could pose a great threat to health and therefore should be avoided. The organisms should also not contain or produce toxic or carcinogenic materials.

(b) *Protein quality and content*: The amount of protein in the organisms should not only be high but should contain as much as possible of the amino acids required by man.

(c) *Digestibility and organoleptic qualities*: The organism should not only be digestible, but it should possess acceptable taste and aroma.

(d) *Growth rate*: It must grow rapidly in a cheap, easily available medium.

(e) *Adaptability to unusual environmental conditions*: In order to eliminate contaminants and hence reduce the cost of production, environmental conditions which are antagonistic to possible contaminants are often advantageous. Thus, strains which grow at low pH conditions or at high temperature are often chosen.

The heterotrophic microorganisms currently used are bacteria (and actinomycetes and fungi (moulds and yeasts); protozoa have not been used in SCP production. Of the substrates currently in use, the gaseous hydrocarbons (methane, propane, butane) are almost exclusively attacked by bacteria. Liquid hydrocarbons (n-paraffins, gas oil, diesel oil) and alcohols are utilized by both bacteria and yeasts. Of the carbohydrates sugar is readily utilized by all the classes of microorganisms; a large number of them can also utilize starch. Cellulose is not generally utilized directly by many microorganisms save after treatment. Cellulose and other materials in peanut shells, carob beans, spoiled fruits, corn and pea wastes, sugarcane bagasse, palm, cassava wastes have been used to make SCP using the moulds *Trichoderma* sp., *Glicladium* sp., *Geotrichum* sp., *Fusarium*, and *Aspergilus. Paecilomyces variotii* is used in the Pekilo sulfite liquor SCP method. Fungi have the advantage that they are lower in RNA content and are easily harvested.

15.3 USE OF AUTOTROPHIC MICROORGANISMS IN SCP PRODUCTION

Autotrophic organisms include the photosynthetic bacteria and algae. Most of the work on SCP production by autotrophic microbes seems to be limited to the algae. It does not

Table 15.3   Organisms and substrates which have been used for single cell protein production

Gaseous hydrocarbons	Bacteria	Fungi
i) Methane	*Methanomones* sp.	
	Methylococius capsulatus	
	Pseudononas sp.	
	Hyphomisobium sp. Mixed	
	Acinetobacter sp.	
ii) Propane	*Flavobacterium* sp.	
	Arthrobacter simplex	
iii) Butane	*Nocardia paraffinica*	
	Nocardia paraffinica	
Liquid Hydrocarbons:		
i) n-Alkanes (C_{10}–C_{30})	*Mycobacterium phlei*	*Candida guillermondi*
	Nocardia sp.	*Candida lipolytica*
carbon length		
ii) n-Alkanes (unspecified)		*Candida Kofuensis*
		Candida tropicals
iii) Liquified petroleum gas		*Candida lipolylica*
		Candida rigida
iv) Gas oil	*Acinetobacter*	*Candida tropicalis*
	Pseudomonas	*Candida lipolytica*
v) Diesel oil	*Acromobacter delcavate*	
Alcohols	*Methylomonas Methanolica*	*Torulopsis methansoba*
Methanol	*Methyliphilus (Pseudomonas)*	*Toralopsis methanolove*
	methylotrophus	*Candida boidini*
Ethanol	*Methylomonas clara*	*Hansemula polymorpha*
		Candid ethanomorphium
		Candida tropicalis
Plant/Wood Wastes		
Sulphite liquor	*Thermomonespore fusca*	*Paecylomyces variotii*
Cellulose pulping fines	*Brevibacterium* sp.	*Candida utilis*
Mesquite wood		
Wheat bran	*Rhodopseudomonas glatinosa*	
Wastes from carb,		*Fusarium sp*
Palms, papaya, etc.		*Aspergillus sp*
Starch Wastes		
Potato hydrolysate		*Rhodotromla rubra*
Tapioca (Cassava) starch		*Candida tropicalis*
Diary Wastes		*Endomycopsis*
		Libuligera
Whey		*Kluyveromyces fragilis*
		Trichosporon cutaneum
Sugar Wastes		
Molasses		*Candida utilis*
Chemical Industries Wastes:		
Oxanone Waste Water		*Trichosporon cutaneum*
		Candida pseudotropicalis
Waste polyethylene		

appear that photosynthetic bacteria hold out much hope for use as SCP sources because they require anaerobic conditions for photosynthesis. These conditions are difficult to provide and maintain.

The annual production of the oceans and seas of the world which harbor the bulk of the world's algae is very high – some 550×10^9 tons – about 100 tons per annum for every human being alive. Man consumes algae through fish for which the algae serve as food. The algae themselves are rich in protein and could be harvested for this purpose. However, the concentration of algae in marine water is only 3 mg/litre whereas for economic viability the harvest should be at least 250 mg/litre. This is the first reason why algae need to be cultivated and produced in high concentrations.

The second reason is that, when given adequate conditions, 20 tons (dry weight) of algae having a protein concentration of 50% can be produced per acre of pond per annum. In terms of the yield of digestible protein, this is 10-15 times greater than the soya bean and 25-50 times more than the same area planted with corn. From the view point of good energy the yield per acre of algae in terms of dietary energy is eight times as great as sugar beet, and between 22 and 45 times as great as that of corn and potato respectively.

Third, in terms of water use for the same amount of protein, much less water is required (Table 15.4).

Table 15.4   Comparison of water use for food production by some conventional crops and by algae

	Annual Protein Yield (Ib/Acre)	Annual Water Consumption Acre/Feet/Acre	Ib Protein/ Acre Food
Soya bean	576	2.0	288
Corn	240	2.0	120
Wheat	135	1.5	90
Algae	20,000	4.0	5,000

Capital investment involving various facets of the development of land, energy consumption, and manpower utilization are about the same when conventional and algal farming are compared.

Feeding tests in animals (using the green algae *Scendemus* and *Cholorella*) showed that in general algae had beneficial effects if fed to animals in small amounts at a time. Feeding in large amounts for long periods was more successful if they were supplemented by proteins from other sources. Humans consuming foods containing algae, did so for periods of up to 20 days with only minor abdominal upset due, probably, to the novelty of the foods. However, due to the peculiar taste of the algae, such foods would probably be more immediately acceptable in communities such as in Chad or in Mexico which have developed a taste for the blue-green algae *Spirulina* through generations of consumption. For the highest algal yields carbon dioxide is supplied to algae growing in day light; where natural saline water rich in bicarbonates is available, such as is found in Chad Republic or in Lake Texicoco in Mexico, supplementation with CO_2 is not necessary.

Effluents emanating from sewage treatment with their rich content of minerals would be ideal for growing algae for animal feeding. The resulting algae should be heat-treated to avoid any possibility of pathogen transmission.

With the advantages of algae cultivation mentioned earlier, particularly the comparatively low capital investment, the digestibility by ruminants and other animals the possibility of integrating waste disposal with algae production and above all the availability of sunlight and warm temperatures throughout the year, the tropics should be the place for algae cultivation for animal feeding.

As can also be seen from Table 15.5 the production cost of protein per kilogram is very favorable for SCP when compared with other protein sources.

Table 15.5   Production of various proteins (1980 figures, USA, $ dollars)

	$ kg^{-1}	% Protein	$ per kg protein
Beef	1.54	15	20.3
Pork	1.10	12	19.1
Poultry	0.66	20	6.6
Cheese	0.78	24	6.5
Soy flour	0.15	52	0.6
Peanut flour	0.15	59	0.5
Yeast from n-alkanes	0.42	53	1.4
Yeast from molasses	0.33	53	1.3
Fungi from celluloses	0.15	43	0.7
Algae	0.66	46	2.8

15.4   SAFETY OF SINGLE CELL PROTEIN

Probably on account of the novelty of SCP as food it has met with very strong opposition especially in some countries, notably Japan and Italy. The public in the former country had become aware of the health hazards of environmental pollution and in particular the 'minimata disease' which was due to the consumption of mercury from sea food contaminated with it. The government was concerned with the possibility of the presence of carcinogenic compounds in petroleum-grown SCP, with its limited and non-renewable nature and because it was not conventional. The oil companies which had been working on the SCP from petroleum-derived substrates switched over to working with non-petroleum substrates. In Italy the concern was over the safe content of nucleic acid in SCP, the polycyclic aromatic hydrocarbons, fatty acids containing odd-numbered carbon skeletons and the presence of n-paraffins carried over from protein-grown yeasts fed to farm animals. Evidence in support of the overall safety of SCP has been however, presented and it is likely that SCP will eventually receive official approval.

The two examples given above are typical of the concern shown by the public and organizations in many parts of the world, including some specialized agencies of the United Nations, namely the World Health Organization (WHO), the Food and Agriculture Organization (FAO) and the United Nations International Children's Emergency Fund (UNICEF). The concern for the nutritional completeness and toxicological safety of novel protein foods designed for developing countries (solvent and heat-extracted soy proteins, synthetic amino acids, flours from ground nut and cotton seed, fish and leaf proteins, and microbial protein, etc.) led the WHO in 1955 to form the Protein Advisory Group (PAG). WHO was joined by the two other above-mentioned bodies in sponsoring the Protein Advisory Group (PAG) in 1960. The PAG appointed a number of ad hoc working groups including an Ad hoc Working Group of SCP which was formed in 1969. The PAG on SCP concluded that low levels of residual alkanes, and the presence of odd-number fatty acids, or polycyclic hydrocarbons which are all derived from petroleum do not constitute a danger in terms of carcinogenicity or toxicity. It has also developed guidelines for the production, and nutritional and safety standards, of SCP for human consumption. The International Union of Pure and Applied Chemists (IUPAC) has a Fermentation Section which has prepared a set of standards and specifications relating to the feeding of SCP to farm animals since these are ultimately consumed by man. The two groups mentioned have similar protocols for determining safety. These include microbiological examination for pathogens and toxin producers, chemical analyses for heavy metals, nucleic acid content, presence of hydrocarbons, safety tests on animals and protein quality studies.

15.4.1 Nucleic Acids and their Removal from SCP

Apart from the fears of carcinogenicity and toxicity from petroleum derivatives mentioned above, both of which fears have been allayed in extensive studies, another area of concern in SCP feeding is the consumption of high levels of nucleic acid. Man has lost the enzyme uricase which oxidizes uric acid to the soluble and excretable allantoin. When nucleic acid is eaten by man, it is broken up by nucleases present in the pancreatic juice, and converted into nucleosides by intestinal juices before absorption. Guanine and adenine are converted to uric acid, which as had been pointed out earlier cannot be converted to the soluble and excretable allantoin. As a result when foods rich in nucleic acid are consumed in large amounts, an unusually high level of uric acid occurs in the blood plasma. Owing to the low solubility of uric acid, uricates may be deposited in various tissues in the body including the kidneys and the joints when the diseases known as kidney stones and gout may respectively result. In June, 1970 the PAG working group of SCP established the upper limit of 2 gm nucleic acid per day in addition to the quantity present in the usual diet for adults.

Some ordinary foods are high in nucleic acid (mostly RNA): liver, sardines, and fish roe (caviar) contain 2.2. and 5.7 gm of nucleic acids per 100 gm of proteins respectively. With SCP, comparable figures vary from 8 to 25. The proportion of nucleic acids in total cell content of various micro-organisms is as follows: moulds, 2.5-6%; algae, 4-6%; yeasts, 6-11%; bacteria, up to 16%.

Various ways have been devised for the removal of nucleic acids from SCP.

(a) *Growth and cell physiology method*: The RNA content of cell is dependent on growth rate: the higher the dilution rate (in continuous cultures) the higher the RNA/protein ratio. In other words the higher the growth rate the higher the RNA content. The growth rate is therefore reduced as a means of reducing nucleic acid. It must however be borne in mind that high growth is one of the requirements of reducing costs in SCP, hence the method may have only limited usefulness.

(b) *Extraction with chemicals*: Dilute bases such as NaOH or KOH will hydrolyze RNA easily. Hot 10% sodium chloride may also be used to extract RNA. The cells usually have to be disrupted before using these methods. In some cases the protein may then be extracted, purified and concentrated.

(c) *Use of pancreatic juice*: RNAase from bovine pancreatic juice, which is heat-stable, has been used to hydrolyze yeast RNA at 80°C at which temperature the cells are more permeable.

(d) *Activation of endogenous RNA*: The RNAase of the organism itself may be activated by heat-shock or by chemicals. The RNA content of yeasts have been reduced in this way.

15.5   NUTRITIONAL VALUE OF SINGLE CELL PROTEIN

The nutritional value of SCP depends on the composition of the microbial cells used especially their protein, amino acid, vitamin, and mineral contents. These to some extent also depend on the conditions of growth of the organism. The FAO has set up reference values for the amino acid content of proteins. On this basis, SCP derived from bacteria and yeasts is deficient in methionine. Glycine and methionine are sometimes deficient in molds. These can be improved by supplementation with small amounts of animal proteins.

SUGGESTED READINGS

Boze, H., Moulin, G., Galzy, P. 1991. Production of Microbila Biomass. In: Biotechnology G., Reed T.W. Nagodawitana, (eds). VCH Weinheim Germany. pp. 167-220.

Caron, C. 1991. Commercial Production of Bakers Yeast and Wine Yeast In: Biotechnology G. Reed, T.W. Nagodawitana (eds). VCH Weinheim Germany. pp. 321-350.

Flickinger, M.C., Drew, S.W. (eds) 1999. Encyclopedia of Bioprocess Technology - Fermentation, Biocatalysis, and Bioseparation, Vol 1-5. John Wiley.

Litch Field, J. 1994. Foods, Nonconventional. Kirk-Othmer Encyclopedia of Chemical Technology 10.

Scrimshaw, N.S., Murray, E.B. 1991. Nutritional Value and Safety of "Single Cell Protein". In: H.J. Rehm, (ed) Biotechnology. 2nd Edit VCH Weinheim Germany. pp. 221-240.

CHAPTER 16

Yeast Production

Yeasts have interacted with man from time immemorial – from the time when he first learnt that fruit juices developed into intoxicating drinks and that the dough produced from his ground cereal can be leavened, although he did not understand these two phenomena. Today yeasts which are produced and used in all the six continents of the world form the single most produced micro-organisms in terms of weight. The estimated world production (excluding the Eastern European countries) for 1977 is given in Table 16.1, which are clearly underestimates in such areas as Middle East/Asia and Africa. In the United States.

It would be safe to double the figures in the table as today's estimated production. Baker's yeast is manufactured by six major companies in the United States. These companies are Universal Foods (Red Star Yeast), Fleischmanns, Gist-brocades, Lallemand (American Yeast), Minn-dak, and Columbia. There are 13 manufacturing plants owned by these companies. Table 16.2 lists the locations of these plants by manufacturer.

Table 16.1   Estimated yeast production (dry weight, tons) in 1977

Region	Baker's Yeast	Food and Fodder Yeast
Europe (excluding Eastern)	74,000	160,000
North America	73,000	53,000
Middle east/Asia Countries	15,000	25,000
United Kingdom	15,000	-
South America	7,500	2,000
Africa	2,700	2,500

The purpose for which yeasts are used and the types of yeasts employed for each purpose are given in Table 16.2. Of these, the production of baker's yeasts has received the greatest attention, followed by food and fodder yeasts. Due to recent interest in the production of single cell protein, food and fodder yeasts may become as important as baker's yeasts in terms of total quantity produced. The chapter will discuss the large-scale production of baker's, food and fodder yeasts.

16.1   PRODUCTION OF BAKER'S YEAST

The use of yeasts in bread making is an ancient art, although man did not always recognize the mechanism of the rise of dough. It is of interest to give a brief historical

Table 16.2   Yeast manufacturing plants in the United States

Company	Location of Plant
Lallemand (American yeast)	Baltimore, Maryland
Columbia	Headland, Alabama
Fleischmanns	Gastonia, North Carolina
	Memphis, Tennessee
	Oakland, California
	Sumner, Washington
Gist-brocades	Bakersfield, California
	East Brunswick, New Jersey
Minn-dak	Wahpeton, North Dakota

Table 16.3   Major uses of yeasts

Use	Yeasts involved
1. Bakery	*Saccheromyces cerevisiae*
2. Beer brewing	*Sacch. uvarum (Sacch cerevisiae) Scch. cerevisiae*
3. Food yeasts and feed yeasts	*Candida tropicalis Candida pseudotropicalis*
	Candida utilis Sacch. cerevisiae
	Kluyveromyces fragilis (Sacch. fragilis
4. Feed yeasts	*Candida lipolytica*
5. Wine making	*Saccharomyces cerevisiae var. elliposoides*
6. Wine making (sparkling wines)	*Sacch. bayanus*
7. Industrial alcohol/spirits	*Sacch. cerevisiae (Sacch. fragilis with whey)*
8. Yeast-products (antolysates, biochemicals)	*Sacch. cerevisiae*

account of the development of the yeast industry. The dough of leavened bread whose antiquity is testified by biblical records, was probably raised by a mixture of yeasts and lactic acid bacteria. A small piece of successful dough was used as the inoculum for the next batch, providing a type of early continuous culture. This system of course has been largely abandoned except in a special type of sour bread produced in San Francisco, California, United States.

From about the Middle Ages, bakery yeasts were obtained from winemaking and brewing. But the quality of the yeast was variable and in the case of yeast obtained from beer the product was bitter because of the hops in the beer. This period lasted until the latter part of the 19th century when the work of Pasteur from 1855 to 1857 elucidated the nature of yeasts.

The first major step in the development of baker's yeast technology can be said to be the so-called Vienna process introduced about 1860 in which grain mash meant for anaerobic alcohol production was gently aerated so that a good quantity of yeast was obtained. Thus two birds were killed with one stone – yeasts and alcohol being obtained in one operation even if the quantities were less than would be optimal if either one or the other alone were sought. The work of Pasteur later led to more vigorous aeration, thus yielding more cells and less alcohol. As a result of grain shortages resulting from World War I a shift was made from the use of grain to the use of beet molasses, supplemented by ammonia and phosphate.

The next major step in the development of baker's yeast technology was the introduction of *fed-batch* or incremental addition of nutrients rather than the introduction of all nutrients at the beginning, as is the case in the classical batch method. The essence of this system known as the *Zulaut* method is still used today in baker's yeast manufacture and ensures that an excess of molasses sugar which might lead to alcohol production is avoided.

Another important development, the production of dried active yeast, was necessitated by the need to provide troops fighting in far off lands a means of producing bread, instead of the compressed yeast normally used in temperate countries.

Today's production methods for baker's yeast do not allow alcohol production because of the vigorous aeration used. Furthermore the yield has increased from 3% in the mid-19th century through 13% early in this century to the present-day yield of over 50% dry weight of yeast.

16.1.1   Yeast Strain Used

Non-sporulating 'torula' yeasts have occasionally been used for baking; nowadays however specially selected strains of *Saccharomyces cerevisiae* are used. For some time two strains of baker's yeasts were available: one was highly active but had poor stability during storage; the other had poor activity but was highly stable in storage. Successful breeding program were then undertaken to produce new strains from them.

New large-scale factory process for bread-making used in Western countries (the Chorleywood Bread Process) involving sophisticated plants and computerization have led to new demands on traditionally used yeasts. These new demands include the fermentation of more complex sugars, high initial fermentation ability, faster adaptation to maltose fermentation, and ability to reconstitute rapidly when prepared in the active dry form. Yeast strains used for the modern fast-rising dough have been developed with the following traditional and new physiological properties in mind.

(a) ability to grow rapidly at room temperature of about 20-25°C;

(b) easy dispensability in water;

(c) ability to produce large amounts of CO_2 in flour dough, rather than alcohol;

(d) good keeping quality, i.e., ability to resist autolysis when stored at 20°C;

(e) high potential glycolytic activity;

(f) ability to adapt rapidly to changing substrates;

(g) high invertase and other enzyme activity to hydrolyze the higher glucofructans rapidly;

(h) ability to grow and synthesize enzymes and coenzymes under the anaerobic conditions of the dough;

(i) ability to resist the osmotic effect of salts and sugars in the dough;

(j) high competitiveness i.e. high yielding in terms of dry weight per unit of substrate used.

In many Eastern European countries no special yeasts are produced to cope with the newer baking techniques mentioned above. Baking yeasts are not therefore produced specially and a type of Vienna process is used, the yeast being obtained from grain mash fermented for alcohol distillation.

16.1.2   Culture Maintenance

The specially selected baking strains of *Saccharomyces cerevisiae* are apt to mutate and therefore proper storage is most important. Of the various methods used, storage in liquid nitrogen and the oil culture method in which sterile oil is placed over a slant of yeast and refrigerated at 4°C are most widely used. Freeze drying is not highly used as it induces loss of viability in many yeasts and a tendency towards mutation.

16.1.3   Factory Production

16.1.3.1   Substrate

The substrate usually used for baker' yeast production is molasses. Where these are not available or are too expensive any suitable sugar-containing substrate e.g. corn steep liquor may of course be used. In the Soviet Union for example sulphite liquor is used for both alcohol and baker's yeast production. Ethanol has been used in laboratory studies, but is yet to be used on a large scale.

Beet and cane molasses, when they are simultaneously available, are treated separately: clarified, pH adjusted and sterilized. They are then mixed in equal amounts so that the nutritional deficiency of one type is made up by the other (Chapter 4). Cane sugar is particularly richer in biotin, panthothenic acid, thiamin and magnesium and calcium; while beet molasses is much richer in nitrogen. Molasses composition is however not constant and varies with the geographical area of growth, the factory extraction of the sugar and other factors. When only one type of molasses is available, deficiencies are made up by adding appropriate nutrients.

The molasses is clarified to remove inert colored material arising from colloidal particles and which can impart undesirable color to the yeast. Clarification may be achieved by precipitation with alum or calcium phosphate or by poly-electrolyte flocculating agents such as alginates and polyacrylamides. Clarification also helps reduce foaming. 'Sterilization' is achieved by heating at 100°–110°C for about an hour, after the pH has been adjusted to pH 6-8 to prevent caramelization of the sugar.

Phosphorous, ammonium and smaller amounts of magnesium, potassium, zinc, and thiamin are added for maximum productivity to the mixed molasses. Antifoam is sometimes added.

16.1.3.2   Fermentor processes

The fermentor for baker's yeast propagation is nowadays made of stainless steel. The trade fermentor (i.e. the final fermentor) may be anything from 75 to 225 cubic meters. Of this about 75% is occupied by the medium, the unused space being allowed for foaming. The typical stirred tank fermentor with agitator baffles and sparging is not often used in yeast growth because of the high initial and operating cost. Generally, baker's yeast fermentors are aerated only by spargers which are so arranged that large volumes of air pass through per unit time: about one volume of air per volume of broth per minute. Spargers of different types are available. It is most important that aeration be high and constant. When the oxygen falls below 0.2 ppm anaerobic conditions set in and alcohol is formed.

The aeration through sparger holes is started as soon as mixing begins in the steam sterilized fermentor. Water, mineral nutrients, yeasts and the blended molasses containing 1% glucose are mixed. The amount of blended molasses added is calculated so that the total sugar in the fermentor does not exceed 0.1%. Molasses is added incrementally during the course of the fermentation as it is used up by the yeast, beyond the 0.1% ceiling. The pH is maintained at pH 4-6 by the addition of alkali and the temperature at 30°C by cooling. The amount of molasses to be added at predetermined intervals is arrived at by experimentation. Automatic sensoring and self-adjusting equipment for temperature pH, aeration, sugar, etc., are built into some modern fermentors. Large amounts of heat are evolved and the cooling of the fermentor is very important. 1-lb dry wt. of yeast would require 4.3 lb of molasses, 0.9 lb of ammonia, 0.3 lb of $NH_4H_2PO_4$, 1.1 lb of $(NH_4)_2SO_4$ and 60 lb of air. Continuous fermentation has not been widely used in baker's yeast production. It is used (see below) for feed yeast production from sulfite liquor.

16.1.3.3 Harvesting the yeast

The period of fermentation in the trade or production fermentor varies from 10 to 20 hours depending on how much yeast is pitched into it; cells form from 3.5% to 5% dry wt. of the broth. In some processes aeration is allowed to continue for 30-60 minutes at the end of the feeding to allow unused nutrients to be used up, budding cells to divide so that most cells are 'resting' at the beginning of the budding cycle. This ensures that the cells divide somewhat 'synchronously' when growth resumes.

The fermentation broth is cooled and cells concentrated in centrifugal separators; they are washed by resuspension in water and centrifugation until they are lighter in color. The yeast cream resulting from this treatment contains 15-20% yeast cells. It is further concentrated by passing over a rotary vacuum filter or through a filter press. Sometimes the Mautner process is used to ensure a friable dry cream during vacuum filtration. This latter process consists of adding before filtration 0.2-0.6% (w/v) sodium chloride, which causes cell shrinking by osmosis. Excess salt is removed during filtration by spraying water over the filtered yeast, so that the cells swell again. The resulting product has a dry matter content of 28-30%.

The yeast may then be packaged as compressed yeast or active dry yeast. It may also be converted into *dried yeast* for human or animal feeding as described further on in this chapter.

16.1.3.4 Packaging

Baker's yeasts may be packaged as moist (compressed) yeasts or as dried active yeast.

(i) *Compressed yeast*: The yeast product obtained after harvesting, is mixed with fine particles of ice, starch, fungal inhibitors and processed vegetable oils (e.g. glyceryl monostearate) which all help to stabilize it. It is then compressed into blocks of small (1-5 lb) blocks for household use or large (up to 50 lb) for factory bakery operations, stored at – 7 to 0°C and transported in refrigerated vans.

(ii) *Active dry yeast*: Dry yeast is more stable in that it can be used in areas or countries where refrigeration is not available. In many developing countries baker's yeast is imported from abroad in the form of active dry yeast. For active dry yeast production special strains better suited for use and dry conditions may be used. It has been found that when regular strains are used they perform better as dry yeasts when they are subjected to a number of treatments. These treatments include raising the temperature to 36°C (from about 30°C) towards the end of the fermentation, addition of alcohol-containing spent broth (resulting from centrifugation or finished yeast fermentation), synchronization of budding by alternate feeding and starving. The reason for the benefit is not known.

Yeast cream of 30-38% content from filter pressing is extruded through a screen to form continuous thread-like forms. These are then chopped fine and dried, using a variety of driers: tray driers, rotary drum driers, or fluidized bed driers. The final product has a moisture content of about 8% and may be packaged in nitrogen-filled tins. Sometimes anti-oxidants may be added to the yeast emulsion to further ensure stability.

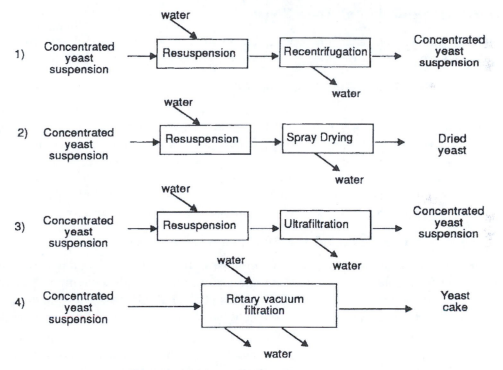

Fig. 16.1 Various Methods for Packaging Yeasts

16.2 FOOD YEASTS

Yeasts are used for food by man for the following reasons: to provide protein; to impart flavor and to supply vitamins especially B-vitamins. Food yeasts are sometimes prescribed medically when a deficiency of B-vitamins exists in a patient. Food yeasts have several synonyms: dried yeast, inactive dried yeast, dry yeast, dry inactive yeast, dried torula yeast, *Saccaromyces siccum*, sulfite yeast, wood sugar yeast, and xylose yeast. The link between wood and the production of yeast for animal consumption is shown by the last three names.

As food yeasts are consumed by man, stringent standards are imposed on the product by governmental agencies, professional bodies and manufacturers. The standards of the International Union of Pure and Applied Chemists (IUPAC) will be quoted here only as an example, because they are the most comprehensive. The IUPAC requires that any organism to be used as a food yeast belong to the family *Cryptococcaceae,* should be unextracted, should have a fat content of not more than 20%, should contain no inert fillers not indigenous to the yeasts, and should be free of *Salmonella.* The IUPAC has also set upper limits for bacterial and fungal counts, lead, arsenic and lower limits for protein, thiamin, riboflavin and niacin.

Too high a consumption of yeasts is detrimental to health because of the high RNA content of yeasts, which the kidneys are unable to dispose of. This was discussed in the previous chapter.

16.2.1   Production of Food Yeast

While baker's yeasts are usually produced from molasses using special strains of *Sacch. cerevisiae* food yeasts are produced from a wide variety of yeasts and substrates.

16.2.1.1   Yeasts used as food yeasts

Yeasts used as food yeasts are *Saccharomyces cerevisiae, Saccharomyces carlbergiensis, Saccaromyces fragilis, Candida utilis,* and *Candida tropicalis.* Only *Saccharomyces fragilis* (imperfect stage *Candida pseudotropicalis*) can utilize lactose hence it is used for the fermentation of whey. Ethanol may be used as substrate for food yeast production; it is however used only by *Saccharomyces fragilis* and *Candida utilis. Candida utilis* is the most versatile of all the yeasts and will utilize a wider range of carbon and nitrogen sources than any other, hence it is most widely used in food yeast preparations.

16.2.1.2   Substrates used for food yeast production

The most commonly used substrates are molasses, sulphite liquor, wood hydrolysate, and whey. Since interest developed in single cell protein other unconventional sources have been developed. These include hydrocarbons, alcohol and wastes of various types. These have been discussed in the previous chapter. Only molasses, sulphite liquor, wood hydrolysate and whey will be discussed.

(i) *Molasses*: Bakers yeast grown on molasses as described above may, after separation from the spent liquor by centrifugation, be dried to yield food yeast. Drum-drying, spray-drying or fluidized bed drying may be used to reduce the moisture content to only about 5%. Sometimes food yeast is grown on molasses for that purpose per se. Thus *Candida utilis* is grown fed-batch in Taiwan in Waldhof fermentors. The fed-batch method using molasses is also used in South Africa.

   Recently food yeasts using *Candida utilis* in continuous culture in molasses has been grown in Cuba and Eastern Europe.

(ii) *Sulfite liquor*: The impetus to produce food and fodder yeast from sulfite liquor (Chapter 4) derived from an attempt to reduce the pollution which would arise if the wastes containing fermentable substrates were discharged directly into a stream. The use of continuous fermentation was attractive because the sulfite is produced almost continuously in the operation of the pulp factory. In general a Waldhof-type fermentor is used for the continuous production of yeasts from sulfite waste. Liquors from various sources are usually blended. Thereafter, the sulfite containing compounds are removed either by precipation with lime, by aeration or by passing steam through it (steam stripping). The pH is adjusted from about 2 to 5.5 using ammonia. The lowest pH consistent with high yield is usually preferred in order to lessen the chances of contamination.

Ammonium, phosphate and potassium are monitored and supplied continuously. The versatile and hardy yeast *Candida utilis* is usually used so that biotin is not added. The yeast is harvested continuously and recovered by removing liquor at the same rate as it is introduced. The effluent liquor containing about 1% cell is concentrated to an 8% concentration by centrifuging. It is usually washed with water by diluting and centrifuging to remove lignosulfnic acid. Yield from sulfite liquor, whose assimilable matter content is usually low may be increased by the addition of new carbon sources e.g. acetic acid and ethanol. The liquor may be re-hydrolysed with H_2SO_4 thereby increasing the sugar content from 4% to about 24%. In some cases, addition of nutrients to the liquor e.g. yeast hydrolysate or corn steep liquor leads to an increased yield of about 5%. Simultaneously, more efficient organisms are usually also sought.

(iii) *Production of food yeast from whey*: The effluent which drains from the coagulum from milk during cheese manufacture is known as whey. It contains approximately 4% sugar (lactose), 1% mineral and some of the lactic acid which enabled the coagulation of the milk protein. In countries where a lot of cheese is produced, whey is a waste product but it is sometimes turned into good use in the production of alcohol or yeasts. Very few yeasts metabolize lactose. Those which do include *Saccharomyces lactis, Kluyveromyces fragilis* and its imperfect or asporogeneous stage *Candida pseudotropicalis.* The whey is diluted, fortified with ammonia, phosphate, minerals, yeast extract and then pasteurized at 80°C for about 45 minutes. It is then inoculated with yeasts at pH 4.5 at an incubation temperature of 30°. Any of the above yeasts could be used but in the United States the preference is for *K. fragilis.* In many establishments the fermentation is continuous and sugar, pH and minerals are monitored automatically. The yeast is recovered by centrifugation and may be drum or spray dried.

16.3 FEED YEASTS

Feed yeasts are the same as food yeasts described above. The only difference is that less rigid standards are imposed on the production of feed yeasts. Thus, feed yeasts intended for animal feeding are usually obtained by drying out the whole fermentation broth, often without washing.

Several thousands tons of yeasts are recovered from breweries around the world annually. To be used as food yeasts, such yeast is 'debittered' of hop resins by repeated washing with dilute alkali until the bitterness no longer exists. It is then slightly acidified to about pH 5.5. Cells are recovered by centrifugation and spray – or drum-dried.

16.4   ALCOHOL YEASTS

Alcohol yeasts are those to be used in beer brewing wineries and distilleries for spirits of industrial alcohol. In the production of alcohol yeasts, the aim is cell production. The methods are generally similar to those already described for baker's yeasts. Beginning from a lypholized vial or tube, contamination is checked in a plate. A single colony (or prefereably a single spore by micromanipulation) is picked and multiplied in sequentially increasing amounts.

The yeasts used are specially selected strains of the following:

Brewing: *Saccaromyces cerevisiae, Saccharomyces uvarum carlbergensis S. uvarum.*

Wine: *Saccharomyces cerevisiae, Sacch. bayanus, Sacch. beticus, Sacch. elipsoides.*

Distillery Yeasts: *Saccharomyces cerevisiae.*

The medium used in the multiplication of the yeast is made of materials to be found in the final fermentation. Thus for growing brewery yeasts wort is used, for distiller's yeast a rye-malt medium is used, and for wine grape juice is used.

Alcohol yeasts are usually recovered and reused for several rounds of fermentation before being discarded.

16.5   YEAST PRODUCTS

Various products used in the food, pharmaceutical and related industries may be produced from yeasts.

Yeast extracts are used in the preparation of soups, sausages, gravies, to which they impart a meaty flavor. A well-known example is marketed in certain parts of the world as 'marmite'. The extracts may be obtained by autolysing the yeasts and thereafter spray-drying or drum-drying with or without extracting soluble materials from the autolysate.

The extract may also be obtained by hydrolyzing the yeast cells in acid solution. It is neutralysed with sodium hydroxide, filtered, decolorized through charcoal and concentrated to a syrup or spray-dried. Yeast products are usually fortified with the flavoring compound, mono-sodium glutamate, extracts of animal or vegetable protein or with yeast cells.

Yeast extracts are consumed for dietary purposes on pharmaceutical grounds as a source of vitamins, mainly vitamin B_{12}.

SUGGESTED READINGS

Boze, H., Moulin, G., Galzy, P. 1991. Production of Microbila Biomass. In: Biotechnology G. Reed, T.W. Nagodawitana, (eds). VCH Weinheim Germany. pp. 167-220.

Burrows, S. 1979. In: Microbial Biomass. A.H. Rose, (ed.) Academic Press, New York, USA. pp. 32-64.

Caron, C. 1991. Commercial Production of Bakers Yeast and Wine Yeast In: Biotechnology G. Reed, T.W. Nagodawitana, (eds). VCH Weinheim Germany. pp. 321-350.

Flickinger, M.C., Drew, S.W. (eds) 1999. Encyclopedia of Bioprocess Technology - Fermentation, Biocatalysis, and Bioseparation, Vol 1-5. John Wiley, New York, USA.

Scrimshaw, N.S., Murray, E.B. 1991. Nutritional value and Safety of "Single Cell Protein". In: Biotechnology G. Reed, T.W. Nagodawitana, (eds). VCH Weinheim Germany. pp. 221-240.

CHAPTER 17

Production of
Microbial Insecticides

Insects are major pests of crops. Enormous losses occur when they attack various plant parts, often transmitting disease in the process. Even after harvest insects attack stored foods; this attack of stored foods is not limited to plant foods, but also includes animal foods such as dried fish. Besides the loss they cause in agriculture and food, insects are also vectors of various animal and human diseases.

In modern times insects have been controlled mainly with the use of chemicals. Over the past decade or so there has been a move away from the sole use of chemical control, and towards integrated control, which employs other methods as well as chemical control. The reasons for this include non-specificity of chemical insecticides leading to the destruction of pests as well as their natural predators, resistance to chemical insecticides, concern for the environment and human health since the insecticides enter drinking water from soil, and since some are toxic or carcinogenic. Finally due to increased cost of petroleum on which many of these insecticides are based, their cost has also increased.

17.1 ALTERNATIVES TO CHEMICAL INSECTICIDES

The alternatives to the use of chemicals include the following:

(a) *Predators*: Among vertebrates one of the best known is the use of fish especially *Gambusia affinis* to eat mosquito larvae. Invertebrate predators include other larger insects e.g. wasps, while plant predators include *Utricularia* (a bladder wort).

(b) *Genetic manipulations*: These include the production (by chemicals or by irradiation) of large numbers of sterile males, whose mating does not result in fertile eggs.

(c) *The use of hormones or hormone analogs*: Pheromones are synthetic compounds which act as sex attractants. The insects attracted are destroyed.

(d) *The use of pathogens*: Pathogens of insects are found among bacteria, fungi, protozoa, viruses and nematodes. The idea of using pathogens to control insects originated from studies of the diseases of the silkworm *Bombyx mori*. The pioneer work of Bassi was followed by those of Le Conte, Pasteur, Hagen until Metchnikoff actually tested the control of sugar beet pests with the fungus *Metarrhizium anisopliae* in South Russia.

17.2 BIOLOGICAL CONTROL OF INSECTS

Biological control has been studied or practiced to a large extent in relation to agriculture, food production and forestry. Its study and use in the control of insect vectors of disease such as mosquitoes has been small in comparison. In 1976 the World Bank in collaboration with the United Nations Development Programme (UNDP) and the World Health Organization (WHO) put into operation a Special Programme for Research and Training in Tropical Diseases. The diseases were malaria, trypanosomiasis, filariasis, leishmaniasis, schistosomiasis and leprosy. Of these the first four are transmitted by insect vectors. The WHO which administered the Programme also studied biological control with respect to the insect-borne disease and ranked the organisms to be used in order of effectiveness from time to time (Table 17.1). The organisms included bacterial, fungi, nematodes, and fish, and targets are mainly mosquitoes-vectors of malaria and yellow fever and black flies *(Simulium* spp), vectors of oncocerchiasis (river blindness). For agricultural biological control however some viruses pathogenic to insects are also used (Table 17.2). In this chapter only control using microorganisms will be discussed.

Table 17.1 Biological control agents as ranked in order of priority by the special program of the World Health Organization 1980

Priority 1	*Bacillus thuringiensis,* serotype H-14 (bacterium)
Priority 2	*Bacillus sphaericus,* strain 1593 (bacterium)
	Culicinomyces sp. (fungus)
	Poecilia reticulata (fish)
	Romanomermis culicivorax (nematode)
	Toxorhynchites (predatory mosquitoes)
	Zacco platypus (fish)
Priority 3	*Aphanius dispar* (fish)
	Coelomomyces (fungi)
	Lagenidium (fungus)
	Leptolegnia (fungus)
	Metarrhizium anisopliae (fungus)
	Parasitoids in general (insects)
	Romanomermis iyengari (nematode)
	Stenopharyngodon idella (fish)
Priority 4	**Group A**
	Aplocheilus (fish)
	Baculoviruses (viruses)
	Dimorphic Microspordia (protozoa)
	Dugesia (Planaria)
	Lutzia (predatory mosquito)
	Octomyomermis muspratt (nematode)
	Protozoa of snails (protozoan)
	Tolypocladium
	Group B
	Entomophthorales (fungi)
	Nosema algerae (protozoan)
	Vavraia culicis (protozoa)

Table 17.2 Pathogenic viruses found in insects

	Family	Nucleic Acid[a]	Particle Symmetry	Vertebrate and plant viruses resembling families of insect viruses	
				Vertebrate viruses	Plant viruses
1.	Baculoviridae (Baculovirus groups A,B,C)	DNA	Rod (occluded)	None	None
2.	Poxviridae (Entomopox viruses)	DNA	'Brick' (occluded)	Orthopoxvirus Avipoxvirus Capripoxvirus Leporipoxvirus Parapoxvirus	None Reoviruses
3.	Reoviridae (Cytoplasmic polyhedrosis viruses)	dsRNA	Isometric (occluded)	Reovirus Orbivirus	Plant
4.	Irodioviridae (Iridovirus)	DNA	Isometric	African Swine Fever Frog Viruses 1-3 Lumphocystis virus	Fungal Algal
5.	Parvoviridae (Densovirus)	ssDNA	Isometric	Parvovirus Adeno-associated Group	None
6.	Picornaviridae (Enterovirus; unclassified groups)	RNA	Isometric	Enterovirus	Small RNA Viruses (single polypeptide)
7.	Rhabdoviridae (Sigmavirus)	RNA	Bullet/ bacciliform	Vesiculovirus Lyssavirus	Plant rhab-dovirus

[a]ds RNA, double stranded RNA, ss, single stranded DNA

17.2.1   Desirable Properties in Organisms to be Used for Biological Control

The following are desirable in microorganisms to be used in the biological control of insects:

(a) The agent should be highly virulent for the target insect, but should kill no other insects.

(b) The killing should be done quickly so that in the case of crops, damage is kept as low as possible, and in the case of vectors of disease before extensive transmission of the disease occurs.

(c) The killing ability should be predictable.

(d) The agent should not be harmful to man, animals or crops; in other words it should be safe to use.

(e) It should be technically amenable to cheap industrial production.

(f) When produced, it should be stable under the conditions of use such as under the high temperature and ultra violet light of ordinary sunlight.

(g) It should be viable over reasonably long periods to permit storage and transportation as necessary.

(h) It should ideally persist or recycle and/or be able to search for its host.

17.2.2 Candidates Which have been Considered as Biological Control Agents

(i) *Bacteria* A large number of bacteria are pathogenic to insects including *Bacillus* spp., *Pseudomonas* sp. *Klebsiella* sp., *Serratia marcescens*. In practice, spore formers have been developed commercially because they survive more easily in the environment then vegetative cells, but especially because they are amenable mass production. The four bacilli which have been produced for control purposes are:

(a) *Bacillus thuringiensis*: *B. thuringiensis* (commonly known as 'Bt') is an insecticidal bacterium, marketed worldwide for control of many important plant pests–mainly caterpillars of the Lepidoptera (butterflies and moths) but also mosquito larvae, and simuliid blackflies that vector river blindness in Africa. Bt products represent about 1% of the total 'agrochemical' market (fungicides, herbicides, and insecticides) across the world. The commercial Bt products are powders containing a mixture of dried spores and toxin crystals. They are applied to leaves or other environments where the insect larvae feed. The toxin genes have also been genetically engineered into several crop plants. The method of use, mode of action, and host range of this biocontrol agent differ markedly from those of *Bacillus popilliae*.

It is a complex of several organisms regarded by some as being variants of *B. cereus*. There are 19 serotypes based on some flagellar or H-antigens. Serotype H3 and H3A are used in the United States on alfalfa, cotton, tobacco, spinach, potatoes, tomatoes, oranges, and grapes. Serotype H14 attacks mosquitoes and blackflies and will be discussed below. *Bacilus thuringiensis* produces at least three toxins, a Phospholipase C, a water-soluble heat stable B-exotoxin potentially toxic to mammals, and a crystalline, d-toxin or the parasporal body which is enclosed within the sporangium (this will be discussed further below).

The crystalline d-toxin is the active principle against most insects. The spores and crystals are released into the medium in most strains of *B. thuringiensis* following the lysis of the sporangium.

(b) *Bacillus moritai*: This is used in Japan for the same purpose as *B. thuringiensis* serotypes H3 and H3A.

(c) *Bacillus popilliae*: This is an obligate pathogen of the Japanese beetle *Popilla japonica* against which it is used. Since it is an obligate parasite it is produced in the larvae of the beetle.

(d) *Bacillus thuringinensis var. israelensis* (also known as serotype H14). This was isolated in 1976 by Goldberg and Margalit from a mosquito breeding site in Israel. It has proved very effective in killing mosquito larvae and the black fly (*Simulium* spp). So promising is it from results of various projects sponsored by the Special Program of the WHO that it was expected that it would be produced on a large scale in the US and Europe and probably on smaller scales in tropical countries. It has a (nearly 100%) kill of mosquito larvae and shows no adverse effect on non-target organisms. Unlike classical *Bacillus thuringiensis* it does not produce a beta-toxin. Its killing effect is therefore based principally on its crystalline delta-toxin, (d-toxin) which is resistant to both heat (surviving 80°C for 10 minutes and 60°C for 20 minutes) and ultra violet light.

(e) *Bacillus sphericus*: *Bacillus sphericus* is an highly specific for mosquito larvae as *Bacillus thuringiensis*, var *israelensis (B.t.i.)*. However, whereas the lethality of *B.t.i.* resides in toxic protein crystals formed during the spores of the organism, the toxin of *B. sphericus* resides in the cell wall of the organism. The toxin of *B. sphericus* works slowly (8-40 hours) compared with that of *B.t.i.* (2-10 hours). *Bacillus sphericus* however, has the advantage of being able to lay dormant in muds or sewage ponds and to recycle as susceptible mosquito larvae appear. Like *B.t.i.*, it had reached stage 4 of had WHO scheme for screening and evaluating biological agents for control of disease vectors shown in Table 17.3.

(ii) *Viruses*: A large number of viruses has been isolated from insects. The advantages of viruses as biological control agents is that they are specific. Seven groups of insect-pathogentic viruses have been identified (Table 17.2). The most useful of them for biological control purposes are the baculoviruses, which are easily recognizable because the virus particles are included within a proteinaceous inclusion body large enough to be seen under a light microscope. (These inclusion bodies, polyhedrons and granules, are found in the nucleus of the host cell – hence they are nuclear polyhedrosis and granuloses).

The baculoviruses are the best candidates for insect control because they are (a) effective in controlling insect populations, (b) restricted to a host range of invertebrates, (c) relatively easy to produce in large quantities and (d) stable under specific conditions because of the inclusion bodies.

Several experimental preparations are available and at least two (one each in the USA and Japan) have been produced on a commercial scale. The preparations are ingested when the insects consume leaves and other plant parts on which the virus particles have been sprayed. After ingestion the polyhedral inclusion bodies dissolve within the mid-gut; the released virions pass through the mid-gut epithelial cells into the haemocoel. Death of the larvae occurs four to nine days after ingestion.

(iii) *Fungi*: All the four major groups of fungi, *Phycomycetes*, *Ascomycetes*, *Fungi Imperfecti* and *Basidiomycetes* contain members pathogenic to insects. The great difficulty with using fungi for biological control is that environmental conditions including temperature and humidity must be adequate for spore germination and insect cuticle penetration by the hyphae. Since these environmental conditions are not always assured the result is that fungi are used for biological control only in a few countries especially the USSR. Fungi which have been most widely used as *Beauvaria bassiana* and *Metarrhizium anisopliae*. Others are *Hirsutella thompsonii Verticillium* and *Aschersonia aleyrodis*. *H. thompsonii* is being developed commercially as acaricide, for killing mites which attack plants, although a large number of other fungi attack mites. *H. thomposonii* has been

Table 17.3 Scheme for screening and evaluating the efficacy, safety, and environmental impact of biological agents for control of disease vectors designed by the WHO

Stage I	Stage II	Stage III	Stage IV	Stage V
Laboratory A. Identification and charaterization	*Laboratory* A. Mammalian infectivity tests to ensure safet to laboratory and field personnel	*Preliminary field trials* Results of Review of stages I and II Strictly regulated ponds tests under WHO supervision to determine efficacy against disease vectors under natural conditions	*Laboratory* A. More detailed tests on mammalian infectivity using appropriate techniques *Laboratory and field trials*	*Large-scale field trials* To be conducted under WHO auspices. Not presently defined, and will vary according to target vector, habitat(s), mode of application, etc. Review of stages I, II, III and IV by informal consultation group
B. Assessment against selected target vectors species	B. Preliminary assessment against certain non-target		B. Detailed studies on non-target range especially other fauna in habitats where stage V trials may be conducted	
C. Preliminary evaluation of ease of rearing in			*Formulation* C. Studies on stability of suitable formulations and delivery system	

found particularly active against mites which attack citrus. It is applied as the conidial powder and maximum effectiveness occurs at 27°C and under moist conditions or at relative humidities of 79-100%. *Coelomomyces* sp. is very effective against mosquitoes but its production is difficult because of the need of a secondary host. Most effective and specific against mosquitoes are *Culicinomyces* sp. which was isolated in Australia and produced a mortality rate on mosquitoes of 90-100%. *Tolypocladium cylindrosporum* is essentially like *Culicinomyces* in being highly specific for mosquitoes. *Lagenidium giganteum* and *Leptolegnia* sp have been shown to have high mortality for mosquitoes. All the above (except *Coelomomyces*) can be mass-produced by fermentation. *Beauvaria* and *Metarrhizium* already discussed have broad activity against mosquitoes.

(iv) *Protozoa*: In constrast to the rapid action of viruses and spore-forming bacteria, killing by protozoa is slow and may take weeks. Furthermore they are difficult to produce, being accomplished only *in vivo*. Nevertheless they have been produced and successfully used experimentally for stored-product pests *(Matosia trogoderina)* mosquitoes *(Nosema algerae)* and grasshoppers *(Nosema pyrasta)*.

Vavra vilivis is also effective against mosquitoes and has properties similar to those of *Nogema algerae*. Studies sponsored by the WHO have shown that *N. algerae* does not seem to constitute a safety hazard for man. Factors favoring the use of *N. algerae* are spore-longevity, ease of spore-production under laboratory and especially cottage industry conditions and the probable impact on disease transmission by reducing the longevity of infected female mosquitoes.

So far however protozoa have not been produced on an industrial scale for biological control.

17.2.3 *Bacillus thuringiensis* Insecticidal Toxin

B. thuringiensis strains produce two types of toxin. The main types are the **Cry** (crystal) toxins, encoded by different **cry genes**, and this is how different types of Bt are classified. The second types are the **Cyt** (cytolytic) toxins, which can augment the Cry toxins, enhancing the effectiveness of insect control. Over 50 of the genes that encode the Cry toxins have now been sequenced and enable the toxins to be assigned to more than 15 groups on the basis of sequence similarities. The table below shows the state of such a classification in 1995. An alternative classification has recently been proposed based on the degree of evolutionary diversity of the amino acid sequences of the toxins, but this has not yet been widely adopted.

Cry toxins are encoded by genes on plasmids of *B. thuringiensis*. There can be five or six different plasmids in a single Bt strain, and these plasmids can encode different toxin genes. The plasmids can be exchanged between Bt strains by a conjugation-like process, so there is a potentially wide variety of strains with different combinations of Cry toxins. In addition to this, Bt contains transposons (transposable genetic elements that flank genes and that can be excised from one part of the genome and inserted elsewhere). All these properties increase the variety of toxins produced naturally by Bt strains, and provide the basis for commercial companies to create genetically engineered strains with novel toxin combinations.

Table 17.4   Bt toxins and their classification

Gene	Crystal shape	Protein size (kDa)	Insect activity
cry I [several subgroups: A(a), A(b), A(c), B, C, D, E, F, G]	bipyramidal	130-138	lepidoptera larvae
cry II [subgroups A, B, C]	cuboidal	69-71	lepidoptera and diptera
cry III [subgroups A, B, C]	flat/irregular	73-74	coleoptera
cry IV [subgroups A, B, C, D]	bipyramidal	73-134	diptera
cry V-IX	various	35-129	various

Mode of Action of Bt Toxin

The toxin of Bt is lodges in a large structure, the parasporal structure, which is produced during sporulation. The parasporal crystal is not the toxin. However, once it is solubulized a protoxin is released.

The crystals are aggregates of a large protein (about 130-140 kDa) that is actually a **protoxin**, which must be activated before it has any effect. The crystal protein is highly insoluble in normal conditions, so it is entirely safe to humans, higher animals and most insects. However, it is solubilised in reducing conditions of high pH (above about pH 9.5), the conditions commonly found in the mid-gut of lepidopteran larvae. For this reason, Bt is a highly specific insecticidal agent.

Once it has been solubilized in the insect gut, the protoxin is cleaved by a gut protease to produce an active toxin of about 60 kDa. This toxin is termed **delta-endotoxin**. It binds to the mid-gut epithelial cells, creating pores in the cell membranes and leading to equilibration of ions. As a result, the gut is rapidly immobilized, the epithelial cells lyse, the larva stops feeding, and the gut pH is lowered by equilibration with the blood pH. This lower pH enables the bacterial spores to germinate, and the bacterium can then invade the host, causing lethal septicaemia.

17.3   PRODUCTION OF BIOLOGICAL INSECTICIDES

Microbiological insecticides are produced in one of three ways: submerged fermentation; surface or semi-solid fermentation; and *in vivo* production. The first two are for facultative pathogens and the third is for obligate pathogens.

17.3.1   Submerged Fermentations

These have been used for the production of *Bacillus* spp. (excluding production of *B. poppillae* which is produced *in vivo*) and to a lesser extent, fungi.

Medium: In fermentation for *Bacillus thuringiensis* the active principle sought is the delta toxin found in the crystals. Media for submerged fermentation have been compounded by various workers in a number of patents. In one such preparation, the initial growth in a shake flask occurred in nutrient broth; in the second shake flask, and in the seed fermentor best molasses (1%), corn steep liquour (0.85%) and $CaCO_3$ (0.1%) were used. A typical medium for production would be beet molasses (1.86%), pharmamedia (1.4%) and $CaCO_3$ (0.1%). Other production media contain corn starch (6.8%), sucrose (0.64%), casein (9.94%), corn steep liquor (4.7%), yeast extract (0.6%) and phosphate buffer (0.6%).

A third medium contained soya bean meal (15%), dextrose (5%), corn starch (5%), $MgSO_4$ (0.3%), $FeSO_4$ (0.02%), $ZnSO_4$ (0.02%) and $CaCO_3$ (1.0%).

The above media were used for agricultural strains of *B. thuringiensis* but could no doubt be used also for *B. thuringiensis var israelensis*.

Bacillus thuringiensis var insraelensis and *Bacillus sphericus* do not require carbohydrates for growth and can grow well and produce materials which will kill the larvae of mosquitoes in a variety of proteinaceous materials such as commercial powders of soy products, dried milk products, blood and even materials from primary sewage tanks. Effective powders of *B. sphericus* 1593 and *B. thuringiensis var israelensis* have been produced using discarded cow blood from abattoirs and various legumes.

Extraction: At the end of the fermentation, the active components of the broth are recovered by centrifugation, vacuum filtration with filter aid or by precipitation. Precipitation has been done with $CaCl_2$ and the acetone method yields products of very high potency. The fermentation beer may readily be diluted and used directly.

17.3.2   Surface Culture

Surface culture techniques are used for fungi and for spore formers. The organisms after shake-flask growth are cultured in a seed tank from where the broth is transferred to flat bins with perforated bottoms. The semi-solid medium is a mixture of an agricultural by-product such as bran, an inert product such as kisselghur, soy bean meal, dextrose, and mineral salts. The use of this medium increases the surface area and hence aeration because of the thinness of its spread in the bins. Hot air is passed through the perforations to dry the material. It is ground, assayed and compounded to any required strength with inert material. Submerged, culture in which the hyphae are used have been carried out with good results in the United States using *Hirsutella thompsonii*.

17.3.3   *In vivo* Culture

In vivo culture methods are used for producing caterpillar viruses, mosquito protozoa and *Bacillus popillae*. The method is labor-intensive and could be easily applied for suitable candidates in developing countries where expertise for submerged culture production is usually lacking.

Once the organism has been obtained in a sufficient quantity to last for several years it is lyophilized and stored at low temperature. The viruses are introduced into the food of the larvae and the dead larvae are crushed, centrifuged to remove large particles and the rest are dried. The amount of viruses in each larva is variable but the virus content of between one and one hundred caterpillars should be sufficient to treat one acre in the case of cotton moths. Usually separate facilities are used for rearing the caterpillars, for infecting them and for the extraction of the virus particles. The preparation is then bio-assayed and mixed with a suitable carrier.

17.4   BIOASSAY OF BIOLOGICAL INSECTICIDES

It is obvious that a reference standard must be set up against which various preparations can be compared. The standard will differ with each particular bioinsecticide. Thus, standards do exist for *Bacillus thuringiensis* serotypes H3 and H3A used against caterpillars and a standard for *B. thuringiensis* var *israelensis* against mosquito 'IPS82' exists. Both standards are prepared and deposited at the Institute Pasteur in Paris. In the simplest terms a standard is based on the LD_{50}, the dose of the insecticide which will kill 50% of the population must be clearly defined; the age and type of insect to be used; the food of the insect; the temperature conditions and a host of other parameters.

17.5 FORMULATION AND USE OF BIOINSECTICIDES

The formulation of the bioinsecticides is extremely important. An insecticide shown to be highly potent under laboratory experimental conditions may prove valueless in the field unless the formulation has been correctly done. Since microorganisms cannot by themselves be patented, industrial firms producing bioinsecticides depend for their profits on the efficiency of their formulation (i.e., the inert material which ensures adequate presentation of the larvicide to the target insect). The inert material is referred to as a carrier or an extender. Carriers or extenders are the *solids* or *liquids* in which the active principle is diluted. When the carrier is a liquid and the active principle is suitable in it the application is a spray. There are thus two types of formulation: (a) powders and dusts (b) flowable liquid; which of the two is manufactured depends to a large extent on the method of production and intended use of the insecticide.

17.5.1 Dusts

Semi-solid preparations based on waste plant products usually are compounded as dusts or powders because making them into liquid causes the bran to absorb water and prevent free flow thus leading to the clogging of conventional liquid applicators. The advantage of dusts is greater stability of the preparation. They are also useful when the insecticide is intended to reach the underside of low lying crops such as cabbages. Heavy rains unfortunately wash off dusts. They may also lead to inhalation of the bioinsecticides by the persons applying them. Diluents which have been used in commercial dust of *Bacillus thuringiensis* are celite, chalk, kaolin, bentonite, starch, and lactose. Lactose has also been used for diluting virus insecticide dusts. When the active principle is *absorbed* on to the extender (or filler), the extender is referred to as a *carrier*.

If the extender or carrier is attractive to the insect as a food, oviposition site etc., then the extender or filler is known as a *bait*. Baits for *Bacillus thuringiensis* include ground corn meal, and for protozoa, cotton seed oil, honey, hydroxyethyl cellulose.

17.5.2 Liquid Formulation

Liquid formulations are usually made from water in which both the crystal and spores are stable. Sometimes oils and water/oil emulsions may be used. When liquids other than water are used it must be ascertained that they do not inactivate the active agent. Emulsifiers may be added to stabilize emulsions when these are used. Some emulsifiers which have been used for *B. thuringiensis* and viruses are Tween 80, Triton B 1956, and Span 20.

The nature of the surface on which the insecticide is applied and which may be oily, smooth or waxy may prevent the liquid from wetting the sprayed surface. *Spreaders* or wetting agents which are surface-tension reducers may be added. Wetting agents may be added to dusts to produce *wettable-powders* which are more easily suspended in water. Some wetting agents and spreaders which have been used for agricultural *Bacillus thuringienses* include alkyl phenols Tween 20, Triton X114 and for viruses Triton X100 and Arlacel 'C' which are all commercial surface-tension reducing agents.

To prevent run-off of liquids or wettable powders, *stickers* or *adhesives* are added to hold the insecticide to the surface. Stickers which have been used for bacteria and viruses include skim milk, dried blood, corn syrup, casein, molasses, and polyvinyl chloride latexes.

Protectants are often added to insecticides which protect the active agent from the effect of ultra violet light, oxidation, desiccation, heat and other environmental factors which reduce the effectiveness of the active agent. These are usually trade secrets and their composition is not disclosed. Dyes combined with proteins such as brewers yeast plus charcoal, skin milk plus charcoal, and albumin plus charcoal have also proved effective in protecting virus preparations from the effect of the ultraviolet light of the sun. Micro-encapsulation of bioinsecticides with carbon also affords protection.

17.6   SAFETY TESTING OF BIOINSECTICIDES

Many individuals on first learning of the use of microorganisms to control insect pests and vectors of disease express fear about the effect of these entomopathogens or their effective components (e.g., crystals of *B. thuringiensis*). For this reason animal tests including feeding by mouth, inhalation, intraperitoneal, intradermal and intravenous inoculations, and teratogenicity and carcinogenicity tests are done. Test animals include rats, mice, monkeys, rabbits, fish, and sometimes when appropriate, human volunteers.

Tests conducted on the following agricultural entomopathogens in the United States, Russia and Japan have shown them non-toxic for man, other animals or plants: Bacteria (*Bacillus popillae, B. thuringiensis, B. moritai*), five viruses (*Heliothus, Orgyia, Lymantria, Autographa, Dendrolimus*), three protozoa (*Nosema locustae, N. algerae, N. troqodermae*) and two fungi (*Beauveria bassiana, Hirsutella thompsonii*).

Tests sponsored by the WHO and carried out in France and the United State have shown the following useful or potentially useful entomopathogens to be safe. *Bacillus sphericus* stain SS11-1, *B. sphericus* strain 1593-4; *B. sphericus* strain 1404-9, *Bacillus thuringiensis, var. israelensis* (serotype H14) strain WHO/CCBC 1897; *Metarrhuzium anisopliae, Nosema algerae*.

17.7   SEARCH AND DEVELOPMENT OF NEW
BIOINSECTICIDES

There are a number of stages in the development of a new bioinsecticide. The World Health Organization has for some years followed the scheme given in Table 17.3 for the screening, evaluation, safety, and environmental impact of entomopathogens to be used for biological control.

Except where the material can be produced on a small scale, cottage industry, level, production and sale of the final material will have to be done by industry, with its experience of formulation and sale distribution. It has been estimated that it will take five to seven years to develop an entomopathogen into a biological insecticide; it will take less than five years if some information on safety already exists on safety of a related bioinsecticide. The cost of developing a biological insecticide is far less than that of developing a chemical insecticide by between 20% and 50%.

SUGGESTED READINGS

Glare, R.E., Callaghan, M. 2000. *Bacillus thuringiensis:* Biology, Ecology and Safety. Wiley Chichester UK.

Knowles, B.H. 1994. Mechanism of action of *Bacillus thuringiensis* insecticidal delta-endotoxins. In: Advances in Insect Physiology, P.D. Evans, (ed.) Vol 24 Academic Press. London: UK. pp. 275-308.

Obeta, J.A.N., Okafor, N. 1984. Medium for the production of the primary powder of *Bacillus thuringiensis* sub-species *israelensis*. Appl. Environ. Microbiol. 47, 863-867.

Obeta, J.A.N., Okafor, N. 1983. Production of *Bacillus sphaericus* strain 1593 primary powder on media made from locally obtainable Nigerian agricultural products. Canadian J. Microbiol. 29, 704-709.

World Health Organization. 1979a. Report of a meeting on standardization and industrial development of Microbial Control Agents. TDR/BCV/79.01.

World Health Organization. 1979b. Biological Control Data Sheet: *Bacillus thuringiensis de* Barjac, 1978. VBC/BCDS/79.01.

World Health Organization. 1979c. Progress Report: Mammalian Safety Tests on the Entomocidal Microbials Contract V2/181/113/(A): WHO/TDR/VBC.

World Health Organization. 1979d. Proposals for the adoption of a standardized bioassay method for the evaluation of insecticidal formulations derived from serotype H14 of *Bacillus thuringiensis* (Barjac de. H., and Larget, I.) WHO/VBC/79.744. Geneva.

World Health Organization. 1980. Annual Rept. Scientific Working Group on Biological Control of Vectors, July 1979-June, 1980. WHO Geneva, Switzerland.

Yousten, A.Y., Federici, B., Roberts, D.W. 1991. Microbial Insecticides. In: Encyclopedia of Microbiology Vol 2, Academic Press Sandiego, USA. pp. 521-531.

CHAPTER 18

The Manufacture of *Rhizobium* Inoculants

Nitrogen is a key element in the nutrition of living things because of its importance in nucleic acids, which are concerned with heredity, and in proteins, which *inter alia* provide the bases for enzymes. Gaseous nitrogen is present in abundance in the Earth's atmosphere forming about 80% of atmospheric gases. Indeed it has been estimated that each acre of land has about 3,500 tons of N_2 above it. Unfortunately, most living organisms cannot utilize gaseous nitrogen but require it in a fixed form; that is, when it forms a compound with other elements. Nitrogen can be fixed both chemically and biologically. Chemical fixation is employed in the production of nitrogenous chemical fertilizers, which are used to replace nitrogen removed from the soil by plants. The ability to carry out biological fixation is found only in the bacteria and blue-green algae. Some of these organisms fix nitrogen in the free-living state and thereby contribute to the improvement of the nitrogen status of the soil. Others do so closely associated (in symbiosis) with higher plants. In some of these associations such as with some tropical cereals, the organisms live on the surface of the plant roots and fix the nitrogen there. In some others the microorganism penetrates the roots and forms outgrowths known as nodules within which the nitrogen is fixed. Of the nodule-forming nitrogen fixing associations between plant and micro-organisms, the most important are the legume-bacteria associations. There are about 1,200 legumes species and their nodulation is important because about 100 of them are used for food in various parts of the world.

Apart from serving as food for man and animals the legumes provide nitrogenous fertilization to the soil for the better growth of crops in general; the importance of soil fertilization can be seen from the fact that 100 kg of symbiotically-fixed nitrogen has been estimated to be equivalent to an application of 50 kg of ammonium sulfate fertilizer.

The bacteria which form nitrogen-fixing nodules with legumes are members of the genus *Rhizobium*. Inoculation of legumes with rhizobia started as long ago as 1896, eight years after the beneficial association between the legumes and rhizobia was discovered. Today there are thriving industries producing rhizobia inoculants in most parts of the world. The inoculation of rhizobia into soils or on seeds is done where the specific rhizobia which will form nodules with a given legume are absent in the soil because the legume is new to the area or where because of a lapse of many years without the legume the soil may become deficient in effective strains of rhizobia.

The need for legume inoculation has become more urgent in recent years because of the rise in the cost of chemical fertilizers, the inefficient use of chemical fertilizers by agricultural crops and the short-term and long-term environmental consequences of unused nitrate fertilizers which find their way into, and cause the pollution of, drinking water.

Finally the problem of providing more protein for an ever-rising world population has been compounded by the fact that areas of the world where protein shortage is most acute are just those least able to afford the plants for the manufacture of chemical nitrogenous fertilizers. By contrast investment in rhizobium inoculant production is relatively cheap and the manufacture unsophisticated in comparison with chemical factories.

18.1 BIOLOGY OF *RHIZOBIUM*

18.1.1 General Properties

Members of the genus *Rhizobium* are aerobic, Gram-negative relatively large rods which are motile and have peritrichuous flagellation. Older cells contain prominent granules of poly-β-hydroxybutyrate which give them a banded appearance. Within nodules they form irregularly shaped cells known as 'bacteroids"; no fixation occurs in the absence of bacteroids. At the end of the growing season, the nodules occur in the absence of bacteroids. At this time, the nodules disintegrate, releasing bacterioids which, once in the soil, revert to rod-shaped cells and can survive there, sometimes for years. They invade roots of appropriate legumes and form new modules when appropriate legumes are planted.

18.1.2 Cross-inoculation Groups of *Rhizobium*

The most distinctive feature of *Rhizobium* spp. is their ability to form nodules with legumes. Different rhizobia will form nodules only with some legumes. A group of legumes with which a particular rhizobia bacterium will form nodules is known as a cross-inoculation group. Over 20 cross-inoculation groups have been set-up out of which seven (Table 18.1) are well-known. As will be seen from Table 18.1, rhizobia which lack a more suitable group get lumped into the 'cowpea rhizobia' group. The group has also been reclassified using numerical taxonomy, DNA base composition and homology, serology phage susceptibility, patterns of isoenzymes, etc.

18.1.3 Properties Desirable in Strains to be Selected for use as *Rhizobium* Inoculants

Before the production of a rhizobial inoculant is done the strain of the organism which will yield the highest amount of fixed nitrogen in a given legume and a given environment (or the most effective strain) is selected on the following basis.

(i) *Effectiveness*: 'Effectiveness' is a term used to describe the overall ability of a given rhizobial strain to form nodules with a particular legume in a given environment and fix useful quantities of nitrogen. Effectiveness itself is based on nitrogen-fixing ability, invasiveness and competitiveness.

Table 18.1 Cross-inoculation groups of the genus *Rhizobium*

Rhizobium *species*	Cross Inoculation Group	Legume host included Trigonella (Fenugreak)
1. Clover group	*R. Trifolii*	*Trifolium* (clover)
2. Alfalfa group	*Rhizobium meliloti*	*Medicago* (alfalfa)
		Melilotus (sweet clover)
3. Bean group	*R. Phaseoli*	*Phaseolus* (Beans)
4. Pea group	*R. leguminosarum*	*Pisum* (pea); *Vivia* (Vetch)
		Lathyrus (sweet pea)
5. Lupine group	*R. lupini*	*Lens* (Lentil)
		Lupinus (lupines)
6. Soybean group	*R. japonium*	*Ornithopus* (Serradella)
7. Cowpea group		*Glycine* (soy bean)
		Vigna (cowpea); *Grotolaria* (Crotolaria) *Pueraria* (Kudzu) *Arachis* (Peanut) *Phaseolus* (lima beans)

Nitrogen-fixing ability is an important factor as not all nodules fix nitrogen, or do so to the same extent. Nitrogen-fixing nodules are usually comparatively few in number, relatively large, and are pink in color. *Invasiveness* (also referred to as infectiveness) is a measure of the ability of the legume to invade and nodulate the roots of a high proportion of the plants to which it is applied. *Competitiveness* is related to invasiness and is a measure of the ability of the rhizobium strain to produce a large number of nodules in the presence of other infective or invasive strains.

(ii) *Ability to perform in the environment of a particular soil*: When effectiveness is a measure of properties inherent in the bacterium itself other factors relate to performance in the soil. These are:

(a) ability to fix nitrogen in the presence of fertilizers used in the soil;

(b) tolerance of insecticide and seed disinfectants;

(c) resistance to bacteriophages common in the soil;

(d) adequate performance under the pH, aeration, and the mineral status of the soil.

(iii) *Growth and Survival in the carrier*: The ability of the organism to survive and multiply in the carrier (i.e., the inert support for distributing the bacteria) is important; an otherwise adequate organism may be inhibited by the carrier.

18.1.4 Selection of Strains for Use as Rhizobial Inoculants

Methods available for assessing the performance of a rhizobium strain before choosing it as an inoculant are discussed below:

(i) *The agar method*: Seeds sterilized with mercuric chloride, sulphuric acid and alcohol are washed with sterile water are allowed to germinate in contact with the rhizobia being assessed. One seedling each is placed on nitrogen free slants of agar containing appropriate mineral salts in a large test tube. (approx. 5 cm x 20 cm). The surface of the agar is flooded with a heavy suspension of the rhizobium to be tested. Control slants with nitrate and without rhizobia are prepared. The nitrogen-fixing ability of the system is assessed by harvesting the seedlings and determining their dry weight and nitrogen contents over the control after an appropriate period of growth. Alternatively all the nodules may be excised from the various test groups and assessed indirectly for nitrogen fixation by the acetylene reduction method. This is a rapid and convenient method for assessing nitrogen fixation. Nodules are exposed to acetylene and thereafter the gases are checked in a gas chromatograph for ethylene production. The ability of nodules, free-living bacteria or other biological systems to fix nitrogen is determined from the extent of acetylene-ethylene conversion.

(ii) *Soil cores*: Undisturbed soil cores from the field may be planted with germinated seedlings and inoculated with a culture of rhizobium in a glass house. Assessment of nitrogen fixation may then be done as described above.

(iii) *Field assessment*: For field assessment the material in which the rhizobium is carried should have been selected and the experiment laid out in such a way that; (a) uninoculated plots are tested for the presence of naturally occurring rhizobia; (b) plots are inoculated with the organism being tested; (c) plots are inoculated with the test organism and simultaneously supplied with nitrogen fertilizers. The plants are then harvested and checked for dry weight and nitrogen content.

While soil agar and soil tube tests are useful for rapid screening, there is no substitute for field testing using the expected final carrier in soil as similar as possible to that in which the rhizobium is destined to grow.

18.2 FERMENTATION FOR RHIZOBIA

(i) *The inoculum*: The inoculum is prepared from a stock culture preferably stored in sterile soil or on agar overlain with oil. The organisms used in preparing the inoculum are preferably those able to form nitrogen-fixing nodules with several legumes (i.e., the so-called 'broad-spectrum' strains). Where these are not available a mixture of several strains effective in a wide range of legumes are used. In this way the need to prepare a large number of small amounts of the inoculant is obviated. The inoculum added is preferably about 1.0% of the total volume and should have a density of 10^6-10^7 organisms per ml.

(ii) *Medium*: Rhizobia are not very demanding of nutritional requirements. Most industrial media used consist of yeast extract or yeast hydrolysate, a carbohydrate source and mineral salts. In some media yeast extracts supply all the nitrogen, vitamins (especially biotin) and minerals required by the bacteria. Corn steep liquour and hydrolyzed casein are sometimes used to supplement yeast extract. In some media, one or more of potassium phosphate, magnesim sulfate, ferric chloride, and sodium chloride may be added. For the fast growers the carbohydrate source is usually sucrose; for the slow growers it may be mannitol, galactose or arabinose. Fast growers (e.g. *Rhizobium meliloti)* have large gummy colonies which form in three to five days. Slow growers (e.g. *Rhizobium japonium*) have small colonies taking 7 to 10 days to develop. The former are petrichuously flagellated while the latter are sub-polarly flagellated.

(iii) *Aeration*: Rhizobia are aerobic organisms; nevertheless the type of intensive agitation and aeration used for the production of yeasts or some antibiotics does not seem necessary. Indeed rhizobia will grow quite well in an unaerated fermentor if there is a broad enough surface area to permit oxygen diffusion. Very low aeration, as low as 0.5 liters/hour, has been found satisfactory. In large fermentors air sparging without agitation is usually satisfactory.

(iv) *Time and temperature*: The temperature used is about 20°C and while fast growers attain high numbers (in excess of 140×10^9/ml) in about two to three days, slow growers in medium specially designed to facilitate their growth attain slightly less than this number in three to five days.

(v) *The fermentor*: Fermentors used for rhizobium culture are small in comparison with those used for antibiotics. The larger sizes range from 1,000 to 2,000 liters. Ordinarily they range from 5 liters through 40 liters to about 200 liters and are also usually quite unsophisticated compared with those used for producing antibiotics.

18.3   INOCULANT PACKAGING FOR USE

After the fermentation of the organism it is packaged for delivery in either of two forms: (a) as a coating on the seeds, (or seed inoculants) from which the rhizobia develop at planting and invade the roots; (b) direct application into soil with the seeds introduced shortly before or after soil inoculants.

18.3.1   Seed Inoculants

Seed inoculants are more commonly used than soil inoculants. In seed inoculation the rhizobia may be offered as a liquid or broth, frozen concentrates, freeze-dried or oil-dried preparation. No specific carrier is used for these preparations. Although gum Arabic, milk, and sucrose are sometimes added to these essentially liquid preparations, they still do not offer sufficient protection to the bacterial from the environment. The bacteria therefore die out quickly. By far the commonest preparations are offered with carriers.

A carrier is the material which binds the rhizobium to the seed. Carriers should have a high water-holding capacity provided a nutritive medium for the growth of rhizobia protect the bacteria from harmful environmental effect e.g. sunlight and favor their survival on the seeds and in the soil, and in particular they should not be toxic to the bacteria.

Agar may sometimes be used as a carrier but by far the most widely used carrier is peat. Other locally available materials may be used.

(i) The use of peat as a carrier

Peat is the first stage in the formation of coal, which when freshly obtained is moist. It must be dried and milled or shredded. Peats vary in their properties and each must be studied and undesirable properties rectified before final use. For example peats mined in some parts of the world have a high content of sodium chloride, which must be removed by leaching with water before the peat is used. Peats are also usually acid, and finely ground $CaCO_3$ up to about 5% is used to raise the pH to about 6.8.

The survival rate of rhizobia in unsterilized carriers is low because of competition and antagonism from resident organisms. The peat is therefore usually sterilized by hot air, steam (including autoclaving), gamma irradiation and chemical sterilants. Hot air and steam seem to be favored, as gumma-irradiation facilities are not always easily accessible and the post-treatment removal of chemical sterilants may sometimes be difficult. Care must be taken to ensure that materials toxic to the rhizobia are not released by the use of too high a temperature.

In order to introduce the organism into a carrier two methods are used. In the United States the broth containing a heavy growth of rhizobium in excess of 10^9 /ml, is mixed with $CaCO_3$ and sprayed onto sterilized ground peat. It is then incubated in thin layers at 26-28°C for two to three days to allow the heat generated during the wetting of the peat to dissipate. Thereafter it is ground and bagged: adequate numbers are reached in the peat in three to five weeks for fast growers and in about twice as long for slow growers. At the end of this period the preparation is refrigerated till used. In other parts of the world, the broth is inoculated into bottles or polythene bags containing sterile peat or a peat/soil mixture. They are shaken after 24 hours and allowed to grow for one to two weeks at 26-30°C and thereafter stored at 2-4°C. For seed inoculation the rhizobium in broth or with a carrier is brought in contact with the seeds, which then become coated with the organisms.

(ii) Use of other carriers

Peat is not available in some parts of the world. Any carrier which meets the requirements indicated above would do. A wide range of materials have in fact been tried with success including lignite, coal, charcoal, bagasse, coir dust, composted straw plus charcoal, ground wheat straw, rice husk, and ground talc.

The essential thing is that the carrier, be it peat or any other substance(s) must be shown to be suitable both in laboratory experimentation, and also in the field.

18.3.2 Soil Inoculants

When seed inoculation is not practicable, soil inoculation is resorted to. The following are conditions when seed inoculation is not efficient:

(a) In some epigeal legumes for example soybean, the seed coat of the seedling is lifted out of the ground during the emergence of the cotyledons (seed leaves). Under this condition, the rhizobia clinging to the seed coat are not deposited in the soil.

(b) Seeds coated with fungicides or insecticides cannot be successfully inoculated with rhizobia.

(c) With legumes having small seeds and therefore only a limited number of rhizobia are introduced, heavy soil inoculation is practiced, especially if very aggressive but ineffective rhizobia exist in the soil.

(d) Finally fragile seeds such as peanuts (groundnuts) may break following prior wetting during seed inoculation.

Besides the above difficulties soil inoculation has the additional advantage of relative ease of application; furthermore, it may be used to inoculate the growing plant in situations where earlier inoculation failed.

Rhizobia may be inoculated into soil using one of two ways:

(a) Frozen concentrate obtained from centrifugation is thawed and diluted with water and applied to soil;

(b) Free flowing granules made from peat-rhizobium preparations may be applied to soil.

18.4 QUALITY CONTROL

It is important that the number and nitrogen-fixing qualities of the inoculant meet the expectation of the user at the time of application. The number of rhizobia required for effective nodulation depends on a large number of factors including the size of the seed, the presence or absence of competing rhizobia in the soil weather conditions; the temperature moisture, and type of soil. Nevertheless the accepted standard is a minimum of 1,000 cells of *Rhizobium* per seed, except in the case of soybean where the minimum is 10.

In some countries, such as Australia a government or university body supervises all stages of the production of inoculants including the testing and selection of the strains, maintenance and issuing of stock cultures to manufacturers, assessing the quality of the broth and the final preparation of the peat-carried inoculum. In other countries notably the United States the control is left to the producing companies. In either case the performance of the inoculants in the hands of the farmer is the ultimate test. Wherever rhizobia are inoculated some means of ensuring the quality of the product must exist.

Once the material has been prepared and the maximum number possible in the preparation attained quality is most easily maintained by storing it under refrigeration until use, provided that adequate content and some aeration is maintained in the package. A system of expiry date based on experience is often indicated on the package.

SUGGESTED READINGS

Brockwell, J. 1977. In: A Treatise on Dinitrogen Fixation: Section IV: Agronomy and Ecology. R.W.F. Hardy, A.H. Gibson, (eds). Wiley, New York, USA. pp. 277-309.

Flickinger, M.C., Drew, S.W. (eds), 1999. Encyclopedia of Bioprocess Technology - Fermentation, Biocatalysis, and Bioseparation, Vol 1-5. John Wiley, New York, USA.

Temprano, F.J., Albareda, M., Camacho, M., Daza, A., Santamaria, C., Rodrýguez-Navarro. 2002. Survival of several *Rhizobium/Bradyrhizobium* strains on different inoculant formulations and inoculated seeds. International Microbioliology. 5, 81–86.

CHAPTER 19

Production of Fermented Foods

19.1 INTRODUCTION

Fermented foods may be defined as foods which are processed through the activities of microorganisms but in which the weight of the microorganisms in the food is usually small. The influence of microbial activity on the nature of the food, especially in terms of flavor and other organoleptic properties, is profound. In terms of this definition, mushrooms cannot properly be described as fermented foods as they form the bulk of the food and do not act on a substrate which is consumed along with the organism. In contrast, yeasts form a small proportion by weight on bread, but are responsible for the flavor of bread; hence bread is a fermented food.

Fermented foods have been known from the earliest period of human existence, and exist in all societies. Fermented foods have several advantages:

(a) Fermentation serves as a means of preserving foods in a low cost manner; thus cheese keeps longer than the milk from which it is produced;

(b) The organoleptic properties of fermented foods are improved in comparison with the raw materials from which they are prepared; cheese for example, tastes very different from milk from which it is produced;

(c) Fermentation sometimes removes unwanted or harmful properties in the raw material; thus fermentation removes flatulence factors in soybeans, and reduces the poisonous cyanide content of cassava during garri preparation (see below);

(d) The nutritive content of the food is improved in many items by the presence of the microorganisms; thus the lactic acid bacteria and yeasts in garri and the yeasts in bread add to the nutritive quality of these foods;

(e) Fermentation often reduces the cooking time of the food as in the case of fermented soy bean products, or ogi the weaning West African food produced from fermented maize.

Fermented foods are influenced mainly by the nature of the substrate and the organisms involved in the fermentation, the length of the fermentation and the treatment of the food during the processing.

The fermented foods discussed in this chapter are arranged according to the substrates used:

Wheat
  Bread

Milk
  Cheese
  Yoghurt

Maize
  Ogi, Akamu, Kokonte

Cassava
  Garri
  Foo-foo, Akpu, Lafun

Vegetables
  Sauerkraut
  Pickled cucumbers

Stimulant beverages
  Coffee, Tea and Cocoa

Legumes and oil seeds
  Soy sauce, Miso, Sufu
  Oncom. idli
  Ogili, Dawa dawa, Ugba

Fish
  Fish sauce

19.2 FERMENTED FOOD FROM WHEAT: BREAD

Bread has been known to man for many centuries and excavations have revealed that bakers' ovens were in use by the Babylonians, about 4,000 B.C. Today, bread supplies over half of the caloric intake of the world's population including a high proportion of the intake of Vitamins B and E. Bread is therefore a major food of the world.

19.2.1 Ingredients for Modern Bread-making

The basic ingredients in bread-making are flour, water, salt, and yeasts. In modern bread-making however a large number of other components and additives are used as knowledge of the baking process has grown. These components depend on the type of bread and on the practice and regulations operating in a country. They include 'yeast food', sugar, milk, eggs, shortening (fat) emulsifiers, anti-fungal agents, anti-oxidants, enzymes, flavoring, and enriching ingredients. The ingredients are mixed together to form dough which is then baked.

19.2.1.1   Flour

Flour is the chief ingredient of bread and is produced by milling the grains of wheat, various species and varieties of which are known. For flour production most countries use *Triticum vulgare*. A few countries use *T. durum*, but this yellow colored variety is more familiarly used for semolina and macaroni in many countries. The chief constituents of flour are starch (70%), protein (7-15%), sugar (1%), and lipids (1%).

In bread-making from *T. vulgare* the quality of the flour depends on the quality and quantity of its proteins. Flour proteins are of two types. The first type forming about 15% of the total is soluble in water and dilute salt solutions and is non-dough forming. It consists of albumins, globulins, peptides, amino acids, and enzymes. The remaining 85% are insoluble in aqueous media and are responsible for dough formation. They are collectively known as gluten. It also contains lipids.

Gluten has the unique property of forming an elastic structure when moistened with water. It forms the skeleton which holds the starch, yeasts, gases and other components of dough. Gluten can be easily extracted, by adding enough water to flour and kneading it into dough. After allowing the dough to stand for an hour the starch can be washed off under a running tap water leaving a tough, elastic, sticky and viscous material which is the gluten. Gluten is separable into an alcohol soluble fraction which forms one third of the total and known as gladilins and a fraction (two thirds) that is not alcohol-soluble and known as the glutenins. Gladilins are of lower molecules weight than glutenins; they are more extensible, but less, elastic than glutenins. Glutelins are soluble in acids and bases whereas glutenins are not. The latter will also complex with lipids, whereas glutelins do not. 'Hard' wheat with a high content of protein (over 12%) are best for making bread because the high content of glutenins enables a firm skeleton for holding the gases released curing fermentation. 'Soft' wheat with low protein contents (9-11%) are best for making cakes.

19.2.1.2   Yeast

The yeasts used for baking are strains of *Saccharomyces cerevisiae*. The ideal properties of yeasts used in modern bakeries are as follows:

(a) Ability to grow rapidly at room temperature of about 20-25°C;

(b) Easy dispersability in water;

(c) Ability to produce large amounts of CO_2 rather than alcohol in flour dough;

(d) Good keeping quality i.e., ability to resist autolysis when stored at 20°C;

(e) Ability to adapt rapidly to changing substrates such as are available to the yeasts during dough making.

(f) High invertase and other enzyme activity to hydrolyze sucrose to higher glucofructans rapidly;

(g) Ability to grow and synthesize enzymes and coenzymes under the anaerobic conditions of the dough;

(h) Ability to resist the osmotic effect of salts and sugars in the dough;

(i) High competitiveness i.e., high yielding in terms of dry weight per unit of substrate used.

The amount of yeasts used during baking depends on the flour type, the ingredients used in the baking, and the system of baking used. Very 'strong' flours (i.e., with high protein levels) require more yeast than softer ones. High amount of components inhibitory to yeasts e.g., sugar (over 2%), antifungal agents and fat) usually require high yeast additions. Baking systems which involve short periods for dough formation, need more yeast than others. In general however yeast amounts vary from 2-2.75% (and exceptionally to 3.0%) of flour weight. The roles of yeasts in bread-making are leavening, flavor development and increased nutritiveness. These roles and the factors affecting them are discussed more fully below.

Yeast 'food' The name yeast 'food' is something of a misnomer, because these ingredients serve purposes outside merely nourishing the yeasts. In general the 'foods' contain a calcium salt, an ammonium salt and an oxidizing agent. The bivalent calcium ion has a beneficial strengthening effect on the colloidal structure of the wheat gluten. The ammonium is a nitrogen source for the yeast. The oxidizing agent strengthens gluten by its reaction with the proteins' sulfydryl groups to provide cross-links between protein molecules and thus enhances its ability to hold gas releases during dough formation. Oxidizing agents which have been used include iodates, bromates and peroxide. A well-used yeast food has the following composition: calcium sulfate, 30%, ammonium chloride, 9.4%, sodium chloride, 35%, potassium bromate, 0.3%; starch (25.3%) is used as a filler.

19.2.1.3 Sugar

Sugar is added (a) to provide carbon nourishment for the yeasts additional to the amount available in flour sugar (b) to sweeten the bread; (c) to afford more rapid browning (through sugar caramelization) of the crust and hence greater moisture retention within the bread. Sugar is supplied by the use of sucrose, fructose corn syrups (regular and high fructose), depending on availability.

19.2.1.4 Shortening (Fat)

Animal and vegetable fats are added as shortenings in bread-making at about 3% (w/w) of flour in order to yield (a) increased loaf size; (b) a more tender crumb; and c) enhanced slicing properties. While the desirable effects of fats have been clearly demonstrated their mode of action is as yet a matter of controversy among bakery scientists and cereal chemists. Butter is used only in the most expensive breads; lard (fat from pork) may be used, but vegetable fats especially soy bean oil, because of its most assured supply is now common.

19.2.1.5 Emulsifiers (Surfactants)

Emulsifiers are used in conjunction with shortening and ensure a better distribution of the latter in the dough. Emulsifiers contain a fatty acid, palmitic, or stearic acid, which is bound to one or more poly functional molecules with carboxylic, hydroxyl, and/or amino groups e.g., glycerol, lactic acid, sorbic acid, or tartaric acid. Sometimes the carboxylic group is converted to its sodium or calcium salt. Emulsifiers are added as 0.5% flour weight. Commonly used surfactants include: calcium stearyl- 2-lactylate, lactylic stearate, sodium stearyl fumarate.

19.2.1.6 Milk

Milk to be used in bread-making must be heated to high temperatures before being dried; otherwise for reasons not yet known the dough becomes sticky. Milk is added to make the bread more nutritious, to help improve the crust color, presumably by sugar cearamelization and because of its buffering value. Due to the rising cost of milk, skim milk and blends made from various components including whey, buttermilk solids, sodium or potassium caseinate, soy flour and/or corn flour are used. The milk substitutes are added in the ratio of 1-2 parts per 100 parts of flour.

19.2.1.7 Salt

About 2% sodium chloride is usually added to bread. It serves the following purposes:

   (a) It improves taste;
   (b) It stabilizes yeast fermentation;
   (c) As a toughening effect on gluten;
   (d) Helps retard proteolytic activity, which may be related to its effect on gluten;
   (e) It participates in the lipid binding of dough.

  Due to the retarding effect on fermentation, salt is preferably added towards the end of the mixing. For this reason flake-salt which has enhanced solubility is used and is added towards the end of the mixing. Fat-coated salt may also be used; the salt becomes available only at the later stages of dough or at the early stages of baking.

19.2.1.8 Water

Water is needed to form gluten, to permit swelling of the starch, and to provide a medium for the various reactions that take place in dough formation. Water is not softened for bread-making because, as has been seen, calcium is even added for reasons already discussed. Water with high sulphide content is undesirable because gluten is softened by the sulphydryl groups.

19.2.1.9 Enzymes

Sufficient amylolytic enzymes must be present during bread-making to breakdown the starch in flour into fermentable sugars. Since most flours are deficient in alpha-amylase flour is supplemented during the milling of the wheat with malted barley or wheat to provide this enzyme. Fungal or bacterial amylase preparations may be added during dough mixing. Bacterial amy1ase from *Bacillus subtilis* is particularly useful because it is heat-stable and partly survives the baking process. Proteolytic enzymes from *Aspergillus oryzae* are used in dough making, particularly in flours with excessively high protein contents. Ordinarily however, proteases have the effect of reducing the mixing time of the dough.

19.2.1.10 Mold-inhibitors (antimycotics) and enriching additives

The spoilage of bread is caused mainly by the fungi *Rhizopus, Mucor, Aspergillus* and *Penincillium*. Spoilage by *Bacillus mesenteroides* (ropes) rarely occurs. The chief anti-mycotic agent added to bread is calcium propionate. Others used to a much lesser extent are sodium diacetate, vinegar, mono-calcium phosphate, and lactic acid.

Bread is also often enriched with various vitamins and minerals including thiamin, riboflavin, niacin and iron.

19.2.2   Systems of Bread-making

Large-scale bread-making is mechanized. The processes of yeast-leavened bread-making may be divided into:

(a) *Pre-fermentation (or sponge mixing)*: At this stage a portion of the ingredients is mixed with yeast and with or without flour to produce an inoculum. During this the yeast becomes adapted to the growth conditions of the dough and rapidly multiplies. Gluten development is not sought at this stage.

(b) *Dough mixing*: The balance of the ingredients is mixed together with the inoculum to form the dough. This is the stage when maximum gluten development is sought.

(c) *Cutting and rounding*: The dough formed above is cut into specific weights and rounded by machines.

(d) *First (intermediate) proofing*: The dough is allowed to rest for about 15 minutes usually at the same temperature as it has been previous to this time i.e., at about 27°C. This is done in equipment known as an overhead proofer.

(e) *Molding*: The dough is flattened to a sheet and then moulded into a spherical body and placed in a baking pan which will confer shape to the loaf.

(f) *Second proofing*: This consists of holding the dough for about 1 hour at 35-43°C and in an atmosphere of high humidity (89-95°C)

(g) *Baking*: During baking the proofed dough is transferred, still in the final pan, to the oven where it is subjected to an average temperature of 215-225°C for 17-23 minutes. Baking is the final of the various baking processes. It is the point at which the success or otherwise of all the previous inputs is determined.

(h) *Cooling, slicing, and wrapping*: The bread is depanned, cooled to 4-5°C sliced (optional in some countries) and wrapped in waxed paper, or plastic bags.

The Three Basic Systems of Bread-making

There are three basic systems of baking. All three are essentially similar and differ only in the presence or absence of a pre-fermentation. Where pre-fermentation is present, the formulation of the pre-ferment may consists of a broth or it may be a sponge (i.e., includes flour). All three basic types may be sponge i.e includes flour. All three basic types may also be batch or continuous.

(i) Sponge doughs: This system or modification of it is the most widely used worldwide. It has consequently been the most widely described. In the sponge-dough system of baking a portion (60-70%) of the flour is mixed with water, yeast and yeast food in a slurry tank (or 'ingridator') during the pre-fermentation to yield a spongy material due to bubbles caused by alcohol and CO_2 (hence the name). If enzymes are used they may be added at this stage. The sponge is allowed to rest at about 27°C and a relative humidity of 75-80% for 3.5 to 5 hours. During this period the sponges rises five to six times because of the volatile products released by this yeast and usually collapses spontaneously. During the next (or dough) stage the sponge is mixed with the other ingredients. The result is a dough which follows the rest of the scheme described above. The heat of the oven causes the metabolic products of the yeast – CO_2, alcohol, and water vapor to expand to the final size of the loaf. The protein becomes denatured beginning from about 70°C; the denatured protein soon sets, and imposes fixed sizes to the air vesicles. The enzymes alpha and B amylases are active for a while as the temperature passes through their optimum temperatures, which are 55-65°C and 65-70°C respectively. At temperatures of about 10°C beyond their optima, these two enzymes become denatured. The temperature of the outside of the bread is about 195°C but the internal temperature never exceeds 100°C. At about 65-70°C the yeasts are killed. The higher outside temperature leads to browning of the crust, a result of reactions between the reducing sugars and the free amino acids in the dough. The starch granules which have become hydrated are broken down only slightly by the amylolytic enzymes before they become denatured to dextrin and maltose by alpha amylase and B amylase respectively.

(ii) The liquid ferment system. In this system water, yeast, food, malt, sugar, salt and, sometimes, milk are mixed during the pre-fermentation at about 30°C and left for about 6 hours. After that, flour and other ingredients are added in mixed to form a dough. The rest is as described above.

(iii) The straight dough system: In this system, all the components are mixed at the same time until a dough is formed. The dough is then allowed to ferment at about 28-30°C for 2-4 hours. During this period .the risen dough is occasionally knocked down to cause it to collapse. Thereafter, it follows the same process as those already described. The straight dough is usually used for home bread making.

The Chorleywood Bread Process

The Chorleywood Bread Process is a unique modification of the straight dough process, which is used in most bakeries in the United Kingdom and Australia. The process, also know as CBP (Chorleywood Bread Process) was developed at the laboratories of the Flour Milling & Baking Research Association (Chorleywood, Herefordshire, UK) as a means of cutting down baking time. The essential components of the system are that:

(a)   All the components are mixed together with a finite amount of energy at so high a rate that mixing is complete in 3-5 minutes.

(b)   Fast-acting oxidizing agents (potassium iodate or bromate, or more usually ascorbic acid) are used.

(c)   The level of yeast added is 50-100% of the normal level; often specially-developed fast-acting yeasts are employed.

(d)   No pre-fermentation time is allowed and the time required to produce bread from flour is shortened from 6-7 hours to 1½-2 hours.

19.2.3   Role of Yeasts in Bread-making

Methods of Leavening: Leavening is the increase in the size of the dough induced by gases during bread-making. Leavening may be brought about in a number of ways.

(a)   Air or carbon dioxide may be forced into the dough; this method has not become popular.

(b)   Water vapor or steam which develops during baking has a leavening effect. This has not been used in baking; it is however the major leavening gas in crackers.

(c) Oxygen has been used for leavening bread. Hydrogen peroxide was added to the dough and oxygen was then released with catalase.

(d) It has been suggested that carbon-dioxide can be released in the dough by the use of decarboxylases, enzymes which cleave off carbon dioxide from carboxylic acids. This has not been tried in practice.

(e) The use of baking powder has been suggested. Baking powder consists of about 30% sodium bicarbonate mixed in the dry state with one of a number of leavening acids, including sodium acid pyrophosphate, monocalcium phosphate, sodium aluminum phosphate, monocalcium phosphate, glucono-delta-lactone. CO_2 is evolved on contact of the components with water: part of the CO_2 is evolved during dough making, but the bulk is evolved during baking. Baking powder is suitable for cakes and other high-sugar leavened foods, whose osmotic pressure would be too high for yeasts. Furthermore, weight for weight yeasts are vastly superior to baking powder for leavening.

(f) Leavening by microorganisms, may be done by any facultative organism releasing gas under anaerobic conditions such as heterofermentative lactic acid bacteria, including *Lactobacillus plantarum* or pseudolactics such as *Escherichia coli*. In practice however yeasts are used; even when it is desirable to produce bread quickly such as for the military or for sportsmen and for other emergency conditions the use of yeasts recommends itself over the use of baking powder.

The Process of Leavening: The events taking place in dough during primary fermentation i.e. fermentation before the dough is introduced into the oven may be summarized as follows. During bread making, yeasts ferment hexose sugars mainly into alcohol (0.48 gm) carbon dioxide (0.48 gm) and smaller amounts of glycerol (0.002-0.003 gm) and trace compounds (0.0005 gm) of various other alcohols, esters aldehydes, and organic acids. The figure given in parenthesis indicate the amount of the respective compounds produced from 1 gm of hexose sugars. The CO_2 dissolves continuously in the dough, until the latter becomes saturated. Subsequently the excess CO_2 in the gaseous state begins to form bubbles in the dough. It is this formation of bubbles which causes the dough to rise or to leaven. The total time taken for the yeast to act upon the dough varies from 2-6 hours or longer depending on the method of baking used.

19.2.3.1   Factors which effect the leavening action of yeasts

(i) *The nature of the sugar available*: When no sugar is added to the dough such as in the traditional method of bread-making, or in sponge of sponge-doughs and some liquid ferments, the yeast utilizes the maltose in the flour. Such maltose is produced by the action of the amylases of the wheat. When however glucose, fructose, or sucrose are added these are utilized in preference to maltose. The formation of 'Malto-zymase' or the group of enzymes responsible for maltose utilization is repressed by the presence of these sugars. Malto-zymase is produced only at the exhaustion of the more easily utilizable sugars. Malto-zymase is inducible and is produced readily in yeasts grown on grain and which contain maltose. Sucrose is inverted into glucose and fructose by the saccharase of the cell surface of bakers yeasts. While fructose and glucose are rather similarly fermented, glucose is the preferred substrate. Fermentation of the fructose moeity of sucrose is initiated after an induction period of about 1 hour. It is clear from the above that the most rapid leavening is achievable by the use of glucose.

(ii) *Osmotic pressure*: High osmotic pressures inhibit yeast action. Baker's yeast will produce CO_2 rapidly in doughs up to a maximum of about 5% glucose, sucrose or fructose or in solutions of about 10%. Beyond that gas production drops off rapidly. Salt at levels beyond about 2% (based on flour weight) is inhibitory on yeasts. In dough the amount used is 2.0-2.5% (based on flour weight) and this is inhibitory on yeasts. The level of salt addition is maintained as a compromise on account of its role in gluten formation. Salt is therefore added as late as possible in the dough formation process.

(iii) *Effect of nitrogen and other nutrients*: Short fermentations require no nutrients but for longer fermentation, the addition of minerals and a nitrogen source increases gas production. Ammonium normally added as yeast food is rapidly utilized. Flour also supplies amino acids and peptides and thiamine. Thiamine is required for the growth of yeasts. When liquid pre-ferments containing no flour are prepared therefore thiamine is added.

(iv) *Effect on fungal inhibitors (anti-mycotic agents)*: Anti-mycotics added to bread are all inhibitory to yeast. In all cases therefore a compromise must be worked between the maximum level permitted by government regulations, the minimum level inhibitory to yeasts and the minimum level inhibitory to fungi. A compromise level for calcium propionate which is the most widely used anti-mycotic, is 0.19% (based on flour weight).

(v) *Yeast concentration*: The weight of yeast for baking rarely exceeds 3% of the flour weight. A balance exists between the sugar concentration, the length of the fermentation and the yeast concentration. Provided that enough sugar is available the higher the yeast concentration the more rapid is the leavening. However, although the loaf may be bigger the taste and in particular the texture may be adversely affected. Experimentation is necessary before the optimum concentration of a new strain of yeast is chosen.

19.2.3.2 Flavor development

The aroma of fermented materials such as beer, wine, fruit wines, and dough exhibit some resemblance. However, the aroma of bread is distinct from those of the substances mentioned earlier because of the baking process. During baking the lower boiling point materials escape with the oven gases; furthermore, new compounds result from the chemical reactions taking place at the high temperature. The flavor compound found in bread are organic acids, esters, alcohols, aldehydes, ketones and other carbonyl compounds. The organic acids include formic, acetic, propionic, n-butyric, isobutyric, isocapric, heptanoic, caprylic, pelargonic, capric, lactic, and pyruvic acids. The esters include the ethyl esters of most of these acids as would be expected in their reaction with ethanol. Beside ethanol, amyl alcohols, and isobutanol are the most abundant alcohols. In oven vapor condensates ethanol constitutes 11-12 % while other alcohols collectively make up only about 0.04%. Besides the three earlier-mentioned alcohols, others are n-propanol, 2-3 butanediol, β-phenyl ethyl alcohol. At least one worker has found a correlation between the concentration of amyl alcohols with the aroma of bread. Of the aldehydes and ketones acetaldehyde appears to be the major component of pre-fermentation. Formaldehyde, acetone, propinaldehyde, isobutyraldehyde and methyl

ethyl ketone, 2-methyl butanol and isovaleradehyde are others. A good proportion of many of these is lost during baking.

19.2.3.3 Baking

Bread is baked at a temperature of about 235°C for 45–60 minutes. As the baking progresses and temperature rises gas production rises and various events occur as below:

- At about 45°C the undamaged starch granules begin to gelatinize and are attacked by alpha-amylase, yielding fermentable sugars;
- Between 50 and 60°C the yeast is killed;
- At about 65°C the beta-amylase is thermally inactivated;
- At about 75°C the fungal amylase is inactivated;
- At about 87°C the cereal alpha-amylase is inactivated;
- Finally, the gluten is denatured and coagulates, stabilizing the shape and size of the loaf.

19.3 FERMENTED FOODS MADE FROM MILK

19.3.1 Composition of Milk

Milk is the fluid from the mammary glands of animals which is meant for feeding the young of mammals. It is a complex liquid consisting of several hundred components of which the most important are proteins, lactose, fat, minerals, enzymes, and vitamins in which emusified fat globules and casein micelles are present. Its composition varies from breed to breed, as shown in Table 19.1.

Proteins: Milk proteins are divided into two: caseins and whey proteins. Caseins consist of carbohydrate, phosphorus, and protein (glyco-phspho-protein) and make up 85% of the total milk proteins. Casein exists in milk as the calcium salt, ie, as calcium caseinate in globules (micelles) ranging from 40-300 mµ in diameter. Casein exists in four types designated, α β, κ and γ and depending on their electric charges. The proportion of the various types in milk depends on the breed of the cow producing the milk. The letter 's' after α_s - caseins indicates its sensitivity to precipitation by calcium.

Table 19.1   Chemical composition (%) of milk from various mammals

Animal	Water	Fat	Protein	Lactose	Ash
Ass	89.0	2.5	2.0	6.0	0.5
Buffalo	82.1	8.0	4.2	4.9	0.8
Camel	87.1	4.2	3.7	4.1	0.9
Cow	87.6	3.8	3.3	4.7	0.6
Goat	87.0	4.5	3.3	4.6	0.6
Mare	89.0	1.5	2.6	6.2	0.7
Reindeer	63.3	22.5	10.3	2.5	1.4
Sheep	81.6	7.5	5.6	4.4	0.9

Whey proteins consist of different components which are normally stable to acid, but very sensitive to heat. β-Lactoglobulin forms about 66% of the total whey proteins, followed by α-Lactabumin (22%). The immune globulins from about 10% of the the total, and contribute towards the immunity derived by the young from the consumption of colostrum.

Lactose: The main carbohydrate in milk is lactose, which is found only in milk. It is a dissacharide of glucose and galactose and has a low sweetening ability, as well as low solubility in water.

Fat: Fat consists of one molecule of glycerol and three of fatty acids. Over 60 different acids are known in butter, many of them, being of low molecular weight of about 10 carbon atoms or less, and include saturated and unsaturated fatty acids.

Enzymes: Enzymes found in milk include proteases, carbohydrases, esterases, oxidases/reductases.

Minerals: Milk is a major source of calcium; other minerals in milk are phosphorous, magnesium, sodium, potassium, as well as sulphate and chloride ions.

When fat is removed from milk such as during butter making, the remnant is *skim milk*. On the other hand, when casein is removed such as during cheese manufacture, the remnant is known as *whey*. Whey is high in lactose and its disposal sometimes poses some problem as not all microorganisms can break down whey. It is however used in the production of yeasts to be used as food or fodder.

19.3.2   Cheese

Cheese is a highly proteinaceous food made from the milk of some herbivores. Cheese is believed to have originated in the warm climates of the Middle East some thousands of years ago, and is said to have evolved when milk placed in goat stomach was found to have curdled. The scientific study and manipulation of milk for cheese manufacture is however just over a hundred years old. Most cheese in the temperate countries of the world such as Western Europe and the USA is made from cow's milk, the composition of which varies according to the breed of the cattle, the stage of lactation, the adequacy of its nutrition, the age of the cow, and the presence or absence of disease in the breasts (udders), known as mastitis. In some subtropical countries milk from sheep, goats, the lama, yak, or ass is also used. Sheep milk is used specifically for the production of certain special cheese types in some parts of Europe (e.g. Roquefort in France, and Brinsen in Hungary). Milk from the water buffalo may be used in India and other countries, while milk from the reindeer and the mare may be used in northern parts of Scandinavia and in Russia, respectively. Cheese made from the milk of goat and sheep has a much stronger flavor than that made from cow's milk. This is because the fat in goat and sheep milk contain much lower amounts of the lower fatty acids, caproic, capryllic, and capric acids. These acids confer a sharp taste (similar to that of Roquefort cheese) to cheese made from these mammals. Future discussion of cheese in this chapter will however refer to that made from cow's milk.

About a thousand types of cheese have been described depending on the properties and treatment of the milk, the method of production, conditions such as temperature, and the properties of the coagulum, and the local preferences.

19.3.2.1   Stages in the manufacture of cheese

The manufacture of all the types of cheeses include all or some of the following processes:

(a) *Standardization of milk*

The quality of the milk has a decided effect on the nature of cheese. Cheese made from skim milk is hard and leathery; the more fat a cheese contains the smoother its feel to the palate. The fat/protein ratio is often adjusted through fat addition in order to yield a cheese of consistent quality. In the US, pasteurization (High Temperature Short Time) or (Long Temperature Short Time) must be given to milk to be in certain types of cheeses, such as cottage or cream cheese. For others the milk need not be pasteurized. but must be stored at for at least 60 days at 2°C. If however the 'starter' is slow acting or souring is delayed, food-poisoning staphylococci could develop and produce toxins in the cheese. Sub-pasteurization temperatures are often the legal compromise. Pasteurization gives a better control over the processes of cheese production. However, the organisms present in raw milk are important during the ripening processes.

The milk may also be homogenized by forcing it at high speed through small orifices to reduce the milk fat globules for use in producing soft cheeses.

(b) *Inoculation of pure cultures of lactic acid bacteria as starter cultures*

In the past, lactic acid was produced by naturally occurring bacteria. Nowadays they are inoculated artificially, by specially selected bacteria termed starters. Indeed lactic acid formation is necessary in all kinds of cheese. The propagation and distribution of lactic acid bacteria for use in cheese manufacture is an industry in its own right in the United States. For cheese prepared at temperatures less than 40°C strains of *Lactococcus lactis* are used. For those prepared at higher temperatures the more thermophilic *Streptococcus thermophilus, Lactobacillus bulgaricus,* and *Lact. helveticus are used.*

Lactic acid has the following effects:

  (i) It causes the coagulation of casein at pH 4.6, the isoelectric point of that protein, which is used in the manufacture of some cheeses, e.g. cottage cheese.
 (ii) It provides a favorably low pH for the action of rennin the enzyme which forms the curd from casein in other types of cheeses.
(iii) The low pH eliminates proteolytic and other undesirable bacteria.
 (iv) It causes the curd to shrink and thus promotes the drainage of whey.
  (v) Metabolic products from the lactic acid bacteria such as ketones, esters and aldehydes contribute to the flavor of the cheese.

Problems of lactic acid bacteria in cheese-making

  (i) *Attack by bacteriophages*: Bacteriophages sometimes attack the lactic acid starters and besides choosing strains that are resistant to phages, rotations (i.e., using different lactic mixtures every three or four days) helps eliminate them.
 (ii) *Inhibition by penicillin and other antibiotics*: Lactic acid bacteria, being Gram-positive are particularly susceptible to penicillin used to treat diseased udder in mastitis if the antibiotic finds its way into the milk; other antiobiotics also have an inhibitory effect on them.

(iii) *Undesirable strains*: Some strains of lactic acid bacteria are undesirable in cheese making because they produce too much gas, undesirable flavors, or produce antibiotics against other lactic acid bacteria. They arise by mutation.

(iv) *Sterilant and detergent residues*: Sterilant and detergent residues may inhibit the growth of starter bacteria. The minimum concentration required for inhibition varies with the different anti-microbial agents and between different strains of starter bacteria. Residues gain entry to milk at the (a) farm, (b) during transportation to the factory, and (c) the factory due to careless use of sterilants or detergents, incomplete draining or inadequate rinsing of equipment. The inhibitory effects of sterilant and detergent residues are prevented by the correct and ethical use of these materials. Proper use includes the use of the chemical at the correct concentration and adequate rinsing and draining. Their presence is mitigated by dilution with uncontaminated milk.

(c) *Adding of rennet for coagulum formation*

The classical material used in the formation of the coagulum is 'rennet' which is derived from the fourth stomach, abomasum or vell of freshly slaughtered milk-fed calves. Besides those of calves, the abomasum of kids (young goats), lamb or other young mammals have been used. Rennet is produced by soaking and/or shredding air-dried vells under acid conditions with 12-20% salt. Extracts from young calves contain 94% rennin and 6% pepsin and from older cows, 40% rennin and 60% pepsin. Rennin (chymosin) is the enzyme responsible for the coagulation of the milk. Pepsin is proteolytic and too high an amount of pepsin can result in the hydrolysis of the coagulum and a resulting low yield of cheese, and a bitter taste may result from the amino acids. Due to the high cost of animal rennet, other sources, mostly of microbial origins, have been found (Table 19.2).

Table 19.2 Some commercial microbial rennets and their microbial sources

Commercial Rennet	*Microbial source*
Harmilase	*Mucor miehei*
Rermilase	*Mucor miehei*
Fromase	*Mucor miehei*
Emposase	*Mucor pusillus*
Meito	*Mucor pusillus*
Suparen	*Endothia parasitica*
Surd curd	*Endothia parasitica*
Mikrozyme	*Bacillus subtilis*

The major effect of the milk-clotting enzymes is the conversion of casein from a colloidal to a fibrous form. First the pH of the milk is brought down from pH 6.8 -7 to pH 5.5 by the action of lactic acid bacteria which produce lactic acid from lactose in the milk. On addition of rennet, the active component, rennin, catalyses the hydrolyses of k-casein to release para-k-casein and k-casein macropeptide. The latter goes into whey, while the para-k-casein remains part of casein micelles, which now bind together to form the curd following the removal of carbohydrates with the k-casein macropeptide and the exposure of binding surfaces. The events up this coagulation are aided by lowered pH and by increasing temperatures up to 45°C. Most of the bacteria, fat, and other particulate matter are entrapped in the curd. When casein is removed the remaining liquid containing proteins, lactalbumin, globulin, and yellow-green riboflavin (vitamin B_2) is whey. The whey proteins may be precipitated by heat, but not acid or rennin and they are used in making whey cheese. The enzymes used in cheese making are now obtained from microorganisms, mainly fungi.

(d) *Shrinkage of the curd*

The removal of whey and further shrinkage of the curd is greatly facilitated by heating it, cutting it into smaller pieces, applying some pressure on it and lowering the pH. In many types of cheeses, such as Parmesan, Emmenthal and Gruyere, there is a stage known as 'scalding' in which the temperature can be as high as 56°C in the preparation. Acid produced by the lactic starters introduce elasticity in the curd, a property desirable in the final qualities of cheese.

(e) *Salting of the curd and pressing into shape*

Salt is added to most cheese varieties at some stage in their manufacture. Salt is important not only for the taste, but it also contributes to moisture and acidity control. Most importantly however it helps limit the growth of proteiolytic bacteria which are undesirable. The curd is pressed into shape before being allowed to mature.

(f) *Cheese ripening*

The ripening or maturing of cheese is a slow joint microbiological and biochemical process which converts the brittle white curd or raw cheese to the final full-flavored cheese. The agents responsible for the final change are enzymes in the milk, in the rennet and those from the added starter microorganisms as well as other micro-organisms which confer the special character of the cheese to it. Among the cheese whose peculiar characteristics are dependent on particular microorganisms are the blue-veined cheese Roqueforti, Gorgonzola, Stilton, conferred by *Penicillium roquefort,* Swiss cheese, with its characteristic flavor and holes produced by the fermentation products and gases from *Propionibacterium* spp. Yeasts, micrococci, and *Brevibacterium linens* impart the characteristic flavor of Limburger cheese. In soft cheese, such as Camembert, the protein is completely broken down to almost amino acids, whereas in the hard cheese, the protein remains intact.

19.3.3   Yoghurt and Fermented Milk Foods

Many types of fermented milks are produced and drunk around the world (Table 19.3) Yoghurt is a fermented milk traditionally believed to be an invention of the Turks of Central Asia, in whose language the word yoghurt means to blend, a reference to how the milk product is made. Although accidentally invented thousands of years ago, yoghurt has only recently gained popularity in the United States. While yoghurt has been present for many years, it is only recently (within the last 30-40 years) that it has become popular. This is due to many factors including the introduction of fruit and other flavorings into yoghurt, the convenience of it as a ready-made breakfast food and the image of yoghurt as a low fat healthy food.

Table 19.3 Fermented milks and their presumed countries of origin

Name	Presumed country of origin	Description	Cultures
Yoghurt	Asia, Balkans	Acidic, set or stirred, characteristic aroma	*S. thermophilus, Lb. bulgaricus, (and Lb. acidophilus, Bifidobacterium* spp.) *
Acido-philus milk	USA	Set, stirred or liquid, mild flavor	*Lb. acidophilus*
Kafir	Caucasus	Stirred beverage, creamy consistency, characteristic taste and aroma (CO_2)	*Lc. lactis, Lc. cremoris, Lb. kefir, Lb. casei,Lb.*
Kumiss	Mongolia	Frothy beverage, acid, refreshing taste	*Lb. bulgaticus, Lb. acidophilus, yeasts*
Lassi	India	Sour milk drink diluted with water, consumed salted, spicy or sweet	*Lactococcus* spp., *Lactobacillus* spp., *Leuconostoc*
Dahi	India	Set, stirred, or liquid beverage pleasant	*S. thermophilus, Lb. bulgaricus, Lc. diacetylactis,*
Leben	Middle East	Set or stirred product, pleasant taste and aroma	*S. thermophilus, Lb. bulgaricus, Lb. acidophilus,*
Filmjilk	Sweden	Viscous stirred beverage, clean acid taste	*Lc. lactis, Lc. cremoris, Lc. diacetylactis, Ln.*
Villi	Finland	Viscous stirred product, mildly sour	*Lc. lactis, Lc. cremoris, Lc. diacetylactis, Lc.*

* Depending on the country

In the manufacture of yoghurt, two kinds of lactic acid bacteria, *Lactococcus* spp. and *Lactobacillus* spp., are generally used with usually unpasteurized milk. Most commonly used are *Lactococcus salivarius* and *thermophilus*, and *Lactobacillus* spp., such as *Lacto. acidophilus, bulgaricus* and *bifidus*.

The bacteria produce lactic acid from lactose in the milk causing the pH to drop to about 4-5 from about 7.0. This drop in pH causes the milk to coagulate. The lactic acid gives yoghurt its sour taste and limits the growth of spoilage bacteria. Yoghurt is flavored usually with fruits.

19.4 FERMENTED FOODS FROM CORN

Corn is a tropical crop, but grows in the summer in temperate climates, which have at least 90 days which are frost free. It is known as maize in some parts of the world. Scientifically known as *Zea mays* L, it is used to make important fermented foods in west Africa and it is sometimes mixed with sorghum, *Sorghum bicolor* Linn for this purpose.

19.4.1   Ogi, Koko, Mahewu

Ogi, also known as *akamu,* is a Nigerian sour gruel made from maize. In Ghana the equivalent foods are known as koko. For ogi preparation, corn is soaked in water for about two days. Thereafter the cereal is wet-milled and sieved to remove the fibrous portions of the maize. The starchy sediment is allowed to settle and to ferment for another two days. The water is decanted off and the starchy sediment is ogi. It is prepared by boiling it to form a thick gruel which can be consumed sweetened with sugar, or eaten with foods made from beans. It may also be heated to form a stiff gel, known as *agidi* or *eko* when cool. Ogi or the stiff gel made from it are popular weaning and convalescent foods in Nigeria.

Studies of the microbiology of ogi production show that in the early stages of fermentation, fungi such as *Fusarium* and *Cephalosporium* which are acquired from the field, and which form the bulk of the organisms in the first 24 hours, soon disappear to be replaced with lactic acid bacteria especially *(Lactobacillus plantarum* and *Lact mesenteroides)* and yeasts *(Saccharomyces cerevisiae, Rhodotorula* spp., and *Candida mycoderma* which become the dominant organisms at the time of the milling.

The flavor of ogi has been shown to be due to the activities of lactic acid bacteria and occasionally to yeasts and acetic acid bacteria. The following acids were identified in that quantitative order by gas chromatography: acetic, butyric, pentanoic, isohexanoic, and isobutyric.

The nutritional quality of ogi suffers from its method of preparation: the soaking of the grains and the discarding of the overtails before milling leads to loss of minerals as well as a portion of the already low quantity of protein and amino acids present in the cereals. The flow charts for ogi production and that of a similar food, *koko,* from Ghana are in Fig. 19.1

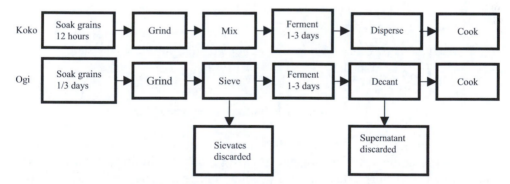

Fig. 19.1   Flow Charts for *Koko* and *Ogi* Production

Mahewu, also known as *mogou,* is a South African sour food. Although the main organism found in locally-made mahewu is *Streptococcus lactis,* mahewu is made on an industrial scale by inoculating *Leuconostoc delbruckii* into autoclaved 8-10% maize slurry, fermenting the mixture for about 12 hours and spray-drying the slurry. It is an acid food of about pH 3.5, and in order to ensure that the proper level of lactic acid is produced to attain this pH level, buffering salts such as $CaHPO_4$ are somtimes added. It is a convenient food consumed by miners and the dry powder needs only to be reconstituted in cold water to get the food ready for consumption. The food is sometimes enriched with vitamin and protein rich additives such as yeast extract.

19.5 FERMENTED FOODS FROM CASSAVA: GARRI, FOO-FOO, CHIKWUANGE, KOKONTE, BIKEDI, AND CINGUADA

Cassava is an important source of food all over the tropical world in South America, Africa, India and the Far East. Botanically it is a member of the family *Euphorbiaceae* and is classified as *Manihot esculenta* Crantz (formerly *Manihot utillisaima* Pohl). It has a number of synonyms around the world: *manioc* (Madagascar and French-speaking Africa) *tapioca* (India, Malaysia), *ubi scetlela* (Indonesia) *manioca* or *yucca* (Latin America).The plant tolerates low soil fertility and drought, better than most crops and needs little maintenance once planted. It has also been claimed to be a higher producer of carbohydrate then commonly cultivated cereals and tuber crops and under favorable conditions will yield above 90 ton/hectare. It is therefore not surprising that it is the staple food in densely populated areas of the tropics such as Central Java in Indonesia, the State of Kerala in India, south-eastern Nigeria and north-eastern Brazil and is consumes by an estimated 400 million people around the world. Nigeria is currently the world's largest producer followed by Brazil, Indonesia, Zaire, Thailand, and Tanzania.

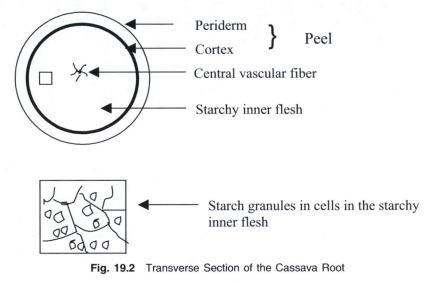

Fig. 19.2 Transverse Section of the Cassava Root

A major shortcoming of cassava roots is that they are very low in protein. In addition many varieties contain the cyanogenic glucosides, linamarin and (to a lesser extent) lotaustralin. These glucosides can give rise to fatalities if cassava roots are consumed unprocessed. Cassava may be processed by boiling, roasting, drying, leaching with cold water, or by fermentation. By far the most popular method of processing cassava is by fermentation. In producing fermented cassava products, the roots may first be grated before fermentation, the whole root may be cut into large pieces and fermented in water (retted). The best known example of foods produced from cassava pulp is garri, while those produced from the retting of whole roots include foo-foo, chikwuangue, kokonte, and cinguada.

19.5.1   Garri

Garri is a popular food for about 100 miilion people in West Africa; in Cote d'Ivoire, *atieke*, a food very similar to garri, produced from cassava but is is not fried like garri. A similar food known as *farinha de manioc* or *farinha de mega* is consumed in parts of Brazil.

Preparation of Garri: Garri is currently prepared mostly on small, house-hold scales. The first stage is the peeling to remove the brownish thin outer covering (Fig. 19.2) to reveal the white fleshy inner portion which is grated on a hand-held rasper or crushed in a grating machine. The central pith and primary xylem provide some fibers in the grated material some of which is removed by sieving, but which is appreciated by some garri consumers.

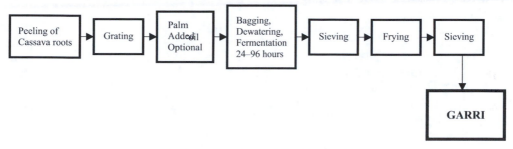

Fig. 19.3   Flow Chart of Garri Production

The mash resulting from the grating is placed in cloth bags for between 18 and 48 hours and fermented. During the period of fermentation, the mash is dewatered by placing heavy objects on the cloth bags. At the end of the fermentation period, the mash is sieved through a coarse sieve and heated, sometimes with a little palm oil, in a flat iron pot with stirring.

Microbiology of the fermentation of Garri: In 1959 Collard and Levi published their study on the two-stage fermentation of cassava pulp. In the first stage which lasted for the first 48 hours a yellow-pigmented bacterium, *Corynebacterium manihot*, proliferated. This organism broke down starch eventually to organic acids including lactic acid. The resulting drop in pH led to the spontaneous breakdown of the linamarin and the proliferation of the fungus *Geotrichum candidum* which produced the flavoring aldehydes, ketones and other compounds. In 1977 Okafor re-examined fermenting pulp and while there he found some *Corynebacterium*, the bulk of the organisms were lactic acid bacteria, especially *Lactobacillus*, *Leuconostoc* and and yeasts. He suggested the absence of lactic acid bacteria in the work of Collard and Levi was probably because they used nutrient agar, a medium lacking sugars in which lactic acid bacteria grow better. Furthermore, linamarin is fairly stable and the glucoside is probably broken down more by the indigenous linamarase of the enlarged roots, which is released when the roots are crushed than by a change of pH. Other workers have since confirmed the importance of lactic acid bacteria and yeasts in the fermentation of cassava during garri production.

The current thinking on the microbiological processes of cassava fermentation for garri production is that when cassava roots are grated to produce the mash which is bagged and fried to produce garri, the indigenous linamarase present in the roots is released and makes contact with the cyanogenic glucosides in the roots. The glucosides, mostly linamarin and some lotaustralin (about 5% of the total glucosides) are then broken down into glucose and HCN (Fig. 19.4). The HCN release is characteristically noticed by the pungent smell in evidence whenever cassava is being grated. As the amount of the linamarase is insufficient to hydrolyze all the cyanogenic glucosides or because the particles are not fine enough to ensure complete contact between the enzyme and the substrate, there is always residual glucoside which enters the garri as cyanide. Many of the organisms encountered in fermenting cassava mash are lactic acid bacteria and yeasts. *Lactobacillus, Leuconostoc,* the yeast *Candida* and various other yeasts are encountered in fermenting cassava mash, and many strains of these have been found to produce linamarase. By inoculating one of such linamarase producing organisms into fermenting cassava mash the group was able to almost totally remove the residual cyanide in garri.

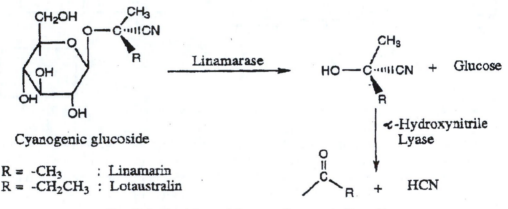

Fig. 19.4   Breakdown of Cassava Cyanogenic Glucosides

19.5.2   Foo-Foo, Chikwuangue, Lafun, Kokonte, Bikedi, and Cinguada

The preparation of these foods, although eaten in different parts of Africa, is similar. Foo-foo is eaten in parts of eastern Nigeria, while lafun is eaten in western Nigeria. Chikwuangue is eaten in Democratic Republic of the Congo, bikedi in Congo (Brazzaville), kokonte and cinguada are eaten in Ghana and East Africa respectively. In the preparation of these foods, cassava roots are cut into large pieces and immersed in still water in pots or in running stream water and allowed to ret for one to five days; for foo-foo or chikwuangue fermentation retting takes between three and six days so that the starch can be extracted from the retted roots by macerating with the hands. For kokonte and cinguada, retting is only partial and hardly lasts more than two days; the material is then sun-dried and pounded into a flour when it is known as lafun in Nigeria. For foo-foo and bikedi the retted roots are macerated to extract the starch. The retting is a result of the breakdown of the pectin in the cell walls of the cassava root brought about by pectinases produced by bacteria of the genus *Bacillus* spp., while the lactic acid bacteria are responsible for the flavor of these foods.

19.6   FERMENTED VEGETABLES

Like the fermentation of other foods, vegetables have been preserved by fermentation from time immemorial by lactic bacterial action. A wide range of vegetables and fruits including cabbages, olives, cucumber, onions, peppers, green tomatoes, carrots, okra, celery, and cauliflower have been preserved. Only sauerkraut and cucumbers will be discussed, as the same general principles apply to the fermentation of all vegetables and fruits. In general they are fermented in brine, which eliminates other organisms and encourages the lactic acid bacteria.

19.6.1   Sauerkraut

Sauerkraut is produced by the fermentation of cabbages, *Brassica oleracea*, and has been known for a long time. Specially selected varieties which are mild-flavored are used. The cabbage is sliced into thin pieces known as slaw and preserved in salt water or brine containing about 2.5% salt. The slaw must be completely immersed in brine to prevent it from darkening. Kraut fermentation is initiated by *Leuconostoc mesenteroides*, a heterofermentative lactic acid bacterium (i.e., it produces lactic acid as well as acetic acid and CO_2.) It grows over a wide range of pH and temperature conditions. CO_2 creates anaerobic conditions and eliminates organisms which might produce enzymes which can cause the softening of the slaw. CO_2 also encourages the growth of other lactic acid bacteria. Gram negative coliforms and pseudomonads soon disappear, and give way to a rapid proliferation of other lactic acid bacteria, including *L. brevis*, which is heterofermentative, and the homofermentative *L. plantarum*; sometimes *Pediococcus cerevisiae* also occurs. Compounds which contribute to the flavor of sauerkraut begin to appear with the increasing growth of the lactics. These compounds include lactic and acetic acids, ethanol, and volatile compounds such as diacetyl, acetaldehyde, acetal, isoamyl alcohol, n-hexanol, ethyl lactate, ethyl butarate, and iso amyl acetate. Besides the 2.5% salt, it is important that a temperature of about 15°C be used. Higher temperatures cause a deterioration of the kraut.

19.6.2   Cucumbers (pickling)

Cucumber (*Cucumis sativus*) is eaten raw as well after fermentation or pickling. Cucumbers for pickling are best harvested before they are mature. Mature cucumbers are too large, ripen easily and are full of mature seeds. Cucumbers may be pickled by *dry salting* or by *brine salting*.

Dry salting is also generally used for cauliflower, peppers, okra, and carrots. It consists of adding 10 to 12% salt to the water before the cucumbers are placed in the tank. This prevents bruising or other damage to the vegetables.

Brine salting is more widely used. A lower amount of salt is added, between 5 and 8% salt being used. Higher amounts were previously used to prevent spoilage. It has been found that at this salt concentration, the succession of bacteria is similar to that in kraut. However *Leuconostoc* spp. never dominate. During the primary fermentation lasting two or three days, most of the unwanted bacteria disappear allowing the lactics and yeasts to proliferate. In the final stages, after 10 to 14 days, *Lactobacillus plantarum* and *L. brevis*, followed by *Pediococcus*, are the major organisms.

19.7 FERMENTATIONS FOR THE PRODUCTION OF THE STIMULANT BEVERAGES: TEA, COFFEE, AND COCOA

Tea, coffee, and cocoa are produced mainly in the rainforest zones of the Indian sub-continent and in South America and West Africa respectively. Tea can also grow in the cooler temperatures of mountains. The beverages are stimulating on account of their content of either one or the other of two chemically similar stimulants, caffeine and theobromine (Fig. 19.5). Of the three, only cocoa and coffee are produced by some form of fermentation; the production of tea is strictly speaking a chemical reaction, but it is included for completeness.

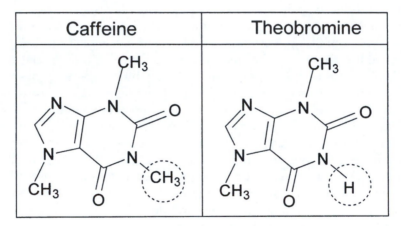

Fig. 19.5 Structure of Caffeine and Theobromine

19.7.1 Tea Production

Tea (*Camellia sinensis*; previously *Thea*) is believed to have originated from south-east China. It has now spread to many parts of the world including India, Sri Lanka, Malaysia, Kenya, Georgia in the former USSR, Turkey, Iran, Malawi, Cameroon, Thailand, Vietnam, Mexico, and Argentina, to name some of the countries where varieties of tea grow. Young tea leaves are harvested by hand and spread on trays to wither. Thereafter the leaves are rolled to squeeze out juices from the leaves and spread the juices over the surface of the leaves. This exposes the polyphenols to oxidation, and the green color gradually begins to turn brownish. Rolling also breaks the leaves into smaller pieces. The 'fermentation' stage follows, but this is a chemical reaction involving polyphenols. After fermentation, the tea is 'fired', i.e. subjected to hot air of between 80 and 90°C. After firing, the tea is sorted and graded.

19.7.2 Coffee Fermentation

Coffee (*Coffea arabica* and *C robusta*) originated from Ethiopia. The main producers of coffee today are Colombia, Brazil, Angola, and Indonesia, in that order. It takes from three to five years of growth before the coffee tree is ready to bear fruit. The fruits grow slowly, taking from 8 to 12 months to reach maturity (when they are bright red in color). Each coffee fruit or berry contains two seeds covered by pulp.

There are two methods of processing coffee: the wet method and the dry method. In the wet method, the fruits are passed through a pulping machine which removes the pulp leaving by mucilage which is removed by pectinolytic enzymes of microbiological origin. The coffee may also be dried by exposure to sunlight. When dry, the fruits are dehulled to remove the dry outer portions. The studies carried on the microbiology of the coffee fermentation showed that many of the organisms were pectinolytic organisms, including spore-forming and non-spore forming ones. Other workers found lactic acid bacteria (*Leuconostoc* spp. and *Lactobacillus* spp.) and yeasts (*Saccharomyces* spp and *Schizosaccharomyces* spp.), and it would appear that these developed from the release of the pectinolytic organisms.

19.7.3   Cocoa Fermentation

Cocoa (*Theobroma cacao*) is a native of South America, but today the major producers are Ghana, Nigeria, Ivory Coast, Cameroon, and Malaysia. The tree produces pods which contain from 40 to 60 seeds. The pods are opened and the seeds heaped and allowed to ferment, often in baskets which permit liquid to drain out. During fermentation the mucilagenous outer covering of the seeds is broken down by microbial action, while the seeds themselves change from pinkish to black. It is believed that the lactic acid bacteria play important roles in the development of the aroma of cocoa.

19.8   FERMENTED FOODS DERIVED FROM LEGUMES AND OIL SEEDS

Legumes are members of the *Leguminosae*. Their seeds are rich in proteins and they are fermented in various parts of the world for flavoring condiments or as major meals. Fermented seeds of soybeans, beans (*Phaeseolus*) and the African oil bean, *Pentaclethra macrophylla* Benth will be discussed.

19.8.1   Fermented Foods from Soybeans

Fermented soybean products have been made and consumed in large amounts in countries of the Orient for thousands of years. It has been suggested that the Buddhist religion which emphasizes the absence of meat from the diet may have been responsible for the development of soy-based foods in China, Japan, Korea, and other oriental countries. Table 19.4 shows some soy foods and where they are consumed. The use of some of them has spread to other parts of the world including the US and parts of Europe.

The soybean plant itself *Glycine max* is a legume believed to have originated from Eastern Asia. It is now grown around the world.

The soybean seed has an unusual composition. It is rich in protein and oil, and comparatively low in carbohydrates. Its average composition is 42% protein 17% carbohydrate, 18% oil, and 4.6 ash. Sucrose, raffinose, stachyose and pentosans are among the carbohydrates. The beans are rich in phospholipids, nucleic acids, and vitamins especially thiamin, riboflavin, and niacin. It should be noted that the composition of soybeans varies from place to place. The amino acid composition of its

Table 19.4   Fermented products of soybeans and countries of origin or of greatest use

Product	Country
Soy sauce	China, Japan
Miso	China, Japan, Philippines
Natto	Japan, China
Fermented soy sauce	Japan
Sufu	China, Taiwan
Tempeh	Indonesia

protein is also unusual among plant proteins in that it contains high amounts of methionine which is more characteristic of animal than plant proteins.

Soybean is a very nutritious food. However it has shortcomings which are ameliorated by fermentation. Soybeans contain compounds which make the legume unattractive until they are removed by the various stages involved in their processing by fermentation. First, they contain carbohydrates, which are not absorbed until they reach the colon, where the gases produced when they are broken down by microorganisms give rise to flatulence. These carbohydrates include the oligosacharides, raffinose and stachyose and the polysaccharide, arabinogalactan. Second, soybeans have a bitter and 'beany' taste when crushed. This is because the lipoxygenase enzyme which helps produce this taste and the substrate (oil) are held in separate compartments in the tissues of the seeds until the latter are broken or crushed.

Third, soybeans contain anti-nutritional factors such as trypsin inhibitor, hemagglutinins and saponins. Finally even after cooking, about 1/3 of the protein of soybeans cannot be digested.

The soaking of the soybean preceding cooking leaches out a large proportion of the flatulence producing carbohydrates. The 'beany' flavor is due to the presence of several carbonyl compounds such as hexanol and pentanol. These are removed by the action of microorganisms. Fermentation also reduces the carbohydrates of rice and proteins of the bean to lower molecular weights, hence rendering them more digestible. Finally, the anti nutritional factors are destroyed by boiling. In addition to all this, fermentation by *R oligosporous* produces an anti-oxidative compound (4^1, 6^1, 7 trihydroxy-bisoflavane) which is absent from raw soybean, and which helps preserve the fermented foods. The fermented foods derived from soybean (soy sauce, miso, natto) will be considered in this section.

19.8.1.1 Soy sauce

Soy sauce known as shoyu in Japan is a salty pleasantly tasting liquid with a distinct aroma and which is made by fermenting soybeans, wheat, salt with a mixture of molds, yeasts and bacteria. Five different types of shoyu are recognized by the Japanese Government, depending on the proportions of the ingredients used and the method of preparation. Koikuchi-shoyu is the most produced, forming 85% of the total produced. It is this type which is also best known in countries outside the orient. Koikuchi-shoyu is deep red-brown in color and is an all purpose seasoning, with a strong aroma and myriad flavor. It is the only type to be discussed.

Soy sauce manaufacture: The manufacture of koikuchi-shoyu can be divided into four sections: i) the preparation of the ingredients; ii) koji preparation; iii) brine fermentation; iv) refining process

Preparation of the ingredients: Whole wheat is roasted and then coarsely ground. Roasting adds color and flavor to the resulting sauce and kill surface organisms as well as facilitates enzymatic hydrolysis of the grain. Soybeans, usually defatted, are cooked under high pressure and temperature for a short time after a previous soaking in water.

Koji preparation: Whole wheat and soy prepared as described above are used for the preparation of koji. A koji starter, or seed mold or inoculum is first prepared from the spores of several different strains of *Aspergillus oryzae* or *Asp soyae* by inoculating the spores of the fungi on to a mixture of boiled rice and wood ash or mineral salts and spreading the mixture thinly at 30°C for up to five days. The koji starter (also known as *tane koji*) is used to inoculate equal amounts of the wheat and soy prepared as above. This used to be turned manually in shallow trays, but is now also being done mechanically. The mixture is put into large vats and aerated by forced aeration. The important requirements of koji are that it should have high protease and amylase activities. As these are dependent on temperature and humidity, the latter are strictly controlled. After two to three days koji is harvestod as a greenish-yellow material due to the spores of *Aspergillus*.

Brine Fermentation: Koji is introduced into deep fermentation tanks to which an equal volume of salt solution 20-23% is added. The resulting mixture, known as moroni is allowed to ferment for 6-8 months. It is frequently mixed to distribute the material and to eliminate undesirable anaerobic organisms.

During the period, koji enzymes hydrolyze proteins to amino acids and low molecular weight peptides; much of the starch is converted to simple sugars which are then ferrmented to lactic acid, alcohol and CO_2. The pH drops from around 6.5-7.0 to 4.7-4.8. The effective salt concentration is about 18 % (because of the dilution with added koji); it is never allowed to fall below 16% otherwise putrefactive organisms might develop.

There are three stages in the fermentation of moromi, which is brought about by osmophilic strains of microorganisms, after the release of simpler substances by the fungi of the koji. In the first stage, *Pediococcus halophilus* produces lactic acid, causing a drop of the pH. In the second stage *Saccharomyces rouxii* develops and produces alcohol. In the last stage, *Torulopsis* yeasts develop. These produce phenolic compounds which are important components of koichuki-shoyu flavor. The organisms are selected by the conditions of the fermentation, but pure cultures as used more and more nowadays to ensure a more consistent flavor.

Refining: The final state consists of pressing the fermented moromi to release the soy sauce. Hydraulic presses are used in modern production. The raw soy sauce is heated to 70-.80°C to pasteurize it, to develop color and flavor and to inactivate the enzymes. After clarification by sedimentation the sauce is bottled under aseptic conditions, sometimes with the addition of preservatives as well.

In China 'tamari-shoyu' which forms less than 3% of Japanese sauce, is the main type of shoyu. The two differ in that tamari has a higher proportion of soybeans (90% instead of 50%). Furthermore, tamari sauce is not pasteurized. Due to the low quantity of rice, little alcoholic fermentation occurs in tamari because of the paucity of sugars.

19.8.1.2 Miso

Miso, a fermented paste of soybean, wheat and salt is the most important of the soy fermented products in Japan. There are many types of bean pastes. They are also popular in China, Korea and other parts of the Orient, where the different types of paste produced vary according to the proportions of wheat, soybean and salt used, and the lengths of the fermentation and ageing. In Korea they are known as 'jang'; 'miso', and 'shoyu' in Japan, 'tao-tjo' in Indonesia and Thailand and 'tao-si' in the Phillipines. In Japan the average annual consumption is 7.2 kg per person, 80-85% being used in the miso group and the rest as seasonings for various types of foods. Most miso in Japan has a consistency like peanut butter, the color varying from **a** creamy yellowish white to very dark brown. The darker the color in general, the stronger the flavor. It is distinctively salty and has a pleasant aroma.

Manufacture of Miso: Miso production is basically similar to that of shoyu or soy sauce. There are however two basic differences in the production of the two foods. First, the koji or shoyu is made by using a mixture of soybeans and wheat. In koji-making for miso, only the carbohydrate material (rice or barley) is used. Soybeans are not used for making koji miso except in the case of soybean miso. Second, no pressing is done after miso fermentation. Since the material is a paste; the absence of pressing affects the cost of miso. The organisms involved in the fermentation are the same, but *Streptococcus faecalis* is also included. After fermentation, the resulting koji is mixed with salt, cooked soybean, pure cultured yeasts, and lactic acid bacteria and then fermented for a second time. It is then aged and packaged as miso; sometimes it may be freeze-dried before packaging.

19.8.1.3 Natto (Fermented whole soybean)

Whereas soy sauce (shoyu) and miso (bean paste) originated from China, nato, fermented whole soybean is an indigenous Japanese food, originating there more than 1,000 years ago. There are two types of natto, itohiki-natto and hamma-natto. Hama-natto is produced by the action of *Aspergillus*. It is produced only in limited quantities. Natto therefore usually refers to the second and commoner type itohiki-natto. This second type is fermented by *Bacillus natto*. The shape of cooked whole soybean grains is kept, but the surface of each grain is covered with a viscous material consisting of glutamic acid polymers produced by *B natto*.

The manufacture is uncomplicated. Cooked soybean grains are inoculated with the *Bacillus* and put into a small tray, covered, and incubated at 40°C. After 14-18 hour, the packed tray cooled to 2-7°C and then shipped to the market. It is cheap and nutritious and natto is usually served with shoyu and mustard.

19.8.1.4 Tempeh: Oncom and related foods

Tempeh is a popular Indonesian food made by fermenting soybean with strains of *Rhizopus*. Especially in the Indonesian Island of Java, tempeh is a key protein source and 30-120 gm is consumed daily per person. It therefore replaces meat in the grain-centered local meal. It is also eaten in Surinam and New Guinea, but not in the colder regions of the Orient.

Traditional tempeh preparation varies in minor details. Essentially air-dried soybeans are soaked in water and the seed coats are removed. The dehuued beans are boiled in water, drained, cooled, and inoculated with one of the traditional mold inocula. The beans are then packed in small parcels and incubated at room temperature of about 25°C for approximately 40 hour. Fermentation is regarded as complete when the beans have become bound tightly by the mold mycelium into compact white cakes, which are usually consumed within a day or two. It can then, after fermentation be deep-fried for 3-4 minutes or boiled for 10 minutes.

Although several species of *Rhizopus* may be used, *Rhizopus oligosporous* Saito has been shown to be the species producing tempeh. The fungus is strongly proteolytic but has only weak amylase activity, desirable qualities since soybean is high in proteins, but relatively low in carbohydrate content. The proteolytic enzyme of *Rhisopus oligosporous* is not inhibited by inhibitory factors in soy bean.

19.8.2   Fermented Foods from Beans: Idli

Idli is a popular fermented breakfast and hospital food which has been eaten in South India for many years. It is prepared from rice grains and the seeds of the leguminous mung grain, *Phaeseolus mungo*, or from black beans, *Vigna mungo*, which are also known as dahl. When the material contains Bengal grain, *Circer orientium*, the product is known as khaman. It has a spongy texture and a pleasant sour taste due to the lactic acid in the food. It is often embellished with flavoring ingredients such as cashew nuts, pepper and ginger.

Production of Idli

The seeds of the dahl (black gram) are soaked in water for 1-3 hours to soften them and to facilitate decortication, after which the seeds are mixed and pounded with rice in a proportion of three parts of the beans and one of rice. The mixture is allowed to ferment overnight (20-22 hours). In the traditional system the fermentation is spontaneous and the mixture is leavened up to approximately 2 or 3 times. The organisms involved in the acidification have been identified as *Streptococcus faecalis,* and *Pediococcus* spp. The leavening is brought about by *Leuconostoc mesenteroides,* although the yeasts, *Torulopsis candida* and *Trichosporon pulluloma* have also been found in traditional Idli. The fermented batter is steamed and served hot. Idli is highly nutritious, being rich in nicotinic acid, thiamine, riboflavin, and methionine.

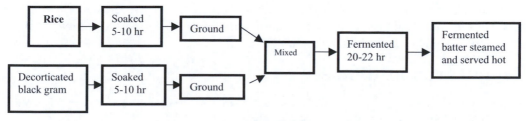

Fig. 19.6 Flow chart of Idli Production

19.8.3 Fermented Foods from Protein-rich Oil-seeds

Stew condiments made from oil-rich seeds are eaten in parts of West Africa, while condiments from fish and fish products appear to be common in parts of Asia.

Stew condiments eaten in parts of Nigeria include *dawadawa*, know as *iru* in the Southwest geopolitical zone of Nigeria which is produced from the seeds of *Parkia biglobosa*. Cadbury Plc Nigeria now makes and markets *dawadawa as Dadawa*. A condiment very similar to *dawawa* is *okpeyi* and comes from the Nsukka area of the Southeast and is made from the seeds of *Prosopsis africana*. Another major soup condiment popular in the Southeast zone is *ogili* which may be made from the seeds of castor-oil seeds (*Ricinus communis*) or egusi (*Citrullus lanatus sub*-species *colocynthoides*). Egusi, pumpkins and squashes are members of the family *Cucurbitaceae* or cucurbits and their seeds contain about 50% oil and 35% protein after dehulling. Besides egusi, another well-known cucurbit in Nigeria is *ugu* or *Telfaria* sp.; its seeds and those of soybeans are sometimes used for making *ogili*.

A fermented delicacy and meat substitute, *ugba* or *ukpaka* is made from the seeds of the African locust bean, *Pentaclethra macrophylla* Benth). This is generally eaten with stockfish and it is very popular in the South-east zone of Nigeria. *Bacillus* spp. have been implicated in the fermentation of the various oilseeds discussed above.

19.8.4 Food Condiments Made from Fish

Fish sauce is eaten in many parts of Asia including Japan, Thailand, Vietnam, the Philippines, Indonesia and Malaysia, and some parts of Northern Europe, including France.

Table 19.5 Methods of fish sauce preparation and countries of use

S/No	Country	Fish used	Method
1	France	*Gobius* spp.	2-8 weeks in 4:1 salt solution
2	Indonesia	*Clupea* spp.	6 months in 6 :1 salt solution
3	Japan	*Clupeapilchardus*	5-6 months in 5 :1 salt to rice
4	Malaysia	*Stelophorus* spp.	4–12 months; 5 :1 salt to sugar
5	The Philippines	*Stelophorus* spp.	3–12 months; 4: 1 salt solution
6	Thailand	*Stelophorus* spp.	6-12 months; 4:1 salt solution

The fish is fermented in a solution of salt. Sometimes, the fish is fermented whole, or sometimes only the viscera are fermented. Sometimes carbohydrate sources such as malted rice may be added, as in Japan.

SUGGESTED READINGS

Amund, O.O., Ogunsina, O.A. 1987. Extracellular amylase production by cassava-fermenting bacteria. Journal of Industrial Microbiology. 2, 123-127.

Bamforth, C.W. 2005. Food, Fermentation and Microorganisms. Blackwell Oxford, UK.

Banigo, E.O.I., de Man, J.M., Duitschaever, C.L. 1974. Utilization of high-lysine corn for the manufacture of ogi, using a new improved proceessing system. Cereal Chemistry 51: 559–573.

Bankole, M.O., Okagbue, R.N. 1992. Properties of "nono", a Nigerian fermented milk food. Ecology of Food and Nuitrition. 27, 145-149.

Batock, M., Azam-Ali, S. 1998. Fermented Frutis and Vegetables. A Global Perspective. FAO. Agricultural Services Bulletin No. 134. Food and Agriculture Organization of the United Nations, Rome, Italy.

Enujiugha, V.N., Akanbi, C.T. 2005. Compositional Changes in African Oil Bean (*Pentaclethra macrophylla* Benth) Pakistan Journal of Nutrition, 4. 27-31.

Ikediobi, C.O., Onyike, E. 1982. The use of linamarase in gari production. Process Biochemistry 17, 2-5.

Kobawila, S.C. Louembe, D., Keleke, S., Hounhouigan, J., Gamba, C. 2005. Reduction of the cyanide content during fermentation of cassava roots and leaves to produce bikedi and ntoba mbodi, two food products from Congo. African Journal of Biotechnology. 4 689-696.

Lei, V., Amoa-Awua, W.K.A., Brimer, L. 1999. Degradation of cyanogenic glycosides by Lactobacillus plantarum strains from spontaneous *cassava fermentation* and other microorganisms International Journal of Food Microbiology 53, 169-184.

Nwachukwu, S.U., Edwards, A.W.A. 1987. Microorganisms associated with cassava fermentation for lafun production. Journal of Food and Agriculture. 1, 39-42.

Okafor, N., Okeke, B., Umeh, C., Ibenegbu, C. 1999. Secretion of Lysine by Lactic Bacteria and Yeasts Associated with Garri Production Using a Synthetic Gene. Letters in Applied Microbiology, 28, 419-422.

Okafor, N., Uzuegbu., J.O. 1993. Studies on the Contributions of Factors other than micro-organisms to the Flavour of Garri. Journal of Agricultural Technology, 1, 36-38.

Okafor, N., Umeh, C., Ibenegbu, C. 1998. Carriers for starter cultures for the production of garri, a fermented food derived from cassava. World Journal of Microbiology and Biotechnology. 15, 231-234.

Okafor, N., Umeh, C., Ibenegbu, C. 1998 Amelioration of garri, a fermented food derived from cassava, *Manihot esculenta* Crantz, by the inoculation into cassava mash, of micro-organisms simultaneously producing amylase, linamarase, and lysine. World Journal of Microbiology and Biotechnology, 14, 835-838.

Okafor, N., Umeh, C., Ibenegbu, C, Obizoba., I.C., Nnam, P. 1998. Improvement in garri quality by the inoculation of micro-organisms into cassava mash. International Journal of Food Microbiology. 40, 43-49.

Okafor, N., Ejiofor, M.A.N. 1985. Microbial breakdown of linamarin in fermenting cassava pulp. *MIRCEN* Journal of Applied Microbiology and Biotechnology. 2, 269-276.

Okafor, N., Ejiofor, M.A.N. 1985. The linamarase of *Leuconostoc mesenteroides:* Production, isolation and some properties. J. Sci. Food Agric. 36, 668-678.

Okafor, N., Ejiofor, A.O. 1990. Rapid detoxification of cassava mash fermenting for garri production following inoculation with a yeast simultaneously producing linamarase and amylase Process Biochemistry International. 25, 82-86.

Okafor, N., Uzuegbu, J. 1987. Studies on the contributions of micro organisms on the organoleptic properties of garri, a fermented food derived from cassava (*Manihot esculenta* Crantz). J. Food Agric. 2, 99-105.

Okafor, N. 1977. Microorganisms associated cassava fermentation for garri production. J. Appl. Bact. 42, 279-284.

Okafor, N., Oyolu, C., Ijioma, B.C. 1984. Microbiology and biochemistry of foo-foo production. J. Appl. Microbiol., 55, 1-13.

Omofuvbe, B.O. Abiose, S.H., Shonukan, O.O. 2002. Fermentation of soybean (*Glycine max*) for soy-daddawa production by starter culturesof *Bacillus*. Food Microbiology, 19, 187-190.

Oteng-Gyang, K., Anuonye, C.C. 1987. Biochemical studies on the fermentation of *cassava* (Manihot utilissima Pohl). Acta Biotechnologica. 7, 289-292.

Oyewole, O.B. 2001. Characteristics and significance of yeasts' involvement in *cassava fermentation* for 'fufu' production. International Journal of Food Microbiology. 65, 213-218.

Oyewole, O.B., Odunfas 1990. Characterization and distribution of lactic acid bacteria in *cassava fermentation* during fufu production. Journal of Applied Bacteriology, 68, 145-152.

Popoola, T.O.S., Akueshi, C.O. 1986. Nutritional value of dawadawa, a local spice made from soybean *(Glycine max)* MIRCEN Journal. 2, 405-409.

Sanni, A.I., Onilude, A.A., Fadahunsi. S.T., Ogunbanwo, S.T., Afolabi, R.O. 2002. Selection of starter cultures for the production of ugba, a fermented soup condiment. European Food Research and Technology, 215, 176-180.

Tamime, A.Y., Robinson, R.K. 1999. Yoghurt Science and Technology. CRC Press.

Wood, B.J.R. (ed) 1985. Microbiology of Fermented Foods, Vol 1 Elsevier Applied Science Publishers. London and New York.

Wood, B.J.R. (ed) 1985. Microbiology of Fermented Foods, Vol 2 Elsevier Applied Science Publishers. London and New York.

Production of Metabolites as Bulk Chemicals or as Inputs in Other Processes

CHAPTER *20*

Production of Organic Acids and Industrial Alcohol

20.1 ORGANIC ACIDS

A large number of organic acids with actual or potential uses are produced by micro-organisms. Citric, itaconic, lactic, malic, tartaric, gluconic, mevalonic, salicyclic, gibberelic, diamino-pimelic, and propionic acids are some of the acids whose microbial production have been patented. In this chapter the production of only citric and lactic acids will be discussed.

20.1.1 Production of Citric Acid

Citric acid is a tribasic acid with the structure shown in Fig. 20.1.

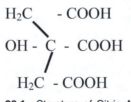

Fig. 20.1 Structure of Citric Acid

It crystallizes with the large rhombic crystals containing one molecule of water of crystallization, which is lost when it is heated to 130°C. At temperatures as high as 175°C it is converted to itaconic acid, aconitic acid, and other compounds.

20.1.2 Uses of Citric Acid

Citric acid is used in the food industry, in medicine, pharmacy and in various other industries.

Uses in the food industry
  (i) Citric acid is the major food acidulant used in the manufacture of jellies, jams, sweets, and soft drinks.

(ii) It is used for artificial flavoring in various foods including soft drinks.

(iii) Sodium citrate is employed in processed cheese manufacture.

Uses in medicine and pharmacy

(iv) Sodium citrate is used in blood transfusion and bacteriology for the prevention of blood clotting.

(v) The acid is used in efferverscent powers which depend for their efferverscence on the CO_2 produced from the reaction between citric acid and sodium bicarbonate.

(vi) Since it is almost universally present in living things, it is rapidly and completely metabolized in the human body and can therefore serve as a source of energy.

Uses in the cosmetic industry

(vii) It is used in astringent lotions such as aftershave lotions because of its low pH.

(viii) Citric acid is used in hair rinses and hair and wig setting fluids.

Miscellaneous uses in industry

(ix) In neutral or low pH conditions the acid has a strong tendency to form complexes hence it is widely used in electroplating, leather tanning, and in the removal of iron clogging the pores of the sand face in old oil wells.

(x) Citric acid has recently formed the basis of manufacture of detergents in place of phosphates, because the presence of the latter in effluents gives rise to eutrophication (an increase in nutrients which encourages aquatic flora development).

20.1.3 Biochemical Basis of the Production of Citric Acid

Citric acid is an intermediate in the citric acid cycle (TCA) (Fig. 20.2). The acid can therefore be caused to accumulate by one of the following methods:

(a) By mutation – giving rise to mutant organisms which may only use part of a metabolic pathway, or regulatory mutants; that is using a mutant lacking an enzyme of the cycle.

(b) By inhibiting the free-flow of the cycle through altering the environmental conditions, e.g. temperature, pH, medium composition (especially the elimination of ions and cofactors considered essential for particular enzymes). The following are some of such environmental conditions which are applied to increase citric acid production:

   (i) The concentrations of iron, manganese, magnesium, zinc, and phosphate must be limited. To ensure their removal the medium is treated with ferro-cyanide or by ion exchange fresins. These metal ions are required as prosthetic groups in the following enzymes of the TCA: Mn^{++} or Mg^{++} by oxalosuccinic decarboxylase, Fe^{+++} is required for succinic dehydrogenase, while phosphate is required for the conversion of GDP to GTP (Fig. 20.2).

   (i) The dehydrogenases, especially isocitrate dehydrogenase, are inhibited by anaerobiosis, hence limited aeration is done on the fermentation so as to increase the yield of citric acid.

   (ii) Low pH and especially the presence of citric acid itself inhibits the TCA and hence encourages the production of more citric acid; the pH of the fermentation

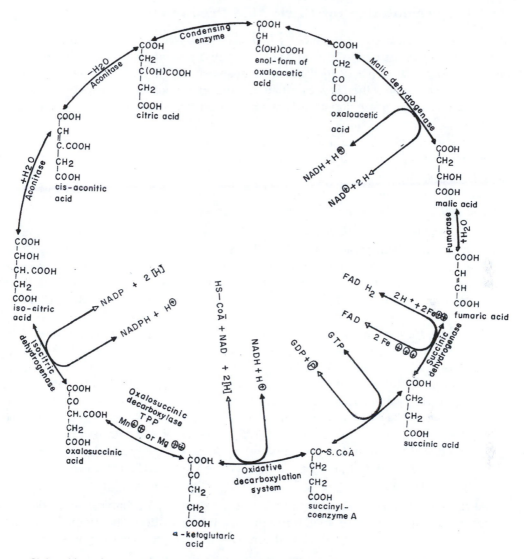

Citric acid can be caused to accumulate by using a mutant lacking an enzyme of the cycle or
by inhibiting the flow of the cycle

Fig. 20.2 The Tricarboxylic Acid Cycle

must therefore be kept low throughout the fermentation by preventing the
precipitation of the citric acid formed.

(iii) Many of the enzymes of the TCA can be directly inhibited by various
compounds and this phenomenon is exploited to increase citric acid
production. Thus, isocitric dehydrogenase is inhibited by ferrocyanide as well
as citric acid; aconitase is inhibited by fluorocitrate and succinic
dehydrogenase by malonate. These at enzyme antagonists may be added to the
fermentation.

20.1.4   Fermentation for Citric Acid Production

For a long time the production of citric acid has been based on the use of molasses and various strains of *Aspergillus niger* and occasionally *Asp. wenti*. Although several reports of citric acid production by *Penicillium* are available, in practice, organisms in this group are not used because of their low productivity. In recent times yeasts, especially *Candida* spp. (including *Candida quillermondi*) have been used to produce the acid from sugar.

Paraffins became used as substrate from about 1970. In the processes described mainly by Japanese workers bacteria and yeasts have been used. Among the bacteria were *Arthrobacter paraffineus* and corynebacteria; the yeasts include *Candida lipolytica* and *Candida oleiphila*.

Fermentation with molasses and other sugar sources can be either surface or submerged. Fermentation with paraffins however is submerged.

(a) *Surface fermentation*: Surface fermentation using *Aspergillus niger* may be done on rice bran as is the case in Japan, or in liquid solution in flat aluminium or stainless steel pans. Special strains of *Asp. niger* which can produce citric acid despite the high content of trace metals in rice bran are used. The citric acid is extracted from the bran by leaching and is then precipitated from the resulting solution as calcium citrate.

(b) *Submerged fermentation*: As in all other processes where citric acid is made the fermentation the fermentor is made of acid-resistant materials such as stainless steel. The carbohydrate sources are molasses decationized by ion exchange, sucrose or glucose. $MgSO_4$, $7H_2O$ and KH_2PO_4 at about 1% and 0.05-2% respectively are added (in submerged fermentation phosphate restriction is not necessary). The pH is never allowed higher than 3.5. Copper is used at up to 500 ppm as an antagonist of the enzyme aconitase which requires iron. 1-5% of methanol, isopranol or ethanol when added to fermentations containing unpurified materials increase the yield; the yields are reduced in media with purified materials.

As high aeration is deleterious to citric acid production, mechanical agitation is not necessary and air may be bubbled through. Anti-form is added. The fungus occurs as a uniform dispersal of pellets in the medium. The fermentation lasts for five to fourteen days.

20.1.5   Extraction

The broth is filtered until clear. Calcium citrate is precipitated by the addition of magnesium-free $(Ca(OH)_2$. Since magnesium is more soluble than calcium, some acid may be lost in the solution as magnesium citrate if magnesium is added. Calcium citrate is filtered and the filter cake is treated with sulfuric acid to precipitate the calcium. The dilute solution containing citric acid is purified by treatment with activated carbon and passing through iron exchange beds. The purified dilute acid is evaporated to yield crystals of citric acid. Further purification may be required to meet pharmaceutical stipulations.

20.1.6 Lactic Acid

Lactic acid is produced by many organisms: animals including man produce the acid in muscle during work.

20.1.6.1 Properties and chemical reactions of lactic acid

(i) Lactic acid is a three carbon organic acid: one terminal carbon atom is part of an acid or carboxyl group; the other terminal carbon atom is part of a methyl or hydrocarbon group; and a central carbon atom having an alcohol carbon group. Lactic acid exists in two optically active isomeric forms (Fig. 20.3).

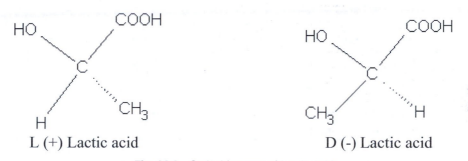

L (+) Lactic acid   D (-) Lactic acid

Fig. 20.3 Optical Isomers of Lactic Acid

(ii) Lactic acid is soluble in water and water miscible organic solvents but insoluble in other organic solvents.

(iii) It exhibits low volatility. Other properties of lactic acid are summarized in Table 20.1.

(iv) The various reactions characteristic of an alcohol which lactic acid (or it esters or amides) may undergo are xanthation with carbon bisulphide, esterification with organic acids and dehdrogenation or oxygenation to form pyruvic acid or its derivatives.

(v) The acid reactions of lactic acid are those that form salts and undergo esterification with various alcohols.

(v) Liquid chromatography and its various techniques can be used for quantitative analysis and separation of its optical isomers

Technical grade lactic acid is used as an acidulant in vegetable and leather tanning industries. Various textile finishing operations and acid dyeing of food require low cost technical grade lactic acid to compete with cheaper inorganic acid. Lactic acid is being used in many small scale applications like pH adjustment, hardening baths for cellophanes used in food packaging, terminating agent for phenol formaldehyde resins, alkyl resin modifier, solder flux, lithographic and textile printing developers, adhesive formulations, electroplating and electropolishing baths, detergent builders.

Lactic acid has many pharmaceutical and cosmetic applications and formulations in topical ointments, lotions, anti acne solutions, humectants, parenteral solutions and dialysis applications, and anti carries agents. Calcium lactate can be used for calcium deficiency therapy, and as an anti caries agent. Its biodegradable polymer has medical applications as sutures, orthopedic implants, controlled drug release, etc. Polymers of lactic acids are biodegradable thermoplastics. These polymers are transparent and their degradation can be controlled by adjusting the composition, and the molecular weight. Their properties approach those of petroleum derived plastics. Lactic acid esters like ethyl/butyl lactate can be used as green solvents. They are high boiling, non-toxic and degradable components. Poly L-lactic acid with low degree of polymerization can help in controlled release or degradable mulch films for large-scale agricultural applications.

Lactic acid was among the earliest materials to be produced commercially by fermentation and the first organic acid to be produced by fermentation.

Table 20.1   Physical properties of ethyl alcohol

Boiling point	78.2
Explosive limit in air, vol %	43-19.0
Freezing point	114.1°C
Specific gravity at 20/20°C	0.7905
Surface tension at 20°C dynes/cm	22.3
Vapor pressure at 20°C mg/HG	44

Chemical processing has offered and continues to offer stiff competition to fermentation lactic acid. Very few firms around the world produce it fermentatively, but this could change when the hydrocarbon-based raw material, lactonitrile, used in the chemical preparation becomes too expensive because of the increase in petroleum prices.

Lactic acid exists in two forms, the D-form and the L-form. When the symbols (+) or (-) are used, they refer to the optical rotation of the acid in a refractometer. However optical rotation in lactic acid is difficult to determine because the pure acid has low optical properties. The acid also spontaneously polymerizes in aqueous solutions; furthermore, salts, esters, and polymers have rotational properties opposite to that of the pure acid from which they are derived. All this makes it difficult to use optical rotation for characterizing lactic acid.

Many organisms produce either the D-or the L-form of the acid. However, a few organisms such as *Lactobacillus plantarum* produce both. When both the D- and L- form of lactic acid are mixed it is a racemic mixture. The DL form which is optically inactive is the form in which lactic acid is commercially marketed.

20.1.6.2   Uses of lactic acid

(i)   It is used in the baking industry. Originally fermentation lactic acid was produced to replace tartarates in baking powder with calcium lactate. Later it was used to produce calcium stearyl 2- lactylate, a bread additive.

(ii)   In medicine it is sometimes used to introduce calcium in to the body in the form of calcium lactate, in diseases of calcium deficiency.

(iii)   Esters of lactic acid are also used in the food industry as emulsifiers.

(iv)   Lactic acid is used in the manufacture of rye bread.

(v)   It is used in the manufacture of plastics.

(vi) Lactic acid is used as acidulant/ flavoring/ pH buffering agent or inhibitor of bacterial spoilage in a wide variety of processed foods. It has the advantage, in contrast to other food acids in having a mild acidic taste.

(vii) It is non-volatile odorless and is classified as GRAS (generally regarded as safe) by the FDA.

(viii) It is a very good preservative and pickling agent. Addition of lactic acid aqueous solution to the packaging of poultry and fish increases their shelf life.

(ix) The esters of lactic acid are used as emulsifying agents in baking foods (stearoyl-2-lactylate, glyceryl lactostearate, glyceryl lactopalmitate). The manufacture of these emulsifiers requires heat stable lactic acid, hence only the synthetic or the heat stable fermentation grades can be used for this application.

(x) Lactic acid has many pharmaceutical and cosmetic applications and formulations in topical ointments, lotions, anti acne solutions, humectants, parenteral solutions and dialysis applications, for anti carries agent.

(xi) Calcium lactate can be used for calcium deficiency therapy and as anti caries agent.

(xii) Its biodegradable polymer has medical applications as sutures, orthopaedic implants, controlled drug release, etc.

(xiii) Polymers of lactic acids are biodegradable thermoplastics. These polymers are transparent and their degradation can be controlled by adjusting the composition, and the molecular weight. Their properties approach those of petroleum derived plastics.

(xiv) Lactic acid esters like ethyl/butyl lactate can be used as environment-friendly solvents. They are high boiling, non-toxic and degradable components.

(xv) Poly L-lactic acid with low degree of polymerization can help in controlled release or degradable mulch films for large-scale agricultural applications.

Table 20.2   Physical properties of lactic acid

Appearance	Yellow to colorless crystals or syrupy 50% liquid
Melting point	16.8°C
Relative density	1.249 at 15°C
Boiling point	122° @ 15 millimeter
Flash point	110°C
Solubility	Soluble in water, alcohol, furfurol
	Slightly soluble in ether
	Insoluble in chloroform, petroleum ether, and carbon disulfide

20.1.6.3   Fermentation for lactic acid

Although many organisms can produce lactic acid, the amount so produced is small: the organisms which produce adequate amounts and are therefore used in industry are the homofermentative lactic acid bacteria, *Lactobacillus* spp., especially *L. delbruckii*. In recent times *Rhizopus oryzae* has been used. Both organisms produce the L- form of the acid, but

Rhizopus fermentation has the advantage of being much shorter in duration; further, the isolation of the acid is much easier when the fungus is used.

Lactic acid is very corrosive and the fermentor, which is usually between 25,000 and 110,000 liters in capacity is made of wood. Alternatively special stainless steel (type 316) may be used. They are sterilized by steaming before the introduction of the broth as contamination with thermophilic clostridia yielding butanol and butyric acid is common. Such contamination drastically reduces the value of the product.

During the step-wise preparation of the inoculum, which forms about 5% of the total beer, calcium carbonate is added to the medium to maintain the pH at around 5.5-6.5. The carbon source used in the broth has varied widely and have included whey, sugars in potato and corn hydrolysates, sulfite liquour, and molasses. However, because of the problems of recovery for high quality lactic acid, purified sugar and a minimum of other nutrients are used.

Lactobacillus requires the addition of vitamins and growth factors for growth. These requirements along with that of nitrogen are often met with ground vegetable materials such as ground malt sprouts or malt rootlets. To aid recovery the initial sugar content of the broth is not more than 12% to enable its exhaustion at the end of 72 hours. Fermentation with *Lactobacillus delbruckii* is usually for 5 to 10 days whereas with *Rhizopus oryzae*, it is about two days.

Although lactic fermentation is anaerobic, the organisms involved are facultative and while air is excluded as much as possible, complete anaerobiosis is not necessary.

The temperature of the fermentation is high in comparison with other fermentation, and is around 45°C. Contamination is therefore not a problem, except by thermophilic clostridia.

20.1.6.4   Extraction

The main problem in lactic acid production is not fermentation but the recovery of the acid. Lactic acid is crystallized with great difficulty and in low yield. The purest forms are usually colorless syrups which readily absorb water.

At the end of the fermentation when the sugar content is about 0.1%, the beer is pumped into settling tanks. Calcium hydroxide at pH 10 is mixed in and the mixture is allowed to settle. The clear calcium lactate is decanted off and combined with the filtrate from the slurry. It is then treated with sodium sulfide, decolorized by adsorption with activated charcoal, acidified to pH 6.2 with lactic acid and filtered. The calcium lactate liquor may then be spray-dried.

For *technical grade* lactic acid the calcium is precipitated as $CaSO_4.2H_2O$ which is filtered off. It is 44-45% total acidity. *Food grade* acid has a total acidity of about 50%. It is made from the fermentation of higher grade sugar and bleached with activated carbon. Metals especially iron and copper are removed by treatment with ferrocyanide. It is then filtered. *Plastic grade* is obtained by esterification with methanol after concentration. High-grade lactic acid is made by various methods: steam distillation under high vacuum, solvent extraction etc.

20.2   INDUSTRIAL ALCOHOL PRODUCTION

Ethyl alcohol, CH_3CH_2OH (synonyms: ethanol, methyl carbinol, grain alcohol, molasses alcohol, grain neutral spirits, cologne spirit, wine spirit), is a colorless, neutral, mobile flammable liquid with a molecular weight of 46.47, a boiling point of 78.3 and a sharp burning taste. Although known from antiquity as the intoxicating component of alcoholic beverages, its formula was worked out in 1808. It is rarely found in nature, being found only in the unripe seeds of *Heracleum giganteun* and *H. spondylium*.

20.2.1   Properties of Ethanol

Some of the physical properties of ethanol are given in Table 20.1

Ethyl alcohol undergoes a wide range of reactions, which makes it useful as a raw material in the chemical industry. Some of the reactions are as follow:

(i) *Oxidation*: Ethanol may be oxidized to acetaldehyde by oxidation with copper or silver as a catalyst:

$$CH_3CH_2OH \xrightarrow{\text{Cu, Ag}} CH_3CHO + H_2$$

(ii) *Halogenation*: Halides of hydrogen, phosphorous and other compounds react with ethanol to replace the $-OH$ group with a halogen:

$$3CH_3CH_2OH + PCl_3 \longrightarrow 3CH_3CH_2Cl + P(OH)_3$$

(iii) *Reaction with metals*: Ethanol reacts with sodium, potassium and calcium to give the alcoholates (alkoxides) of these metals:

$$2CH_3CH_2OH + 2Na \longrightarrow 2CH_3ONa + H_2$$

(iv) *Haloform Reaction*: Hypohalides will react with ethanol to yield first acetaldehyde and finally the haloform reaction:

$$CH_3CH_2OH + NaOCl \longrightarrow CH_3CHO + NaCl + H_2O$$
$$CH_3CHO + 2NaOCl \longrightarrow CCl_3CHO + 3NaOH$$
$$CCl_3CHO + NaOH \longrightarrow \underset{\text{Chloroform}}{CHCl_3} + HCOONa$$

(v) *Esters*: Ethanol reacts with organic and inorganic acids to give esters:

$$CH_3CH_2OH + HCl \longrightarrow \underset{\text{Ethylchoride}}{CH_3CH_2Cl} + H20$$

(vi) *Ethers*: Ethanol may be dehydrated to give ethers:

$$2CH_3CH_2OH \xrightarrow{\text{Catalyst}} CH_3CN_2OCH_2CH_3 + H_2O$$

(vii) *Alkylation*: Ethanol alkylates (adds alkyl-group to) a large number of compounds:

H_3SO_4: $CH_3CH_2HSO_4$ (ethyl hydrogen sulfate)

NH_3: $CH_3CH_2NH_2$ (ethyl amine)

20.2.2 Uses of Ethanol

(i) *Use as a chemical feed stock*: In the chemical industry, ethanol is an intermediate in many chemical processes because of its great reactivity as shown above. It is thus a very important chemical feed stock.

(ii) *Solvent use*: Ethanol is widely used in industry as a solvent for dyes, oils, waxes, explosives, cosmetics etc.

(iii) *General utility*: Alcohol is used as a disinfectant in hospitals, for cleaning and lighting in the home, and in the laboratory second only to water as a solvent.

(iv) *Fuel*: Ethanol is mixed with petrol or gasoline up to 10% and known as gasohol and used in automobiles.

20.2.3 Denatured Alcohol

All over the world and even in ancient times, governments have derived revenue from potable alcohol. For this reason when alcohol is used in large quantities it is denatured or rendered unpleasant to drink. The base of denatured alcohol is usually 95% alcohol with 5% water; for domestic burning or hospital use denatured alcohol is dispened as methylated spirit, which contains a 10% solution of methanol, pyridine and coloring material. For industrial purpose methanol is used as the denaturant. In the United States alcohol may be completely denatured (C.D.A. – completely denatured alcohol) when it cannot be used orally because of a foul taste or four smelling additives. It may be specially denatured (S.D.A. – specially denatured alcohol) when it can still be used for special purposes such as vinegar manufacture without being suitable for consumption.

20.2.4 Manufacture of Ethanol

Ethanol may be produced by either synthetic chemical method or by fermentation. Fermentation was until about 1930 the main means of alcohol production. In 1939, for example 75% of the ethanol produced in the US was by fermentation, in 1968 over 90% was made by synthesis from ethylene. The production of alcohol from ethylene is discussed in Chapter 13.

Due to the increase in price of crude petroleum, the source of ethylene used for alcohol production, attention has turned worldwide to the production of alcohol by fermentation. Fermentation alcohol has the potential to replace two important needs currently satisfied by petroleum, namely the provision of fuel and that of feedstock in the chemical industry.

The production of gasohol (gasoline – alcohol blend) appears to have received more attention than alcohol use as a feed stock. Nevertheless, the latter will also surely assume more importance if petroleum price continues to ride. Governments the world over have set up programs designed to conserve petroleum and to seek other energy sources. One of the most widely publicized programs designed to utilize a new source of energy is the Brazilian National Ethanol Program. Set-up in 1975, the first phase of this program aims at extending gasoline by blending it with ethanol to the extent of 20% by volume. The United States government also introduced the gasoline programme based on corn fermentation in 1980 following the embargo on grain sales to the then Soviet Union.

20.2.4.1 Substrates

The various substrates which may be used for ethanol production have been discussed in Chapter 2. It is clear that the substrate used will vary among countries. Thus, in Brazil sugar cane, already widely grown in the country, is the major source of fermentation alcohol, while it is planned to use cassava and sweet sorghum. In the United States enormous quantities of corn and other cereals are grown and these are the obvious substrates. Cassava grows in many tropical countries and since it is high yielding it is an important source in tropical countries where sugar cane is not grown. It is recognized that two important conditions must be met before fermentation alcohol can play a major role in the economy either as gasohol or as a chemical feedstock. First, the production of the crop to be used must be available to produce the crop without extensive and excessive deforestation. Secondly, the substrate should not compete with human food.

20.2.4.2 Fermentation

The conditions of fermentation for alcohol production are similar to those already described for whisky or rum production. Alcohol-resistant yeasts, strains of *Saccharomyces cerevisiae* are used, and nutrients such as nitrogen and phosphate lacking in the broth are added.

20.2.4.3 Distillation

After fermentation the fermented liquor or 'beer' contains alcohol as well as low boiling point volatile compounds such as acetaldeydes, esters and the higher boiling, fusel oils. The alcohol is obtained by several operations. First, steam is passed through the beer which is said to be steam-stripped. The result is a dilute alcohol solution which still contains part of the undesirable volatile compounds. Secondly, the dilute alcohol solution is passed into the center of a multi-plate aldehyde column in which the following fractions are separated: esters and aldehydes, fusel oil, water, and an ethanol solution containing about 25% ethanol. Thirdly, the dilute alcohol solution is passed into a rectifying column where a constant boiling mixture, an azeotrope, distils off at 95.6% alcohol concentration.

To obtain 200° proof alcohol, such as is used in gasohol blending, the 96.58% alcohol is obtained by azeotropic distillation. The principle of this method is to add an organic solvent which will form a ternary (three-membered) azeotrope with most of the water, but with only a small proportion of the alcohol. Benzene, carbon tetrachloride, chloroform, and cyclohezane may be used, but in practice, benzene is used. Azeotropes usually have lower boiling point than their individual components and that of benzene-ethanol-water is 64.6°C. On condensation, it separates into two layers. The upper layer, which has about 84% of the condensate, has the following percentage composition: benzene 85%, ethanol 18%, water 1%. The heavier, lower portion, constituting 16% of the condensate, has the following composition: benzene 11%, ethanol 53%, and water 36%.

In practice, the condensate is not allowed to separate out, but the arrangement of plates within the columns enable separation of the alcohol. Four columns are usually used. The first and second columns remove aldehydes and fusel oils, respectively, while the last two towers are for the concentration of the alcohol. A flow diagram of conventional absolute alcohol production from molasses is given in Fig. 20.4

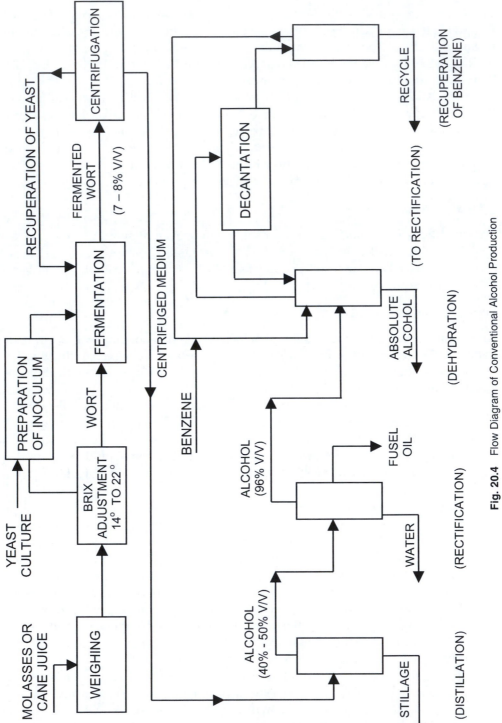

Fig. 20.4 Flow Diagram of Conventional Alcohol Production

20.2.5   Some Developments in Alcohol Production

Due to the current interest in the potential of ethanol as a fuel and a chemical feedstock, research aimed at improving the conventional method of production has been undertaken, and more will, most certainly, be undertaken. Some of the techniques aimed at improving productivity are the following:

(i) Developments of new strains of yeast of *Saccharomyces uvarum* able to ferment sugar rapidly, to tolerate high alcohol concentrations, flocculate rapidly, and whose regulatory system permits it to produce alcohol during growth.

(ii) The use of continuous fermentation with recycle using the rapidly flocculating yeasts.

(iii) Continuous vacuum fermentation in which alcohol is continuously evaporated under low pressure from the fermentation broth.

(iv) The use of immobilized *Saccharomyces cerevisiae* in a packed column, instead of in a conventional stirred tank fermentor. Higher productivity consequent on a higher cell concentration was said to be the advantage.

(v) In the 'Ex-ferm' process sugar cane chips are fermented directly with a yeast without first expressing the cane juice. The chips may be dried and used in the off-season period of cane production. It is claimed that there is no need to add nutrients as would be the case with molasses, since these are derived from the cane itself. A more complete extraction of the sugar, resulting in a 10% increase in alcohol yield, is also claimed.

(vi) The use of *Zymomonas mobilis,* a Gram-negative bacterium which is found in some tropical alcoholic beverages, rather than yeast is advocated. The advantages claimed for the use of *Zymomonas* are the following:

(a) Higher specific rates of glucose uptake and ethanol production than reported for yeasts. Up to 300% more ethanol is claimed for Zymomonas than for yeasts in continuous fermentation with all recycle.

(b) Higher ethanol yields and lower biomass than with yeasts. This deduction is based on Fig. 20.5 where, although the same quantity of alcohol is produced by the two organisms in 30-40 hours, the biomass of *Zymomonas* required for this level of production is much less than with yeast. The lower biomass appears to be due to the lower energy available for growth. *Zymomonas* utilized glucose by the Enthner-Duodoroff pathway (Fig. 5.4) which yields one mole of ATP/mole glucose, whereas yeasts utilize glucose anaerobically via the glycolytic pathway (Fig. 5.1) to give two ATP/mole glucose. Its use does not appear to have gained general acceptance.

(c) Ethanol tolerance is at least as high or even higher [up to 16% (v/v)] in some strains of the bacterium than with yeast.

(d) *Zymomonas* also tolerates high glocuse concentration and many cultures grow in sugar solutions of up to 40% (w/v) glucose which should lead to high ethanol production.

(e) *Zymomonas* grows anaerobically and, unlike yeasts, does not require the controlled addition of oxygen for viability at the high cell concentrations used in cell recycle.

(f) The many techniques for genetic engineering already worked out in bacteria can be easily applied to *Zymomonas* for greater productivity.

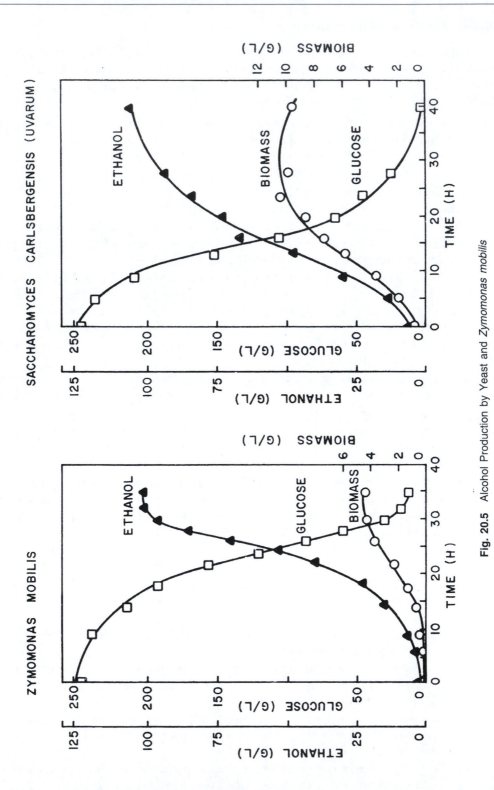

Fig. 20.5 Alcohol Production by Yeast and *Zymomonas mobilis*

SUGGESTED READINGS

Charrington, C.A, Hinton, M., Mead, G.C., Chopra, I. 1991. Organic Acids: Chemistry, Antibacterial Activity and Practical Applications. Advances in Microbial Physiology. 32, 87 – 108.

Ho, N.W.Y. 1980. Ann. Repts. Ferm. Proc. 4, 235-266.

Kosaric, D.C.M., Ng, I.R., Steart, G.S. 1980. Adv. Appl. Microbiol. 26, 137-227.

Lockwood, L.B. 1979. In: Microbial Technology. H.J. Peppler, D. Perlman, (eds.) 2nd Edit. Academic Press, New York, USA, pp. 256-288.

Narayanan, N., Pradip, K., Roychoudhury, P.K., Srivastava, A. 2004. L (+) lactic acid fermentation and its product polymerization. Electronic Journal of Biotechnology 7, Electronic Journal of Biotechnology [online]. 15 August 2004, vol. 7, no. 3 [cited 23 March 2006]. Available from: http://www.ejbiotechnology.info/content/vol2/issue3/full/3/index.html. ISSN 0717-3458.

Ward, W.P., Singh, A. 2003. Bioethanol Technology: Developments and Perspectives. Advances In Applied Microbiology, 51, 53-80.

CHAPTER 21

Production of Amino Acids by Fermentation

Amino acids have the general formula R. CH—COOH and are the main components of
$$\text{R. CH—COOH}$$
$$| $$
$$\text{NH}_2$$
of which proteins are made. The amino acids found in proteins number 20. Of these eight are essential for animals and must be supplied in their food, since animals cannot synthesize them.

Each of the 20 amino acids found in proteins can be distinguished by the R-group substitution on the carbon atom and can be divided into the following groups on that basis: amino acids with aliphatic R groups, non-aromatic amino acids with hydroxyl R groups, amino acids with sulfur containing R groups, amino acids with acidic R groups, amino acids with basic R groups, amino acids with aromatic R groups and amino acids with imino acids as the R groups. The nature of the R group influences the activity of the amino acid. Thus, the hydrophilic amino acids which have –OH groups in their R substituent (e.g. serine) tend to interact with the aqueous environment, are often involved in the formation of H-bonds and are predominantly found on the exterior surfaces proteins or in the reactive centers of enzymes. On the other hand, the hydrophobic amino acids (without –OH groups in the R substituent, for example methionine) tend to repel the aqueous environment and, therefore, reside predominantly in the interior of proteins. This class of amino acids does not ionize nor participate in the formation of H-bonds (Table 21.1).

All the amino acids, except glycine have two optically active isomers, the D – or the L-form. Natural proteins are usually made up of L- (or the so-called natural amino acids.) Outside the 20 amino acids found in protein, many other rare amino acids have been reported in various metabolites such as some antibiotics, other microbiological products and in non-proteinaceous materials in plants and animals.

21.1 USES OF AMINO ACIDS

Amino acids find use in a large number of activities, including human and animal nutrition, medicine, cosmetics, and in the synthesis of chemicals.

Table 21.1   Amino acids found in proteins

Amino Acid	Symbol	Structure
Amino Acids with Aliphatic R-Groups		
Glycine	Gly - G	$H{-}CH{-}COOH$ $\quad\ NH_2$
Alanine	Ala - A	$CH_3{-}CH_2{-}COOH$ $\qquad\quad NH_2$
* Valine	Val - V	H_3C $\quad\ \ \diagdown CH{-}CH{-}COOH$ $H_3C\quad\ \ NH_2$
* Leucine	Leu - L	H_3C $\quad\ \ \diagdown CH{-}CH_2{-}CH{-}COOH$ $H_3C\qquad\qquad NH_2$
* Isoleucine	Ile - I	$H_3C{-}CH_2$ $\qquad\quad\ \diagdown CH{-}CH{-}COOH$ $\qquad H_3C\quad\ NH_2$
Non-Aromatic Amino Acids with Hydroxyl R-Groups		
Serine	Ser - S	$HO{-}CH_2{-}CH{-}COOH$ $\qquad\qquad\ NH_2$
* Threonine	Thr - T	H_3C $\quad\ \ \diagdown CH{-}CH{-}COOH$ $HO\qquad\ NH_2$
Amino Acids with Sulfur-Containing R-Groups		
Cysteine	Cys - C	$HS{-}CH_2{-}CH{-}COOH$ $\qquad\qquad NH_2$
* Methionine	Met-M	$H_3C{-}S{-}(CH_2)_2{-}CH{-}COOH$ $\qquad\qquad\qquad\ NH_2$
Acidic Amino Acids and their Amides		
Aspartic Acid	Asp - D	$HOOC{-}CH_2{-}CH{-}COOH$ $\qquad\qquad\ NH_2$
Asparagine	Asn - N	$H_2N{-}C{-}CH_2{-}CH{-}COOH$ $\qquad\ \ O\qquad\ NH_2$
Glutamic Acid	Glu - E	$HOOC{-}CH_2{-}CH_2{-}CH{-}COOH$ $\qquad\qquad\qquad NH_2$
Glutamine	Gln - Q	$H_2N{-}C{-}CH_2{-}CH_2{-}CH{-}COOH$ $\qquad\ \ O\qquad\qquad NH_2$
Basic Amino Acids		
Arginine	Arg - R	$HN{-}CH_2{-}CH_2{-}CH_2{-}CH{-}COOH$ $C{=}NH\qquad\qquad\qquad NH_2$ NH_2
* Lysine	Lys - K	$H_2N{-}(CH_2)_4{-}CH{-}COOH$ $\qquad\qquad\quad NH_2$
Histidine	His - H	$\qquad{-}CH_2{-}CH{-}COOH$ $\qquad\qquad\quad NH_2$ $HN{\diagdown}\ \diagup N:$

Contd

Table 21.1 Contd.

Amino Acid	Symbol	Structure
Amino Acids with Aromatic Rings		
* Phenylalanine	Phe - F	$\text{—CH}_2\text{—CH—COOH}$, NH_2
Tyrosine	Tyr - Y	HO— ring $\text{—CH}_2\text{—CH—COOH}$, NH_2
* Tryptophan	Trp-W	$\text{—CH}_2\text{—CH—COOH}$, NH_2
Imino Acids		
Proline	Pro - P	COOH

*Essential

(i) *Use in human and animal nutritional supplementation*: Proteins are metabolized constantly in the body and most of the amino acids absorbed are used to replace body proteins. The remaining are metabolized into various body components including hormones, and nucleic acid bases. Of the amino acids in protein eight are essential and the diet must contain them. Foods such as plant proteins lacking in some essential amino acids are fortified with their addition. Most cereals are particularly low in lysine, the addition of which greatly improves the quality of the food as determined by the PER (protein equivalency ration). The PER is a means of comparing the amino acid content of protein with that of hen's egg or human milk. It is usually best determined by feeding tests to rats and mice.

Animal feeds made from inexpensive plant proteins can be greatly improved with only a small quantity of the limiting amino acids, resulting in higher growth rates in the animals. L – lysine and DL methionine are widely used as feed improvers.

(ii) *Flavor and taste enhancement in foods*: Amino acids are important in deciding the taste of meats and such foods. Mono-sodium glutamate well-known as a flavoring agent, will be discussed later.

Amino acids influence the taste of foods. Some are very sweet; for example glycine is as sweet as sugar and is sometimes used in soft drinks and soups. The amino acid is present in large amounts in shrimps to which it confers its sweet taste. The peptide, L- aspartyl – L – phenylalanine methyl ester is particularly sweet. Well-known sweetners such as Aspartame, Sweet and Low, and Splenda contain a dipetide formed from aspartic acid and phenylalanine Other amino acids e.g. valine are bitter. It is interesting that while the L- isomers of leucine, phenylalanine, tyrosine, and tryptophane are bitter, the D-isomers are sweet. The combination of various amino acids influence the taste and flavor of foods. Thus cheese flavor derives from the combined effect of glutamic acid as well as that of bitter amino acids such as valine, leucine and methionine.

(iii) *Medical uses*: The greatest application of amino acids in medicine is in transfusion; which is administered when the oral consumption of proteinaceous food is not possible such as after an operation. In the past only essential amino acids were used in transfusion; nowadays non-essential ones are added.

Various amino acids are used for ammonia detoxification in blood in liver diseases, in the treatment of heart failure, in cases of peptic ulcer and male sterility, etc. Table 21.2 summarizes some of these uses.

In addition derivatives of amino acids are widely used in medicine as discussed below. Methyldopa (L-methy1-3, 4 dihydroxy-phenylalanin) is widely used as an anti-hypertensive with relatively few side effects. Dopa is used in treating Parkinson's disease (Fig. 21.1):

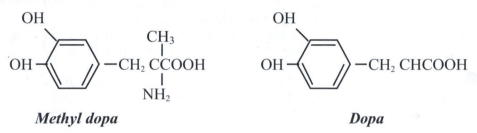

Methyl dopa   **Dopa**

Fig. 21.1   Methyl Dopa and Dopa

A derivative of serine, cycloserine is an antibiotic produced by a streptomycete; it is used for the treatment of tuberculosis.

(iv) *Use as an industrial synthetic raw materials*: Although numerous studies have been concluded on the use of amino acids as raw materials in the chemical industry, very few of these have been put into actual practice. Thus, the manufacture of artificial silk was considered and dropped. However amino acids are used in the following:

(a) *Surface-active agents:* A surface-active agent has a water solube (or hydrophilic end) as well as a water-repellent) or hydrophobic end. The hydrophilic end is dissolved in the water and as a consequence the surface tension of the water is lowered. Surface action agents can be prepared from amino acids by introducing long-chain lipophilic groups to one of the two hydrophilic groups (- COOH, or – NH_2) of amino acids. The resulting surface-active agents is either cationic (if it has a positive charge) or anionic (negative charge). As they lower the surface tension like soap, they foam just like soap. Some are even more effective than soap as cleansing agents. Many of them also have strong bacteriostatic action. Thus sodium lauryl sarcosinate is used in toothpaste and shampoo because it has a bacteriocidal as well as foaming action. These derivatives are also used as fungicides and pesticides.

(b) *Production of polymers from amino acids*: Polymers derived from amino acids are used in making synthetic leather, fire-resistant fabrics and anti-static materials.

(c) *Use as cosmetics:* Amino acids exhibit a buffering action that help maintain normal skin function by regulating pH and a protective action against bacteria. Detergents (surface action agents) derived from amino acids are less irritating than soaps

Table 21.2 Therapeutic uses of some amino acids

Amino Acid	Use
Ornithine	Used for treating cases of hyperammonemia (excessive ammonia in blood) because it increases urease activity in the liver, thus enhancing urease production.
Arginine	More common than ornithine. 1. Use for ammonia detoxication described above. 2. Arginine is the main component of spermatic proteins and hence administered in cases of male sterility due to low number or weak spermatozoa *(Sterilitas virilis)*.
Aspartic acid	1. Used for ammonia detoxification combined with ornithine. 2. Used as carrier for K^+ and Mg^{++} in form of potassium or magnesium aspartate in cases of heart failure, fatigue, etc.
Cysteine and cystine	1. Protect SH-enzymes in the liver from enzyme inhibitors; used for dealing with poisons generally including cyanide poisoning. 2. L- cysteine is used in bronchitis and nasal catarrh.
Gluatamic acid	Important in brain metabolism hence various analogues of glutamic acid are used in treating various neuropathic diseases.
Glycine	Rarely used as a drug, but only as a sweetener in medicines.
Histidine	An amino acid essential for infants but not for adults, it is used in adults for gastric and duodenal ulcers and is administered in cases of anaemia because it helps in haemoglobin regeneration.
Methionine	A sulfur-containing amino acid, it is important in the metabolism of various sulfur-containing compounds in the body. It is also used for detoxification in poisoning by arsenic, chloroform, and benzene derivaties. A derivative of methionine, Vitamin U, is used as an anti-ulcer drug, because it neutralizes histamine which is known to induce ulcer formation.
Tryptophan	Used as an anti-depressant.

because the pH of 5.5-6.0 is closer to that of the skin, whereas soap is slightly alkaline. The addition of different amino acids to shampoo is practiced to achieve different ends: anti-dandruff shampoos contain cysteine; thioglycolic acid is employed as a reducing agent for the cold waving of hair.

21.2 METHODS FOR THE MANUFACTURE OF AMINO ACIDS

The beginning of the development of the amino acid industry can be put at 1908 when Kikunae Ikeda identified and isolated monosodium glutamate (MSG), the sodium salt of glutamic acid, as the flavoring agent in 'kombu', a traditional seasoning agent used in Japan, and derived from some marine algae. The Ajinomoto company the following year started producing MSG by extraction from the acid hydrolysate of wheat gluten or defatted soy. Glutamic acid, lysine and methionine are the most produced amino acids globally (Fig. 21.2). Today amino acids are produced by a number of methods.

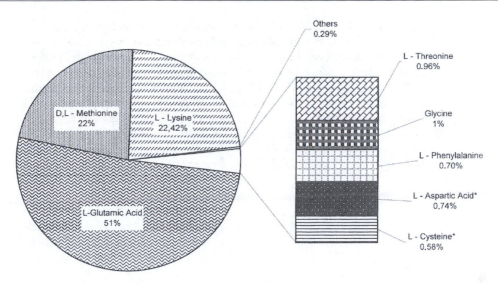

Fig. 21. 2 World Production of Amino Acids, 1996 (from Mueller and Huebner, 2003)

(i) *Protein hydrolysis:* Protein hydrolysis was the original method of amino manufacture. Hair, keratin, blood meal and feathers are hydrolyzed using acid and the amino acid extracted. It is not very popular because it depends on the availability of hair, feathers and the other raw materials. However, cysteine and cystine are still produced by isolating them from chemically hydrolyzed keratin protein in hair and feathers while proline and hydroxyproline are precipitated from gelatin hydrolysates.

(ii) *Chemical synthesis:* Glycine, L-alanine, and DL- methionine are produced by chemical synthesis. Chemical synthesis can only produce the D,L- (recemic) forms of amino acids and an additional step involving the use of an immobilized enzyme, aminoacylase, produced by *Aspergillus niger* is necessary to obtain the biologically active L-form. (Fig. 21. 3). This step is expensive and on account of this, few amino acids are prepared by chemical synthesis. Amino acids produced by chemical synthesis are glycine and methionine; methionine is said to have the same effect as an animal feed additive whether in the L- or in the D, L- form.

(iii) *Microbiological methods:* Microbiological methods are of three types:

1. Semi-fermentation;
2. Use of microbial enzymes or immobilized cells;
3. Direct fermentation.

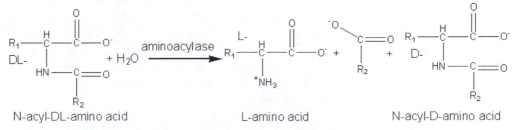

Fig. 21.3 Action of Aminoacylase on Racemic Mixtures of Amino Acids

L-Alanine

$$HOOCCH_2CH(NH_2)COOH \xrightarrow{\textit{Aspartic acid }\beta\textit{-decarboxylase}} CH3CH(NH_2)COOH + CO_2$$

L-Asp   L-Ala

D-Alanine

$$CH_3CH(NH_2)CONH2 \xrightarrow{\textit{D-Alanine amino peptidase}} CH3CHNH2COOH$$

DL-Alanine amide   D-Ala

$$+$$
$$CH_3CHNH_2CONH_2$$
L-Alanine amide

L-Aspartic acid

$$HOOCHC=CHCOOH \xrightarrow{\textit{Aspartase}} HOOCCH_2CH(NH_2)COOH$$

Fumaric acid   L-Asp

Fig. 21.4 Examples of Enzyme Conversions for Producing Amino Acids

21.2.1 Semi-fermentation

In this process, the metabolic intermediate in the amino acid biosynthesis or its precursor is added to the medium, which contains carbon and nitrogen sources, and other nutrients required for growth and production; the metabolite is converted to the amino acid during fermentation. Sometimes the intermediate could be another amino acid. Examples of the commercial production of amino acids by semi-fermentation are L-serine production from glycine and methanol using the methane-utilizing bacterium *Hyphomicrobium* sp. or *Pseudomonas* sp. Other examples are the production of L-tryptophan from anthranillic acid or indole using *E. coli* and *B. subtilis* L-isoleucine production from DL-α-aminobutyric acid and ethanol by *Brevibacterium* sp. has been done commercially by this process (Table 21. 3).

21.2.2 Enzymatic Process

Chemically synthesized substrates can be converted to the corresponding amino acids by the catalytic action of an enzyme or microbial cells as an enzyme source. Often the enzymes or the cells may be immobilized. L-alanine production from L-aspartic acid, L-aspartic acid production from fumaric acid, L-cysteine production from DL-2-aminothiazoline-4-carboxylic acid (Fig. 21. 5). Others are D-phenylglycine (and D-*p*-hydroxyphenylglycine) production from DL-phenylhydantoin (and DL-*p*-hydroxyphenylhydantoin), and L-tryptophan production from indole and DL-serine have been in operation as commercial processes.

Table 21.3   Amino acid production by semi-fermentation process (from Araki, 2003)

Amino acid produced	Precursor added to the medium	Amount, mg/mL
D-alanine	DL-alanine	48.8
L-isoleucine	D-threonine	15
	DL-α-aminobutyric acid	15.7
	DL-α-hydroxybutyric acid	
	DL-α-bromobutyric acid	
L-methionine	L-hydroxy-4-methylthiobutyric acid	10.9
	DL-5-(2-methylthioethyl)hydantoin	34
	D-threonine	
L-phenylalanine	acetoamidocinnamic acid	75.9
L-proline	L-glutamic acid	108.3
L-serine	glycine	16
		54.5
L-threonine	L-homoserine	16
L-tryptophan	anthranilic acid	40
	indole	16.7

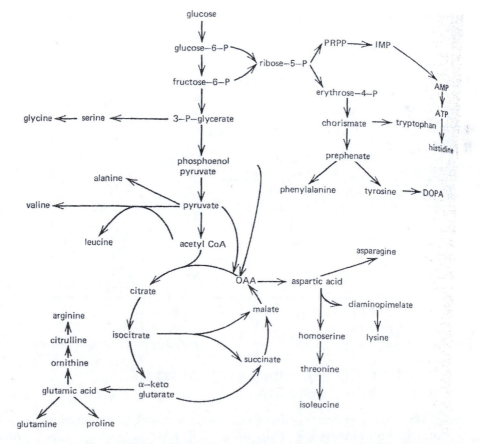

Fig. 21.5   Metabolic Pathways Involved in the Biosynthesis of Amino Acids from Glucose

21.2.3   Production of Amino Acids by the Direct Fermentation

Although other microbiological methods for the commercial production of amino acids exist, such as the biosynthesis of amino acids using intermediates, and the use of enzymes, by far the most important method for producing amino acids microbiologically is by direct fermentation. What method is used in any particular situation depends on factors such as process economics, the available raw materials, market size, the environmental regulation operating in the place of production, etc. Nevertheless, the fermentation method appears to be the dominant one and will be discussed.

The production of amino acids by fermentation was stimulated by the discovery of an efficient L- glutamic acid producer *Corynebacterium glutamincum*. Many microorganisms have been reported to produce amino acids. They are mainly bacteria, but they also include some molds and yeasts. The four most widely reported bacteria belong to the following four genera, the typical species of which are given in parenthesis.

> *Corynebacterium* spp. (*C. glutamicum; C. lilum*)
> *Brevibacterium* spp. (*B. divericartum: B. alanicum*)
> *Microbacterium* spp. (*M. flavum* var. *glutamicum*)
> *Arthrobacter* spp. (*A. globiformis; A. aminofaciens*)

Auxotrophic and regulatory mutants of glutamic acid producing bacteria are used for the commercial production of all amino acids outside L- glutamic acid and L-glutamine, which are produced by the wild type of these organisms (Table 21.4).

Table 21.4   Amino acids produced from wild type and mutant strains of bacteria

Wild-type		Auxotrophic Mutants		Regulatory Mutants	
L- glutamic acid	1	L- citruline	1	L- arginine	1,2,3
L-valine,	1	L- leucine	1	L- histidine	1,3,5
		L- lysine	1	L- isoleucine	1,3,5
		L- ornithine	1	L- leucine	5,6
		L- proline	3	L- lysine	3
		L- threonine	4	L- methionine	1
		L- tyrosine	1	L- phenylalanine	1,3
				L- thereonine	1,3
				L- tryptophane	1,3
				L- tyrosine	1,3
				L- valine	6

1 = *Corynebacterium glutamicum*   2 = *Bacillus subtilis*
3 = *Brevibacterium flavum*   4 = *Escherichia coli*
5 = *Serratia marcescens*   6 = *Brevibacterium lactofermentum*

21.3   PRODUCTION OF GLUTAMIC ACID BY WILD TYPE BACTERIA

(i) *Organisms*: Wild type strains of the organisms of the four genera mentioned above are now used for the production of glutamic acid. The preferred organism is however *Corynebacterium glutamicum*. The properties common to the glutamic acid bacteria are: (a) they are all Gram-positive and non-motile; (b) they require biotin to grow; (c) they lack or have very low amounts of the enzyme α-ketoglutarate, which is formed by removal of CO_2 from isocitrate formed in TCA cycle (citric acid cylce). Since α-ketoglutarate is not dehydrogenated it is available to form glutarate by reacting with ammonia (Fig. 21.1).

(ii) *Conditions of the fermentation*: The composition of a medium which has been used for the production of glutamic acid is as follows (%): glucose, 10; corn steep liquor 0.25; enzymatic casein hydrolysate 0.25; K_2HPO_4 0.1, Mg. SO_4, $7H_2O$, 0.25; urea, 0.5. It should be noted that besides glucose, hydrocarbons have served as carbon sources for glutamic acid production. The optimal temperature is 30° to 35° and a high degree of aeration is necessary.

(iii) *Biochemical basis for glutamic acid production*: Studies by several workers have clarified the basis for glutamic acid production as summarized below.

(a) Glutamic acid production is greatest when biotin is limiting; that is, when it is sub-optimal. When biotin is optimal, growth is luxuriant and lactic acid, not glutamic acid, is excreted. The optimal level of biotin is 0.5 mg per gm of dry cells; with higher amounts glutamic acid production falls.

(b) The isocitrate-succinate part of the TCA cycle (Fig. 21.5) is needed for growth. It is only after the growth phase that glutamic acid production becomes optimal.

(c) An increase in the permeability of the cell is necessary so as to permit the outward diffusion of glutamic acid, essential for high glutamic acid productivity. This increased permeability to the acid can be achieved in the following ways: (i) ensuring biotin deficiency in the medium (ii) treatment with fatty acid derivatives, (iii) ensuring oleic acid deficiency in mutants requiring oleic acid (C_{16} - C_{18}). (iv) addition of penicillin during growth of glutamic acid bacteria, Cells treated in one of the first three ways above have cell membranes in which the saturated to unsaturated fatty acid ratio is abnormal, therefore the permeability barrier is destroyed and glutamic acid accumulates in the medium. The major factor in glutamic acid production by wild type organism is thus altered permeability. Treatment with penicillin prevents cell-wall formation. Cell wall inhibiting antibiotics such as penicillin and cephalosporin have enabled the use of molasses which are rich in biotin for glutamic acid production.

21.4   PRODUCTION OF AMINO ACIDS BY MUTANTS

After wild type strains of *C. glutamicum* and of other bacteria were found to accumulate glutamic acid, efforts to find in nature bacteria able to yield high amounts of other amino acids failed. The reason for this is that microorganisms avoid over-production of amino acids, producing only the quantity they require. To induce the organism to over produce, regulatory mechanisms must be disorganized as discussed in Chapter 6. Two major means of regulating amino acid synthesis are feedback inhibition and repression. Auxotrotrophic mutants and regulatory mutants are two means by which the organisms' tendency not to overproduce can be disorganized. In order to over produce an amino acid which is an intermediate in a synthetic pathway, a mutant auxotroph is produced whose pathway in the synthesis is blocked. When this mutant is cultivated, limiting nutrient feedback and/or repression would have been removed and an overproduction of the amino acid will occur. The mutants used for the production of aminoacids other than glutamic acid are produced from L- glutamic acid producing bacteria. This is because these bacteria assimilate carbon efficiently and also because they do not degrade the amino acid which they excrete.

21.4.1 Production of Amino Acids by Auxotrophic Mutants

Table 21.4 shows the amino acids produced by the use of auxotrophic mutants. The first to be produced was L- lysine using limiting concentrations of either L- homoserine or L-methionine plus L- threonine with a mutant strain of *Corynebacterium glutamicum*. In the wild type of this organism concerted feed back inhibition is by both lysine and threonine. Inhibition does not occur when only one is present. In this particular mutant absence of biosynthetic homoserine derived from aspartic acid causes lysine to accumulate. This is illustrated in Fig. 21.6.

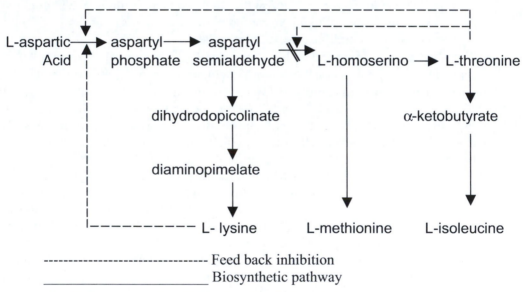

Fig. 21.6   Accumulation of Lysine in a Mutant Strain of *Corynebacterium glutamicum*

21.4.2 Production of Amino Acids by Regulatory Mutants

Regulatory mutants have a feed-back insensitive key enzyme and hence continues to over produce the required amino acid. Examples are given in Table 21.4. In order to obtain such mutants mutations are induced to produce organisms whose growth is not inhibited by analogues of the amino-acid to be overproduced. A good example is the case of lysine production by *Brevibacterium flavum*. In this organism the L- lysine pathway is regulated at aspartate kinase which is the only enzyme sensitive to feed back inhibitation by lysine. Mutants resistant to lysine analogues therefore over produce the amino acid (Fig. 21.7).

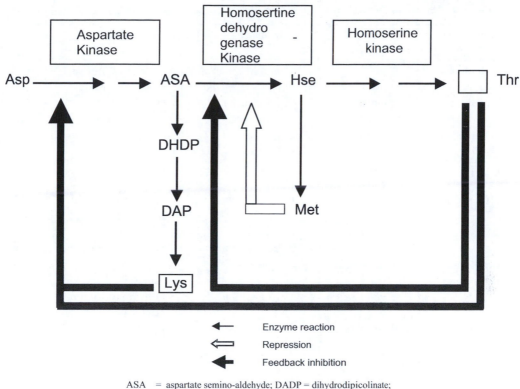

←	Enzyme reaction
⇐	Repression
◄	Feedback inhibition

ASA = aspartate semino-aldehyde; DADP = dihydrodipicolinate;
Hse = Homoserne; DAP = diaminopimelate

Fig. 21.7   Lysine Biosynthesis in *Brevibacteium flavum* and *Corynebacterium glutamicum*

21.5   IMPROVEMENTS IN THE PRODUCTION OF AMINO ACIDS USING METABOLICALLY ENGINEERED ORGANISMS

The improvements of the microorganisms discussed above used classical mutation techniques and screening procedures which relied on deleting competing pathways and eliminating feed back regulations on the biosynthetic pathways. The mutagenic procedure cannot however totally eliminate deregulation. The use of recombinant DNA technology has enabled genetic modifications which have further improved existing production strains through metabolic engineering (Chapter 7). As indicated in Chapter 7, metabolic engineering involves the introduction of genes which will enhance the production of a metabolic pathway. The pathways through which amino acids are made by the organism are shown in Fig. 21.5. The genes limiting the production of the amino acid are enhanced by gene amplication thus leading to a more rational improvement of the organism.

Many examples exist of improvements in amino acid production through cloning of genes (Table 21.5). Among the pathways which have been targeted for improvement through gene cloning are:

Table 21.5 Improvements in amino acid production through the cloning of different genes

Amino acid produced	Microorganisms	Gene donor	Cloned gene or enzyme	Yield mg/mL
L-alanine	E. coli	B. stearothermophilus	Ala dehydrogenase	
D-alanine	E. coli	Ochrobactrum anthropi	D-aminopeptidase	200
L-histidine	C. glutamicum	C. glutamicum	His G, D, C, B	15
	S. marcescens	S. marcescens	His G, D, B	43
L-isoleucine	C. glutamicum	C. glutamicum	Hom dehydrogenase	11
	B. flavum	E. coli	ilv A	21
L-lysine	C. glutamicum	C. glutamicum	Lys A, dap A, B, D, Y	
	C. glutamicum	E. coli	Asp A	
L-phenylalanine	C. glutamicum	C. glutamicum	aro F, chorismate mutase, PRDH	28
	C. glutamicum	E. coli	aro G, Phe A	
	B. lactofermentum	B. lactofermentum	aro F, E, L, PRDH	21
L-proline	S. marcescens	S. marcescens	Pro A, B	75
L-threonine	E. coli	E. coli	Thr A, B, C	55
	C. glutamicum	C. glutamicum	hom dehydrogenase, hom kinase, Thr C	51
	B. lactofermentum	B. lactofermentum	ppc, hom dehydrogenase, hom kinase	33
	B. flavum	E. coli	Thr B, C	27
	S. marcescens	E. coli	ppc	60
L-tryptophan	E. coli	E. coli	Trp A, E, R, tna A	40
	C. glutamicum	C. glutamicum	Trp E, aro F, chorismate mutase, PRDH	45
L-tyrosine	C. glutamicum	E. coli	Aro F	9

(i)   the terminal pathways of the amino acid synthesis
(ii) the central metabolic pathway for producing the amino acid
(iii) the transport process for secreting amino acid

21.5.1 Strategies to Modify the Terminal Pathways

The strategies for modifying the terminal pathways are indicated in Fig. 21.8.

1. *Amplification of rate limiting enzyme*: The gene coding for the rate limting enzyme in the biosynthetic pathway is amplified. Large increases have been observed when this technique was applied to L-phenyl alanine production in *Corynebacterium glutamicum*.

2. *Amplification of branch-point enzyme*: The gene coding for the branch-point enzyme is amplified to redirect the common intermediate to another amino acid. It has been used successfully in converting L-lysine to L-tryptophane and L-tyrosine to L-phenylalanine.

(1) Amplification of rate-limiting enzyme (Removal of bottleneck)	
(2) Amplification of branch-point enzyme (Metabolic conversion)	
(3) Introduction of heterologous enzyme with different control architecture (Bypass of bottleneck reaction)	
(4) Introduction of heterologous enzyme with different catalytic mechanism (Acceleration of key reaction)	
(5) Amplification of the first enzyme in terminal pathways (Augmentation of carbon flow and identification of potential bottleneck)	

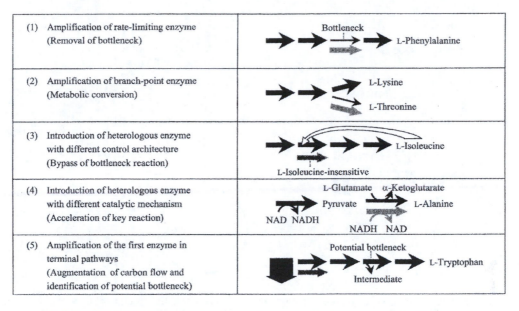

Fig. 21.8 Strategies to Modify Terminal Pathways for the Improved Production of Amino Acids, (from Ikeda, 2003)

3. *Introduction of a different enzyme able to produce the same end amino acid*: The gene for a different enzyme for the same end amino acid is introduced. The enzyme creating the bottle neck is thus bypassed. This has been used for increased L-isoleucine production in *Corynebacterium glutamicum*.

4. *Introduction of a more functional enzyme than the native one*: Introduction of an enzyme which is more active than the native one thereby enhancing the production of the amino acid. This has enhanced the production of L-alanine production by *Corynebacterium glutamicum* when L-alanine dehydrogenase from *Arthrobacter oxydans* was engineered into it .

5. *Amplification of the first enzyme in the terminal pathway*: The first enzyme in a pathway diverging from central metabolism is amplified to increase the flow in that pathway; any bottleneck is removed by the increased down the pathway. This strategy has been applied to obtain increased yield of L-tryptophan by *Corynebacterium glutamicum*.

21.5.2   Strategies for Increasing Precursor Availability

A major aim of metabolic engineering for increased amino acid production is to channel as much carbon as possible from sugar into the production of a desired amino acid. After bottlenecks in the terminal pathway are removed, the main factor limiting increased production is the shifting of intermediates to the central metabolic pathway. The complete genetic sequence of *Corynebacterium glumicum* is available. One strategy is to amplify the genes for the enzymes leading to the formation of aromatic amino acids erythrose 4-P and to L histidine through 5-P (Fig. 21.9).

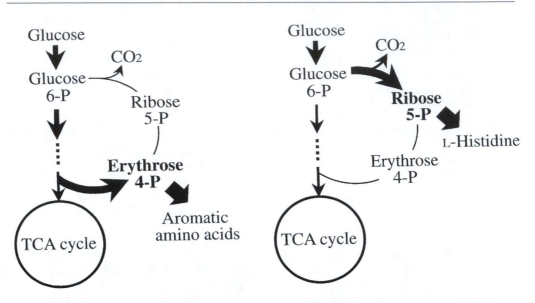

Fig. 21.9   Strategies for Increasing Precursor Availability for the Production of Aromatic Amino Acid and L-Histidine in *Corynebacterium glutamicum*

21.5.3   Metabolic Engineering to Improve Transport of Amino Acids Outside the Cell

The aim of strain improvement is to prevent feedback inhibition when the amino acid accumulates intracellularly. One manner in which feedback inhibition can be avoided is through increased efflux of the amino acid. A gene which codes for increased efflux has been introduced into *E.coli* resulting in a vastly increased production of L-cysteine (Fig. 21.10).

21.6   FERMENTOR PRODUCTION OF AMINO ACID

21.6.1   Fermentor Procedure

Starting from shake flasks the inoculum culture is grown in shake flasks and transferred to the first seed tank (1,000–2,000 liters) in size. After suitable growth the inoculum is

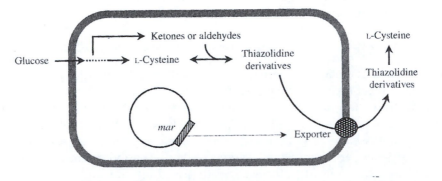

Fig. 21.10   Increased Efflux of Amino Acid in *E. coli* through Metabolic Engineering

transferred to the second seed tank (10,000–20,000 liters), which serves as inoculum for the production tank (50,000–500,000 liters).

The fermentation is usually batch or fed-batch (Chapter 9). In batch cultivation all the nutrients are added at once at the beginning of the fermentation, except for ammonia which is added intermittently to help adjust the pH, and fermentation continues until sugar is exhausted. In a fed-batch process, the fermentor is only partially filled with medium and additional nutrients added either intermittently or continuously until an optimum yield is obtained. The fed-batch appears preferable for the following reasons:

(a) Most amino acid production requires high sugar concentrations of up to 10%. If all were added immediately, acid would be quickly produced which will inhibit the growth of the microorganisms and hence reduce yield.

(b) Where auxotrophic mutants are used, excess supply of nutrients leads to reduced production due to overgrowth of cells or feed back regulation by the nutrient.

(c) During the lag phase of growth, the oxgen demand of the organism may exceed that of the organism leading to reduced growth.

21.6.2   Raw Materials

The main raw materials used are cane or beet molasses and starch hydrolysates from corn or cassava as glucose. In the US, the preferred carbon source is corn syrup from corn, whereas in Europe and South America it is beet molasses.

As nitrogen source, inorganic sources such as ammonia or ammonium sulfate is generally used.

Phosphates, vitamins and other necessary supplements are usually provided with corn steep liquour.

21.6.3   Production Strains

Apart from the glutamatic acid bacteria already discussed, *E. coli* and *Bacillus subtilis* are also good amino acid producing organisms. The glutamic acid bacteria previously classified as four different species are now regarded one species. The optimum temperature of *C glutamicum* is 30°C, whereas that of *E. coli* is higher. Hence, *E. coli* may be prefered for production in tropical countries.

Production strains for amino acids are generally classified as wild-type, capable of producing amino acids under defined conditions, but generally low-yielding in quantity, auxotrophic or regulatory mutants, in which feedback regulations are bypassed by partially starving them of their requirements or by removal of metabolic controls through mutation, and by gentically modifying the organism by amplifying genes coding for rate-limiting enzymes. The strains used belong to the last two categories and have been developed using classical mutation methods or through genetic engineering (Chapter 7). The selection of the strain is not only for high yields but also for those least producing undesirable side products. For instance, when branched chain amino acids are produced, it is essential that other branched chain amino acids do not occur as this increases the cost of separation and extraction.

21.6.4 Down Stream Processing

After fermentation, the cells may be filtered using a rotary vacuum filter (Chapter 10). Sometimes filtration can be improved by using filteraids. These filteraids, usually kiesselghur, which are based on diatomaceous earth, improve the porosity of a resulting filter cake leading to a faster flow rate. Before filtration a thin layer is used as a precoat of the filter (normally standard filters).

The extraction method of the amino acid from the filtrate, depends on the level of purity desired in the product. However two methods are generally used: the chromatographic (ion exchange) method or the concentration-crystallization method.

Crystallization is often used as a method to recover the amino acid. Due to the amphoteric character (contains both acidic and basic groups) of amino acids, their solubility is greatly influenced by the pH of the solution and usually show minima at the isoelectric point (zero net charge). Since temperature also influences the solubility of amino acids and their salts, lowering the temperature can be used in advance as a means of obtaining the required product. Precipitation of amino acids with salts, like ammonium and calcium salts, and with metals like zinc are also commonly used. This is followed by acid (or alkali) treatment to obtain the free or acid form of the amino acid.

Ion exchange resins have been widely used for the extraction and purification of amino acids from the fermentation broth. The adsorption of amino acids by ion exchange resins is strongly affected by the pH of the solution and by the presence of contaminant ions. There are two types of ion exchange resins; cation exchange resins and anion exchange resins. Cation exchange resins bind positively charged amino acids (this is in the situation where the pH of the solution is lower then the isoelectric point (IEP) of the amino acid), whereas anion exchange resins bind negatively charged amino acids (pH of the solution is higher than IEP). Elution of the bound amino acid(s) is done by introducing a solution containing the counterion of the resin. Anion exchange resins are generally lower in their exchange capacity and durability than cation exchange resins and are seldom used for industrial separation. In general, ion exchange as a tool for separation is only used when other steps fail, because of its tedious operation, small capacity and high costs.

SUGGESTED READINGS

Araki, K. 2003. Amino Acids Kirk-Othmer Encyclopedia of Chemical Technology. 2, 554-618.

Currell, B.R.C., Mieras, V.D., Biotol Partners. 1997. Biotechnological Innovations in Chemical Synthesis Elsevier.

Ikeda, M. 2003. Amino Acid Production Processes. Advances in Biochemical Engineering/ Biotechnology, 79, 1–35.

Kelle, R., Hermann, T., Bathe, B. 2005. L-Lysine. In: Handbook of Corynebacterium glutamicm. L Eggelin, and M Bott, (eds). Taylor and Francis, Boca Raton Fl, USA, pp. 465-488.

Kimura, E. 2003. Metabolic Enginering of Glutamate Production. Advances in Biochemical Engineering/Biotechnology, 79, 37–57.

Mueller, U., Huebner, S. 2003. Economic Aspects of Amino Acid Production. Advances in Biochemical Engineering/Biotechnology, 79, 137–170.

Pfefferle, W., Mockel, B., Bathe, B., Marx, A. 2003. Advances in Biochemical Engineering/Biotechnology, 79, 59–112.

Sano, K. 1994. Host – Vector Systems for Amino Acid-Producing Coryneform Bacteria. Improvement of Useful Enzymes by Protein Engineering. In: Recombinant Microbes for Industrial and Agricultural Applications. Y Murooka, T. Imanaka, (eds). Marcel and Dekker, New York, USA. pp. 485-507.

CHAPTER 22

Biocatalysts: Immobilized Enzymes and Immobilized Cells

22.1 RATIONALE FOR USE OF ENZYMES FROM MICROORGANISMS

Enzymes are organic compounds which catalyze all the chemical reactions of living things – plants, animals and microorganisms. They contain mainly protein; some of them however contain non-protein components, prosthetic groups. When excreted or extracted from the producing organism they are capable of acting independently of their source. It is this property of independent action which drew early attention to their industrial use.

All enzymes have infrastructural backbones of protein. In some enzymes only proteins exist, while in others, covalently attached carbohydrate groups may be present; often these carbohydrate groups may play no part in the catalytic activity of the enzyme, though they may contribute to the stability and solubility of the enzyme. Metal ions known as co-factors and low molecular weight organic compounds, known as co-enzymes may also be present. Co-factors and co-enzymes are important for the stability and activity of the enzyme. They have a tendency to be detached and it is important to provide conditions which ensure their retention.

Most industrial enzymes are obtainable from microorganisms. The advantages of using microorganisms are numerous, in contrast with their production from plants (e.g. malt diastase) and animals (e.g. pepsin) and are as follows:

(a) Plants and animals grow slowly in comparison with microorganisms;

(b) Enzymes form only small portions of the total plant or animal and large tracts of land as well as huge numbers of animals would be necessary for substantial productions. These limitations make plant and animal enzymes expensive. Microbial enzymes on the other hand are not subject to the above constraints and may be produced at will in any desired amount.

(c) By far the greatest attraction for the production of microbial enzymes, however, is the great diversity of enzymes which reflects the diversity of microbial types in nature. Thus largely, though not entirely, because of the widely varying environmental conditions in nature, microbial enzymes have been isolated which operate under extreme environmental conditions. For example microorganisms produce amylases functioning at temperatures as high as 110°C and proteases operating at pH values as high as 11 or as low as 3.

(d) Finally, following from greater understanding of the genetic basis for the control of physiological function in micro-organisms it is now possible to manipulate microorganisms to produce virtually any desired metabolic product, including enzymes.

22.2 CLASSIFICATION OF ENZYMES

Based on catalyzed reactions, the enzyme committee (EC) of the International Union of Biochemistry and Molecular Biology (IUBMB) recommended the classification of enzymes into six groups. The nomenclature of enzymes is based on the number assigned to these six major groups, and the sub-groups found within the major groups. Enzymes are also known by long-standing common names which are also widely used.

The IUBMB committee also defines subclasses and sub-subclasses. Each enzyme is assigned an EC (Enzyme Commission) number. For example, the EC number of catalase is EC 1.11.1.6. The first digit indicates that the enzyme belongs to oxidoreductase (class 1). Subsequent digits represent subclasses and sub-subclasses. Thus the enzyme rennet used in cheese manufacture and also known as chymosin, has the number of EC 5.3.1.5. The six major EC groups are as follows.

1. Oxidoreductases catalyze a variety of oxidation-reduction reactions. Common names include dehydrogenase, oxidase, reductase and catalase.

2. Transferases catalyze transfers of groups (acetyl, methyl, phosphate, etc.). Common names include acetyltransferase, methylase, protein kinase, and polymerase. The first three subclasses play major roles in the regulation of cellular processes.

3. Hydrolases catalyze hydrolysis reactions where a molecule is split into two or more smaller molecules by the addition of water. Some examples are:

Proteases: Proteases split protein molecules. They are further classified by their optimum pH as acid, alkaline or neutral. They may also be classified on the basis of their active centers into the following:

(i) *Serine proteases*: These have a residue in their active center and are specifically inhibited by diisopropyl phosphofluoridate and other organophosphorus derivates.

(ii) *Thiol proteases*: The activity of these depends on the presence of an intact-SH group in their active center. They are specifically inhibited by thiol reagents such as heavy metal ions and their derivatives, as well as alkylating and oxidizing agents.

(iii) *Metal proteases*: These depend on the presence of more of less tightly bound divalent cations for their activity.

(iv) *Acid proteases*: Acid proteases contain one or more side chain carboxyl groups in their active center.

Nucleases split nucleic acids (DNA and RNA). Based on the substrate type, they are divided into RNase and DNase. RNase catalyzes the hydrolysis of RNA and DNase acts on DNA. They may also be divided into **exonuclease** and **endonuclease**. The exonuclease progressively splits off single nucleotides from one end of DNA or RNA. The endonuclease splits DNA or RNA at internal sites.

Phosphatase catalyzes dephosphorylation (removal of phosphate groups).

4. Lyases catalyze the cleavage of C-C, C-O, C-S and C-N bonds by means other than hydrolysis or oxidation. Common names include decarboxylase and aldolase.

5. Isomerases catalyze atomic rearrangements within a molecule. Examples include rotamase, protein disulfide isomerase (PDI), epimerase and racemase.

6. Ligases catalyze the reaction which joins two molecules. Examples include peptide synthase, aminoacyl-tRNA synthetase, DNA ligase and RNA ligase.

22.3   USES OF ENZYMES IN INDUSTRY

Most of the enzymes used in industry are hydrolases (i.e., those which hydrolyze large molecules). In particular amylases, proteases, pectinases, and to a lesser extents lipases have been most commonly used. Enzymes are used in a wide range of industries and some uses are discussed below.

(i) *Production of nutritive sweeteners from starch*: Enzymic hydrolysis has now almost completely replaced the use of acid in starch hydrolysis (Chapter 4). The sweeteners which have been produced from starch are high conversion (or high DE) syrup, high maltose syrup, glucose syrup, dextrose crystals and high fructose syrup. These sweeteners are often called corn syrups because they are produced from maize, although starch from any source (e.g. cassava, sorghum, or potatoes) may be used. The processes of production of sweeteners from corn consists of the gelatinization of starch production of water-soluble dextrins with α-amylase, the subsequent application of a de-branching enzyme (e.g. pullulanase) and, depending on the sugar sought, the application of a third enzyme. α-Amylase from *B. licheniformis* is particularly suitable for dextrinization because its optimum temperature is 110°C, a convenient temperature on account of the need to boil starch to gelatinize it. If high maltose syrup is sought α-amylase is applied, while gluco-amylase is applied if glucose syrup is sought (Fig. 22.1). Dextrose crystals are usually produced by removing minerals with ion exchange resins and then crystallizing the liquid after concentration.

Nowadays, most sweeteners produced from starch are in the form of high fructose corn syrup (HFCS), whose production is discussed below. Glucose has a rather bland taste and is not as sweet as sucrose. Fructose, on the other hand, is about 1.7 as sweet as sucrose. In the confectionary industry therefore glucose resulting from the hydrolysis of starch is converted to fructose by the enzyme glucose isomerase which rearranges the glucose molecule to yield fructose and the syrup itself into a glucose/fructose syrup.

Glucose isomerase is completely specific for monomeric D-glucose. The maltose, maltotriose and higher maltooligosaccharides present in glucose syrup are untouched by the enzyme. An acceptable composition of high-fructose glucose syrups in commerce is: fructose 42%, glucose, 50%, maltose, 6%, and maltotriose, 2%. These high fructose

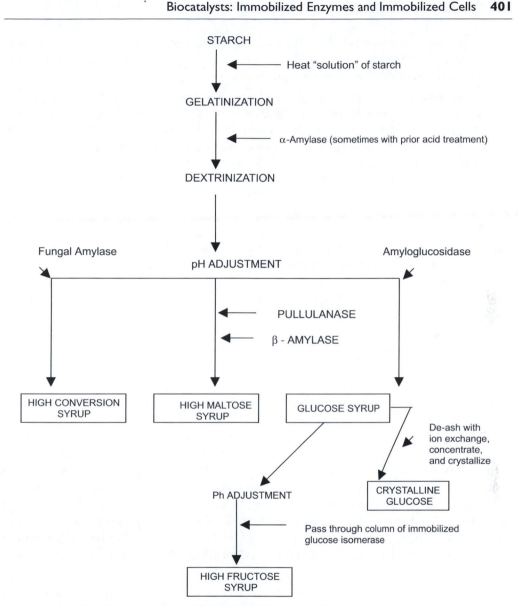

Fig. 22.1 Hydrolysis of Starch for High Fructose Corn Syrup Production

mixtures are used in place of sucrose and invert sugar in the baking and beverages industries. HFCS production indeed represents one of the major uses of enzymes.

Besides its role as a sweetener, fructose has other qualities which make it superior to glucose. It is regarded as a low-calorie sweetener, since it is so much sweeter than sucrose. Furthermore, fructose is favored for intravenous infusions or drip over glucose because the body tolerates it better. Twice as much fructose as of glucose can therefore be permitted in drips, representing a greater caloric intake in as much fluid. It is therefore the preferred sugar for patients in a state of shock. Finally it is more quickly absorbed than glucose and does not need the enzyme and hormonal system required for absorbing the former.

Organisms which have served as sources for glucose isomerase include more than two dozen strains of *Streptomyces*, several of *Arthobacter*, *Nocardia*, *Micromonospora* as well as *Lactobacillus brevis* and *Pseudomonas hydrophilla*.

(ii) *Proteolytic enzymes in the detergent industry*: The detergent industry is at present one of the greatest consumers of enzymes, and uses mostly proteases. Blood and pus stains from hospital linen and other protein dirt precipitate and coagulate on clothes and are ordinarily difficult to remove. The inclusion of proteolytic enzymes in a detergent or washing soap greatly facilitates the removal of such stains. The proteolytic enzymes used for this purpose should have a high pH optimum of 9-11, which is the pH of detergents, and a high temperature optimum of 65-70°C since hot water facilitates laundering. Furthermore, the enzyme should be able to cleave peptide bonds randomly and facilitate the dissolution of the protein. Such proteolytic enzymes have been produced mainly by alkalophilic and aerobic spore-formers such as strains of *Bacillus licheniforms* and *Bacillus amyloliquifaciencs*. The latter has the advantage of producing α-amylases as well.

It is worth mentioning that the history of the use of enzymes in detergents has not always been smooth. Soon after their introduction in the early 1960s, some factory workers handling the enzymes suffered from allergic reactions. Strong public protests led to the withdrawal of enzymes in detergents and, although a commission of inquiry showed that there was no danger to the user, the use of enzymes in detergents suffered a temporary setback. Subsequently, enzymes are added in dust-free encapsulated preparations, to avoid inhalation by producers and users.

(iii) *Microbial rennets*: Rennin is an acid protease found in gastric juice of young mammals where it helps to digest milk. It is used in the manufacture of cheese and functions by hydrolyzing a polypeptide fragment from milk protein-kappa casein to leave paracasein; this then forms an insoluble complex with cations to give a firm curd. The commercial form of rennin known as rennets is obtained from the fourth stomachs of young calves. It is therefore expensive and tedious to produce since it involves the maturation, gestation and delivery of cows. Due to this, a search for substitutes ensued. Strains of *Mucor miehi*, *M. pusillus*, *Endothia parasitica*, *Bacillus polymyxa*, *B. subtilis* and *Aspergilus* are used to produce acid proteases which successfully substitute for rennin. Indeed, microbial rennets constitute about the third largest use of microbial enzymes.

(iv) *Lactase*: Lactase hydrolyzes the disaccharide lactose into its component galactose and glucose, both of which are sweeter than lactose and correspond to the addition of 0.9% sucrose. Thus, dairy products containing lactose, such as yoghurt, and ice cream, can be sweeter and more acceptable to consumers without the extra expense of extraneously added sugar. Galactose and glucose are also metabolized by a far wider range or organisms than can attack lactose. The result is that lactase-hydrolyzed whey can be used to produce alcohol or soft drinks. Furthermore, milk in which lactose is hydrolyzed is preferred by individuals in some parts of the world where intestinal lactase is low. Finally, when lactose occurs in high concentrations, such as is in ice cream, it tends to crystallize out giving the impression during consumption that grains of sand are present in the product. The addition of lactase prevents such crystallization. Lactase is now produced commercially from *Kluyveromyces fragilis*, *Saccharomyces lactis* or *Aspergillus niger*.

(v) *The textile industry*: In the textile industry large amounts of starch, gelatin and their derivatives such as glue are used to strengthen threads (yarns) of synthetic and natural materials (e.g. cotton) to enable them to stand abrasion during weaving and also to polish sewing thread. Starch and its derivatives are furthermore used to restrict dye stuffs and prevent their diffusion to other portions of the fabric.

At the end of the manufacturing the starch is removed with thermostable α-amylases from *Bacillus licheniformis*. The cloth is passed through hot water to remove inorganic salts and to raise the temperature. It is then passed into an enzyme solution where it is allowed to remain from 15 minutes to several hours depending on the enzyme concentration and other factors.

Natural silk threads are proteinaceous in nature and consist of a highly resistant protein known as fibroin, but they are held together by a semi-soluble protein gum known as sericin. Sericin is removed with a neutral proteolytic enzyme; the silk threads are then sized and woven into cloth after which they are de-sized in an amylase or protease depending on the sizing material.

(vi) *Pectinases for use in fruit juice and wine manufacture:* Pectinases are enzymes which attack pectic substances, a group of complex acidic polysaccharides. Pectic substances are high molecular weight substances made up of poly – D – galacturonic acid. As the carboxylic acid groups of the sugar units are partially esterified with methanol, they are regarded as poly-uronides. They are the cementing material holding plant cells together.

The Agricultural and Food Chemistry section of the American Chemical Society has proposed the following nomenclature for the various pectic substances:

Pectic Substances: a group designation for those complex carbohydrate derivatives which occur in or are prepared from plants and contain a large proportion of anhydrogalacturonic acid units. The carboxyl groups of polygalactoronic acids may be partly esterified by methyl groups and partly or completely neutralized by one or more bases.

Protepectin: The water-insoluble parent pectic substance which occurs in plants and which, upon restricted hydrolysis, yields pectin or pectinic acids.

Pectinic acids: are colloidal polygalacturonic acids containing more than a negligible proportion of methyl ester groups.

Pectic substances may be regarded as polygalacturonides composed of unbranched α–1, 4 galacturonic acid residues (Fig. 22.2) but with other non-uronides bound to the chain. However, pectic materials are not uniform because of the great variations which have been observed by many authors in the molecular weights, degree of esterification and acetylation, amount and type of neutral sugars and non-uronide residues found in various preparations of pectic substances.

Pectinolytic enzymes or pectinases are widely distributed in plants and among microorganisms. These enzymes vary greatly in their mechanisms of action, but may be grouped into two: esterases which de-esterify pectin to pectic acid and depolymerases which depolymerize pectin, pectin acid or short-chain galacturonic acid (ligo-D-galacturonates) derived from pectin and pectic acids (Table 22.1). *Aspergilli* (*A.niger, A. oryzae, A. wentii*, and *A. flavus*) and other fungi (Table 22.1) are used for industrial enzyme production. The industrial enzymes themselves are a mixture of various pectinolytic enzymes.

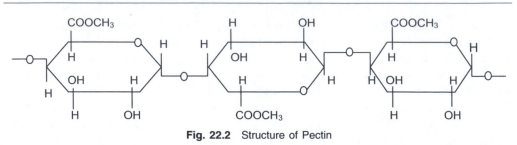

Fig. 22.2 Structure of Pectin

Table 22.1 Distribution of pectinolytic enzymes in some microorganisms

Source	PMGE	PG	PGL	PMG	PMGL	OG	OGL
Bacillus sp.			+			+	
Erwinia aroideae		+	+				+
Pseudomonas sp.			+				+
Ps. Marginalis	+	+	+				
Xanthomonas campestris	+		+				
Clostridium multifermentans	+		+				
Aspergillus niger	+	+		+	+		
Penicillium digitatum	+			+	+		

Key:

PMGE = Polymethylgalacturonase esterase (de-esterify pectin to pectic acid by removal of methoxy residues)

PG = Polygalacturonases (hydrolyze pectic acids randomly, successive bonds or alternate bonds)

PGL = Polygalactorunate lyase (hydrolyzes pectic acids by transelimination)

PMG = Polymethylgalacturonase (hydrolyzes pectin in random or sequential fashion)

PMGL = Polymethylgalacturanate lyase (cleaves pectic acid randomly or sequentially)

OG = Oligogalacturonase (hydrolyzes oligo-galactosiduronates i.e. breakdown products of pectin and pectic acids)

OGL = Oligogalacturonate lyase (acts as OG but by transelimination)

Pectinolytic enzymes are used principally in the fruit juice, fruit processing and the wine industries. In the fruit juice and fruit industry, pectinases are used to disintegrate the fruits, and to clarify the resulting juices to give a clear sparkling liquid after the filtration of the debris.

Another application of pectinases which does not involve the isolation of the enzymes but which deserves mention is the retting of plants for flax (linen) from (*Linum usitatissimum*) and hemp (*Cannabis sativa*) and of jute from *Corchorus* sp. Retting has not been studied intensively probably because of the advent of man-made fibres. Pectinolytic enzymes are produced anaerobically by *Clostridium* spp. when the plants are immersed in water. When aerobic conditions prevail as in 'dew-retting' the organisms which have been isolated include *Bacillus comesii*, and the fungi, *Cladosporium*, *Aspergillus*, and *Penicillium*.

(vii) *Naringinase* is used for removing the bitter tasting substance from citrus fruits, especially grape fruits. Naringin is a flavonoid found in grapefruits, and gives grapefruit its characteristic bitter flavor. Flavonoids are a group of polyphenolic secondary metabolites secreted by plants and found widely among plants. They are present in many plant-based foods such as tea and soybeans, and are generally believed to be beneficial to health. Although naringin is supposed to have some beneficial effect such as stimulating our perception of taste by stimulating the taste buds (for which reason some people eat grape fruits before a meal), the bitter taste is undesirable in fruit juices. Therefore, grapefruit processors attempt to select fruits with a low naringin content, or use naralginase produced by strains of *Aspergillus* spp. to remove it.

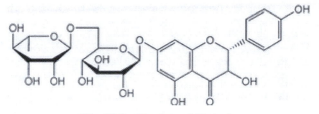

Fig. 22.3 Structure of Naringin

(viii) *Enzymes in the baking industry*: Flour contains 72-75% starch, 11-13% protein, and 0.04-0.4% minerals. It also contains amylases and proteases derived from wheat. These enzymes play major roles in the nature of the final baked product. When the flour is deficient in amylases, unusually low amounts of sugar and intermediate products are produced, giving rise to low volume, dry texture and other undesirable characteristics in the finished bread. Similarly, while wheat brands rich in gluten (or 'hard' wheat) are suitable for bread making because they are highly elastic, they are not suitable for cakes. For the latter, 'soft' wheat with low gluten contents is required. 'Hard' wheat is rendered 'soft' however by the hydrolysis of some of the gluten using proteases. Bakery amylases and proteases are derived from fungi.

(ix) *Enzymes in the alcoholic beverages industry*: Amylases derived from fungi or bacilli may be employed in the distilled alcoholic beverages to hydrolyze starch to sugars prior to fermentation by yeasts. The enzymes may also be used to hydrolyze unmalted barley and other starchy adjuncts converting them to maltose and thus reducing the cost of beer brewing. Although it is unusual, turbidities due to starch may arise in beer due to the destruction of the amylases following the use of too high a temperature during malt kilning. Amylases are used to remove turbidities due to starch. In beer, chill-hazes are due mainly to protein-tannin (protein-polyphenol) precipitates. Chill-hazes may be removed in several ways, one of which is the addition of proteases. Proteases from *Aspergillus niger* are often used (Chapter 12).

(x) *Leather baiting*: After animal skin has been trimmed of flesh, it is de-haired with lime or a proteolytic enzyme and it is then baited. The purpose of baiting, is to prepare the de-haired skins and hides for tanning. Baiting used to be done by keeping the skin in a warm suspension of chicken and dog dung, which probably yielded proteolytic bacteria. Nowadays proteolytic enzymes from *Bacillus* spp. and *Aspergillus* spp. are used. The effect of baiting is to make the fibrous protein collagen of which leather is composed more amenable to the subsequent processes of leather manufacture.

(xi) *Some medical uses of microbial enzymes:* At present, the most successful medical applications of enzymes are the use of proteolytic enzymes from *Bacillus* spp. and other bacteria for the treatment of burns and skin cancers, and the treatment of life-threatening disorders within the blood circulation using hemolytic enzymes produced from B-hemolytic streptococci. Other uses are given below:

(a) Fungal acid proteases may be used to treat alimentary dyspepsia, because of the acid resistance of the enzyme. Fungal amylases may also be used to help digestion.

(b) Dextrans deposited on the teeth by *Streptococcus mutans* may be removed with the use of fungal dextranase often introduced into the toothpaste, thus helping to fight dental decay.

(c) L-asparaginase produced from *E. coli* and other gram-negative bacteria may be used in the treatment of certain kinds of leukemia.

(d) Penicillinases produced by many organisms are sometimes used in emergency cases of penicillin hypersensitivity.

(e) Rhodanase which catalyses the reaction, $S_2O_3^{2-} + CN^- \longrightarrow SO_3^{2-} + SCN^-$ has been used to combat cyanide poisoning. Rhodanase is produced by the thermoacidophilic bacterium *Sulfobacillus sibiricus*.

The above are only a few of some of the uses to which microbial enzymes have been put in the medical area.

22.4 PRODUCTION OF ENZYMES

22.4.1 Fermentation for Enzyme Production

Most enzyme production is carried out in deep submerged fermentation; a few are best produced in semi-solid media.

22.4.1.1 Semi solid medium

This system, also known as the 'Koji' or 'moldy bran' method of 'solid state' fermentation is still widely used in Japan. The medium consists of moist sterile wheat or rice bran acidified with HCl; mineral salts including trace minerals are added. An inducer is also usually added; 10% starch is used for amylase, and gelatin and pectin for protein and pectinase production respectively. The organisms used are fungi, which appear amenable to high enzyme production because of the low moisture condition and high degree of aeration of the semi-soluble medium.

The moist bran, inoculated with spores of the appropriate fungi, is distributed either in flat trays or placed in a revolving drum. Moisture (about 8%) is maintained by occasionally spraying water on the trays and by circulating moist air over the preparation. The temperature of the bran is kept at about 30°C by the circulating cool air.

The production period is usually 30-40 hours, but could be as long as seven days. The optimum production is determined by withdrawing the growth from time to time and assaying for enzyme. The material is dried with hot air at about 37°C–40°C and ground. The enzyme is usually preserved in this manner. If it is desired, the enzyme can be extracted. Growth in a semi-solid medium seems sometimes to encourage an enzyme range different from that produced in submerged growth. Thus, *Aspergillus oryzae* on semi-solid medium will produce a large number of enzymes, primarily amylase, glucoamylose, and protease. In submerged culture amylase production rises at the expense of the other enzymes. Similarly, if *Aspergillus oryzae* producing takadiastase (a commercial powder containing amylase and some protease) is grown in submerged culture four protease components are formed whereas on semi-solid medium not only are two proteases formed, but these are less heat resistant than those produced in submerged fermentation.

22.4.1.2 Submerged production

Most enzyme production is in fact by submerged cultivation in a deep fermentor (Chapter 9). Submerged production has replaced semi-solid production wherever possible because the latter is labor intensive and therefore expensive where labor is scarce, and because of the risk of infection and the generally greater ease of controlling temperature, pH and other environmental factors in a fermentor.

The medium must contain all the requirements for growth, including adequate sources of carbon, nitrogen, various metals, trace elements, growth substances, etc. However, a medium adequate for growth may not be satisfactory for enzyme production.

For the production of inducible enzymes, the inducers must be present. Thus, pectic substances need to be in the medium when pectinolytoc enzymes are being sought. Similarly, in the production of microbial rennets soy bean proteins are added into the medium to induce protease production by most fungi. The inducer may not always be the substrate but sometimes a breakdown or end-product may serve. For example, cellobiose may stimulate cellulose production.

Sometimes some easily metabolizable components of the medium may repress enzyme production by catabolite repression. Strong repression is often seen in media containing glucose. Thus, α-amylase synthesis is repressed by glucose in *Bacillus licheniformis* and *B. subtilis*. Fructose on the other hand represses the synthesis of the enzyme in *B. stearothermophilus*. In many organisms protease synthesis is repressed by amino acids as well as by glucose. It is therefore usual to replace glucose by more slowly metabolized carbohydrates such as partly hydrolyzed starch. High enzyme yield may also be obtained by adding constantly, low amounts of the inducer.

End-product inhibition has also been widely observed. Some specific amino acids inhibit protease production in some organisms. Thus, isoleucine and proline are involved in the case of *B. megaterium* while sulphur amino acids inhibit protease formation in *Aspergillus niger*.

Temperature and pH requirements have to be worked out for each organism and each desired product. The temperature and pH requirements for optimum growth, enzyme production, and stability of the enzyme once it is produced are not necessarily the same for all enzymes. The temperature adopted for the fermentation is usually a compromise taking all three requirements into account.

The oxygen requirement is usually high as most of the organisms employed in enzyme production are aerobic. Vigorous aeration and agitation are therefore done in the submerged fermentations for enzyme production. Batch fermentation is usually employed in commercial enzyme fermentation and lasts from one to seven days.

Continuous fermentation, while successful experimentally, does not appear to have been used in industry.

In a few cases the enzyme production is highest during the exponential phase of growth. In most others, however, it occurs post-exponentially. Furthermore, different enzymes are produced at different stages of the growth cycle. Thus *Asp. niger* produces mostly α–amylases in the first 72 hours but mainly maltase thereafter.

22.4.2 Enzyme Extraction

The procedures for the extraction of fermentation products described in Chapter 10 are applicable to enzyme extraction. Care is taken to avoid contamination. In order to limit contamination and degradation of the enzyme the broth is cooled to about 20°C as soon as the fermentation is over. Stabilizers such as calcium salts, proteins, sugar, and starch hydrolysates may be added and destabilizing metals may be removed with EDTA. Anti-microbials if used at all are those that are normally allowed in food such as benzoates and sorbate. Most industrial enzymes are extra-cellular in nature. In the case of cell bound enzymes, the cells are disrupted before centrifugation and/or vacuum filtration.

The extent of the purification after the clarification depends on the purpose for which the enzyme is to be used. Sometimes enzymes may be precipitated using a variety of chemicals such as methanol, acetone, ethyl alcohol or ammonium sulfate. The precipitate may be further purified by dialysis, chromatography, etc., before being dried in a drum drier or a low temperature vacuum drier depending on the stability of the enzymes to high temperature. Ultra-filtration separation technique based on molecular size may be used.

22.4.3 Packaging and Finishing

The packing of enzymes has become extremely important since the experience of the allergic effect of enzyme dust inhalation by detergent works. Nowadays, enzymes are packaged preferably in liquid form but where solids are used, the enzyme is mixed with a filler and it is now common practice to coat the particles with wax so that enzyme dusts are not formed.

22.4.4 Toxicity Testing and Standardization

The enzyme preparation should be tested by animal feeding to show that it is not toxic. This test not only assays the enzyme itself but any toxic side-product released by the microorganisms. For a new product extensive testing should be undertaken, but only spot checks need to done for a proven non-toxic enzyme in production. The potency of the enzyme preparation, based on tests carried out with the substrate should be determined. The shelf life and conditions of storage for optimal activity should also be determined.

22.5 IMMOBILIZED BIOCATALYSTS: ENZYMES AND CELLS

The major handicap in the traditional use of enzymes is that they are used but once. This is mainly because the enzymes are unstable in the soluble form in which they are used and because recovery would be expensive, even if it were possible. It is not surprising that influenced by the idea of the catalyst in the chemical industry ways should be sought to re-use biological catalysts. The immobilization of enzymes and cells provides a basis for the re-use of enzymes and cells. Interest in immobilized enzymes has grown since the 1960s and numerous conferences and papers have been held and given on them. Immobilized cells have received a great deal but perhaps slightly less attention judging from the literature, since their study came a little later than that of immobilized enzymes.

(i) *Imobilized enzymes*: An immobilized enzyme may be defined as an isolated or purified enzyme confined or localized in a defined volume of space.

(ii) *Immobilized cells*: Immobilized cells, also referred to as controlled biological catalysts, may be defined as a high density of cells physically confined on a solid phase or in pellets or clumps and in which cell movement is restricted for the period of their use as biological agents. This definition excludes cells in a chemostat, or cells which are recovered by centrifugation in a batch culture and returned to the fermentation. It is not a completely satisfactory definition as the term 'high density' can be elastic. Cell immobilization has existed or been exploited long before it became recognized as potentially valuable in industry. Thus, microorganisms in natural habitats such as soil, marine, alimentary canal, dental plaque or in the 'Orleans' process of vinegar production where cells are immobilized on wood shavings, the activated and trickling filter treatment of wastewater, may be seen as examples of immobilized cells.

22.5.1   Advantages of Immobilized Biocatalysts in General

The advantages of immobilized enzymes beside reuse are as follows:

(i) They can be easily separated from the reaction mixture containing any residual reactants and reused in subsequent conservations.

(ii) Immobilized enzymes are more stable over broad ranges of pH and temperature.

(iii) Enzymes are absent in the waste-stream

(iv) Immobilized systems specially lend themselves to continuous processes.

(v) Reduced costs in industrial production.

(vi) Greater control of the catalytic effect.

(vii) Greater ease of new applications for industrial and medical purposes.

(viii) Immobilized enzymes permit the use of enzymes from organisms which would not normally be regarded as safe (i.e. non-GRAS).

22.5.2   Methods of Immobilizing Enzymes

Immobilized enzymes have been classified in a number of ways. The classification method adopted here is the one published in 1995 by the International Union of Pure and Applied Chemistry (IUPAC), and which divides methods of immobilized enzymes into four broad groups, based on:

(a) covalent bonding of the enzyme to a derivatized water-insoluble matrix,

(b) intermolecular cross-linking of enzyme molecules using multifunctional reagents,

(c) adsorption of enzyme onto a water-insoluble matrix, and

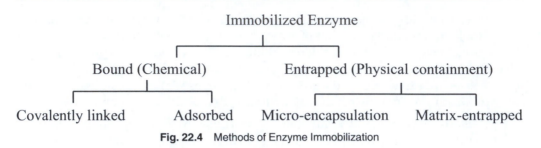

Fig. 22.4 Methods of Enzyme Immobilization

(d) entrapment of the enzyme molecule inside a water-insoluble polymer lattice or semi-permeable membrane.

The IUPAC groups can be divided into two basic groups, the chemical and the physical methods as shown in Fig. 22.4.

The IUPAC emphasizes that in dealing with immobilized enzymes, the properties of the free enzyme, the type of support used and the methods of support activation and enzyme attachment must be specified.

22.5.2.1 Immobilization by covalent linkage

This is by far the most widely studied method. The covalent linkage is achieved between a functional group on the enzyme not essential for catalytic activity and a reactive group on a solid water-insoluble support. The functional groups available on enzymes for linkage are amino and carboxyl groups, hydroxyl groups, imidazole groups, indole groups, phenolic groups and sulphydryl groups. The nature of the enzyme's functional group through which immobilization is to be effected determines the reaction which will be used to bind the enzyme to the support. Some of the reactions which have been used are acylation, amination and arylation and alkylation.

Some supports which have been used for immobilization include agarose, celluose, dextran, chitin, starch, polygalacturonic acid (pectin), polyacrylamide, polyvivyl alcohol, polystyrene, polyprpylene, polyamino acids, polyamide, glass and metal aides and bentonite. Many of these are organic, but recently the use has been advocated of inorganic support on the grounds of reuse of inorganic materials, non-toxicity, good half-life of enzymes immobilized on inorganic supports, and the ease with which inorganic materials can be fashioned to suit any particular enzyme system.

In the above description the reaction has been a direct one between functional groups on the enzyme and reactive sites on the support. However, in some cases intermolecular linkage may occur through the mediation of a crosslinking reagent, In such cases the cross linking agent has a number of different active sites. Some of these react with the support and others with the enzyme. One of the most commonly used cross-linking agent is glutaraldehyde which has two cross-linking sites. Several others are available. The advantage of the covalent bonding method of enzyme immobilization are:

(i) The coupling of the enzyme to the support is easy to conduct and consists of allowing support and enzyme to interact and therefore facilitates centrifuging and washing off any enzymes not bound.

(ii) The enzyme-support derivative is easy to manipulate and adapt because of the great physical and chemical variation in the available support: they can be used in a variety of reactors including stirred tank, fluidized bed-reactors and can also be modified into flat sheets fiber.

(iii) Covalent coupling has been widely described and methods for carrying it out are readily available in the literature.

(iv) The supports themselves are widely available commercially.

The disadvantages are that some preparations are tedious to make; the chemical bonding may inactivate the enzyme in some cases; and finally covalently-bound water-insoluble enzyme-substrate derivatives act poorly on high molecular weight substrates.

22.5.2.2 Immobilization by adsorption

This method is both simple and inexpensive and consists of bringing an enzyme solution in contact with a water-insoluble solvent surface and washing off the unadsorbed enzyme. The extent of the adsorption depends on a number of factors including the nature of the support, pH, temperature, time, enzyme concentration. In principle, though not always in practice, adsorption is reversible. Adsorbents which have been used include alumina, bentonite, calcium carbonate, calcium phosphate, carbon, cellulose, charcoal, clay, collagen, diatomaceous earth, glass, ion-exchange resins, sephadex, and silica gel. Apart from the ease of the operation, the other advantage is that the enzymes are unlikely to be inactivated because the system is mild. The disadvantage is that in cases of weak binding the enzyme may be easily washed away.

22.5.2.3 Immobilization by micro-encapsulation

Micro-encapsulation consists in packaging the enzyme in tiny usually spherical capsules ranging from 5-300 μ in diameter in semi-permeable (permanent) or liquid (non-permanent) membranes. The former are more commonly employed. To prepare micro-capsules a high aqueous concentration of the enzyme is first prepared. The aqueous enzymes solution is then emulsified in an organic solvent or solvents with a surfactant which is soluble in the organic solvents. Two methods are then used to form micro-capsules from this enzyme-surfactant emulsion.

In one method known as the interfacial polymerization technique, the enzyme solution contains the enzyme as well as one component of the membrane that will form round the micro-capsule. The emulsion is stirred vigorously and more of the organic solvent(s) containing the rest of the capsule-forming reagent is added.

In the second method, the coacervation-dependent method, the added organic solvents contain all the components of the polymer. In both cases the enzyme droplets are formed during the vigorous stirring. The semi-permeable membrane is allowed to harden around the micro-droplets; the micro-capsules are then washed and then transferred.

Semi-permeable membranes have been made of cellulose nitrate, polystyrene, etc., with the coacervation method.

Micro-capsule formation by the interfacial method has been produced from nylon and widely investigated because of its application in medicine e.g. in urease immobilization in artifical kidneys. The organic solvent usually used for the polyamide Nylon-6,10 semi-permeable membrane from hexamethylenediamine and sabacoyl chloride is a chloroform-cyclohezane mixture.

$$H_2N-(CH_2)_6-NH_2 + Cl-\overset{\overset{\displaystyle O}{\|}}{C}-(CH_2)-\overset{\overset{\displaystyle O}{\|}}{C}-Cl-HCI$$

Hexamethlenediamine   Sabacoyl chloride

$$-NH-(CH_2)_6-NH-\overset{\overset{\displaystyle O}{\|}}{C}-(CH_2)_8-\overset{\overset{\displaystyle O}{\|}}{C}(CH_2)_6-NH\,\overset{\overset{\displaystyle O}{\|}}{G}-(CH_2)_8-\overset{\overset{\displaystyle O}{\|}}{C}-NH-$$

Non-permanent liquid membranes are prepared by emulsifying the aqueous enzyme solution in a surfactant to form the liquid membrane-encapsulated enzyme. It is not commonly used.

The advantages of the permanent (semi-permeable) micro-capsulation method are:

(i) An extremely large surface is provided by the tiny bubbles of enzymes. A micro-capsule with 20μm diameter would for example have a surface area of 2,500 cm^2/ml.

(ii) The specificity of the micro-capsule is increased by the possibility of using a membrane which will favor the diffusion of substrates of certain types.

The disadvantage is that a high concentration of enzyme is needed and only low molecular weight substances pass through. With the non-permanent liquid membranes, the same advantages accrue; the disadvantage however is leakage.

22.5.2.4   Immobilization by entrapment

In the entrapment of enzymes, no reaction occurs between support and the enzyme. A cross-linked polymeric network is formed around the enzyme; alternatively the enzyme is placed in a polymeric substance and the polymeric chains cross-linked. Polyacrylamide gels have been widely used for this purpose, although enzymes do leak through the network in some cases.

The advantages of the entrapping method are: (i) its simplicity, (ii) the small amount of enzyme used, (iii) the unlikelihood of damage to the enzyme, (iv) applicability to water insoluble enzymes.

The disadvantages include leakage of enzymes and some chemical and thermal enzyme damage during gel formation.

22.5.3   Methods for the Immobilization of Cells

Three general methods are available for immobilizing microbial cells.

(i) *Ionic binding to water-soluble ion-exchangers*: Cells of E. coli and *Azotobacter agile* bound to Dowex-l resin have been studied while mold spores have been bound to ion-exchange cellulose derivatives. In both cases successful demonstration of succinic acid oxidation and invertase activity were demonstrated. This method is however not entirely satisfactory as enzymes may leak out following autolysis during continuous enzyme reaction.

(ii) *Immobilization in cross-linked chemicals*: Microbial cells have been immobilized by cross-linking each other with bi-functional reagents such as glutaraldehyde. But non-cross-linking agents are equally effective.

(iii) *Entrapment in a polymer matrix:* This appears to be the most widely used method of immobilizing cells. In this method the cells are entrapped in a polymer matrix where they are physically restrained. The following matrixes have been used: polyacrylamide, collagen, cellulose triacetate, agar, alginate and polystyrene. Methods for immobilization of cells and enzymes are given in Fig. 22.5.

22.5.3.1 Advantages of immobilized cells

Immobilized cells have the following advantages over conventional batch fermentation as well as over immobilized enzymes.

(i) In batch fermentation, a significant proportion of the substrate is 'wasted' for the growth of the microbial population and for producing other substances other than enzymes required for the conversion at hand. Once the cells are immobilized however, they need to be offered nutrients for growth.

(ii) When cells are immobilized the reactions are more homogeneous and can be treated more like catalysts.

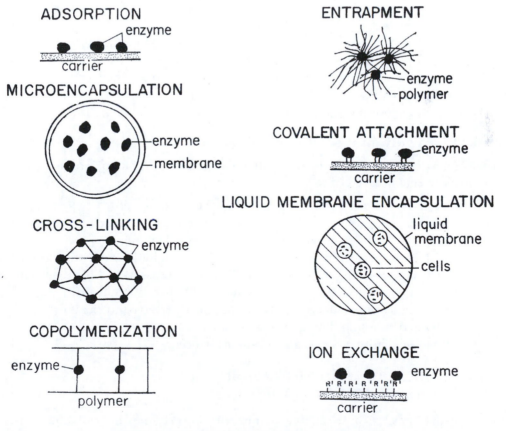

Fig. 22.5 Methods of Immobilizing Enzymes and Cells

(iii) The lag period which occurs in a conventional batch fermentation is eliminated for the accumulation of products associated with non-growth phase of the cells.

(iv) It is more feasible to run immobilized cells continuously at high dilution rates without the risk of washout which would occur in a conventional continuous culture system.

(v) Higher and faster yield is possible because of the greater density of cells; furthermore, toxic materials are continuously removed.

(vi) It is possible to recharge or resuscitate the cells by inducing growth and reproduction among resting cells.

(vii) A high capital cost is involved in installing, and operating a fermentor; in systems where comparison have been made, immobilized cells are cheaper than the conventional batch production.

(viii) The use of immobilized cells eliminates the need for enzyme extraction and purification. Furthermore, systems involving multi-enzyme reactions can occur more easily in intact cells harboring these enzymes.

(ix) Immobilized cells are more suited to multiple step processes

(x) Cofactor regeneration is not a problem

Immobilized cells are particularly appropriate under the following conditions:

(i) When the enzymes are intracellular: the use of immobilized cells would eliminate the need for breaking the cells for enzyme isolation.

(ii) When extracted enzymes are unstable during or after immobilization.

(iii) When the micro-organism does not produce enzymes which can cause undesirable side reactions; or when such side-reaction producing enzymes can be readily inactivated.

(iv) When the substrates and products are not high molecular weight substances.

22.5.3.2 Disadvantages of immobilized cells

Some of the disadvantages of the conventional system of cultivation of organisms spill over to the immobilized cell thus:

(i) The cells may produce enzymes other than the one (s) sought.

(ii) Genetic changes, although with reduced likelihood in comparison with conventional fermentation, can also occur during immobilization.

(iii) Immobilization may result in the loss of a specific catalytic activity due to enzyme inactivation, resulting from the immobilization process or to diffusional barriers hindering substrate access, or product removal from the organisms.

(iv) Cells located in the center of a cell flow may be deprived of nutrients or be inactivated by accumulating toxic wastes.

(v) Contamination by other microorganisms can occur.

22.6 BIOREACTORS DESIGNS FOR USAGE IN BIOCATALYSIS

A variety of bioreactors are available for immobilizing enzymes and cells and these are shown in Fig. 22.6.

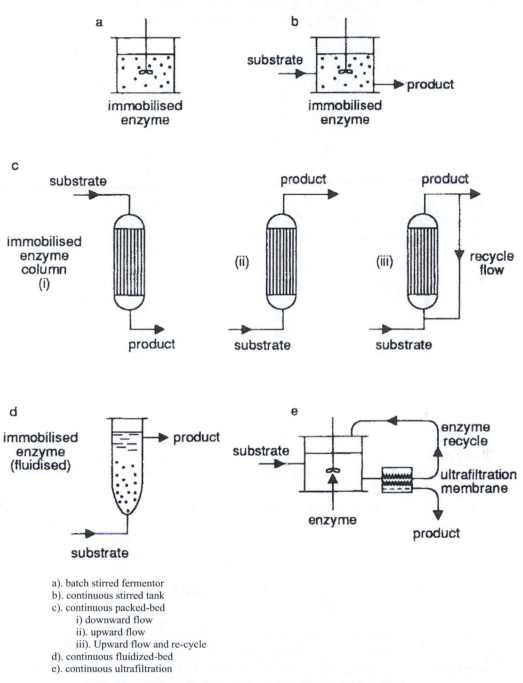

a). batch stirred fermentor
b). continuous stirred tank
c). continuous packed-bed
    i) downward flow
    ii). upward flow
    iii). Upward flow and re-cycle
d). continuous fluidized-bed
e). continuous ultrafiltration

Fig. 22.6 Various Designs of Bioreactors for Use in Biocatalysis.

22.7 PRACTICAL APPLICATION OF IMMOBILIZED BIOLOGICAL CATALYST SYSTEMS

Immobilized enzymes and cells have been intensively studied in the hope that they can be used industrially. Only some of the expectations have been realized because of economic reasons. Soluble 'once only' enzymes marketed in the form of powder or liquids are available at low prices. Amylases and glucoamylases used in the starch industry are so low-priced comparatively that immobilized forms can hardly compete. The only immobilized enzyme currently used on a large scale is glucose isomerase used to produce about 2 million tons of high-fructose syrups around the world, but especially in the USA, Europe, and Japan. High fructose syrup competes successfully with sugar from beet or cane.

Several immobilized enzymes or whole cell processes are being applied in the Japanese pharmaceutical industry. These include L-amino acid from racemic acyl – D – L – amino acids, L-aspartic acid from ammonium fumarate, L-citrulline from L-arginine, and the production of 6-amino penicillanic acid (6-APA) for semi synthetic penicillin production. In the United States and Europe 6-APA is produced with immobilized enzymes. In Italy whole milk lactose hydrolysis is carried out by fiber entrapped lactase.

Many other applications are nearing the point in their development where they are ready for commercialization: saccharification of starch by immobilized glucoamylase; cheese whey lactose hydrolysis by bound β-galactosidase; beer chill-proofing, steroid transformations, protein-hydrolysis to improve digestibility. Industrial processes do not receive publicity rapidly and it is not unlikely that some of these may well have been commercialized already.

Immobilized cells have been used industrially in Japan for the transformations mentioned above except for L-amino acid isolation from racemic mixtures. The most-widely employed use of immobilized cells however are glucose isomerization and the hydrolysis of raffinose in beet sugar using mycelial pellets of the fungus *Mortierrella* sp. Raffinose (in beet molasses) is hydrolyzed to sucrose and galactose by β–galactosidase (mellibiase) produced by the fungus.

The potentials of immobilized enzymes and cells are yet far from realized. When economic conditions permit them to become so, the whole process of fermentation as we know it today may be revolutionized and fermentors may become largely for the growth of cells for subsequent use in immobilized enzyme or cell production.

22.8 MANIPULATION OF MICROORGANISMS FOR HIGHER YIELD OF ENZYMES

Until recently higher yields in enzyme production have been achieved, as has been the case with most other industrial microbial products, by empirical means using selection from natural variants, mutants obtained through treatment with various mutagens, and improvement in the environmental conditions of the fermentor or semi-solid medium. With these methods the rate of enzyme production has increased from two to five-hundred times.

In recent times more knowledge has accumulated about various aspects of enzyme production including those of molecular and other dimensions, and the manipulation of industrial organisms in general. In this section some of these new developments and their use or possible use in increasing enzyme yield will be discussed. Some of them promise the possibility of producing virtually any enzyme extracelluarly and at will.

22.8.1 Some Aspects of the Biology of Extracellular Enzyme Production

(i) *The nature of extra-cellular enzymes secretion*: Extracellular enzymes have been defined as those which are secreted into the medium outside the cell *without involving cell lysis*. This distinction is important because most extraccellular enzyme-producing organisms are Gram-positive organisms. Gram-negative organisms, in general produce enzyme in the medium only when the cell is lysed. Most Gram-negative organisms do not therefore, according to this definition, produce 'true' extra-cellular enzymes. However, it has been found in recent times that some Gram-negative organisms do in fact secrete extracellular enzymes. Furthermore many Gram-negative organisms do in fact produce and secrete enzymes across the cytoplasmic membrane. Such enzymes are however held within the periplasmic zone (Fig. 22.7) of the Gram-negative cell wall and hence do not find their way to the medium. Thirdly mutants of Gram-negative cells defective in the ability to synthesize cell wall components continue to synthesize and secrete polypeptides into the environment. On account of these observations extra-cellular enzymes have been redefined as those which are secreted across the cell membrane. In terms of industrial microbiology it is an apt definition as methods for deranging the molecular arrangements of cell walls exist, which when successfully applied to Gram-negative bacteria secreting into the periplasmic space convert them to extra-cellular secreters. Such methods include the formation of protoplasts by the prevention of cell wall formation using suitable antibiotics, limited digestion by trypsin, solubilization of the cell wall with a combination of a detergent (e.g. laurodeoxycholate and a chelating agent).

(ii) *Some biochemical properties of extra-cellular enzymes*: Bacterial extracellular enzymes vary in molecular weight from 12,000 to 500,000 but in the main they range from 20,000 to 40,000. Secondly, most but not all bacterial exoproteins lack cysteine. It has been suggested, but not entirely accepted, that this absence of cysteine will confer the property of malleability on extra-cellular enzymes thus facilitating their export.

(iii) *Site of synthesis of extra-cellular proteins*: Even in cells actively secreting extra-cellular proteins, an examination of the cytoplasm shows a complete absence or only a trace amount of the enzymes being excreted. Early report for instance claiming that α–amylase of *β–amyloliquifaciens* is first produced as a high molecular weight precursor have been shown to be wrong on the basis of radio-isotope (labeling) experiments. Since no evidence exists for the cytoplasmic synthesis of extracellular proteins it has been suggested they are synthesized on ribosomes associated with the cells in much the same way as in eucaryotic cells. In eucaryotic cells, membrance-bound polysomes are engaged in secreting proteins for export whereas polysomes secrete non-exportable proteins.

According to the currently accepted model synthesis takes places on the cell membrane and is secreted directly into pores in the cell membrane. Indeed synthesis and secretion are one process, following the system in eucaryotic cells. Some evidence for this are as follows. In many bacteria a considerable but variable fraction of the ribosomes is

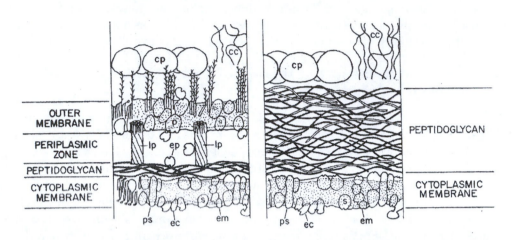

Left = Gram-negative wall; Right = Gram-positive wall; Dotted areas = hydrophobic zones; cc=capsular carbohydrate; cp = capsular polysaccharide; ec = cytoplasmic membrane enzymes in the cytoplasmic which synthesize cell wall macro-molecules; lp = lipoprotein; p = structural and encymic proteins of the outer layers of the Gram-negative wall; s = structural proteins of cytoplasmic membrane; sp = enzymes in the periplasmic zone; ps = permeases.

Fig. 22.7   Generalized Structure of the Bacterial Cell Wall

associated with the cytoplasmic membrane. In exponential phase of *Bacillus licheniformis* for example, 96% of the ribosomes are membrane bound. It is also know that exoenzyme synthesis is more sensitive to antibiotic inhibition than general protein synthesis. This has been interpreted as being so because of the membrane bound ribosomes are more accessible to the antibiotic.

(iv) *Control of extra-cellular enzyme secretion by gene cloning:* When the terminal portion of the β-*galactosidase* gene of *E. coli* was replaced with a gene that codes for a protein of the outer cell wall membrane of the bacteria, the β-*galactosidase* activity which is normally intracellular was formed extracellularly depending on the size of the latter that attached on the β-*galactosidase* gene. This and other similar experiments show that in due course it may be possible to produce virtually any enzyme extracellularly by gene cloning.

(v) *Some methods for increasing enzyme yields*: The increased yields which have been observed in enzyme production is based on strain selection, improved environmental factors, regulatory controls and genetic manipulation.

(a) *Strain selection*: Strain selection from natural variants of the same species or even entirely new species have resulted in the array of enzymes available in some industries e.g. the starch hydrolyzing industries.

  The natural strains may then be mutagenized for increased variation in the gene pool. Strains have been selected in the above two manners for a wide variety of properties including temperature-tolerant enzymes and resistance to feedback regulation.

(b) *Environmental factors:* Exoenzymes may be constitutive, but the majority are inducible or partially so. Inducers are therefore important in increasing the yields of many extracellular enzymes. Since many of the substrates are insoluble they cannot enter the cell, and hence their analogues or gratuitious inducers (those that induce the enzyme but are not substrates or breakdown products) have been used. Inducers are usually cheap in order to bring down costs. Thus, corn cobs are hydrolyzed to produce xylose which act as inducers for glucose isomerase.

Most extracellular enzymes are produced in the idiophase and maximum production is usually in the late log and early stationary phase. This period coincides with the period when the organism is released from catabolite repression. Increased yields may therefore be achieved by feeding low levels of the substrate or feeding them intermittently. Yields may also be increased by increasing cell-wall permeability. Surfactants may be incorporated into the medium for this purpose although how they affect the wall permeability is not fully understood.

(c) *Regulatory control*: Control of regulation is through induction and catabolite and feedback regulations. Mutants resistant to all three have been produced with consequent boost in production. An example of inducer-resistance is in the case of the glucose isomerase producing antinomycete *Streptomyces phaechromogenes*, the wild strain of which will not germinate on L-lyxose, another form of D-xylose. It will grow on lyxose only if germination is first obtained on xylose. Mutants were selected which would germinate directly on lyxose, thus eliminating the need for xylose.

The bypassing of catabolite repression has led to the production of large amounts of enzymes. This has been achieved by using toxic analogues of the substrates. Thus, 2-deoxy-glucose is used as a toxic analogue when seeking for mutants able to over produce glucosidase.

Feedback regulation is not very applicable to enzyme synthesis, although some examples are known.

(d) *Genetic manipulation*: As had been indicated earlier the gene specifying the extra-cellular secretion may be cloned on those controlling the synthesis of particular enzymes, thus causing the enzyme to be secreted extracellularly.

The number of copies of specific genes may be increased by gene amplification methods thus increasing enzyme yield several times. For example, plasmids specifying particular extra-cellular enzymes may continue to replicate while the parent chromosome is inhibited by, for example, chloramphenicol thereby permitting an amplification of the genes – sometimes up to 2,000 copies or up to 40% of the cells total DNA. The result is increased yield of the enzyme.

SUGGESTED READINGS

Anon. 1995. Classification and Chemical Characteristics of Immobilized Enzymes. (Technical Report), International Union of Pure and Applied Chemistry. Pure and Applied Chemistry, 67, 597-600.

Butterfoss, G.L., Kuhlman, B. 2006. Computer-Based Design of Novel Protein Structures. Annual Review of Biophysics and Biomolecular Structure, 35, 49–65.

Chaplin, M.F., Bucke, C. 1990. Enzyme Technology. Cambridge University Press. New York, USA.

Cheetam, S.J. 2004. Bioprocesses for the Manufacture of Ingredients for Food and Cosmetics. Advances in Biochemical Engineering/Biotechnology, 86, 83–158.

Desai, M.A. 2000. Downstream Processing of Proteins. Humana Press Totowa, New Jersey, USA.

Fogarty, M., Kelly, C.T. (eds) 1990. Microbial Enzymes and Biotechnology. 2nd ed. Elsevier Applied Science London and New York.

Imanaka, T. 1994. Improvement of Useful Enzymes by Protein Engineering. In: Rombinant Microbes for Industrial and Agricultural Applications. Y. Murooka, T. Imanaka, (eds). Marcel and Dekker. New York, USA. pp. 449-465.

Kennedy, J.F. 1995. Principles of Immobilization of Enzymes. In: Handbook of Enzyme Biotechnology. A Wiseman, (ed) 3rd ed. Ellis Horwood, London, UK. pp. 235–310.

Mitchel, D.A., Berovic, M., Krieger, N. 2000. Biochemical Engineering Aspects of Solid State Bioprocessing. Advances in Biochemical Engineering/Biotechnology. 68, 61–132.

Pasechnik, V.A. 1995. Practical Aspects of Large-scale Protein Purification. In: Handbook of Enzyme Boiotecnology. A Wiseman, (ed) 3rd ed. Ellis Horwood, London, UK. pp. 379–418.

Puri, M., Kaur, H., Kennedy, J.F. 2005. Covalent immobilization of naringinase for the transformation of a flavonoid. Journal of Chemical Technology & Biotechnology, 80, 1160 - 1165.

Puri, M., Banerjee, U.T. 2002. Production, purification, and characterization of the debittering enzyme naringinase. Biotechnology Advances, 18, 207-217.

Roy, I., Sharma, S., Gupta, M.N. 2004. Smar Biocatalysts: Design and Application. Advances in Biochemical Engineering/Biotechnology, 86, 159–190.

Schugerl, K. 2000. Recovery of Proteins and Microorganisms from Cultivativation Media by Foam Flotation. Advances in Biochemical Engineering/Biotechnology. 68, 191-233.

Tanaka, A, Tosa, T, Kobayashi, T, (1993). Industrial Applications of Immobilized Biocatalysts New York: Dekker.

CHAPTER 23

Mining Microbiology: Ore Leaching (Bioleaching) by Microorganisms

23.1 BIOLEACHING

The term bioleaching refers to the conversion of an insoluble metal (generally a metal sulfide, e.g., CuS, NiS, ZnS) into a soluble form (usually the metal sulfate, e.g., CuSO4, NiSO4, ZnSO4). When this happens, the metal is extracted into water; this process is called bioleaching. As these processes are oxidations, this process may also be termed bio-oxidation. However, the term bio-oxidation is usually used to refer to processes in which the recovery of a metal is enhanced by microbial decomposition of the mineral, but the metal being recovered is not solubilized. An example is the recovery of gold from arsenopyrite ores where the gold remains in the mineral after bio-oxidation and is extracted by cyanide in a subsequent step. The term bioleaching is clearly inappropriate when referring to gold recovery (although arsenic, iron, and sulfur are bioleached from the mineral). Biomining is a general term that may be used to refer to both processes.

Bacterial leaching of metals is a process in which the ores of the metals, usually their sulphides are solubilized by bacterial action. Basically the process is a chemical oxidation following the equation

$$MS + 2O_2 \xrightarrow{\text{Microorganisms}} MSO4$$

where M is a bivalent metal. The solution collected after the solubilization is processed to recover the metal. In the case of insoluble sulfides such as is the case with lead sulfide, the fact of insolubility may be used to separate it from the dissolved metals.

Bacterial leaching has been practiced by man over many centuries without any understanding of the microbiological basis of the process. It was used for mining copper by the Romans in Wales, in Rio Tinto, Spain in the 18th century and in the USA in this century. Indeed about 12% of copper produced in the USA is obtained by bacterial leaching of low grade ores.

23.2 COMMERCIAL LEACHING METHODS

There are two types of processes for commercial microbially-assisted metal recovery: the irrigation-type and the stirred tank type.

23.2.1 Irrigation-Type Processes

The irrigation-type processes can be grouped into three: the dump, the heap and the *in-situ* methods. The most widely used methods are dump and heap leaching. The metal most commonly bioleached metal in the irrigation methods is copper.

23.2.1.1 Dump leaching

In this method large quantities of low-grade ore are placed in valleys with impermeable grounds. The dumps are shaped by bull-dozers into cones which may be as high as 600 ft with 600 ft diameter at the base. Acid solutions usually H_2SO4, known as leach solutions, are then sprayed into, or flooded over, the dumps or injected into it through steel pipes. The acid provides the low pH required by the microorganisms whose activities are responsible for dissolving the ores. Liquid collected at the bottom of the dump contains dissolved metal which is recovered after processing.

In some processes the dumps are subjected to preconditioning, irrigation, rest, and conditioning, each of which may extend for a year. The irrigation is done with an iron- and sulfate-rich recycyled waste-water from which the copper has been removed. Microorganisms growing in the dump bring about the reactions which cause the insoluble copper sulfides to become soluble copper sulfate. The copper sulfate is collected from the bottom of the dump and the copper is recovered by solvent extraction and electrolysis. One of the best-known dump leaching sites is the Kennecot Copper mine in Bingham Canyon, Utah, USA; another is Bala Ley plant in Coledo, Chile.

23.2.1.2 Heap leaching

Heap leaching is very similar to dump leaching, except that steps are taken to make the process more efficient through ores of smaller particles and a smaller scale of operation. The ores are crushed to make them finer, and then staked in much smaller heaps of about 6 to 50 ft high. Aeration pipes may be included to permit forced aeration so as to speed up the process. To further speed up the reaction inorganic ammonium salts and phosphates may be added. The heaps are placed in mounds on drainage pads or in concrete tanks with false bottoms through which liquid collects in a dump. Tanks have capacities of about 12,000 tons. On account of the steps taken to accelerate the process the operation is complete in months rather than in years. An example of a large modern heap plant is to be found in Quabrada Blanca in northern Chile.

Heap bioleaching may also be used to mine gold in low grade gold ores. In this case the ore is initially flooded with acid-ferric iron. Thereafter, it is treated with recycled heap effluent. After the ore is sufficiently broken down, the ore is mixed with lime and the gold extracted with cyanide.

23.2.1.3 *In-situ* Mining

In *in-situ* irrigation bioleaching the ore is not brought to the surface and the ore is extracted *in situ*. In some instances such as in the copper mine at San Manuel, Arizona, injection wells are used to introduce acidified leaching fluids into the mineral deposit. It percolates and collects in disused mines or a specially prepared catchment location. The geology of the location must have a suitable impermeable layer. In spite of this leaching, fluid is lost in this process.

The mining of uranium is generally done *in situ* in the mine. The uranium ore (uranite) is not attacked by acid producing bacteria. However it is solubilized when it is oxidized from the tetravalent form which is insoluble, to the soluble hexavalent form by chemical oxidants such as ferric iron, MnO_2, H_2O_2 chlorates or nitric acid. In uranium leaching mediated by bacteria, the role of the bacteria is to produce ferric iron which then oxidizes the uranite to make it soluble.

23.2.2   Stirred Tank Processes

Stirred tank bioreactors are highly aerated tanks much like the fermentors already discussed except that the sterility maintained in fermentors used, for example for antibiotics, is not necessary in these bioreactors. They are expensive to construct and maintain and hence they are used only for high value minerals such as gold.

They are usually arranged in series and function as continuous fermentors. The tanks in the first stage are usually arranged in parallel so that the fermentation broth is retained long enough for the cell numbers to reach a steady state without being washed out. The mineral ore is suspended in water to which fertilizer grade $(NH_4)_2SO_4$ and KH_2PO_4 are added. The pH is adjusted 1.5–2.0. The bioreactor is vigorously aerated using air spargers and baffles. The released heat is removed by a jacket of cold water. Systems using aerated bioreactors use pretreatments for the recovery of gold, especially where the gold is finely divided in a mixture of pyrite and arsenopyrite (i.e. iron ores mixed with arsenic). Such ores are known as recalcitrant ores. Normally gold is recovered from rich ores by treatment with cyanide. However, because gold forms only a small proportion of recalcitrant aesenopyrite ores, the ores are pretreated in the aerated bioreactors. During the pretreatment which lasts for about four days microbes oxidize the arsenopyrite which is decomposed into iron and sulfate releasing the gold, and is then treated with cyanide. Stirred bioreactors are used to pretreat gold ores in many countries around the world including Australia, Brazil, Ghana, and Peru. The Ghana plant in Sensu is perhaps the largest fermentation in the world with 24 tanks each with a capacity of 1 million liters.

Aerated bioreactors are used for other minerals in France and Uganda (cobalt) and South Africa (nickel).

23.3   MICROBIOLOGY OF THE LEACHING PROCESS

The primary biomining microorganisms involved in ore leaching have several properties in common:

 (i) They are Gram-negative specialized chemolitho-authotrophic able to use ferrous iron and reduced inorganic sulphur compounds or both as electron donors;

 (ii) They are able to fix carbon with energy derived from the oxidation of inorganic compounds such as ferrous iron, sulfur and sulfides according to the following equations:

$$4FeSO4 + 02 + 2H2SO4 \quad \rightarrow \quad 2Fe_2\,(SO4) + 2H_2O \qquad (1)$$
$$(S_8 + 120_2 + 8H_2) \quad \rightarrow \quad 8H_2SO4 \qquad (2)$$
$$H_2S + 20_2 \quad \rightarrow \quad H_2SO4 \qquad (3)$$

(iii) They are acidophilic and will thrive under very acid conditions at pH ranges of 1.5 – 2.0.

(iv) Although they can use electron acceptors other than oxygen they grow best in highly aerated solutions.

(v) Some are able to fix atmospheric nitrogen, when the oxygen supply is limited

(vi) Some are obligately autotrophic, while others are facultative autotrophs and able to grow in the presence of organic matter.

(vii) The organisms are usually found naturally in waters in contact with exposed sulphides or in mines.

The organisms involved are the following:

(a) *Acidothiobacillus*: Organisms belonging to this genus were previously classified as *Thiobacillus*. Following 16S rRNA analysis they have been reclassified *as Acidothiobacillus* to accommodate the very acidophilic members of the genus. These include *Acidothiobacillus ferrooxidans* (formerly *Thiobacillus ferrooxidans*) which is the most intensively studied and until recently was thought to be the sole microorganism involved. It is an obligate autotroph and obtains its energy for the fixation of CO_2 from the oxidation of ferrous irons, sulfides, and other sulfur compounds such as thiosulfate.

(b) *Leptospirillum ferrooxidans*: This grows only in soluble ferrous iron and not on sulfur or mineral sulfides.

(c) *Acidothiobacillus thiooxidans* (formerly *Thiobacillus thiooxidan*): It cannot oxidize iron, but grows on elemental sulfur and soluble compounds including those generated in the leaching systems of *T. ferrooxidans*.

(d) *Other bacteria*: *Thiobacillus organosporus* is facultatively chemolithotrophic (i.e. while it is autotrophic it will also grow when organic compounds are available to it). It oxidizes sulfur, but not iron or sulfides. Mixed with other organisms e.g. *Leptospirriullum ferrooxidans* it will degrade iron sulfides which neither can do alone.

Unlike other industrial microbiology processes there is no conscious attempt to use pure cultures. The highly acidophilic organisms create conditions which are unsuitable for other organisms.

23.4   LEACHING OF SOME METAL SULFIDES

(i) *Copper sulfides*: One of the world's main source of copper is the iron-copper sulfide, chalcopyrite ($CuFe_2S_2$). Others are chalcocite (Cu_2S) and covellite (CuS). The reactions in which these minerals are leached by bacterial action are complex, but the end equations may be given as follows:

Chalcopyrite

$$60CuFeS_2 + 25SO_2 + 90H_2O \xrightarrow{\text{Bacteria}} 60CuSO_4$$
$$+ 20H\,Fe\,(SO_4)_2.\,2Fe\,(OH)_3$$
$$+ 20H_2\,SO4 \tag{4}$$

Chalcocite

$$10\,Cu_2S + 10H_2\,SO4 \xrightarrow{\text{Bacteria}} 20CuSO4 + 10H_2O \tag{5}$$

(ii) *Uranium extraction*: Bacteria especially *T. ferrooxidans*, oxidize ferrous sulphate to ferric sulphate (Equation I). The latter then reacts with uranium ores thus:

Uranite

$$UO_2 + Fe_2\,(SO4)_3 \longrightarrow UO_2\,SO4 + FeSO4 \tag{6}$$

The tetravalent form in the ore is insoluble in the leach solution while the oxidized hexavalent form is. There is thus a two-stage action which seems especially appropriate as the uranyl ion is toxic to most strains of *T. ferrooxidans*.

(iii) *Cobalt and nickel sulfides leaching*: The major nickel bearing sulfide ore is usually pentlandite which contains about 1% of nickel. The reaction for leaching with *T. ferrooxidans* is given as follows:

$$(Ni, Fe)_9\,S_8 + 175/8\,O_2 + 3\tfrac{1}{4}\,H_2SO4 \xrightarrow{\text{Bacteria}} 4\tfrac{1}{2}\,NiSO4$$
$$+ \tfrac{1}{4}\,Fe_2\,(SO4)_3 + 3\tfrac{1}{4}\,H_{20} \tag{7}$$

(iv) *Zinc and lead sulfide leaching*: Zinc and lead sulfides respectively may be oxidized by *T. ferrooxidans* according to the equations:

$$ZnS + 2O_2 \xrightarrow{\text{Bacteria}} Zn\,SO4 \tag{8}$$

$$PbS + 2O_2 \xrightarrow{\text{Bacteria}} Pb\,SO4 \tag{9}$$

23.5 ENVIRONMENTAL CONDITIONS AFFECTING BACTERIAL LEACHING

The following factors affect the efficiency of leaching via bacterial action. Much of this has been studied using *T. ferrooxidans*.

(i) *Temperature*: The optimum temperature for the bacterial solubilization of metals lies between 25°C and 45°C for different strains of *T. ferrooxidans*. Above 55°C the activity is mainly chemical. No minimum temperature has been established, but it is believed that action stops at freezing.

(ii) *pH*: *T. Ferrooxidans* is acidophilic and has been studied at pH values ranging from 1 to 5. The optimum pH values for acting on many minerals lie between pH 2.3 and 2.5. Other organisms outside *T. ferrooxidans* appear to have about the same pH requirements.

(iii) *Nutrient status of the leaching medium*: Like any other bacterium, the iron-oxidizing thiobacilli must have the appropriate nutrients in the leaching medium. The energy source is, as has been stated, ferrous sulfate, but minerals containing iron or sulfur may be utilized. The carbon is obtained from CO_2 of the air. It also needs nitrogen, potassium, magnesium, phosphate, and calcium. Most of these are available from the surrounding rocks; where the deficiency of one or more of them is established, it must be replaced in the leaching solution for optimum productivity. The thiobacilli are strict aerobes and oxygen deficiency leads to limitations in leaching productivity.

(iv) *Particle size*: In general the finer the particle size of the ore the greater the extraction of the leaching solution. Below a certain particle size however, especially in the case of low grade ores, the ratio of the mineral to unwanted part of the ore increases and productivity falls.

SUGGESTED READINGS

Kelly, D.P., Norris, P.R., Brierley, C.L. 1979: In: Microbial Technology: Current State, Future Prospects. A.T., Bull, D.C. Ellwood, C. Ratledge, (eds). Cambridge University Press, Cambridge. UK. pp. 263-308.

Lundgren, D.G. 1980. Ore Leaching by bacteria Ann. Rev. Microbiol. 24, 263-283.

Rawlings, D.E. 2002. Heavy Metal Mining Using Microbes Annual Review of Microbiology, 56, 65-91.

Zajic, J.E. 1969. Microbial Biogeo-chemistry Academic Press, New York, USA.

Production of Commodities of Medical Importance

CHAPTER 24

Production of Antibiotics and Anti-Tumor Agents

As currently defined, antibiotics are chemicals produced by microorganisms and which in low concentrations are capable of inhibiting the growth of, or killing, other microorganisms. Anti-microbial substances are also produced by higher plants and animals. Such substances are however excluded by this definition. Bacteriocins although produced by microorganisms are also not included in this definition because they are not only larger in molecular size than the usual antibiotics, but they are mainly protein in nature; furthermore they affect mainly organisms related to the producing organism. In comparison with bacteriocins, conventional antibiotics however are for more diverse in their chemical nature and attack organisms distantly related to themselves. Most importantly, while the information specifying the formation of 'regular' antibiotics is carried on several genes, that needed for bacteriocins being single proteins need single genes. It will be seen later that in the last few years this definition has been somewhat broadened by some authors to include materials produced by living things – plants, animals or microorganisms – which inhibit any cell activity.

Antibiotics may be wholly produced by fermentation. Nowadays, however, they are increasingly produced by semi-synthetic processes, in which a product obtained by fermentation is modified by the chemical introduction of side chains. Some wholly chemically synthesized compounds are also used for the chemotherapy of infectious diseases e.g. sulfonamides and quinolones. But these will not be considered since they are not produced wholly or partially by fermentation. Some antibiotics e.g. chloramphenicol were originally produced by fermentation, but are now more cheaply produced by chemical means.

Thousands of antibiotics are known; and every year dozens are discovered. However, only a small proportion of known antibiotics is used clinically, because the rest are too toxic.

24.1   Classification and Nomenclature of Antibiotics

Several methods of antibiotic classification have been adopted by various authors. The mode of action has been used, e.g. whether they act on the cell wall, or are protein inhibitors, etc. Several mechanisms of action may operate simultaneously making such a method of classification difficult to sustain. In some cases they have been classified on the basis of the producing organisms, but the same organism may produce several antibiotics, e.g. the production of penicillin N and cephalosporin by a *Streptomyces* sp. The same antibiotics may also be produced by different organisms. Antibiotics have been classified by routes of biosynthesis; however, several different biosynthetic routes often have large areas of similarity. The spectra of organisms attacked have also been used, e.g. those affecting bacteria, fungi, protozoa, etc. Some antibiotics belonging to a well known group e.g. aminoglycosides may have a different spectrum from the others. The classification to be adopted here therefore is based on the chemical structure of the antibiotics and classifies antibiotics into 13 groups. This enables the accommodation of new groups as they are discovered (Table 24.1).

Table 24.1   Grouping of antibiotics based on their chemical structures

Chemical Group	Example
Aminoglycosides	Streptomycin
Ansamacrolides	Rifamycin
Beta-lactams	Penicillin
Chloramphenicol and analogues	Chloramphenicol
Linocosaminides	Linocomycin
Macrolides	Erythromycin
Nucleosides	Puromycin
Puromycin	Curamycin
Peptides	Neomycin
Phenazines	Myxin
Polyenes	Amphothericin B
Polyethers	Nigericin
Tetracyclines	Tetracycline

One well-known example of each group has been given to facilitate recognition of the groups.

The nomenclature of antibiotics is also highly confusing as the same antibiotic may have as many as 13 different trade names depending on the manufacturers. Antibiotics are therefore identified by at least three names: the chemical name, which prove long and is rarely used except in scientific or medical literature; the second is the group, generic, or common name, usually a shorter from of the chemical name or the one given by the discoverer; the third is the trade or brand name given by the manufacturer to distinguish it from the product of other companies.

The production of antibiotics is a very wide subject and because of space limitations only beta-lactam antibiotics will be discussed. Even among them, only penicillin and cephalosprin will be discussed in any detail.

24.2   BETA-LACTAM ANTIBIOTICS

The Beta-lactam antibiotics are so-called because they have in their structure the four-membered lactam ring. Figure 24.1 shows the structures of the various Beta-lactam antibiotics.

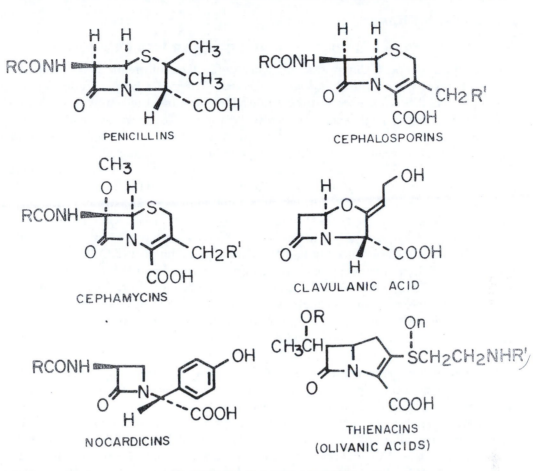

Fig. 24.1 The Beta-lactam Antibiotics

The Beta-lactam structure is not very common in nature and besides the antibiotic groups to be discussed it is only found in some alkaloids and some anti-metabolite toxins including pachystermines from the higher plant, *Pachystradra terminalis,* wild-fire toxin from *Pseudomonas tabici* and the anti-tumor antibiotics, phleomycins and bleomycins from *Streptomyces verticillus.*

The Beta-lactam antibiotics include the well-established and clinically important penicillins and cephalosphorins as well as some relatively newer members: cephamycins, nocardicins, thienamycins, and clavulanic acid. Except in the case of nocardicins these antibiotics are derivatives of bicyclic ring systems in which the lactam ring is fused through a nitrogen atom and a carbon atom to ring compound. This ring compound is five-membered in penicillins (thiazolidine), thienamycins (pyrroline) and clavulanic acid (oxazolidine); it is six-membered (dihydrothiazolidine) in cephalosporins and cephamycins (Fig. 24.1).

The Beta-lactam antibiotics inhibit the formation of the structure-conferring petidoglycan of the bacterial cell wall. As this component is absent in mammalian cells, Beta-lactam antibiotics have very low toxicity towards mammals.

24.2.1 Penicillins

24.2.1.1 Strain of organism used in penicillin fermentation

In the early days of penicillin production, when the surface culture method was used, a variant of the original culture of *Penicillium notatum* discovered by Sir Alexander Fleming was employed. When however the production shifted to submerged cultivation, a strain of *Penicillium chrysogenum* designated NRRL 1951 (after Northern Regional Research Laboratory of the United States Department of Agriculture) discovered in 1943, was introduced. In submerged culture it gave a penicillin yield of up to 250 Oxford Units (1 Oxford Unit = 0.5988 of sodium benzyl penicillin) which was two to three times more than given by *Penicillium notatum*. A 'super strain' was produced from a variant of NRRL 1951 and designated X 1612. By ultraviolet irradiation of X-1612, a strain resulted and was named WISQ 176 after the University of Wisconsin where much of the stain development work was done. On further ultra violet irradiation of WISQ 176, BL3-D10 was produced, which produced only 75% as much penicillin as WISQ 176, but whose product lacked the yellow pigment the removal of which had been difficult. Present-day penicillin producing *P. chrysogenum* strains are far more highly productive than their parents. They were produced through natural selection, and mutation using ultra violet irradiation, x-irradiation or nitrogen mustard treatment. It was soon recognized that there were several naturally occurring penicillins, viz. Penicillins G, X, F, and K (Fig. 24.2).

Penicillin G (benzyl penicillin) was selected because it was markedly more effective against pyogenic cocci. Furthermore, higher yields were achieved by supplementing the medium with phenylacetic acid, analogues (phenylalanine and phenethylaninie) of which are present in corn steep liquor used to grow penicillin in the United States. Present day penicillin-producing strains are highly unstable, as with most industrial organisms, and tend to revert to low-yielding strains especially on repeated agar cultivation. They are therefore commonly stored in liquid nitrogen at $-196°$ or the spores may be lyophilized.

Penicillin has since been shown to be produced by a wide range of organisms including the fungi *Aspergillus, Malbranchea, Cephalosporium, Emericellopsis, Paecilomyces, Trichophyton, Anixiopsis, Epidermophyton, Scopulariopsis, Spiroidium* and the actionomycete, *Streptomyces*. The only type of penicillin produced by actinomycetes however is Penicillin N (with the chemical structure D-α (δ- aminoadipyl) penicillin usually accompanied by cephamycins and/or deacetyl – 3 – 0-carbamoylcephalosporin C.

24.2.1.2 Fermentation for penicillin production

The inoculum is usually built up from lyophilized spores or a frozen culture and developed through vessels of increasing size to a final 5-10% of the fermentation tank. As the antibiotic concentration in the fermentation beer is usually dilute the tanks are generally large for penicillin and most other antibiotic production. The fermentors vary from 38,000 to 380,000 liters in capacity and in modern establishments are worked by computerized automation, which monitor various parameters including oxygen content, Beta-lactam content, pH, etc.

The medium for penicillin production now usually has as carbohydrate source glucose, beet molasses or lactose. The nitrogen is supplied by corn steep liquor. Cotton

GENERAL STRUCTURE OF PENICILLINS

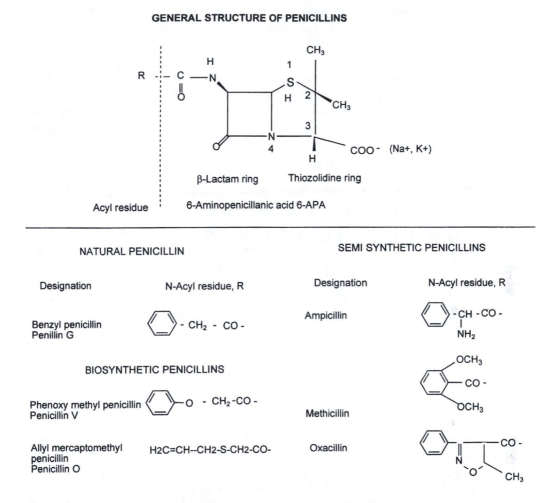

Fig. 24.2   Natural and Biosynthetic Penicillins

seed, peanut, linseed or soybean meals have been used as alternate nitrogen sources. The nitrogen source is sometimes exhausted towards the end of the fermentation and it must then therefore be replenished. Calcium carbonate or phosphates may be added as a buffer. Sulfur compounds are sometimes added for additional yields since penicillin contains sulfur. The practice nowadays is to add the carbohydrate source intermittently, i.e. using fed-batch fermentation. Lactose is more slowly utilized and need not be added intermittently. Glucose suppresses secondary metabolism and excess of it therefore limits penicillin production. The pH is maintained at between 6.8 and 7.4 by the automatic addition of H_2SO_4 or NaOH as necessary.

Precursors of the appropriate side-chain are added to the fermentation. Thus if benzyl penicillin is desired, phenylacetic acid is added. Phenyl acetic acid is nowadays added continuously as too high an amount inhibits the development of the fungus. High yielding strains of *P. chrysogenum* resistant to the precursors have therefore been developed.

Penicillin production is stimulated by the addition of surfactants in a yet unexplained mechanism. The temperature is maintained at about 25°C, but in recent times it has been found that yields were higher if adjusted according to the growth phase. Thus, 30-32°C was found suitable for the trophophase and 24°C for the idiophase. Aeration and agitation are vigorous in order to keep the components of the medium in suspension and to maintain yield in the highly aerobic fungus.

Penicillin fermentation can be divided into three phases. The first phase (trophophase) during which rapid growth occurs, lasts for about 30 hours during which mycelia are produced. The second phase (idiophase) lasts for five to seven days; growth is reduced and penicillin is produced. In the third phase, carbon and nitrogen sources are depleted, antibiotic production ceases, the mycelia lyse releasing ammonia and the pH rises.

24.2.1.3 Extraction of penicillin after fermentation

At the end of the fermentation the broth is transferred to a settling tank. Penicillin is highly reactive and is easily destroyed by alkali conditions (pH 7.5-8.0) or by enzymes. It is therefore cooled rapidly to 5-10°C. A reduction of the pH to 6 with mineral acids sometimes accompanied by cooling helps also to preserve the antibiotic. The fermentation broth contains a large number of other materials and the method used for the separation of penicillin from them is based on the solubility, adsorption and ionic properties of penicillin. Since penicillins are monobasic carboxylic acids they are easily separated by solvent extraction as described below.

The fermentation beer or broth is filtered with a rotary vacuum filter to remove mycelia and other solids and the resulting broth is adjusted to about pH 2 using a mineral acid. It is then extracted with a smaller volume of an organic solvent such as amyl acetate or butyl acetate, keeping it at this very low pH for as short a time as possible. The aqueous phase is separated from the organic solvent usually by centrifugation using Podbielniak centrifugal countercurrent separator (Chapter 9).

The organic solvent containing the penicillin is then typically passed through charcoal to remove impurities, after which it is back extracted with a 2% phosphate buffer at pH 7.5. The buffer solution containing the penicillin is then acidified once again with mineral acid (phosphoric acid) and the penicillin is again extracted into an organic solvent (e.g. amyl acetate). The product is transferred into smaller and smaller volumes of the organic solvent with each successive extraction process and in this way, the penicillin becomes concentrated several times over, up to 80-100 times. When it is sufficiently concentrated the penicillin may be converted to a stable salt form in one of several ways which employ the fact that penicillin is an acid: (a) it can be reacted with a calcium carbonate slurry to give the calcium salt which may be filtered, lyophilized or spray dried. (b) it may be reacted with sodium or potassium buffers to give the salts of these metals which can also be freeze or spray dried; (c) it may be precipitated with an organic base such as triethylamine.

When benzyl penicillin is administered intramuscularly it is given either as the sodium (or potassium) salt or as procaine penicillin. The former gives high blood levels but it quickly excreted. Procaine penicillin gives lower blood levels, but it lasts longer in the body because it is only slowly removed from the blood. It is produced by dissolving sodium or penicillin in procaine hydrochloride.

24.2.1.4 Production of semi-synthetic penicillins

In the late 1940s it was shown by labeling experiments that penylacetamide derivatives were directly incorporated into the benzyl penicillin molecule. The possibility was recognized of inducing the mold to produce new antibiotics antibiotic by the introduction of various precursors. Phenoxymethyl penicillin (penicillin V) which had greater acid stability than penicillin G, allythiomethyl penicillin (Penicillin O) which was less likely to induce allergic reactions and butylthiomethyl penicillin (Penicillin S) were thus produced. The natural penicillins (formed in unsupplemented media) and the biosynthetic (produced by the addition of specific side-chain precursors) are indicated in Fig 24.2 The high expectations of making new penicillins by the introduction of side-chains during fermentation, did not however, result in many new pencillins.

In 1959 6-amino penicillanic acid (6-APA) was isolated from precursor-starved *P. chrysogenum* fermentations and this ushered in the era of semi-synthetic penicillins and indeed other semi-synthetic antibiotics. Today the only 'natural' penicillins used are benzyl penicillin (Penicillin G) and phenoxymethyl penicillin (Penicillin V). All others are semi-synthetic.

In preparing semi-synthetic penicillins, 6-APA is not produced by starving *P. chrysogenum* of precursors, because yields are low. It is prepared by cleaving from penicillin G or penicillin V, the 6-acyl group by chemical means or with enzymes (acylases) produced by a wide range of microorganisms including bacteria, yeasts, and molds and even mammals (hog kidney acylase). The various acylases have different substrates. Actinomycete and mold acylases usually attack penicillins with aliphatic side-chains, e.g. penicillin V or phonoxymetyl penicillin (Penicillin); penicillin G is attacked more slowly. On the other hand, bacterial acylases attack penicillin G rapidly. Immobilized enzymes and cells are being used in these processes.

The introduction of the acyl side chain is done by reacting 6-APA with a suitable derivative of a carboxylic acid, usually a chloride, in organic solvents under anhydrous or aqueous conditions. In the latter system it is done in acetone-water mixtures in the presence of sodium bicarbonate. The resulting penicillins can be extracted by solvent extraction as already described, followed by charcoal treatment.

Semi-synthetic penicillins were developed to meet some of the short-comings of benzyl penicillin. Some of the desired properties were greater intrinsic activity against Gram-positive bacteria, increased antibacterial spectrum including Gram-negative organisms, gastric acid stability and oral absorbability and resistance to beta-lactamases, (i.e. enzymes which open the Beta-lactam ring). All penicillins to some extent bind to the serum albumin and therefore reduce the quantity of the antibiotic available to attack microorganisms. The less binding therefore, for pencillins of otherwise equal activity, the more bactericidal. The final desirable property is reduced ability to induce hypersensitivity, a phenomenon which occurs in 9-10% of the population.

24.2.2 Cephalosporins

Cephalosporins in general have a broader spectrum than penicillins and are less likely to induce allergic reactions among those who react to penicillins. The first cephalosporin was discovered in 1948 and was produced by *Cephalosporium acremonium*. It was later shown that the organism in fact produced five antibiotics. Five of these were steroids (and therefore hydrophobic) and were active only against Gram-positive bacteria. A hydrophilic component active against both Gram-positive and Gram-negative bacterial was also found and named cephalosporin N. While purifying cephalosporin N, another compound, cephalosporin C was discovered. It was latter found that cephalosporin C was stable to acid and to pencillianse; most importantly it was active against Gram-negative bacteria although it had about one-tenth the activity of cephalosporin N against Gram-positives. With further study Cephalosporin N was found to be a penicillin and renamed Penicillin N. Cephalosporin is now known to be produced by as wide a range of micro-organisms as produce penicillins and these include fungi and actinomycetes. The natural cephalosporins and cephamycins are given in Fig 24.3.

24.2.2.1   Production of cephalosporin

While two of the clinically important penicillins (Penicillin G & V) are produced entirely by fermentation the rest being semi-synthetic, all of the cephalosporins in use on the other hand are semi-synthetic. They however have as their starting point Cephalosporin C produced by fermentation, otherwise the production of both antibiotics is similar.

24.2.2.2   Strain of organism used

The original strain of *Cephalosporium acremonium* (C Ml 49, 137 – Common Wealth Mycological Institute, Kew Gardens, London) produced by violet mutagenesis, *C. acremonium* 8650. This latter organism is the parent of most of the various commercially used *C. acremonium*. In passing, Cephamycins, (Fig. 24.3) are produced by *Streptomyces lipmanni* and *S. clavuligenis*.

24.2.2.3   Fermentation

The medium used for cephalosporin fermentation is same as used for penicillin N. Paraffins have however been used to produce several cephalosporins. Methionine, arginine, ornithine, spermine, cadaverine, and lysine have been shown to increase cephalosporin production.

24.2.2.4   Extraction of cephalosporin after fermentation

While penicillins are carboxylic acids, cephalosporin C is amphotheric having both alkaline and acidic properties. For this reason it cannot be extracted directly into organic solvents. Cephalosporin C is more commonly isolated by ion exchange and precipitation. The broth is acidified to a low pH after filter-aid filtration in a rotary vacuum filter. The broth usually contains penicillin N and deactylacephalosporin. The low pH destroys penicillin N and converts the deactylcelphalosporin to cephalosporin G.

While 6-APA can be made during fermentation by starving *P. chrysogenum* of the side-chain precursor, 7-amino cephalosoporanic acid (7-ACA) cannot be produced by fermentation even if precursors are not added; neither can it be produced by the enzymatic side-chain cleavage as with 6-APA. The production of 6-amino cephalosporanic acid is therefore by chemical means. 7-ACA is obtained by the chemical removal of the Alpha-amino adipic side chain of cephalosporin when preparing

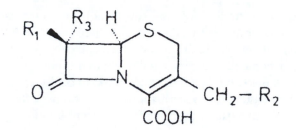

β-Lactam   Dihydro-
ring   thiazine ring

Designation	R_1	R_2	R_3
7-Aminocephalo-sporanic acid (7-ACA)	$- NH_3^{\oplus}$	$-O-CO-CH_3$	$-H$
NATURAL CEPHALOSPORINS			
Cephalosporin C		$-O-CO-CH_3$	$-H$
Deacetyl-3'-carbamoyl-cephalosporin C		$-O-CO-NH_2$	$-H$
7-Methoxy-cephalosporin C		$-O-CO-CH_3$	$-OCH_3$
Cephamycin A			$-OCH_3$
Cephamycin B			$-OCH_3$
Cephamycin C		$-O-CO-NH_2$	$-OCH_3$
SEMI-SYNTHETIC CEPHALOSPORINS			
Cephalotin		$-O-CO-CH_3$	$-H$
Cephalexin		H	$-H$

For the natural cephalosporins R_1 is:
$$- CO - (CH_2)_3 - (D)\,CH - NH_3^{\oplus}$$
$$\qquad\qquad\qquad\quad | \atop COO^{\ominus}$$

Cephamycin A R_2: $-OCO-C{=}CH-\bigcirc\!\!\!\!\!\!\!\!\!\!\!\!-O-SO_3H$

Cephamycin B R_2: $-OCO-C{=}CH-\bigcirc\!\!\!\!\!\!\!\!\!\!\!\!-OH$, with $O-CH_3$

Fig. 24.3 Natural Cephalosporins and Cephamycins

cephalosporins which have Alpha-3 acetoxymethyl group or a derivative from it. However, cephalosporins with a 3-methyl substituent (deacetoxy-cephalosporanic acids) are derived from penicillins. The more complex nature of cephalosporins in comparison with penicillins offers greater opportunities for the production of semi synthetics and besides 6-ACA, 6-ADCA (6 amino decephalosporanic acid) is also used as a basis for the production of semi synthetic cephalosporins. The methods of their preparation are similar to those described for the semi-synthetic penicillins.

24.2.2.5   Use of cephalosporins

Among cephalosporins, cephalothin occupies the same position as benzyl-penicillin, which continues to be widely used despite some of its deficiencies and the presence of newer products. Cephalothin is broad-spectrum although ineffective against some Gram-negative organisms such as *Proteus* and *Pseudomonas*. It is administered preferably intravenously because it is poorly absorbed and because intra-muscular pain is high in some individuals. New cephalosporins have been produced which are effective against Gram-negative organisms e.g. cefazolin and cefenandole. Oral cephalosporins include cefatrizine and cefachlor.

24.2.3   Other Beta-Lactam Antibiotics

24.2.3.1   Cephamycins, (7-Methoxycephalosporins)

Three cepham antibiotics produced by *Streptomyces* were discovered in 1971. A cepham with a methoxy group at position C-7 was produced by *Streptomyces lipmanni* while *Streptomyces clavuligenis* produced cephalosporin (A-16886-A) and 7-methoxycephalosporin (A-16886-B) with carbamoxy-loxymethyl function at C-3 position. Penicillin N was produced simultaneously in both strains. Soon afterwards several species of *Streptomyces* able to produce 7-methoxycephalosporins were found (Fig. 24.1).

As with other cephams and penicillin N, the production of cephamycins is stimulated by methionine. They are generally broad-spectrum though the extent to which they affect Gram-positive and Gram-negative organisms vary among different compounds in the group. Cephamycin C is for instance specially active against *Proteus* and *E. coli*. They are resistant to hydrolysis by cephalosporinases produced by cephalosporin-resistant bacteria. They appear most important for the present as starting points for the synthesis of new semi-synthetic cephams.

24.2.3.2   Nocardicin

The nocardicins were first isolated in 1976 using a super sensitive mutant strain of *E. coli*, strain ES 11, whose minimum inhibitory concentrations on penicillin G (100) Cephalosporin (400) and Nocardium A were 0.8, 0.4, and 0.4 respectively. They are produced by *Nocardia uniformis* subspecies *suyamanensis*. This antibiotic is novel among the Beta-lactams (Fig. 24.1) in that it is monocyclic (i.e., no ring is fused to the Beta-lactam ring). In comparison with the cepham, cephazolin, it is highly effective against the Gram-negative *Proteus* and *Shigella* but has limited activity against *Pseudomonas*. An enhanced activity occurs however when *Pseudomonas* is treated *in vivo*. Against Gram-positive bacteria and yeasts and molds it is completely ineffective. Norcadicin A and B differ slightly in their structures and are very non-toxic to mammals.

24.2.2.3 Clavulanic acid

This antibiotic was first described in 1976 using another novel method of isolation. In this method, agar plates containing 10 mcg/ml of benzylpenicillin are seeded with Beta-lactamase – producing *Klebsiella aerogenes*. Test samples which did not inhibit the organisms without the introduced penicillin give a zone of inhibition on the test plate when a diffusible Beta-lactamase inhibitor was present. Clavulanic acid, cephalosporins and penicillin N are produced by strains of *Streptomyces clavuligerus* using the above method. Clavulanic acid was not however discovered with the classical method. It is a weak antibiotic but has broad-spectrum activity against bacteria. However, it has an obvious potential value if it can be used along with penicillinase-susceptible antibiotics of greater potency than itself. Structurally it resembles cephalosporins.

24.2.3.4 Thienamycins

Thienamycin was first described in 1976. Like the cephalosporins and the cephamicins it was discovered as a fermentation product with broad spectrum activities and was produced by *Streptomyces cattleya*. Thienamycins are reported to be broad spectrum even at lower concentrations while being as non-toxic as the known natural and semi-synthetic Beta-lactams.

24.3 THE SEARCH FOR NEW ANTIBIOTICS

In the 1970s the view was that the fight against communicable diseases was about to be won. The rates of bacterial disease were falling through vaccination and the effectiveness of the available antibiotics. Many pharmaceutical companies decided to focus attention away from anti-microbial drugs production as there seemed to be little need for new compounds. The situation has changed and there now appears an urgent need for new antimicrobial drugs. This section will discuss the need for new antibiotics, the classical methods for searching new antibiotics, and some of the newer methods for prospecting them.

24.3.1 The Need for New Antibiotics

24.3.1.1 The problem of multiple resistance to existing antibiotics

Microorganisms have developed multiple resistance to many of the antibiotics currently in common use. This is due to several factors some inherent in the nature of microorganisms, others relating to use (or misuse) of antibiotics by humans. Some of the factors pertaining to the human use of antibiotics include the wide spread and sometimes unnecessary use of antibiotics, the prophylactic use of antibiotics, and the use of low doses of antibiotics for encouraging the growth of farm animals. In the case of bacteria, their sheer numbers means that there is a potentially large number of genotypes waiting to be selected and their short generation time fuels the rapidity of this selection. Finally their ability to horizontally transfer genetic materials through plasmids, transposons, and by conjugation and transformation set the stage for the (almost) inevitability of the development of resistance among microorganisms, especially bacteria.

24.3.1.2 The development of previously non-pathogenic microorganisms into pathogens

As clinical practice now use more invasive methods and people live longer, more and more people now depend on adequate antimicrobial coverage. Especially in patients who are immunocompromised, microorganisms which were previously non-pathogenic or ordinary commensals have become pathogens due to the widespread use of antibiotics. Thus, *Proteus* sp., *Acinetobacter* sp. and yeasts have all gained new status as pathogens especially in intensive care units.

24.3.1.3 Need to develop anti-fungal antibiotics

Currently there are few satisfactory systemic anti-fungal and anti-viral antibiotics. In the case of anti-fungal agents, currently few satisfactory systemic antifungal antibiotics outside Amphothericin B seem to exist; even Amphothericin B is not always efficacious. There is a similar dearth in the anti-viral antibiotics.

24.3.1.4 Need to develop antibiotics specifically for agricultural purposes

The growing needs for antibiotics used in agriculture for combating plant diseases, in animal feeds and in veterinary practice dictate that antibiotics be found specially for food production, but not for human medicine because of the problem of resistance.

24.3.1.5 Need for anti-tumor and anti-parasitic drugs

Although microbial metabolites for combating tumors, helminthes and parasites cannot be strictly described as antibiotics in the conventional sense, their production by the fermentation of microorganisms may allow the loose use of the term anti-tumor antibiotics and anti-helminthes antibiotics. Antibiotics need to be produced from microorganisms for these purposes.

24.3.2 The classical method for searching for antibiotics: random search in the soil

The classical method for the search for new antibiotics is by random search in the soil. This method will be described briefly below as a setting for more recent methods.

Although the first important commercially produced antibiotic was discovered by chance, most present day antibiotics were discovered by systematic search. The soil is a vast repository of microorganisms and it is to the soil that search is turned when antibiotics are being sought. The stages to be discussed below are not necessarily rigidly followed; they are merely meant to indicate in a general manner some of the activities involved in the development of antibiotics. The most important are:

(i) *The primary screening*: Several methods have been employed in primary screening.

   (a) **The crowded plate**: This method is used to isolate soil organisms able to produce antibiotics against other soil organisms. A heavy aqueous suspension (1:10; 1:100) of soil is plated on agar in such a way as to ensure as much as possible the development of confluent growth. Organisms showing clear zones around themselves are isolated for further study. Different groups of organisms could be encouraged to develop by altering the media used.

This method has the disadvantage that slow-growing antibiotic-producing organisms such as actinomycetes are usually over grown and are therefore hardly isolated. Furthermore, the test organisms used in this method are soil organisms. The susceptibility of soil organisms to the antibiotics produced in the test, may therefore be unrelated to the susceptibility of clinically important organisms.

(b) **The direct-soil-inoculation method**: This method is used when the aim is to isolate antibiotics against a known organism or organisms. Pour plates containing the test organisms are prepared. Soil crumbs or soil dilutions are then placed on the plates. Antibiotic producing organisms develop which then inhibit the growth of the organisms in the plate. They are recognized by the cleared zone which they produce around themselves and they may then be picked out.

(c) **The cross-streak method**: This method is used for testing individual isolates, especially actinomycetes which may be obtained from soil without any previous knowledge of their antibiotic-producing potential. The organism may come from one of the two methods already indicated above.

The purified isolate is streaked across the upper third of plate containing a medium which supports its growth as well as that of the test organisms. A variety of media may be used for streaking the antibiotic producer. It is allowed to grow for up to seven days, in which time any antibiotic produced would have diffused a considerable distance from the streak. Test organisms are streaked at right angles to the original isolates and the extent of the inhibition of the various test organisms observed (Fig. 24.4).

(d) **The agar plug method**: This method is particularly useful when the test organism grows poorly in the medium of the growth of the isolate such as fungi. Plugs about 0.5 cm in diameter are made with a sterile cork borer at progressive distances from the fungus. These plugs are then placed on plates with pure cultures of different organisms. The diameters of zones of clearing are used as a measure of antibiotic production of the isolate. The method may be used with actinomycetes.

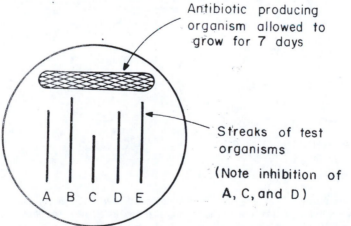

Fig. 24.4   The Cross Streak Method for the Primary Search of Antibiotic Producing Organisms

(e) **The replica plating method**: If a large number of organisms are to undergo primary screening, one rapid method is the use of replica plating. This is a well-known method used in microbial genetics. It was discussed in Chapter 5 dealing with the production of mutants. The method consists of placing a sterile velvet pad on the colonies formed in the crowded plate or soil inoculation plate, or on series of discrete colonies to be tested for antibiotic properties. The pad is thereafter carefully touched on four or five plates seeded with the test organisms. As a landmark is placed on the pad as well as on the plates it is possible to tell which colonies are causing the cleared zones on the tested plates (Fig. 24.5).

(ii) *Secondary screening*: Organisms showing suitably wide zones of clearing against selected target organisms are cultivated in broth culture in shake flasks using components of the solid medium in which the isolate grew best. Crude methods of isolating the active antibiotic are developed by extracting the broth using a wide range of extractive methods. With each extraction the resultant material is assessed for activity against the target organisms at various dilutions. The extract is either spotted on filter paper discs placed on agar seeded with the test organism or introduced into wells dug out from the seeded agar with sterile cork borers. In this manner the most efficient extractive methods and the spectrum of activity of the organisms are determined.

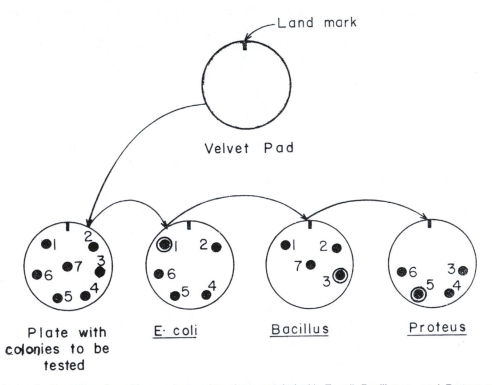

Colonies 1 - 7 are transferred by a velvet pad to plates seeded with *E. coli, Bacillus* sp., and *Proteus* sp. respectively. Note that colonies 1,5,7 produce dear zones in *E. coli, Bacillus* sp., and *Proteus* sp. respectively.

Fig. 24.5   Replica Plating Method of Testing Antibiotic Producing Colonies

Secondary screening is aimed at eliminating at an early stage any antibiotic which does not appear promising either by virtue of low activity, other undesirable properties or because it has been discovered previously.

Antibiotic spectrum: The minimal inhibitory concentration (MIC) is a means of determining the activity of the isolated antibiotic and comparing this activity with those of existing antibiotics. Tests involving agar diffusion such as filter paper discs or agar wells described above are rapid and very useful for initial screening. However, they involve not only the intrinsic anti-microbial potency of the antibiotic produced but also its ability to diffuse through agar. The MIC has the advantage that it is performed in broth thereby eliminating the disadvantage of large-molecule slower-diffusing antibiotics.

(iii) *Other properties*: The other qualities of the antibiotic outside anti-microbial activity depend on its intended use. For antibiotics meant for clinical use, information on a number of the following may be sought at this stage:

(a) Toxicity to mammals, determined by intra peritoneal injection into animals;
(b) Haemolysis is tested by observing the effect on blood agar;
(c) Serum binding is tested by adding serum to the broth before testing against susceptible organisms;
(d) The inactivation of the antibiotic by several enzymes from various organs is tested by exposing the antibiotic to them;
(e) Acid stability is tested if the antibiotic is meant for oral fermentation
(f) Tetragonicity tests, which determine the effect on the unborn are carried out on laboratory animals;
(g) For plant antibiotics, phytotoxicity as shown by damage to leaves in the laboratory and in the green house, is determined;
(h) For feed antibiotics, low absorbability and low toxicity are desirable and are tested.

Several other tests designed to certify the safety of administering the antibiotic may be carried out at this stage or performed later.

As part of the secondary screening, initial studies on the chemical nature of the crude antibiotics is determined using paper chromatography, ultraviolet absorption, solubility in acid and alkali, optical rotation, infra-red absorption, nmr, etc. The aim of this study is to see if it is an already known antibiotic. If it is, then further work on it may be abandoned. It is important to make this decision early. As has been shown previously, the same antibiotic is often produced by a large number of different organisms. This method known as 'finger printing' is more fully discussed when anti-tumor antibiotics are examined.

(iv) *Further laboratory evaluation*: If and after all the above, the antibiotics is promising, then further experimentation is done in shake flasks as a preparation for pilot production. The optimal conditions of growth are determined; the most suitable medium, optimal pH, temperature, length of fermentation etc. are all determined.

(v) *Pilot plant production*: The results obtained in previous experimentation are fed into the pilot plant. The material produced is subjected to further safety tests and chemical analysis. Enough materials are made available for the structure to be determined.

Methods for the isolation are improved and perfected. Clinical tests on a limited scale may be carried out at this stage. *The Journal of Antibiotics* regularly carries several articles describing the procedures for the chemical analysis of many new antibiotics. Unless the organism is already known, it is also described and may be named. The mode of action, the route of biosynthesis and strain improvement are undertaken, but the production of a good antibiotic does not await their successful completion.

(vi) *Plant production*: The production plant utilizes all the information obtained in the pilot experimentation.

(vii) *Certification*: A government agency must approve the antibiotic before it becomes available for general use. In the US, it is the Food and Drug Administration. In the UK and EU countries, it is the European Medicines Agency (EMEA) which was established in 1993 and is based in London. The process for the certification of drugs by the FDA is discussed in greater detail later in Chapter 28.

(viii) *Marketing and financing*: The marketing and financing of the business are of paramount importance since the aim of the producing firm is profit maximization.

24.4 COMBATING RESISTANCE AND EXPANDING THE EFFECTIVENESS OF EXISTING ANTIBIOTICS

Many ways have been devised to combat microbial resistance to existing antibiotics or expand the effectiveness and to discover new ones. These include modifying the procedures for the classical search in soil, searching for antibiotics in novel environments, and chemically manipulating the antibiotics directly or through using mutant microorganisms.

24.4.1 Refinements in the Procedures for the Random Search for New Antibiotics in the Soil

To increase the chances of finding new antibiotic - producing organisms, some refinements of the classical methods of searching for antibiotic producers were introduced as shown below.

24.4.1.1 The use of super-sensitive mutants

By using super-sensitivity strains of test organisms, organisms producing only small amounts of an antibiotic may be detected. Antibiotics so produced are useful because they may have a wider spectrum than those of the same class already in existence. Furthermore they may provide better substrates for semi or muta-synthesis. This method has led to the discovery of novel Beta-lactam antibiotics, thienamycin, olivanic acid, nocardicins, clavulanic acid etc. It is remarkable that several natural clavulanic acid type compounds (e.g. 2-hydroxymethylclavam) have significant antifungal properties. The use of super-sensitive mutants has shown that Beta-lactam antibiotics are produced by a wider spectrum of organisms – ascomycetes, fungi imperfecti and actinomycetes – than was previously thought.

24.4.1.2   The application of criteria other than death or inhibition

Reactions such as irregular growth of the fungal mycelium, inhibition of sporulation, or some readily ascertainable deficiency rather than death may be used to follow the antibiotic effect. When anti-fungal antibiotics are sought, the clear zone principle is employed using yeasts or fungal spore in the test plate. With this method the existing antifungal antibiotics, namely polyenes as well as cycloheximide and actimycins were found. If criteria less drastic than death e.g. abnormal growth of hyphae, or inhibition of zygosphore formation, then a wide range of antibiotics may be found. Thus some new antibiotics have been found in actinomycetes including boromycin, venturicidin and mikkomycin, with this method.

24.4.1.3   Search for antibiotics effective in conjunction with other antibiotics

Some antibiotics while being effective are not permeable through the wall of the test organism or the pathogen. Such antibiotics should be sought in nature by using test-organisms with deficient cell walls, protoplasts, or by the incorporation of detergents or EDTA which enable the permeation of antibiotics into the organism. When they are discovered they can be made permeable by using them in conjunction with wall-inhibiting antibiotics or compounds, or coupling them to compounds which bacteria ingest by active transport.

24.4.1.4   Use of organisms of recent clinical importance as test organisms

The classical search used routine organisms such as *E. coli* and *Bacillus*. The result as has been shown has been that fewer and fewer new antibiotics have been discovered. In recent times new organisms previously of little importance in clinical practice have emerged, due to the widespread use of antibiotics as important medical organisms. These include anaerobes, Gram-negatives e.g. *Proteus*, Beta-lactamase producing gonococci, facultatives, haemolysis, etc. These should be used as the test organisms in place of the previous ones.

24.4.2   Newer Approaches to Searching for Antibiotics

In spite of the above modifications and refinements in the classical methods for searching for new antibiotics, antibiotics with new structures were not discovered and cross-resistance among the available antibiotics continue to occur. In recent times some newer approaches to discovering new antibiotics have been adopted.

24.4.2.1   Search in novel environments

Systematic search for antibiotics is usually from soil. Other natural bodies exist which can provide novel organisms. Two of such habitats will be discussed in this section, namely the sea and white blood cells.

24.4.2.1.1 The sea as a habit for prospecting for micro-organisms producing antibiotics (and other drugs)

The seas and oceans occupy 70% of the earth surface. Until recently they were not exploited as sources of antibiotic-producing organisms. Although they would present new difficulties such as the need for a boat, they are a unique habitat. They are not only twice the land area of the earth, they contain large amounts of salt and other mineral nutrients, have fairly constant temperature, and have a higher hydrostatic pressure and less sunlight in the deeper regions. The coastal area is constantly changing with tides and such areas should be expected to have a wide variety of organisms, peculiar to the littoral environment.

Although deep ocean exploration is still in its infancy, many scientists now believe that the deep sea harbors some of the most diverse ecosystems on Earth. This diversity holds tremendous potential for human benefit. More than 15,000 natural products have been discovered from marine microbes, algae, and invertebrates, and this number continues to grow. The uses of marine-derived compounds are varied, but the most exciting potential uses lie in the medical realm. More than 28 marine natural products are currently being tested in human clinical trials, with many more in various stages of preclinical development. To date, most marketed marine products have come from shallow and often tropical marine organisms, due mainly to the ease of collecting them. But increasing scientific interest is now being focused on the potential medical uses of organisms found in the deep sea, much of which lies in international waters. These organisms have developed unique adaptations that enable them to survive in dark, cold, and highly pressurized environments. Their novel biology offers a wealth of opportunities for pharmaceutical and medical research and a growing body of scientific evidence (Table 24.2) suggests that deep sea biodiversity holds major promise. The medicines in the Table do not include antibiotics, but it could be because search for them was not conducted in this particular study. Nevertheless the search for antibiotics in the sea has indeed led to the discovery of new and unique antibiotics. These include antibiotic SS-228R from *Chainia* sp. effective against Gram-positive bacteria and tumors, bromopyrrole from a marine *Pseudomonas,* and leptosphaenin from marine fungi.

24.4.2.1.2 Antibiotic sources other than microorganisms: bactericidal/permeability increasing protein (BPI)

Antibiotics are produced also by higher organisms – plants and animals – and they also should be screened by the regular method of antibiotic screening. However because of established practice, antibiotics from such higher organisms have been usually screened for anti-tumor and anti-viral activity. Nothing intrinsic in materials from higher organisms should stop them from acting against microorganisms in suitable cases. In this section Bactericidal/permeability increasing protein (BPI) will be discussed as an example of a novel antimicrobial agent derived from a living thing higher than microorganisms.

BPI, is a protein and has been studied for several decades. It is derived from the white blood cells, polymorphonuclear leucocytes, the primary phagocytic white blood cells responsible for part of the body's innate immune response. BPI has great affinity for the lipopolysaccharide layer (LPS) of Gram-negative bacteria. It immediately arrests the

Table 24.2   Deep sea compounds in development for medical use (July, 2005)

Name	Application	Source	Depth/Location	Status	Comments
E7389	Cancer: non-small cell lung and other types	Sponge: Lissodendoryx sp.	330 ft (100 m) New Zealand	Phase I clinical trials	
Discode-morlide	Cancer: solid tumors	Sponge: Discondermia Dissolute	460 ft (140 m) Bahamas	Phase l trials (completed)	Toxicity similar to Taxol ®; works on multi-drug resistant tumors
Diclyos-tatin -1	Cancer	Sponge: Order Lithistida, Family Corallistadae	1,460 ft (442 m) Jamaica	Preclinical Development	Toxicity similar to Taxol ®
Sarcodictyin/ Eleutherobin (related compounds)	Cancer	Coral: Sarcodictyon roseum	330 ft (100 m) Mediterranean	Preclinical Development	Toxicity similar to Taxol ®
Salinospor-amide A	Concer: melanoma, colon, breast, non-small cell lung	Microbe: Selinospora	More than 3,300 ft (1,000 m) North Pacific Ocean	Preclinical Development	Will enter clinical trails in 2005; potency 35x omuralide
Topsentin	Anti-inflammatory: arthritis, skin irritations Cancer: colon (preventive) Alzheimer's	Sponge: Spongosporites ruetzleri	1980 ft (1,000 m) Bahamas	Preclinical Development	
Orthopedic implants	Bone grafting	Coral: Family Isididae	More than 3,300 ft (1,000 m) North Pacific Ocean	Preclinical Development	Resources risk of mammalian disease

growth of Gram-negative bacteria, increases the permeability of the outer and inner membranes of the Gram-negative bacterial wall and eventually kills the organism. It is attractive as a possible antibiotic for several reasons. First, it is highly potent and specific against Gram-negative bacteria, which include many important human pathogens; it is at least 10 times more potent than any known mammalian anti-microbial protein or peptide. Second, it is non-toxic to mammalian cells. Third, it maintains its anti-microbial activity in the complex environment of body fluids, unlike many mammalian antimicrobial agents. Finally it is not only a potent antimicrobial agent, but it also reduces the effect of the inflammatory response known as 'septic shock' which the lipopolysaccharide of the Gram-negative cell wall induces in patients. None of the other proteins able to bind to the Gram-negative lipopolysaccharide is able to neutralize the effect of toxic shock. It is currently at the stage of clinical trials.

24.4.3 Chemically Modifying Existing Antibiotic: The Production of Semi-synthetic Antibiotics

The production of a semi-synthetic antibiotic involves the use of a fermentation-derived antibiotic, which is then modified by the addition of side chain to give rise to an antibiotic with new properties. As has been discussed, a well-known example is the modification of penicillin G to 6-APA and the subsequent use of chemical reaction to produce semi-synthetic penicillins. Another example achieved in a different manner is the chemical alteration of specific sites in an antibiotic in order to render the antibiotic immune to an enzyme which destroys it, as is done with streptomycin.

24.4.4 Modifying an Existing Antibiotic Through Synthesis by Mutant Organisms: Mutasynthesis

This method is one in which a mutant of an antibiotic producing organism is fed different precursors leading to the production of new antibiotics. Since the original production of hybridicins from the neomycin synthesizing *Strep. fridiae*, this method has been used in the production of paromomycin by *Strep. rimogus* forma *paromonycins*, ribostamycin by *Strep. ribosidificus* and butrosin by *Bacillus circulans*.

24.5 ANTI-TUMOR ANTIBIOTICS

24.5.1 Nature of Tumors

Each cell in the animal (and human) body has a definite function which it carries out in cooperation with other cells. Thus the brain, the skin and the intestines are composed of specialized cells which cooperate to carry the functions of these organs. Sometimes however a cell in any part of the body may no longer cooperate with others with which it normally functions in an organ. Such cells divide indiscriminately and independently of the others to form a structure called a *tumor* or a *neoplasm*. Sometimes the body restricts the growth of tumors by forming a capsule round them. Under these conditions they do not spread: they are known as *benign* tumors. Other tumors however grow rapidly and are not restricted by a capsule. Such tumors are *malignant*. The cells in malignant tumors often break off and are carried via blood vessels and lymphatic vessels to other parts of the body where they initiate new tumors. When such secondary growth occurs away from the primary tumor the situation is known as metastasis.

Tumors are further classified according to the type of tissue they attack. Some of these will be mentioned: a malignant tumor composed of epithelial cells is called a *cancer* or a *carcinoma*. *Adenocarcinomas* are tumors formed around the mucous membranes such as in the alimentary canals. *Sarcomas* are connective tissue tumors. The term 'hard' tumor is sometimes used to distinguish neoplasms formed in the solid parts of the body such as the gut, bones, brain etc. from those of blood such as leukemia which is a neoplasm of the white blood cells. Neoplasms are treated by one or more of three methods: (1) by surgery to remove the cancer; (2) by radiation, which aims at selectively destroying the cancer cells and (3) by chemotherapy or the use of chemicals which affect the tumor cells without damaging the normal cells. When such chemicals are produced by microorganisms they are called anti-tumor antibiotics. Chemotherapy is particularly useful when the disease has metastasized to several sites in the body so that it becomes practically impossible to achieve any success by surgery or by radiation. It is also used after treatment by surgery or radiation to attack those cancerous cells missed by the other two treatments. Many of the chemotherapeutic agents used in cancer treatment are secondary metabolites produced by microorganisms, especially of the genus *Streptomyces*. This chapter is concerned with these metabolites known as anti-tumor antibiotics.

24.5.2 Mode of Action of Anti-tumor Antibiotics

The anti-tumor antibiotics are heterogeneous in their chemical natures. Some of the best known groups used in clinical practice include anthracyclines, actinomycins and bleomycins (Fig. 24.7). In terms of their modes of action their common characteristic seems to be interaction in some form with DNA. Daunomycin and adriamycin which are anthracyclines link up base pairs and thus inhibit RNA and DNA synthesis. Mithramycin and chromomycin A_3 which are actinomycins inhibit DNA – dependent RNA synthesis. On the other hand bleomycins which are peptides react with DNA and cause it to break. Other anti-tumor antibiotics operate through alkylation e.g. streptonigrin, mitomycin C, and profiromycin. Still others interfere with membrane functions or interact with the micro-tubules in the cell.

The basis of all chemotherapy whether with anti-bacterial or with anti-tumor drugs is the ability of the drugs to selectively attack the pathogen or the errant tumor cell. In the well-known case of penicillin for example, the absence of mucopeptides in animal cell wall is the key to the operation of the drug. Several mechanisms have been suggested which allow the selective attack of anti-tumor drugs on tumor cells. These include inability of the tumor cell to repair damage by the anti-tumor drug, higher distribution of the drug in tumor than in normal cells, greater ability to inactivate tumor cells. These are based on structural and bio-chemical differences between normal and tumor cells. Unfortunately anti-tumor antibiotics as well as other anti-tumor drugs do not always discriminate successfully between tumor and normal cells. Varying degrees of toxicity therefore usually accompany the use of anti-tumor antibiotics. The most severe of these is damage to the bone marrow which is involved in synthesis of blood components a damage that may be fatal. Toxicity is in many cases being successfully handled clinically by various means including reduced dosage, change of route of administration, etc. Less toxic antibiotics are also being produced by semi-synthesis or by the modification of the antibiotic molecule.

24.5.3 Search for New Anti-tumor Antibiotics

The search for anti-tumor antibiotics is more difficult than that of anti-bacterial or anti-fungal agents in terms of methodology and interpretation. In the search for the latter

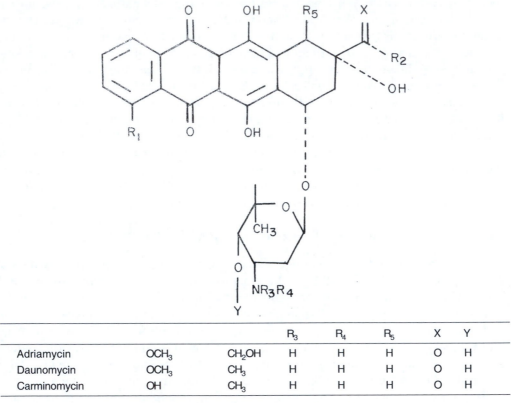

			R_3	R_4	R_5	X	Y
Adriamycin	OCH_3	CH_2OH	H	H	H	O	H
Daunomycin	OCH_3	CH_3	H	H	H	O	H
Carminomycin	OH	CH_3	H	H	H	O	H

Fig. 24.7   General Formula of Anthracyclines

agents it is usually possible to isolate the pathogen from the diseased animal and test the isolate against a wide range of possible antibiotics *in vitro* and *in vivo* in experimental animals. In general an antibiotic successfully tested *in vivo* in experimental animals will be expected to be reasonably efficacious in treating human disease. In the case of anti-tumor agents the relationship is not so straight-forward.

As is the case with anti-microbial antibiotic no order of procedure can be prescribed. The description that follows is built up from published materials from several groups, especially at the National Cancer Institute, and the Cancer Research Laboratories, Kalamazoo, both in the USA. The stages involved in the search and initial development of anti-tumor antibiotics may be enumerated as follows.

24.5.3.1   Screening of potential antibiotic producing organism

A wide variety of microorganisms is obtained from all over the world and their fermentation broth is evaluated for the presence of anti-tumor drugs. Cultures whose identities are established as well as those still to be established are obtained from culture collections and individual scientists around the world. Fresh isolations are also made from natural habitats including soil, aquatic and other environments.

For the isolation (and often the maintenance of the organisms, a wide variety of carbon sources is used. Especially in well-studied groups, such as actinomycetes, novel carbon sources are used in the hope that fermentation broth may contain some anti-tumor antibiotics, as a result of the blockage of certain pathways or the enhancement of others. Such carbon sources include monosaccharides, glycosides, substituted sugars, polyhydric alcohols, oligosaccharides, terpens, and hydrocarbons.

Isolations of microorganisms are done by sprinkling soil on plates containing the above carbon sources. Perfusion technique in which soil is bathed constantly with a solution containing the chosen carbon source may also be used. Isolates are grown in shake flasks using the various carbon sources and a variety of environmental conditions including pH, minerals, temperature, nitrogen sources, and aeration.

24.5.3.1.1   *In vitro* prescreening

Since large numbers of samples are generated, it would be extremely expensive to test them directly in tumor-bearing animals. Prescreens are therefore used. Such prescreens should ideally select the broths successfully containing potentially *in vivo* active components, should be relatively inexpensive in terms of money and time and should require only small quantities of the test broth. *In vitro* screens are also used to follow the course of fermentation. *In vitro* methods are particularly essential because the frequency of occurrence of active anti-tumor components in fermentation beers is low, and when present at all, the concentration especially initially is also low. Some of the *in vitro* screening methods which are currently in use are as follows:

(i) *Use of anti-microbial activity as prescreens:* As most anti-tumor agents will also inhibit micro-organisms, the latter came to be used during the early stages of the search for these drugs. Indeed a number of drugs possessing carcinostatic properties were first isolated as anti-microbial agents. These include actinobolin, cycloheximide, and actinomycin. Azaserine was the first anti-tumor drug isolated following activity against a bacterium, which in this case was *E. coli.* A wide range of microorganisms can and indeed have been used in prescreens. However, many groups favor using a set containing a few microorganisms to facilitate the identification by 'finger printing' of already discovered antibiotics. Finger-printing will be discussed more fully below. A set of microorganisms which has been used include *Bacillus subtilis* ATCC 6633, *Sarcina lutea* ATCC 9341, *Torulopsis albida* NRRL Y1400 and *Escherichia coli* M 1262.

The beer to be tested is spotted on filter paper disks placed on freshly made pour plates of these organisms or in agar wells dug from such plates. The extent of the inhibition is determined by the diameter of the zone of clearing.

(ii) *Anti-metabolite activity*: The use of anti-metabolites as a prescreen was based on the observation that some drugs or broths which did not inhibit microorganisms in complex media such as nutrient agar did so on synthetic minimal media. It was soon shown that by adding various compounds to the minimal medium it was possible to determine the missing metabolite. Broth samples are incorporated into rich agar and minimal agar respectively. Those broths which are more active against susceptible microorganisms in synthetic minimal medium are regarded as potential leads. This methods led to the discovery of 5-Azacytidine and its principal advantage is that it gives an idea at an early stage of the mode of action of the active component of the broth.

(iii) *Inhibition of tumor cells in culture*: The ability of the broth to inhibit animal tumor cells in tissue culture is tested. Among cells which have been used are L1240 (mouse leukemia cells), KB (human carcinoma cells of the nasopharynx) and p388 mouse leukemia cells. L1240 appears to be most widely used. The tumor cells are grown in liquid culture with and without the broth over a period of about three days. They are then counted in a coulter counter. The potency of the same is given as ID_{50} or ID_{90} or the dilution that will cause 50% or 90% inhibition of growth as compared with control.

(iv) *Nuclear cytotoxicity*: Instead of using cell cultures, nuclei from tumor cells can be isolated and the broth tested against these. Isolation may be achieved using citric acid, detergents, organic solvents, and glycerol. The limitation of the system is that it can be used to detect only those agents which in some way affect nuclear synthesis. It however lends some insight into the mode of action of the agent.

24.5.3.2   Finger printing of anti-tumor antibiotics in culture

Broths showing some activity by any of the above prescreening methods are subjected to 'finger-printing' or 'dereplication' in order to avoid the isolation of already known antibiotics. The data used include the following:

  (i) The anti-microbial spectrum of the active components of the broth is determined using a set of microorganisms. Besides bacteria and yeasts some workers use protozoa and algae.
 (ii) The characteristics of the culture used in the fermentation.
(iii) Chromatographic analyses of the broth using paper, thin layer, and high performance liquid chromatography.
(iv) Ultraviolet absorption.

24.5.3.3   *In vivo* assessment

If the results of the finger-printing indicate that the active components are new then *in vitro* testing is done. The one common characteristic of tumors or neoplasms is the uncontrolled growth of cells. Outside this property they are in fact biologically heterogeneous in terms of site of origin, cell type involved, the course of the disease, or response to curative procedures. In assessing active components *in vivo* therefore this diversity is acknowledged by testing various types of artificially induced tumors usually in mice. These include mouse tumors of the colon, breast, lungs, and white blood cells (leukemia). Human cancer from the colon, breast and lungs, are also grafted on these regions of mice whose thymus glands have been removed to avoid rejection of the grafts.

Often it is necessary to have an *in vivo* prescreen before subjecting the broth to the above tests. Some workers have found that mouse leukemia P388 is more sensitive as an *in vivo* prescreen than L1240.

The parameters for *in vivo* tests are increased lifespan of the animal or tumor growth inhibition as measured by tumor weight inhibition over the control.

The *in vivo* test is expensive in time and money. It takes three to four weeks to perform whereas cell or microbial cultures take about three days or less.

24.5.3.4   Extraction and manipulation of the pure drug

Since the active component may often be present in very low concentration it is necessary to obtain the active component in the broth in a reasonably pure form. Subsequent to this it is characterized chemically and then subjected to further animal tests before being assessed clinically. Since many anti-tumor antibiotics have serious side effects, analogues of the drugs are produced and modifications to the molecule are then carried out. Other procedures, e.g. improvement in the environmental conditions of the broth, development of optimum isolation procedures, scale up, sales, etc. are carried out as for anti-microbial antibiotics.

24.5.3.5   Towards a new definition of 'antibiotic'

The current definition of the term 'antibiotic' which restricts them to chemicals produced by microorganisms is credited to Waksman who had won the Nobel Prize for discovering streptomycin. However, when screenings have been done outside microorganisms, the higher organisms so screened have been shown to produce anti-microbial substances. Such substances are low molecular weight secondary metabolites in the same way as regular antibiotics are. Due to this, there is now a tendency to extend the term antibiotic to all secondary metabolites, irrespective of their origin, which are able to inhibit various growth processes at low concentration. Not only that, even wholly synthetic antimicrobials such as ciprofloxacin are now legitimately termed antibiotics. It is not an altogether unreasonable redefinition. After all, the word antibiotic derives from two origins, anti (against) and bios (life). Nothing in the word itself restricts antibiotics both in origin or in use to microbial life

24.6   NEWER METHODS FOR SEARCHING FOR ANTIBIOTIC AND ANTI-TUMOR DRUGS

In recent times newer method have been developed for searching for new antibiotics, anti-tumor agents and other drugs. These methods include computer-aided drug designing, synthesis of new drugs by combinatorial chemistry and genome-based methods. These are further discussed in Chapter 28, where drug discovery is examined.

SUGGESTED READINGS

Allsop, A., Illingworth, R. 2002. The impact of genomics and related technologies on the search for new antibiotics. Journal of Applied Microbiology, 92, 7-12.

Anon, 1993. Congress of the United States, Office of Technology Assessment. Pharmaceutical R&D: Costs, Risks and Rewards: 1993; pp. 4-5.Washington, DC, USA.

Anon, 1999. From Test Tube to Patient: Improving Health Through Human Drugs. Special Report, Center Drug Evaluation and Research. Food and Drug Administration. Rockville, MD, USA.

Austin, C. 2004. The Impact of the Completed Human Genome Sequence on the Development of Novel Therapeutics for Human Disease. Annual Review of Medicine, 55, 1-13.

Bansal, A.K. 2005. Bioinformatics in the microbial biotechnology – a mini review. Microbial Cell Factories, 4, 4-19.

Beamer, L. 2002. Human BPI: One protein's journey from laboratory to clinical trials. ASM News. 68, 543-548.

Behal, V. 2000. Bioactive Products from *Streptomyces*. Advances in Applied Microbiology. 47, 113-156.

Bull, A.T., Ward, A.C., Goodfellow, M. 2000. Search and Discovery Strategies For Biotechnology:. The Paradigm Shift. Microbiology and Molecular Biology Reviews, 64, 573-548.

Dale, E., Wierenga, D.E., Eaton, C.R. 2001. Processes of Product Develpoment. http://www.allpcom/drug-dev.htm. Accessed on September 28, 2005 at 12.05 pm GMT.

Debouck, C., Metcalf, B. 2000. The Impact of Genomics on Drug Discovery. Annual Review of Pharmacology and Toxicology, 40, 193–208.

Fan, F., McDevitt, D. 2002. Microbial Genomics for Antibiotic Target Discovery. In: Methods in Microbiology. Vol 33, Academic Press. Amsterdam: The Netherlands. pp. 272–288.

Feling, R.H., Buchanan, G.O., Mincer, T.J., Kauffman, C.A., Jensen, P.R., Fenical, W. 2003. Salinosporamide A: a highly cytotoxic proteasome inhibitor from a novel microbial source, a marine bacterium of the new genus *Salinospora*. Angewandte Chemie International Edition, 42, 355-357.

Fraser, C.M., Rappuoli, R. 2005. application of microbial Genomic Science To Advanced Therapeutics. Annual Review of Medicine, 56, 459–74.

Handelsman, J., Rondon, M.R., Brady, S.F., Clardy, J., Goodman. R.M. 1998. Molecular biological access to the chemistry of unknown soil microbes: a new frontier for natural products. Chemistry and Biology, 5, 245-249.

Hill, D.C., Wrigley, S.K., Nisbet, L.J. 1998. Novel screen methodologies for identification of new microbial metabolites with pharmacological activity. Advances in Biochemical Engineering and Biotechnology, 59, 75–124.

Maxwell, S., Ehrlich, H., Speer, L., Chandler, W. 2005. Medicines from the Deep Sea. Washington, DC, USA.

Rosamond, J., Allsop, A. 2000. Harnessing the Power of the Genome in the Search for New Antibiotics. Science, 287, 1972-1976.

Wlodawer, A., Vondrasek, J. 1998. Inhibitors of HIV-1 Protease: A Major Success Of Structure-Assisted Drug Design. Annual Review of Biophysics and Biomoecular Structurrs, 27, 249–84.

CHAPTER 25

Production of Ergot Alkaloids

25.1 NATURE OF ERGOT ALKALOIDS

Alkaloids are laevorotatory basic naturally occurring hetereocyclic organic nitrogen containing compounds which are biosynthesized from amino acids by plants, microorganisms and some animals. Many of them are pharmacologically active and are consequently used as drugs. The main precursors for alkaloid biosynthesis are ornithine, lysine, aspartic acid, phenylalanine, tyrosine and tryptophan. For example the alkaloid in tobacco, nicotine, is derived from ornithine while phenylalanine and tyrosine give rise to simple alkaloids such as ephedrine or more complex ones such as morphine.

The name alkaloid (literally alkali like) derives from their basic nature, because of which they readily form salts with acids present in the natural sources from which they are derived. Their chemical classification is based on the carbon-nitrogen skeletons. Ergot alkaloids, the subject of this chapter are of the indole type and are derived from tryptophan.

The ergot alkaloids are so called because they were originally derived from ergot, a sclerotium (twisted mat of fungal hyphae) formed as a disease on the grain of rye *(Secale cereale* L.) a temperate cereal. The dried ergot is known among pharmacists as secale cornutum. The cause of the rye disease is a fungus, an ascomycete, *Claviceps purpurea.* As will be seen later, ergot contains several (more than 40) highly potent alkaloids and the unwitting consumption of grain attacked by fungi producing ergot alkaloid has led to 'ergotism', (previously known as 'Holy fire' or 'St. Anthony's fire' when it was not understood) a disease characterized by among other symptoms, convulsions. . Ingestion of contaminated grain, most often after the grain has been made into bread, causes ergotism, also known as the 'Devil's curse' or 'St. Anthony's fire,' and has been a problem for centuries. It has been noted in writings from China as early as 1100 B.C. and in Assyria in 600 B.C., and Julius Caesar's legions suffered an epidemic of ergotism during one of their campaigns in Gaul (France). In 994 A.D., an epidemic in France killed between 20,000 and 50,000 people, and in 1926, at least 11,000 cases of ergotism occurred in Russia.

Ergotism can cause convulsions, nausea, and diarrhea in mild forms, and there is some thought that an outbreak of ergotism may have been the cause of the 'bewitchings' which led to the Salem witch trials in the United States in 1691. Ergotism may also have caused some of the extreme destruction associated with the French Revolution. In the Middle Ages, ergotism was described as causing victims to die "miserably, their limbs eaten up by the holy fire that blacked like charcoal." People turned to the church for help, assuming that the disease was retribution for their sins. In particular, they prayed to St. Anthony for deliverance, giving rise to the name for the disease. Ergotism takes two forms, gangrenous ergotism, in which tingling effects were felt in fingers and toes followed in many cases by dry gangrene of the limbs and finally loss of the limbs, and convulsive ergotism, in which the tingling was followed by hallucinations and delerium and epileptic-type seizures. In both cases, death was slow and painful. Ergotism has now been recognized as a result of infection by a mycotoxin, and the ergotism plagues have been eliminated.

About 50 ergot alkaloids are known today. While most of these alkaloids are derived from the *Claviceps sclerotium* formed on the rye grain, hundreds of other cereals and grasses can serve as hosts for the fungus. About 50 species of *Claviceps* itself are known. In addition, in recent times the alkaloids have been produced by other fungi including *Aspergillus, Penicillium,* and *Rhizopus.* Some recent fungi shown to produce *alkaloids* are *Balansia epichloe, B. henningsiana, B. strangulans, Myriogenospore atrementose* and *Epichre typhine.* Furthermore, ergot alkaloids have recently been found in the seeds of some higher plants, *Ipomea, Rivea, Agyreis* which belong to *Convulvulaceae* the family to which the flower morning glory belongs.

The life cycle of *Claviceps* is given in Fig. 25.1. The sclerotium forms in the size and shape of the grain which it replaces. These sclerotia fall to the ground at the end of the growing season and remain dormant till the beginning of the next growing season, when they germinate and form ascocarps (perithecia). The ascospores are distributed by wind to the newly formed flowers of grains. The germinated ascospores yield hyphae which produce masses of conidia supported in a sugary liquid which attracts insects. These insects further help distribute the conidia to other plants.

The ergot alkaloids are classified as indole alkaloids, which are derived from tyrptophan. With one exception, chanoclavine, the ergot alkaloids possess the basic tetracyclic (four-ringed) structure known as ergoline (Fig. 25.2).

The naturally occurring ergot alkaloids can be divided into two groups. (a) lysergic acid derivaties (Fig. 25.2) and (b) clavine alkaloids. Although the clavine alkaloids were the first to be prepared in fermentation broths, and were used for biosynthetic studies, they are much less known than the lysergic (Fig. 25.3) acid ones in terms of their fermentation and even in terms of pharmacological activity. Therefore only the lysergic alkaloids will be handled in this discussion.

The lysergic acid derivates can be further divided into two depending on the nature of the amide substituents. First are the simple amide substituents and second are the peptide alkaloids in which a cyclic peptide is attached to lysergic acid. The greatest interest appears to center on the peptide alkaloids. In this group a modified tripeptide containing proline and an α–hydroxy - α-amino acid which has undergone cyclic foundation with the carbonyl atom of proline substitutions at C-2 and C5 of the cyclol peptide moeity creates variability as shown in Fig. 25.4. This group includes ergotamine.

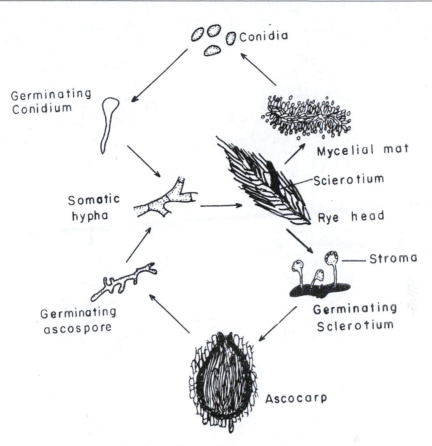

Fig. 25.1   Life cycle of the Ergot Fungus

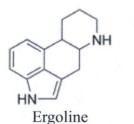

Ergoline

Fig. 25.2   Structure of Ergoline

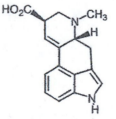

Fig. 25.3   Lysergic Acid

25.2   USES OF ERGOT ALKALOIDS AND THEIR DERIVATES

Ergot alkaloids and their derivates are powerful drugs and may be used as such or may be the basis of semi-synthetic preparations. Almost all of them have one important pharmacological effect or another, depending on the nature of the substituents and on the tissue of the body concerned. Thus ergotamine will cause vessels to constrict while ergometrine has a minimal effect on blood vessels but will cause the uterus (womb) to constrict; LSD on the other hand will excite the brain cells in a manner not achieved by the

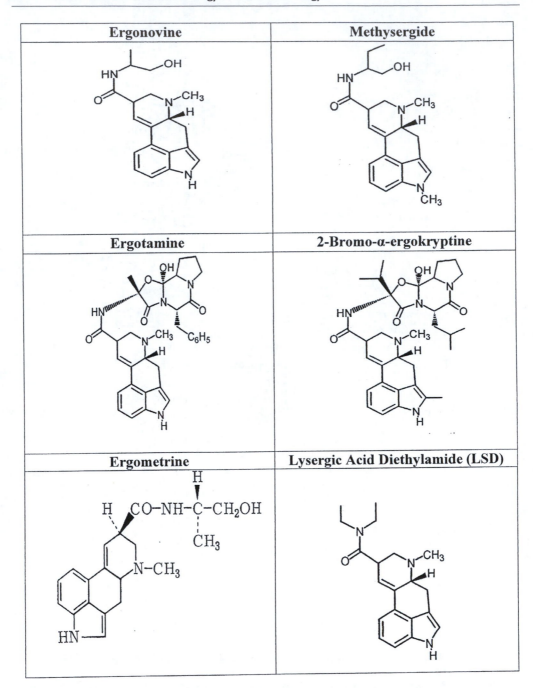

Ergonovine (uterine contraction; treating post-partum hemorrhages**). Methysergide** (cranial vasodilator; treatment of migrane headaches). **Ergotamine** (treatment of severe migrane head aches). **2-Bromo-α-ergokryptine** (semi-synthetic, reduction of lactation in women). **Ergometrine** (used for treating post-partum hemorrhages). **Lysergic Acid Diethylamide** (for treating psychiatric disordrers).

Fig. 25.4   Some Therapeutically Useful Ergot Alkaloids

other three. These various effects are exploited in using ergot alkaloids as drugs in the manner shown below:

   (i) Ergot alkaloids (e.g. ergometrine) have been used in mid-wifery to induce labor for centuries.

  (ii) Ergonovine, (the 2-aminopropanolamide derivative) is used to stop bleeding after birth due to its stimulatory effect on the sympathetic nervous system.

 (iii) Ergotamine blocks the sympathetic system and is used for treating strong headaches such as migraines.

 (iv) The diethylamide derivate of lysergic acid (known as LSD) is a powerful hallucinogenic drug and is often utilized illegally for this purpose in many countries, as well as for experimental psychotherapy.

  (v) Most of the clavine alkaloids do not possess strong pharmaceological properties. However a few of the didydro derivates have found use as strong stimulants of oxytoxic (milk secreting) activity or of uterine contractions.

In recent times newer uses have been found for ergot alkaloids especially semi-synthetic ones. These include:

 (vi) 'Nigericoline', a derivate of lysergic acid which is a receptor-blocking agent and is used for treating peripheral and cerebral circulation disorder.

 (vii) 'Lysenyl' a diethyl derivate of isolysergic acid is used to treat hypertension and migraine, since it is a serotonin antagonist.

(viii) In recent times it has been found that some ergot alkaloids affect functions controlled by the hypothalamic pituitary system particularly the release of prolactin (which deals with milk secretion) from the pituitary gland. Since the prolactin level seems to play a part in the growth and development of certain breast cancers they are used for therapy.

 (ix) Some of newer ergot alkaloids have also been implicated as potential therapeutic agents in the treatment of diseases such as Parkinson's disease, lack of milk production after child birth, and cancer of the prostrate.

  (x) The alkaloids have been used as models for the synthesis of several potent drugs.

25.3   PRODUCTION OF ERGOT ALKALOIDS

A large market based mainly in Europe derives from ergot drugs; this market seems to be expanding as more and more pharmacological properties are discovered in the ergot drugs.

The methods for producing these drugs are three.

 (a) Isolation from field cultivated ergot.

 (b) Fermentation of the ergot fungus

 (c) Partial or total chemical synthesis.

Until now the bulk of the production seems to be the extraction of field inoculated ergot. The pattern seems however to be changing. Fermentation is increasing and as with the antibiotics wholly new drugs are being produced by semi-synthesis with substrates derived from fermentation.

(a) *Isolation from field cultivated ergot (or parasitic production)*: This method is widely used in Europe. Inoculation of the rye plants with conidiospores of *Claviceps* and other fungi takes place two to three weeks before flowering begins and may continue during flowering. Harvesting of ergotized ears of rye takes place about two months later. This method has numerous disadvantages. Firstly, only one crop a year can be obtained. Secondly, the yield of alkaloid in terms of quality (type) and quantity is highly unpredictable. Thirdly, the vicissitudes of the weather attendant on all field operations make the operation highly unstandardizable.

(b) *Fermentation production*: Due to the above problems, efforts have been put in over the years to devise fermentation methods. A good measure of success has been achieved especially for the clavine alkaloids and simple lysergic acid derivates. Ergot alkaloids can be produced in submerged fermentation by *Claviceps* or *Penicillium* species, which are used for their industrial production. Initial work in Japan showed that submerged cultures did not produce the typical alkaloids associated with the sclerotium but instead produced a series of new non-peptide bases (clavines) which did not possess any significant pharmacological action. Attempts were made by many workers to influence alkaloid production by modification of the culture medium and the fungus strain. The first pure ergot alkaloid, ergotamine, was obtained by Stoll in 1920. Subsequently, others reported the discovery of the "water soluble uterotonic principle of ergot" which was subsequently determined to be ergonovine (also called ergometrine). As a result of further successful experiments the commercial manufacture of simple lysergic acid derivatives by fermentative growth of a strain of *Claviceps paspali* became feasible

(i) *Production of clavine alkaloids*: Different species of *Claviceps* parasitizing a variety of grasses have been isolated and grown in liquid medium and the different alkaloids assayed. Clavine alkaloids were obtained from *Cl. litoralis, Cl. microcephals*. One strain *Cl. purpurea* produced a clavine and the other a pepetide alkaloid. The medium used contained mannitol (5%) ammonium citrate (0.7%), KH_2PO4 (0.1%) and MgSO4 (0.03%). The pH was 5.2. The use of 10% sucrose instead of mannitol gave higher yields.

The fermentation lasted from 30 days to 40 days and by classical strain improvement methods yields of up to 1.0-1.5 gm/liter were obtained.

Improvements have since been obtained by other workers who found that a mixture of mannitol (6.5%) and glucose (1%) gave up to 1 gm/liter yield in about 14 days. A high carbon to nitrogen in the medium increased yield.

(ii) *Production of simple lysergic acid derivates*: The production of simple lysergic acid derivates was achieved in 1961 using *Claviceps paspali* isolated from an infected *Paspalum digitatum*.

Higher yields have since been attained by slight modifications of the original mannitol-succinic acid-mineral salts medium. Fumaric acid gave higher yields when it was used in place of succinic acid. The fermentation lasted nine days. The addition of hydrophilic non-ionic surfactants had the greatest effects on yield. Aeration was vigorous.

(iii) *Peptide alkaloids*: The physiology of ergotamine formation has been studied using *Claviceps purpurea, C. litoralis, Elymus mollis*. Improved ergotoxine yields were obtained by the mutation and selection of a strain of *Claviceps purpurea*. This strain on a mannitol-ammonium-succinate medium produced up to 40 gm/liter.

25.4 PHYSIOLOGY OF ALKALOID PRODUCTION

(i) *Induction*: Tryptophan is the central precursor of all ergot alkaloids and it is therefore required in the medium. In addition to being a direct precursor of ergot alkaloids, tryptophan is also a factor in the induction and derepression of enzymes necessary for alkaloid synthesis. If the amino acid is not added to the medium the organism manufactures it. When it is added the organism accumulates it in its hyphae.

(ii) *Feedback regulation*: Feedback regulation studies have been hampered by the cell wall as studies using protoplasts of *Claviceps* sp. have unambiguously demonstrated. It was shown that the addition of elymoclavine inhibited (not repressed) the first enzyme in the synthesis of the alkaloid, namely dimethylollyl tryptophane synthase (DMAT synthase). This was demonstrated by supplying to washed stationary phase cultures of the producing organism, basal culture medium or basal culture medium supplemented with elymoclavine. Cultures with fresh medium synthesized the alkaloid at a slower rate than when supplemented. Both of them however reached the same final alkaloid concentration.

(iii) *Phosphate repression*: Like many secondary metabolites ergot alkaloid formation is inhibited by increasing the level of phosphate. In this case the unfavorable limit was 1.1. gm/liter. The addition of tryptophan helped to nullify the effect of phosphate, showing that phosphate inhibition was mediated through tryptophan probably by preventing its accumulation from exceeding the amount required for alkaloid synthesis induction. In support of this explanation, it is noted that 5-methyltryptophan also overcomes the inhibition of alkaloid synthesis by tryptophan. Phosphate inhibited culture had very low levels of the first enzyme in the synthetic chain, namely DMAT synthase; tryptophan caused more than 10-fold increase in the activity of the enzyme.

(iv) *Catabolite regulation*: High levels of glucose (5%) greatly inhibited enzyme production. The addition of a small amount of camp restored alkaloid synthesis to a little extent.

(v) *Alkaloid formation and morphological structures*: It is interesting to note that alkaloids seem to form in one structure and accumulated in another. Thus, in the plant genus *Ipomea*, *alkaloids* are formed in the leaves and accumulated in the seeds, which do not produce alkaloids at all. In ergotoxine alkaloid fermentation, alkaloids are not elaborated until specific morphological structures, pellets, are formed in the medium. The medium composition appears to influence the formation of these pellets. Sucrose appears to encourage their formation while they are poorly formed in malt. The peptide alkaloids are found in these structures, while clavines and simpler derivates of lysergic acid are found in the medium.

(vi) *Biosynthesis*: Work on biosynthesis of alkaloids has been greatly facilitated by the use of protoplasts. Being secondary products, alkaloids are produced by pathways different from those of general metabolism. Furthermore, synthesis initiates with carbon limitation. The ergot skeleton is derived from tryptophan, mevalonic acid, and methionine. The central precursor of the ergoline skeleton is tryptophan, which is initially dealkylated. The five-carbon unit is obtained from mevalonic acid. Although a component of the ergoline unit, it is neither limiting nor does it participate in the regulation of alkaloid synthesis induction. Mevalonic acid may be formed from the malonyl COA pathway with biotin and decarboxylation.

Clavines are simpler in terms of biosynthesis than lysergic acid derivates. Lysergic acid itself is a key substrate in ergot alkaloid synthesis and the simpler amide as well as the peptide derivates come from it (see Fig. 25.5).

Surprisingly not many enzymes appear to be involved in the synthesis of alkaloids. Two of them have so far been isolated: Dimethyl tryptophan synthase (DMTPS) and chanoclavine synthase. DMTPS catalyses the first specific reaction of the formation of 8-ergoline and introduces the isoprene residue to the C_4 position of tryptophan. Analogues of tryptophan increase the activity of the enzyme. The compound produced, dimethyl tryptophan (4-isoprenyltryptophan) is 5-10 times more effective as a precursor of the clavines than tryptophan.

Chanoclavine cyclase catalyses the cyclization of chanoclavine – 1 to the four-ring 8-ergoline i.e. to elgmoclavine and agroclavine, both individually and simultaneously. Some biogenetic relationships among alkaloids are shown in Fig. 25.5

(vii) *Extraction of the alkaloid*: Alkaloids are readily isolated at an alkaline pH by various organic solvents such as ether, chloroform, and ethyl acetate.

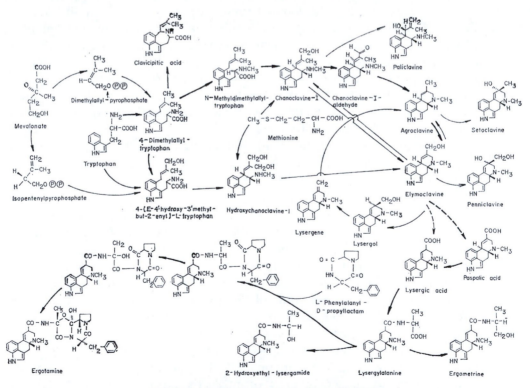

Fig. 25.5 Synthetic Routes for Ergot Alkaloids

The earliest procedures used for extraction were designed to obtain alkaloids from sclerotia. In general the sclerotia would be dried, powdered, alkalinized, and extracted with tartaric acid. For further purification, the tartaric acid solution was made alkaline and extracted with chloroform. This chloroform layer would then contain all the alkaloids. This has led to the terms water-soluble alkaloids which include the simple amide lysergic acid derivatives and the clavines, whereas the water-insoluble alkaloids now include the peptide-type. Nowadays the method adopted for both water-soluble and water-insoluble peptides is to extract the powdered ergot with chloroform to which a small amount of methanolic ammonia has been added. The chloroform extract is concentrated to a small volume, diluted with ether and extrated with concentrated H_2SO_4. On neutralization with ammonia the water-soluble alkaloids can be extracted with water and the water-insoluble ones with carbon-tetrachloride.

SUGGESTED READINGS

Boichenko, L.V., Boichenko, D.M., Vinokurova, N.G., Reshetilova, T.A., Arinbasarov, M.U. 2001. Screening for Ergot Alkaloid Producers among Microscopic Fungi by Means of the Polymerase Chain Reaction, Microbiology, 70: 306-307.

Dongen, van P.W.J., de Groot, A.N.J.A. 1995. History of ergot alkaloids from ergotism to ergometrine, European Journal of Obstetrics & Gynaecology and Reproductive Biology, 60: 109-116.

Evans, W.C. 1996. Pharmacognosy. 14th ed, W B Saunders Company Ltd, London, UK.

Groot, N.J.A. de Akosua, van Dongen, Pieter W.J, Vree, Tom, B., Hekster, Yechiel A., van Roosmalen, Jos. 1998. Ergot Alkaloids - Current Status and Review of Clinical Pharmacology and Therapeutic Use Compared with Other Oxytoxics in Obstretrics and Gynaecology, *Drugs*, 56: 525-8.

Hardman, J.G., Limbird, L.E. (eds) 1996. The Pharmacological Basis of Therapeutics, 9th ed, McGraw-Hill

Kobel, H., Kobel, J. 1986. Ergot alkaloids. In: Biotechnology, H.J., Rehm, G. Reed, (eds) Vol 4,. 2nd Ed. VCH, Weinheim, Germany, pp. 569-609.

Komarova, E.L., Tolkachev, O.N. 2001. The Chemistry of Peptide Ergot Alkaloids, Pharmaceutical Chemistry Journal, 35: 504-506.

Lange, Klaus W. 1998. Clinical Pharmacology of Dopamine Agonists in Parkinson's Disease, Drugs & Aging, 13: 385-386.

Langley, D. 1998. Exploiting the Fungi: Novel Leads to New Medicines, Mycologist, 11: 165-166.

Menge, J.M.M. 2000. Progress and Prospects of Ergot Alkaloid Research. Advances in Biochemical Engineering/Biotechnology. 68, 1-20.

Thompson, F., Muir, A., Stirton, J., Macphee, G., Hudson, S. 2001. Parkinson's Disease, The Pharmaceutical Journal, 267: 600-612.

Votruba, V., Flieger, M. 2000. Separation of Ergot Alkaloids by Adsorption on Silicates, Biotechnology Letters, 22: 1281-1282.

Rehacek, Z., Sajdl, P. 1990. Ergot Alkaloids: Chemistry, Biological Effects, Biotechnology. Academia Praha: Czech Republic.

CHAPTER 26

Microbial Transformation of Steroids and Sterols

26.1 NATURE AND USE OF STEROIDS AND STEROLS

Steroids are a large group of organic compounds with the perhydro- 1, 2-cyclopentano – phenanthrene nucleus, which consists of four fused rings (Fig. 26.1).

Sterols are hydroxylated steroids – that is, they are alcohols derived from steroids. The hydroxyl (OH) group of sterols is usually substituted at position C_3. Unsaturation is usually at C_5 and often as C_7 and C_{22}. The term sterol comes from the Greek (*Steros* = solid) because the earliest members studied were solid alcohols resulting from the unsaponifiable (i.e. could not be broken down by NaOH) fractions of fats of plants and animals. As the variety of known structures increased the general term steroid came into use about 1935. In higher animals the principal sterol is cholesterol but a wider variety exists in lower animals and in plants (Fig. 26.1).

Steroids and sterols are widely distributed in nature and are present in bile salts, adrenal-cortical and sex hormones, insect molting hormones, sapogenins, alkaloids and some antibiotics.

Steroids and sterols differ from each other in two ways: (a) the number, type, and position of the substituents; (b) the number and position of the double bonds in the ring. Steroid molecules are usually flat. However, the substituents at each of the junctions of Rings A and B, Rings B and C, and Rings C and D may be either above or below the plane of the ring. When the substituent group lies above the plane (denoted by a solid line) of the molecule the substituent is denoted by β; when it is below (denoted by a broken line) it is denoted by α. When as is the case in many steroid hormones a double bond exists between C_4 and C_5 the situation is denoted Δ_4. The individual compounds are named systematically as derivatives of steroidal hydrocarbons the more important of which are gonane, estrane, androstane, pregnane, cholane and cholestane. Thus cortisone which is a derivative of pregnane is Δ_4- pregnene – 17α, 24 – diol – 3 11, 20 – trione.

The steroid hormones of the mammalian body have profound effects on the body function and even the behavior of the animal. Thus, the male hormones, androgens secreted by the testes are responsible for the development of the male reproductive organs and the secondary sexual characters such as hairiness among several other functions.

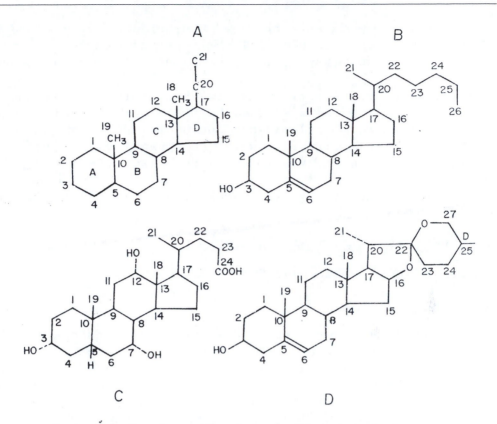

A = Basic skeleton of steroid and sterols. Rings A, B, C, D are fused
to form the perhydro-1, 2, cyclopentano-phenanthrene nucleus.
————————— = α-configuration
- - - - - - - = β-configuration

B = Cholesterol (Cholest-5-en al). An animal sterol; note the - OH
group in the 3 position, a characteristic of sterols

C = Cholic acid (3α, 7α, 12α -Trihydroxy -5 β-cholanic acid)

D = Diosgenin (Spirost -5-en-3 β-ol). A substance used for microbial
transformation of sterols.

Fig. 26.1 Structures of some Steroids and Sterols

The female sexual hormones include estrogens and progesterone. Estrogens are
produced by the ovary – they stimulate the development of the female reproductive
organs and secondary sexual characteristics such as enlarged breasts, etc.

Progesterone is produced by the *corpus luteum,* a body formed by the mature egg in the
female ovary. In association with the oestrogens, progesterone prepares the uterus for the
implanation of the fertilized egg in the uterine wall. Corticosteroid hormones are
produced by the cortex surrounding the adrenal glands, which are themselves located
just above the kidneys. The main steroid hormones produced by the glands are
corticosterone, cortisol, and aldosterone. Aldosterone is involved with mineral
metabolism, mainly of sodium ions and hence indirectly of the blood pressure. Cortisol

and corticosterone help the body handle physiological stress including extreme cold. It is important to underscore, even at the risk of sweeping generalization, the significance of hormones as a background towards appreciating the impetus for the transformation of steroids.

Among insects steroid hormones are also very important in post-embryonic development: juvenile hormones control larval growth; ecdysone controls metarmorphosis of larval-larval larval-pupal and pupal-adult moulting processes; a third hormone affects the brain and controls the production and release of the moulting hormone. These hormones and their laboratory synthesized analogues (pheromones) are used for controlling insects. Bile salts, sterols, oestrogens, progesterone, androgens cortisone and cortisol and other steroids from animal and plants were isolated and studied from 1903.

26.2 USES OF STEROIDS AND STEROLS

The world sale of steroids runs into billions of dollars (see Table 26.1)

26.2.1 Sex Hormones

As will be seen below, many steroids and sterols are manufactured through microbial action. The largest economic impact of synthetic estrogen and progestin production has been for use as contraceptive agents and for treatment and prevention of osteoporosis. Contraceptive steroid mixtures have also been used to treat a variety of related abnormal states including endometriosis, dysmenorrhea, hirsutism, polycystic ovarian disease, dysfunctional uterine bleeding, benign breast disease, and ovarian cyst suppression.

Estrogens are routinely prescribed to post-menopausal women to prevent the development and exacerbation of osteoporosis because it can increase bone density and reduce fractures.

Table 26.1 Total worldwide sales of systemic sex hormones and corticosteroids

	Sales, $x 10^6	
Steroid class	*1990*	*1994*
Systemic sex hormones	3,582	5,436
Corticosteroids		
Topical	1,558	1,891
Systemic	903	1,181
Respiratory	988	2,170
Nasal	382	665
Inhalants, systemic	606	1,505
Steroids for sensory organs	396	507
Total	**7,427**	**11,185**

Testosterone, alkylated testosterone, or testosterone esters are the primary anabolic–androgenic steroid drugs. Most of these synthetic testosterone derivatives were in failed attempts to separate the hormones' masculinizing (androgenic) and skeletal muscle-building (anabolic) effects. The medicinal uses for these drugs include treatment of certain types of anemias, hereditary angioedema, certain gynecological conditions, protein anabolism, certain allergic reactions, and use in replacement therapy in gonadal failure states.

Anabolic–androgenic steroids are best known for their nonmedical, and illegal, use to aid in body-building or to increase skeletal muscle size, strength, and endurance by athletes.

26.2.2 Corticosteroids

The greatest portion of steroid drug production is aimed at the synthesis of glucocorticoids which are highly effective agents for the treatment of chronic inflammation. Glucocorticoids exert their effects by binding to the cytoplasmic glucocorticoid receptor within the target cell and thus either increase or decrease transcription of a number of genes involved in the inflammatory process. Specifically, glucocorticoids down-regulate potential mediators of inflammation such as cytokines (Chapter 28). Typical oral glucocorticoids used to treat rheumatoid arthritis are prednisone and 6 α-methylprednisolone. Corticosteroids are the most efficacious treatment available for the long-term treatment of asthma, and inhaled corticosteroids are considered to be a first-line therapy for asthma. They are also used to treat rhinitis, or nasal congestion and inflammations of the skin.

26.2.3 Saponins

These are used for their hypocholesterolemic (cholesterol lowering) activity. Synthetic steroids that are structurally related to saponins have been shown to lower plasma cholesterol in a variety of different species

26.2.4 Heterocyclic Steroids

Dihydrogentesterone (DHT) is a more potent androgen than testerone. Elevated levels of DHT lead to enlarged prostate (benign hyperplasia), sometimes prostate cancer, and male baldness and the enzyme antagonistic to DHT steroid 5 α-reductase is being developed as treatment for these ailments.

26.3 MANUFACTURE OF STEROIDS

In 1937, the first microbial transformation of steroids was carried out. Testerone was produced from dehydroepiandrosterone using *Corynebacterium* sp. Subsequently, cholesterol was produced from 4-dehydroeticholanic acid and 7-hydroxycholesterol using *Nocardia* spp. These developments were virtually unexploited until 1949, when the dramatic curative effect of cortisone on rheumatoid arthritis, a disease in which painful swellings occur at the joints of the body, was announced. The cortisone used in this work had been prepared by complex and tedious chemical synthesis beginning with deoxycholic acid, a bile acid. So tedious was this that it took 32 chemical steps and two years to produce only 11 gm of cortisone acetate. Additionally, it was difficult to find enough of the starting materials. To meet the demand for cortisone to treat the large number of people suffering from rheumatoid arthritis, a great burst of activity along the four following lines ensured:

(i) The improvements of the original chemical method along lines suitable for commercial production;

(ii) The development of methods of chemical synthesis using other more available and more abundant starting materials including steroids and steroid-containing compounds of plant origin. One of such compounds was diosgenin (a glycoside formed from a steroid and glucose) obtained from a species of yams, *Dioscorea composita* and the South African 'elephant foot' *Testudinaria sylvatica*. The other was a plant sterol, stigmasterol obtained from soy bean, *Glycine max*.

(iii) Total chemical synthesis of cortisone and cortisol from relatively simple materials;

(iv) The use of biological agents to transform readily available steroids by introducing oxygen at carbon C_{11}, a process which took 12 steps in chemical synthesis. The use of biological agents originally consisted of the use of ground or homogenized adrenal tissues and fungi.

The first of the two methods mentioned above had moderate successes and for some time provided steroids for clinical use. The third method remained an academic exercise. The fourth method however gave dramatic results of which microbial transformations eventually became more important. Two of the earliest such microbial transformations were the conversion of progesterone to 11 – a hydroxy progesterone by the introduction of – OH at the position 11 using *Rhizopus nigricans* and the conversion of cortisol to prednisolone by *Corynebacterium simplex* (Fig. 26.2).

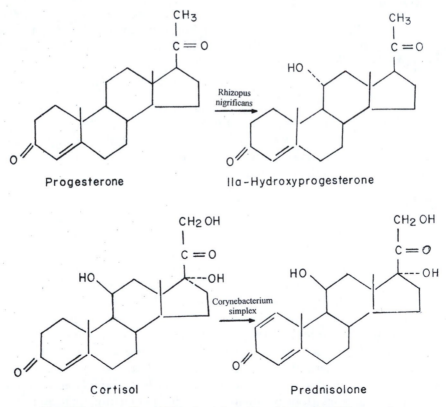

Fig. 26.2 Some Steroid Transformations Brought about by Microorganisms

This latter transformation was notable because the new product was more active than the starting one. Since then a large number of steroid analogues have since been produced using a wide variety of microorganisms. Indeed virtually every steroid is transformable in some way by some microorganism or the other.

From the beginning of 1960, intensive research interest shifted from rheumatoid steroids to the area of sex hormones, especially the progresterone-based drug principally used for birth control pills. This shift of interest was as a result of concern for rising world population. The disclosure about 1965 of the steroidal nature of insect hormones stimulated interest in them as a means of controlling insect pests of agriculture and food and vectors of disease.

A large number of steroids have since been produced and tested for a variety of purposes. Several of them have been found useful as anti-inflammatory, anti-tumor and anti-allergy drugs; as birth control pills; for treatment in heart disease and a vast array of medical and veterinary uses.

The use of microorganisms to transform steroids revolutionized the steroid transformation industry. For example the price of cortisone, a widely used anti-inflammatory drug, fell from US $200 per gm in 1949 to less then US $1.0 per gm in 1979 as a result of this development.

The microbial transformation of steroids differs from the 'traditional' fermentations such as that of penicillin thus:

(i) In many cases steroid transformations are one-step-processes which bring about relatively minor structural changes in the substrate, i.e. the steroid molecule. This differs from the synthesis of penicillin and many other fermentation products in which the product is synthesized entirely from the substrate offered in the medium.

(ii) Whereas in many industrial fermentations, the process of production is completed in the fermentor, in the case of steroid transformations, readily available steroids are micro-biologically transformed into important intermediates which are then converted chemically to the final product. Alternately, the chemical syntheses are first performed and the products transformed microbiologically later.

26.3.1 Types of Microbial Transformations in Steroids and Sterols

Transformations by microorganisms affecting various positions in a wide range of steroids and sterols have been carried out. Although steroid hormones have been most widely studied, the transformation of bile acids, plant and animal sterols, steroid alkaloids have also occurred. The transformation reaction include: hydroxylation, dehydrogenation, reduction, side chain degradation, lactone formation, aromatization, isomerization, epoxidation, hydrolysis, esterification, halogenation, and cleavage of the steroid skeleton.

All of these have been carried out on steroid hormones, but only some of them have been done on the other natural steroids and sterols. Two examples of these reactions have already been described: Progesterone is converted to 11α - hydroxyl progesterone by hydroxylation or the introduction of an OH group at position 11. Similarly, cortisol is converted to prednisolone by dehydrogenation at position 1. Some other examples are given in Fig. 26.2.

A major transformation in which interest has grown sharply in recent times is the cleavage of the C_{17} side chain of sterols. An important source of steroids for the synthesis and production of pharmacologically active steroids used in contraceptives, corticosteroids, geriatic drugs etc. is diosgenin (Fig. 26.1) from *Dioscorea* spp. Due to the shortage of diosgenin, interest has shifted to more abundant sterols from phytosterols (i.e. sterols from plants) and cholesterols from animals. The phytosterols include soy bean sterols mainly β-sitosterol and stigmasterol and tall oil sterols mainly sitosterol and campesterol. For these to be used as starting materials for the production of progesterone and other drugs, the C_{17} side chain must be cleaved hence the interest. The microbial removal of the side chain offers more promise than chemical means. Unfortunately micro-organisms which cleave off the side-chain will also attack the D ring to which the chain is attached. Three methods have therefore been evolved to solve the problem of inhibiting ring degradation, while cleaving the chain.

(i)   The substrate may be modified structurally by chemical means so that the ring is stable while the side-chain is cleaved. Thus while cholesterol rings are degraded when the side-chain is cleaved by *Nocardia* sp. 3-Acetoxy-9- hydroxy-5-cholestene is not. This later compound can be prepared from cholesterol by three chemical steps. The cleavage of the side-chain of cholesterol yields esterone which can then be used for further transformations.

(ii) The enzymes which open the D nucleus may be selectively inhibited. The key stage in the opening of the ring is at the ninth position and since the enzyme for this hydroxylation contains metals, the enzyme and its process may be inhibiting by using chelating agents which remove metals from them.

(iii) Finally, mutants have been developed which will degrade only the side chain. One of the best known is a mutant of *Mycobacterium* sp.

26.3.2   Fermentation Conditions Used in Steroid Transformation

The media used are highly variable, but in the main are not very complex. They are basically mineral salts media containing some carbon source such as glucose, dextrin or glycerol. Nitrogen sources may be ammonium salts, corn steep liquor, soybean, or a protein digest. In some cases yeast extract is added.

Steroid and sterols are lipids; they are not water soluble and therefore must be dissolved in a water-miscible lipid-solvent. Acetone, ethanol, propylene glycol, and methanol are suitable because they dissolve a reasonable amount of the steroid while being relatively non-inhibitory to the enzymes; dimethyl formamide dissolves a reasonable amount of the steroids but has only a minimum of toxicity. Sometimes the steroid is added in small amounts at a time. In this way, any toxic effect of the solvent is minimized.

The level of steroid added is variable and depends both on the transforming ability of the organisms as well as its susceptibility to the toxic effects of the steroid. Normally 200-800 mg/litre are added but much higher amounts are sometimes used. To solve the problem of the insolubility of steroids in water, non-ionic surface-acting agents which reduce surface tension e.g. Tween 80 are often added to the medium. Some poly-saccharides in the medium e.g. yeast cell wall mannan, bind to the steroids and cause them to be more available to the organism.

A wide range of microorganisms, mainly fungi and bacteria, are used in the transformation of steroids. Some of these include the fungi *Rhizopus nigricans, Curvularia lunata, Fusarium* spp. *Cylindrocarpon radicicola* as well as the bacteria *Mycobacterium* spp., *Corynedbacterium simplex,* and *Streptomyces* spp. As has been mentioned, there are organisms to perform just about any conceivable transformation of the steroid molecule. The transformation may occur at different stages of the growth and the steroid may be added to the growing cultures either simultaneously with the inoculation of the culture or the resting or stationary stage of the organism. Fungal spores may sometimes be inoculated as the steroid is introduced into the medium. In recent times immobilized cells have been employed in the transformations of steroids.

Steroid transformations require vigorous aeration and a temperature of about 28°C is usually employed. The fermentation is usually complete in four to five days.

26.4 SCREENING FOR MICROORGANISMS

The screening for microorganisms capable of transforming steroids to yield products of useful pharmacological properties is a continuing one. The processes which are followed in the screening are as follows:

(i) The microorganism is isolated from soil or some suitable source and grown in a suitable medium for 24-28 hours.

(ii) The steroid in a suitable carrier is added to the fermentation and the growth continues for a further period which could be as long as one week.

(iii) The transformation products are extracted with solvents such as methyl acetate and purified by chromatography etc.

(iv) The product is tested for pharmacological properties.

(v) Finally, the structure is elucidated by classical methods of organic chemistry.

SUGGESTED READINGS

Flickinger, Michael C., Drew and Stephen W. 1999. Encyclopedia of Bioprocess Technology - Fermentation, Biocatalysis, and Bioseparation Wiley. Electronic ISBN: 1-59124-457-9.

Martin, C.K.A. 1984. Sterols. In: Biotechnology. Kiesich (ed) Vol 6A Biotransformations Verlag Chemie. Weinheim: Germany. pp. 79–96.

Morgan, B.P., Moynihan, M.S. 1997. Steroids. Kirk-Othmer Encyclopedia of Chemical Technology, 2, 71-113.

Smith, L.L. 1984. Steroids. In: Biotechnology. Kiesich (ed) Vol 6A Biotransformations Verlag Chemie. Weinheim Germany: pp. 31-78.

CHAPTER 27

Vaccines

27.1 NATURE AND IMPORTANCE OF VACCINES

Vaccines are materials which when introduced into the human body help protect the vaccinated person against specified communicable diseases. Communicable diseases are diseases caused by microorganisms, including viruses. Vaccines are preparations of dead or weakened pathogens, or their products, that when introduced into the body, stimulate the production of protective antibodies or T cells without causing the disease.

Vaccination is also called **active** immunization because the immune system of the body is stimulated to actively develop its own immunity against the pathogen. **Passive immunity**, in contrast, results from the injection of antibodies formed by another animal (e.g., horse, human) which provide immediate, but temporary, protection for the recipient.

The name 'vaccine' comes from the Latin *vacca* (for cow). This is because the earliest vaccination was done using the *cow* pox virus (which causes the disease in cow) as a vaccine against small pox in humans. The English physician, Edward Jenner carried out the above vaccination in the late 18th century and published his paper in 1798.

Over the past 200 or so years vaccination has contributed greatly to reducing morbidity and mortality from communicable diseases. The greatest triumph of vaccination is the eradication of smallpox from the earth; no naturally-occurring cases has been reported since 1977. A program to try to eliminate another virus disease, poliomyelitis (polio for short), from the world has been on for some time and the indications are that the number of cases has drastically dropped. Except for the few cases caused by oral polio vaccine (OPV) (see below), in which the live virus reverts, the disease has now been eliminated from the Western hemisphere. Outbreaks of polio still occur in Africa, the Indian subcontinent, and parts of the Near East. Due to the success of vaccination near 100% reduction has been obtained in the cases of many diseases which were previously sources of great mortality and morbidity. These include diphtheria, measles, mumps, pertusis, rubella and tetanus. Table 27.1 gives a list of the most commonly used vaccines today.

27.2 BODY DEFENSES AGAINST COMMUNICABLE DISEASES

In order to better understand the nature of vaccines and their design and production, it is important that the defenses of the human body against communicable diseases be

Table 27.1 Vaccines most commonly used in the world

Disease	Preparation	Notes
Diphtheria	Toxoid	Often given to children in a single preparation (DTP; the
Tetanus	Toxoid	'triple vaccine') or the now-preferred DTaP using acellular pertussis
Pertussis	Killed bacteria ('P') or their purified components (acellular pertussis = 'aP')	
	Inactivated virus previously grown on monkey or human diploid cells	Inactivated polio vaccine: IPV (Salk)
Polio	Attenuated virus inactivated virus previously grown on monkey or human diploid cells	Oral polio vaccine; OPV (Sabin) Both vaccines trivalent (types 1, 2, and 3)
Hepatitis B	Protein (HBsAg) from the surface of the virus	Made by genetic engineering
Diphtheria, tetanus, pertussis, polio, and hepatitis B	uses acellular pertussis and IPV (Salk)	Pediarix®; combination vaccine given in 3 doses to infants
Measles	Attenuated virus	
Mumps	1 Attenuated virus 2 Vaccine: Live in duck cells	Often given as a mixture (MMR) Does **not** increase the risk of autism. (Nor do any vaccines containing thimerosal as a preservative.)
Rubella	Attenuated virus Pig, chick embryo or canine tissue-culture grown	
Chickenpox (Varicella)	Attenuated virus	Caused by the varicella-zoster virus (VZV)
Influenza	Egg-grown virus, formalin inactivated, highly purified by zonal ultracentrifugation Hemagglutinins	Contains hemagglutinins from the type A and type B viruses recently in circulation
Pneumococcal infections	Capsular polysaccharides	A mixture of the capsular polysaccharides of 23 common types. Works poorly in infants.
	7 capsular polysaccharides conjugated to protein	Mobilizes helper T cells; works well in infants.
Staphylococcal infections	2 capsular polysaccharides conjugated to protein	To prevent infection by *Staph. aureus* in patients hospitalized and/or receiving dialysis

Contd.

Table 27.1 Contd.

Disease	Preparation	Notes
Meningococcal disease	Polysaccharides	Used chiefly to prevent outbreaks among the military
Hemophilus influenzae, type b (Hib)	Capsular polysaccharide conjugated to protein	Prevents ear infections in children
Hepatitis A	Inactivated virus	Available in single shot with HBsAg (Twinrix®)
Rabies	Inactivated virus *Active*: (1) β-propiolactone-inactivated virus grown in embryonated duck eggs (2) phenol-inactivated virus grown in rabbit brain *Passive*: equine hyper-immune serum	Vaccine prepared from human diploid cell cultures (HDCV) has replaced the duck vaccine (DEV)
Smallpox	Attenuated live virus: attenuated by passing through calves	Despite the global eradication of smallpox, is used to protect against a possible bioterrorist attack
Anthrax	Extract of attenuated bacteria	Primarily for veterinarians and military personnel
Typhoid	Three available: 1. killed bacteria 2. live, attenuated bacteria (oral) 3. polysaccharide conjugated to protein	
Yellow fever	Live attenuated virus Prepared in chick embryo: Dakar strain or 17D strain	
Tuberculosis	Live attenuated mycobacterium BCG (Bacille calmette Guerin) strain (BCG)	Rarely used in the US

discussed briefly. Ordinarily the human body is surrounded by microorganisms: in the air it breathes, the water it drinks, in the soil around it and on the clothes he wears. Most of these are not normally pathogenic. But even the pathogenic ones do not always cause disease when they come in contact with the human body because the body has evolved ways of dealing with microorganisms and preventing them from causing disease, collectively known as the immune system. The immune system is a complex network of cells and organs which work together to protect the body from communicable diseases. It has two components: the innate or non-specific immunity and the acquired or specific methods. While the innate immunity eliminates the organism no matter the type, acquired or specific immunity specifically recognizes and selectively eliminates the microorganism or foreign molecule.

27.2.1 Innate or Non-specific Immunity

The innate or non-specific defense mechanisms are the first line defense against invading microorganisms. They will act irrespective of the type of microorganism. Briefly they consist of the following:

(a) *Anatomic barriers*: these include mechanical barriers such as skin, which physically keeps out microorganisms and mucous membranes of the alimentary canal respiratory and urinogenital tracts which entrap microorganisms. In addition the mucous membranes harbor a normal set or flora of microorganisms which keep out foreign organisms.

(b) *Physiologic barriers*: The physiology of the human body keeps out some pathogens. Thus the high temperature of the human body, including the fever response keeps out some microorganisms as does the acidic nature of the stomach. Chemical mediators such as lysozyme found in tears breakdown bacterial cell walls.

(c) *Phagocysis and endocytosis*: white blood cells kill and digest whole micro-organisms, while specialized cells engulf and breakdown foreign particles.

(d) *Inflammatory responses*: Tissue damage and infection induce leakage of vascular fluid serum protein with antibacterial fluid and influx of white blood cells leading to pus formation.

27.2.1.1 Acquired or Specific immunity

Acquired or specific immunity has three important properties among others, which are crucial in understanding vaccines and how they function.

 (i) **Antigenic specificity**: An antigen is a material usually a protein which binds specifically to an antibody or to T-cell receptor (see below). Great specifity occurs in the anatigen-antibody or antigen-T-cell receptor relations. Often a small difference of a single amino acid can decide whether or not binding to antibody or T-cell will take place.

 (ii) **Immunologic memory**: Once the acquired immune system has recognized and responded to an antigen, it exhibits immunologic memory: a second encounter with the same antigen induces an increased response.

(iii) **Self/non-self recognition**: Specific immunity recognizes foreign bodies in contrast to those of the body and seeks to destroy the intruders. In rare cases the system of recognition breaks down and the system fails to recognize body cells and proceeds to destroy them, giving rise to auto-immune diseases.

Specific immunity has two components, humoral and cell-mediated immunity which are mediated by white blood cells known as lymphocytes, and as will be seen below both sectors are linked. Humoral immunity is mediated by **B Lymphocytes**, while cell-mediated immunity is brought about by **T Lymphocytes**. White blood cells including lymphocytes and like red blood cells are produced from stem cells in the bone marrow. The B lymphocytes, remain in the bone marrow to mature, while T lymphocytes mature in the thymus, a small organ located above the heart.

27.2.2.2 Specific immunity: humoral immunity

Humoral immunity is also known as antibody immunity. Antibodies are soluble proteins in the blood which bind to foreign agents and mark them for destruction, or neutralize toxins produced by microorganisms. Also known as immunoglobulins, antibodies are glycoproteins by nature (i.e. proteins to which carbohydrates are conjugated). (see Fig. 27.1)

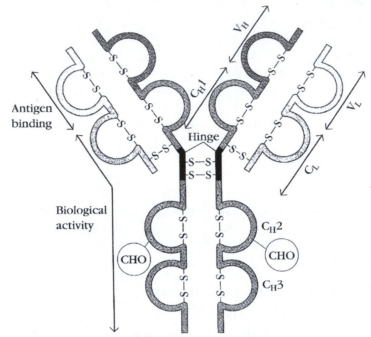

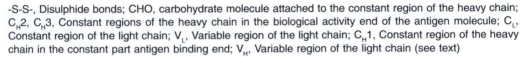

-S-S-, Disulphide bonds; CHO, carbohydrate molecule attached to the constant region of the heavy chain; C_H2, C_H3, Constant regions of the heavy chain in the biological activity end of the antigen molecule; C_L, Constant region of the light chain; V_L, Variable region of the light chain; C_H1, Constant region of the heavy chain in the constant part antigen binding end; V_H, Variable region of the light chain (see text)

Fig. 27.1 General Structure of an Antibody Molecule

When B lymphocytes mature each has one unique antigen-binding molecule, an antibody attached to its membrane; up to 10 different antibody molecules may be carried on the B lymphocytes. When such a mature B lymphocyte which has not encountered any antigen, known as a naïve lymphocyte, encounters an antigen for which its membrane bound antibody is specific, it begins to divide rapidly and differentiates into two types of cells: **memory cells** and **plasma cells**. The memory cells have a longer span of life and continue to express membrane-bound antibody just like the original parent cell. The plasma cells, on the other hand live for four to five days and do not have cell-membrane bound antibodies; instead they produce antibody in a form in which it can be secreted, often in huge amounts, sometimes reaching 2,000 molecules per second. The memory cells are the source of the long-term protection which vaccines confer.

Antibodies (immunoglobulins) are Y- shaped and consist of two identical light chains and two identical heavy chains (Fig. 27.1). The upper end of the Y of the light and heavy chains of the antibody molecule is the variable region. The amino acids in this region vary greatly among different antibodies and this variability confers on antibodies the vast specificity for which they are known. The lower ends of the light and heavy chains are the 'constant' regions and do not show the variability found at the tips of the Y. The variability in protein composition in the 'constant' region of the heavy chains (Fig. 27. 1) leads to antibodies being divided into five major classes, each with a different and distinct property: IgG, IgA, IgD, IgM, and IgE. IgG is the most abundant (80%) of all Igs, and it is the only one able to cross the placenta, helping to confer maternal immunity on the newborn.

An antibody recognizes an antigen in a **specific** manner and the immune system acquires memory towards it. The first encounter with an antigen is known as the **primary response**. Re-encounter with the same antigen causes a **secondary response** that is more rapid and powerful. This is the basis on which vaccines function; they induce the memory lymphocytes to proliferate and the resulting plasma cells to produce soluble antibodies (Fig. 27.2).

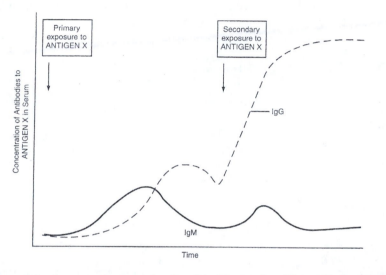

Antibodies are proteins known as immumoglogulins (Ig). The are five different kinds of immunoglobulins IgA, IgD, IgE, IgG, and IgM. The animal body produces antibodies when challemged with materials to which the body can react by producing antibodies (known as antigens). When the animal body is challenged with the same antigen a second time the production of antibodies is not only produced in a shorter time, but the antibody production is more pronounced as shown in the figure above. In the figure above IgM is produced in the first challenge and IgM in the second. (see text).

Fig. 27.2   Antibody Response of the Animal Body to a Second Challenge of an Antigen

27.2.2.3   Specific immunity: cell-mediated immunity

While B lymphocytes mediate antibody or humoral immunity, T lymphocytes are responsible for cell-mediated immunity. T lymphocytes do not have cell membrane bound antibodies, nor do they secrete antibodies. Instead they have **T-cell receptors (TCRs)**. Unlike antibodies which can recognize antigens directly, T-cell receptors can recognize an antigen only if the antigen is associated with cell membrane proteins known as major histocompatibility compatibility (MHC) molecules, of which two classes exist: MHC I and MHC II.

When a naïve T cell encounters an antigen associated with an MHC on a cell, the T cell proliferates and differentiates into **memory T cells, T helper cells (TH)**, and **T cytotoxic cells (TC)**. T_H and T_C cellscarry on their membranes different glycoproteins. T_H cells carry glycoprotein **CD4**, while T_C cells carry **CD8**.

When a T_H cells interacts with an antigen linked to an MHC II compound, it is activated to produce cytokins which also activate B cells to produce antibodies. Cytokins also activate T_C cells when they interact with an antigen linked to an MHC I compound, to differentiate into cytoxic T lymphocytes (**CTLs**). CTLs do not secrete cytokins; instead they monitor the body cells and eliminate any cells which are foreign or contain foreign bodies such as cancer cells or cells containing viruses. To ensure that self cells are not attacked by CTLs, they attack only cells displaying foreign foreigns complexed to an MHC molecule on the surface of cells called antigen presenting cells (**APCs**). Antigen presenting cells adsorb foreign antigens such as viruses, digest them and display the peptides from them on their surfaces. CTLs identify such cells and destroy them. APCS are specialized white blood cells. The relationships between B lymphocytes, T lymphocytes, cytokins, Tc cells and CTLs are depicted in Fig. 27.3. The cell-mediated immune response is important in cases where the pathogen is intracellular as in viruses

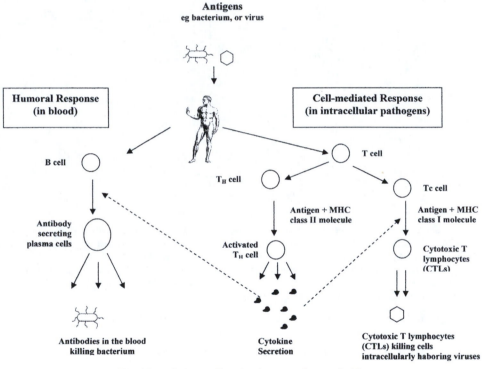

Fig. 27.3 Scheme Showing Immune System in Man

27.2.2.3 Antigens and Epitopes

Antigens are macromolecules that elicit an immune response in the body. Antigens can be proteins, polysaccharides or conjugates of lipids with proteins (lipoproteins) and polysaccharides (glycolipids). Antigens are generally very large and complex and the lymphocytes may not recognize all the sites of a particular antigen. Rather both B and T lymphocytes recognize discreet sites on an antigen known as *epitopes* or *antigenic determinants*. The aim in vaccine production is to ensure that epitopes exist on the vaccine which will elicit humoral or cell-mediated response.

27.3 TRADITIONAL AND MODERN METHODS OF VACCINE PRODUCTION

Traditionally three types of vaccines have been used: attenuated live vaccines, killed vaccines and bacterial toxoids. Recent advances in molecular biology and genomic science have spilled over into vaccine production. This chapter discusses the traditional vaccines, but will also discuss the newer approaches which have been influenced by advances in molecular biology and genomic science.

27.3.1 Traditional Vaccines

27.3.1.1 Live attenuated organisms

In live attenuated vaccines, the organism has been cultured so as to reduce its pathogenicity, but still retains some of the antigens of the virulent form. They consist of the living pathogens whose virulence has been reduced (attenuated) by passaging them through hosts different from the usual. Alternatively, non-virulent strains of the pathogen may be used.

Live agents may be used for one or more of the following reasons:

(i) When the protection-inducing substance is produced as a diffusible product of metabolizing organisms e.g. *Bacillus anthracis.*

(ii) When it is not feasible to produce sufficient amounts of nonviable agents and a small concentration of the living agent can propagate within the vaccinated subject to overcome the deficiency.

(iii) When immunity is induced by the modification of parasitized cells.

Live vaccines in use include those against polio (Sabin oral polio vaccine - OPV), foot and mouth disease of farm animals, mumps, measles, rubella (German measles), tuberculosis, rabies and yellow fever. For tuberculosis the vaccine is derived the Bacillus Calmette-Guérin (BCG) strain of *Mycobacterium tuberculosis*, a weakened version of the bacterium that causes tuberculosis in cows. BCG is used as a vaccine against tuberculosis in many European countries; it is however not commonly used in the U. S.

The OPV has advantages and disadvantages when compared with the (inactivated) Salk polio vaccine (IPV). OPV can be given by mouth rather than by injection, and it can spread to the other members of the vaccinee's family thus immunizing them as well. Its disadvantage is that on rare occasions, the virus regains full virulence and cause the disease. On account of this, the Salk vaccine has gained prominence over the Sabin vaccine in some countries.

27.3.1.2 Killed vaccines

These consist of suspensions of fully virulent organisms (bacteria or viruses) killed as mildly as possible in order not to destroy the antigenic determinants on the organism. Killing can be achieved by heat, (usually about 60°C for 1 hour) chemicals (phenol, alcohol, formalin, β-propiolactone) or ultraviolet irradiation. Killed vaccines do not provide as prolonged antigenic stimuli as living vaccines and two, three or more subcutaneous injections are required to give adequate protection. Examples of killed vaccines include TAB vaccine against typhoid fever and which consists of heat-killed phenol-preserved suspension of *Salmonella typhii* and *Salm. Paratyphii* A & B, whooping cough, cholera, and the Salk IPV.

27.3.1.3 Bacterial toxoids

Toxoids are inactivated bacterial exotoxins. The toxins from *Clostridium botulinum*, *Clostridium tetani* and *Corynebacterium diphtheriae* are inactivated by treatment in formalin. Toxoids induce antibody production when injected into the body, although they are themselves harmless. In some diseases, of which diphtheria and tetanus are good examples, the bacterial metabolite, a protein toxin which they liberate, is the cause of the disease and not the bacteria themselves. Exposing the toxin with formaldehyde, denatures the protein. However, some epitopes on the protein molecule are retained and they elicit antibody production.

27.3.2 Newer Approaches in Vaccinology

The advent of genomics, proteomics, and biotechnology, as well as the increased understanding of pathogenesis and immune responses to various pathogens have led to the development of safer, more effective and cheaper vaccines. Some of these are described below.

27.3.2.1 Sub-unit or surface molecule vaccines

Subunit vaccines contain antigens or epitopes that induce protection rather than the whole organism. The materials usually come from the surface of the organism and hence they are also known as surface molecule vaccines. The potential advantages of using subunits as vaccines are the increased safety, less antigenic competition, since only a few components are included in the vaccine. One of the disadvantages of subunit vaccines is that they generally require strong adjuvants and these adjuvants often induce tissue reactions. (Adjuvants are compounds administered with vaccines so as to increase the immunogenicity of the vaccines.) Second, the duration of immunity is generally shorter than with live vaccines. Sometimes peptide epitopes may be used. Apart from requiring adjuvants, a pathogen can escape immune responses to a single epitope; hence several peptides linked together are used to broaden the immune response to different epitopes. Subunit vaccines are currently available for typhoid and whooping cough. Several vaccines employ purified surface molecules. One of them, the influenza vaccine contains purified hemagglutinins from the viruses currently in circulation around the world. Another example is vaccine against hepatitis B virus. The gene encoding a protein expressed on the surface of the virus, the B surface antigen or HBsAg, can now be expressed in *E. coli* cells and provides the material for an effective vaccine; hepatitis B infection is strongly associated with the development of liver cancer. For the vaccine against *Streptococcus pneumoniae* which causes pneumonia in humans about 80 different strains of the organism are used. They differ in the chemistry of their polysaccharide capsules which surround them and the current vaccine consists of purified capsular polysaccharides of the 23 most common strains.

27.3.2.2 Conjugate vaccines

These are similar to subunit vaccines in the sense that only a part of the organism is used in making the vaccine. Some bacteria which are encapsulated cause important childhood diseases such as septicemia, pneumonia and meningitis. The bacteria are *Hemophilus influenzae* type B (HiB), *Neissseria meningitides* and *Streptococuus pneumoniae*. The capsules of these bacteria are made of carbohydrates which the immune system of adults recognize as foreign, but which that of infants do not and hence cannot make antibodies against them. To solve the problem protein from diphtheria or tetanus toxoids is linked or conjugated to the carbohydrate to make a vaccine. This enables a baby's immune system to respond to the combined vaccine and produce antibodies, initiating an immune response against the disease-causing organism. The licensed conjugate vaccines against *Haemophilus influenzae* type b (Hib), previously the major cause of bacterial meningitis in babies and young children, have virtually eliminated the disease in the United States.

27.3.2.3 Other (Experimental) vaccines

(i) **Polynucleotide (DNA) Vaccines**

A recent development in vaccinology is immunization with polynucleotides. This has been referred to as genetic immunization or DNA immunization. The rationale for this is that cells can take-up DNA and express the genes within the transfected cells. Thus, the animal body itself produces the vaccine. This makes the vaccine relatively inexpensive to produce. Some of the advantages of polynucleotide immunization are that it is extremely safe, induces a broad range of immune responses (cell-mediated and humoral responses), long-lived immunity, and, most importantly, can induce immune responses in the presence of maternal antibodies. Most recently, it has also been used for immunizing animal fetuses. Thus, animals are born immune to the pathogens and at no time in the animal's life are they susceptible to these infectious agents. Although an attractive development, there is a great need to develop better delivery systems to improve the *in vivo* efficiency.

(ii) **Edible Vaccines**

Edible vaccines can safely and effectively trigger an immune response against the *Escherichia coli* bacterium and the Norwalk virus. Attempts are being made to genetically engineer potatoes, bananas, and tomatoes that, when eaten, will initiate an immune response against harmful intestinal bacteria and viruses.

27.3.2.4 Reverse vaccinology

The name reverse vaccinology mimics the established term reverse genetics - meaning the process of identifying a protein or enzyme through its gene product. Despite the numerous successful vaccines produced over the last 200 years of vaccinology and the advances made in vaccine discovery techniques, there are still problems with existing approaches. Traditionally, the initial point in preparing a vaccine is to grow the pathogen, which could often be very fastidious and difficult to grow, then the antigens had to be identified one by one. Sometimes many years might be spent just working one antigen, and traditionally only the most abundant and the easiest to purify and identify (and not necessarily the best) have been studied.. To begin with however, not all pathogens will successfully grow *in vitro* as is required for conventional methodologies, with Hepatitis B and C viruses being prime examples of such organisms. With reverse vaccinology the organism need not even be seen; the genomic constitution of the organism already in databases need only to be accessed *in silico*. Reverse vaccinology utilizes the wealth of information provided by genome sequencing to identify and characterize a whole host of new vaccine targets.

The technology has two major facets, *in silico* and *in vitro/vivo*. The *in silico* aspect is the identification, annotation and then localization of ORFs and their products. Identified targets can then be used for laboratory study (*in vitro/in vivo*) where they are expressed, purified and tested for immunogenicity. Genome-based vaccine discovery was applied for the first time to serogroup B meningococcus, a bacterium which is a major cause of sepsis and meningitis, that had defied all traditional approaches to vaccine development. The *in silico* sequence of the genome predicted 600 potential antigens. Of them 350 were expressed in *Escherichia coli,* purified and used to immunize mice. Twenty nine were found to induce bactericidal antibodies, which will lead to protection. A subgroup of the genome-derived antigens is now being tested in clinical trials. Reverse vaccinology is now a standard technology. Vaccines projects are not now undertaken without knowledge of the the sequence of the pathogen. Successful examples of genome-based vaccine discovery are pneumococcus, group B streptococcus, *Staphylococcus aureus,* and a variety of viruses.

27.3.2.5   Some definitions of terms used in dealing with vaccines

  (i) *Active natural immunity* is immunity arising from a natural disease. Small pox for example, is suffered once in a life-time because of the immunity conferred on the sufferer by the disease, due to antibodies produced during the illness.

 (ii) *Active artificial immunity* is due to the use of vaccines, toxoids, etc., which stimulate the individual to produce his own antibodies.

(iii) *Passive natural immunity* is most easily exemplified by maternal immunity in which new-born animals are immune from certain disease for a short period early in their lives due to the crossing of the placenta by certain antibodies (immunoglobulins). In some animals including man maternal immunity is also acquired by the young's consumption of colostrum (thick cream-colored milk produced during the first few days after childbirth.

(iv) *Passive artificial immunity* occurs when ready-made antibodies are introduced into the body. An example is the use of anti-tetanus serum in which serum from a horse which has been immunized against tetanus is used to protect an individual against tetanus.

27.4   PRODUCTION OF VACCINES

27.4.1   Production of Virus Vaccines

Viruses multiply only in living cells. Viruses to be used for vaccine production must therefore be grown in such cells. In practice they are grown for vaccine purposes in tissue cultures which will first be described briefly below.

27.4.1.1   Tissue cultures and their cultivation

The growth of animal cells *in vitro* in monolayers is known as tissue or cell cultures. Tissue cultures will be discussed briefly here because they are used in an area of industrial microbiology which deals with the manufacture of biological materials used in pharmacy, human medicine and veterinary practice. Such biologicals include vaccines, interferon, hormones, immunological reagents and cellular biochemical such as insulin, enzymes, plasminogen, and plasminogen activators.

(i) **Cells used**: The cells used in tissue cultures for the production of biolgocials are derived from three sources:

(a) *Primary cells* are obtained by treating certain tissues derived from healthy animals with disaggregating enzymes such trypsin and transferred for the first time into an *in vitro* growth environment. Such tissues include decapitated avian embryo, kidneys from virus-free green monkeys widely use for many biologicals (including polio vaccines), rabbit kidney (for rubella and vccinia), calf kidney (for measles in Japan).

When embryo of chicks or ducks are used, the avian embryos are harvested from eggs, obtained from special isolated flocks, and then minced and treated with disaggregating enzymes such as trypsin, collagenase, hyaluronidase, and pronase. The fibroblast cells released by this process are attachment dependent, requiring solid surfaces for growth. The successful commercial manufacture of viral vaccines in attachment-dependent cell systems rely on the establishment and maintenance of healthy cell monolayers. The appropriate growth and maintenance media must be carefully selected, and careful attention must be paid to nutrient depletion, waste accumulation, and changes in pH over time.

To produce measles and mumps vaccines, those viruses are grown and attenuated by passage through cultures of chicken embryo cells. For foot-and-mouth disease which affects farm animals a wide range of primary cells are used: bovine kidney, goat heart, and lung, skin and kidney of camels.

(b) *Diploid cells* are obtained from well-defined human cell lines. Serially passaged diploid strains of cultured human cells were first described in the 1960s and at the concept of using human diploid cell lines for vaccine preparation was hotly debated for fear that cancer-causing DNA or human viral agents might be unknowingly co-administered with the vaccine. Extensive karyological (i.e., chromosomal) characterization and thorough searches for viral contaminants were necessary to ensure that cultures were free of exogenous infectious agents before the first diploid cell product, poliomyelitis vaccine, was licensed in the United States. Such diploid cell lines must have shown a karyology or chromosome characteristics identical with the parent tissue, be free from bacterial, viral or fungal contaminants.). Human diploid cell lines (WI-38 or MRC-5) have since been used to produce a number of licensed vaccine products against poliovirus, adenovirus types 4 and 7, rubella (German measles) virus, rubeola (measles) virus, rabies, hepatitis A, and varicella virus (chickenpox).

(c) *Established cell lines* include those which are capable of growth for an indefinite number of passages. They are used for veterinary rather than human vaccines. A good example is the baby hamster kidney cells. A distinguishing feature of the human diploid cells such as WI-38 or MRC-5 is that they have a finite life span, reaching senescence after 40 to 60 population doublings. In contrast, continuous cell lines exhibit no such constraints and divide indefinitely. An example of a nonhuman continuous cell line that has been used successfully for the manufacturing of vaccines such as polio, rabies, and influenza is Vero. Vero cells are a continuous monkey kidney cell line. Serially cultured or continuous cell lines are advantageous in that each new production batch is derived from a uniform master cell bank, characterized to be free of contaminating infectious agents, contaminating proteins, or nucleic acids. Proper maintenance of a master cell bank is critically important for the consistency of cell culture products.

(ii) **Medium for tissue culture**: Various media are available for the cultivation of cells. They consist essentially of inorganic salts, amino acids, vitamins, nucleotides, and low molecular growth factors such as hormones, steroids, and fatty acids. Another important component is serum and this is often used in conjunction with peptones, tryptic digests and albumin hydrolysates. A mixture of antibiotics is also often added to remove contaminants.

(iii) **Cell Culture fermentors**: Conventionally, animal cells grow on the surface of the glass containers in which they are cultured. Cylindrically shaped roller bottles have been used successfully to establish monolayers of chick fibroblasts on their inner surfaces, which can be monitored microscopically through the clear plastic. The cell sheets are continuously bathed by a growth medium contained within the bottle during slow axial rotation on special roller racks. After the cell sheets are established, they are infected by introduction of the specific virus. After incubation, the cells and virus may be harvested (e.g., varicella). In the case of rubella, viral fluids may be harvested at approximately two- to three-day intervals, through 10 to 12 harvest cycles. Although millions of doses have been successfully manufactured using roller bottles, capacity is limited by the space that a large number of roller bottles requires and by the time-consuming and often labor intensive manipulations for harvesting and pooling the viral fluids from them. Multidisk reactors offer an alternative to roller bottles for attachment-dependent cell lines. They consist of 10-L stainless steel reaction vessels containing approximately 100 parallel titanium disks that slowly rotate through growth media and provide solid surfaces for cell attachment and monolayer formation. For the production of vaccines and other biologicals many thousands of bottles are stacked together. The tendency in recent times is to develop large units of up to 1,000 liters in many small units.

In summary, because cells still require surfaces for growth even on such a large scale, various arrangements are used. In some fermentors, plates or discs made of plastics, glass or metal are supported with a central frame, and which are bathed with a sterile tissue culture solution. The other consist of packed beds of plastic or glass materials over which the medium flows; cells adhere to the surfaces of the support material. In yet others a bank of roller tubes through which medium is circulated support growth on the tubes' internal surfaces.

(iv) **Cell harvest**: Cells may be harvested by the use of trypsin (or other proteolytic enzymes such as papain or pronase), by the use of chelating agents e.g. EDTA, by physical scraping off or a combination of one or more of these methods.

27.4.1.2 Production of salk polio vcaccine

The production of salk polio vaccine will be discussed as an example of the production of a virus vaccine. The cells of the kidneys of rhesus monkeys are caused to separate into individual members by treatment with trypsin. The suspension of cells is then distributed in shallow containers, and covered with a suitable medium. The cells have a tendency to adhere to glass and incubation of the cultures at 37°C for four to six days permits a confluent growth of a monolayer of cells. The culture fluid is removed and is replaced by maintenance medium, which contains no protein as subjects using the vaccine may react adversely to protein if this is present.

Live virus are inoculated into the tissue culture medium and incubated at 37°C for four days. The viruses lyse the kidney cells in a manner characteristic of the particular virus and which is described as the cytopathic effect. The viruses are harvested by centrifuging to remove cell debris. They are then inactivated by treatment with formalin. The success of inactivation is checked by injection into embryonated chick egg or in experimental animals. The inactivating agent is removed before the virus is stored at 4°C sometimes with addition of glycerol.

27.4.2   Production of Bacterial Toxoids

Many clostridia (Gram-positive spore-forming anaerobic rods) cause disease in man and animals by the production of exotoxins. Some examples include:

Cl. tetani	-	tetanus
Cl. botulinum	-	botulism
Cl. welchii	-	gas gangrene

Aerobes may also cause disease by the production of exotoxins e.g. *Corynebacte-rium*. It is possible to collect the toxins so produced in cell-free extracts from *in vitro* cultivation. The toxin can then be inactivated by treatment with formaldehyde. Such inactivated toxin known as a toxoid is antigenic and is able to cause the body to produce antibodies to the original toxin. Toxoids are non-toxic, and are used to artificially induce active specific immunity.

In industrial practice toxoids to clostridial toxins are prepared by inactivating toxins produced from the clostridia grown in large fermentors under anaerobic conditions.

The media used usually consist of hydrolysates of proteins from horse meat, and are sterilized at 15 p.s.i. Nowadays synthetic media containing inorganic components are preferred as toxins therein are easier to isolate from such fermentations. Since the fermentation is anaerobic, satisfactory growth and toxin production are easily obtained in deep fermentors. The only agitation required is to provide uniform temperature. Nitrogen is also blown through to flush away oxygen from the system.

At the end of the incubation, the bulk of the bacterial cells is harvested by centrifuging. The supernatant is further filtered through bacterial filters, before the toxin present therein is converted to toxoid. This is done by incubating the filtrate at 37°C in contact with formalin. The inactivation is tested from time to time by injection into animals.

Protein from the medium is removed by precipitation with ammonium sulphate at 4°C. Excess $(NH_4)_2SO_4$ is removed by dialysis, the product is filter-sterilized and diluted to final strength with buffered saline.

27.4.3   Production of Killed Bacterial Vaccines

A suspension of the organism (usually produced by scrapping from surface cultures on agar) is washed thoroughly with centrifugation. It is then killed usually with heat. For non-spore-forming bacteria, treatment at 60°C for half hour is usually enough. The efficiency of the killing is tested by streaking the killed cells on agar. The density of cells needed to immunize laboratory animals is worked out in experimental animals and that needed for man is obtained by proportion by relating number to weight in both man and animals.

27.5   CONTROL OF VACCINES

(i) **Stringency of standards**: Vaccines produced for man's use must conform to standards laid down by different countries, and manufactures must conform to them. The World Health Organization is also interested in ensuring high standards and has setup its own. The controls are stringent and designed to ensure (a) that the material is potent; (b) it is safe, (c) it will not give rise to unpleasant or undesirable side effects. The control of vaccines is stringent, time-consuming and very expensive because of the expertise involved. Indeed a good deal of the cost of the vaccine is due to the expenses incurred in the tests. No vaccine can be considered ready for use for the health of the general public until it has been extensively tested and conform to the standards of the WHO.

Live vaccines, whether of bacteria or viruses are, understandably more generally stringently tested than killed vaccines or toxoids. Thus, while the killed vaccines/toxins for the bacterial diseases dipheria, tetanus, cholera, typhoid fever and pertussis are merely tested for sterility, toxicity and potency, the live vaccines for tuberculosis, BCG, are tested for contaminations, virulent organisms, identity, skin reactivity, viable counts and stability.

One reason for the stringency of the testing of virus vaccines is that when polio virus was grown on monkey tissue kidney, it was soon found that these monkeys themselves harbored a large number of viruses. Laboratory-grown animals were therefore resorted to, including ducks, chicken, rabbits, or dogs. Even though these contained fewer viruses, the tests had to be gone through because any viruses in the substrate could find its way into man.

(ii) **Potency and field potency testing**: These are based on the ability of the vaccine or toxoid to immunize animals against a lethal, paralytic, skin-test or intracerebral (in the case of pertusis) challenge. In potency testing a set of test animals are protected with the vaccine while the control is not. In a potent vaccine the disease should appear only in the control when the animals are inoculated. A large effort in vaccine production is devoted to potency testing. *Field trials* must be carried out on the vaccine, but this is usually carried out by government regulatory agencies rather than the manufacturer, although he is usually consulted during the tests. The manufacturer usually recommends any adjuvants which may be found necessary. Adjuvants are immunological enhancers and have the following qualities:

   (i)   A smaller quantity of antigen can be used in single and combined vaccines.
   (ii) Owing to the reduced antigen quantity or its slower release, there are fewer local systemic reactions.
   (iii) A better immune response is obtained, an important situation in disease, where a high antibody level is required for protection.
   (iv) Many antigens may be included in a single vaccine.
   (v)   A reduced number of inoculations may be given.

Some adjuvants used are aluminium compounds, groundnut oil and calcium phosphate.

(iii) **Quantity of antigen used in vaccine**: This is to be determined by experimentation by titrating the quantity of antibodies produced against the level of antibody produced. In some cases large quantities of vaccine may be necessary for immunization. It should be noted however that in some cases too high a dose of antigen may paralyze the immune system.

(iv) **The crowding-out effect**: Vaccines may be inactivated when multiple vaccines are administered.

(v) **Choice of test animal**: The test animal in evaluating a vaccine is of paramount importance. The vaccine must be able to induce antibody production in the animal being used.

(vi) **The route of administration**: This may be important in determining the efficiency of the response.

27.6   VACCINE PRODUCTION VERSUS OTHER ASPECTS OF INDUSTRIAL MICROBIOLOGY

It is important that some differences between vaccine production and routine industrial fermentations be discussed in order to emphasize its uniqueness.

   (i) The cells used in vaccine manufacture are usually pathogenic and therefore complete sterility must be maintained. Furthermore, while contaminants may merely hinder production in other industrial fermentations, in vaccine production, contaminants may mean the introduction of an undesirable organism into the human body.

  (ii) The fermentation is usually small (25-1,000 liters) compared to, say, antibiotic fermentation (500,000 liters). This is because the amount required per person is usually small.

 (iii) Potency cannot be determined during cultivation, hence reproducibility during production is viewer the less very essential.

SUGGESTED READINGS

Anon, 1979. Microbial Processes: Promising Technologies for the Developing countries. National Academy of Science, Washington, DC, USA.

Dove, A. 2004. Making Prevention Pay Nature Biotechnology, 72, 387-391.

Fraser, C.M., Rappuoli, R. 2005. Application of Microbial Genomic Science To Advanced Therapeutics. Annual Review of Medicine, 56, 459-474.

Gurunathan, S., Klinman, D.M., Seder, R.A. 2000. DNA VACCINES: Immunology, Application, and Optimization. Annual Review of Immunolology, 18, 927–974.

Kuby, J. 1997. Immunology 3rd Ed W H Freeman and Co. New York, USA.

Meinginer, B., Mongeot, H., Favre, H. 1980. Development in Biological Standardization, 46, 249-256.

Mora, M, Veggi, D., Santini, L., Pizza, M., Rappuoli, R. 2003. Reverse Vaccinology *Drug* Discovery Today 8, 459-464.

Rappuoli, R. 2001. Reverse Vaccinology, a Genome-based Approach to Vaccine Development. Vaccine, 19, 2688-91.

Robertson, B.H., Nicholson, J.K.A. 2005. New microbiology Tools For Public Health And Their Implications. Annual Review of Public Health. 26, 281–302.

Spier, R.E. 1980. Adv. Biochem. Eng. 14, 141-162.

Stoughton, R.B. 2005. Applications of DNAmicroarrays in Biology. Annual Reviews of Biochemistry 74, 53–82.

Telford, J.L., Pizza, M., Grandi, G., Rappuoli, R. 2002. Reverse Vaccinology: From Genome to Vaccine In: Methods in Microbiology. Vol 33, Academic Press. Amsterdam the Netherlands: pp. 258–269.

Yewdell, J.W., Haeryfar, S.M.M. 2005. Understanding Presentation of Viral Antigens To Cd8+ T Cells In Vivo: The Key to Rational Vaccine Design. Annual Review of Immunology 23, 651–82.

Zinkernagel, R.M. 2003. On Natural And Artificial Vaccinations. Annual Review of Immunolology, 21, 515–46.

CHAPTER 28

Drug Discovery in Microbial Metabolites: The Search for Microbial Products with Bioactive Properties

Microorganisms produce a wide array of chemically diverse secondary metabolites which are not necessarily anti-microbial in nature. Many of them have turned out to be very important to the pharmaceutical industry, while some are of importance in the agricultural industry. In Chapter 24 we discussed the production of anti-microbial and anti-tumor agents by microorganisms; in this chapter we will look at other products with pharmaceutical relevance outside antibiotics.

Organized search for microbial metabolites of pharmaceutical and clinical importance began in the late 1960s when methods were developed for the isolation of enzyme inhibitors of microbial origin. This led to the discovery of many drugs of clinical importance. One such enzyme inhibitor is an beta-lactamase inhibitor which is administered with Beta-lactam antibiotics; the other is an inhibitor of cholesterol accumulation, while a third is the immunosupressant, cyclosporin A. This section will discuss the conventional methods for assaying microbial metabolites as a means of discovering those with positive bioactive activities with the potential of resulting in new drugs. Some of the newer methods which have come into being following the recent successes with the human genome project and such developments as the involvement of the computer in biotechnology, or bioinformatics, will also be touched upon. The examples given will illustrate how knowledge of a disease helps us look for drugs against it from among the metabolites of microorganisms.

Prior to the assay for possible drug activities the microbial metabolite must be studied using various chemical methods including solvent extraction, precipitation, chromatography, spectroscopic methods; spectral libraries should be searched to eliminate known compounds. The assays may be cell-based, receptor-binding or enzyme assays. Examples from each type of assay are given below to illustrate the immense diversity of microbial metabolites. Many assays are available and the examples given are a small selection, and designed to expose the student to the general procedure for drug discovery in microbial metabolites. Finally the processes of drug approval by regulatory agencies will be discussed using those of the Food and Drug Administration (FDA) as illustration.

It will be seen that the processes of drug discovery are beyond the ordinary capabilities of the microbiologist working single-handedly, and thus illustrate the team-work nature of industrial microbiology and biotechnology discussed earlier in this book. The human physiologist, the biochemist, the clinician and many others may be involved in the process of drug discovery.

28.1   CONVENTIONAL PROCESSES OF DRUG DISCOVERY

28.1.1   Cell-based Assays

Cell-based tests are used to screen for novel microbial metabolites which inhibit a cellular function, but where a specific molecular target has not been identified. The active compound(s) may interact with the cell at a number of points and the cell may even be killed. Compounds with activity at the cell level need to be further screened to identify the exact mechanism by which they affect the cell. A number of cell reactions are used to determine whether or not microbial metabolites are bioactive and hence their potential to become new drugs.

28.1.1.1   Inflammatory reactions

When a foreign protein such as a bacterium enters the body, the body reacts by developing a non-specific inflammatory reaction in which white blood cells rush to the site; it may also develop antibodies to the foreign protein. The ability to enhance or initiate or to suppress immunologic reaction is used to assess bioactivity in microbial metabolites. One test used is the mixed lymphocyte reaction (MLR) test. This test is an *in vitro* assay of T_H–cell proliferation in a cell-mediated response. In an MLR test cytotoxic lymphocytes (CTLs) are generated by co-culturing spleen cells from two different species, the rat and the mouse (Chapter 27). The T lymphocytes undergo extensive transformation and cell proliferation. The degree of cell proliferation is assessed by adding labeled [3H] thymidine to the culture medium and monitoring the uptake of the label into DNA during cell divisions. Both components proliferate unless one population is rendered unresponsive with an inhibitory antibiotic such as mitomycin C or has been killed by irradiation.

28.1.1.1.1   Immunomodulation

Microbial metabolites with the ability to enhance immunologic reactions are those able to enhance the incorporation of [3H] thymidine into spleen cells and T-cells in the MLR test. Similarly the ability of a microbial metabolite to restore antibody production to mice unable to produce antibodies is an indication of bioactivity. Kifunesine a compound produced by the actinomycete, *Kitasatosporia kifunense* has been found to have immunomodulatory activity.

28.1.1.1.2   Immunosupression

The enhancement of the body's immune system so as to protect it from disease by foreign organisms such as bacteria is normally desirable. However under some conditions it is necessary to suppress the body's immune system. One such situation is when immunosupression is desirable during tissue or organ transplantation when a person receives body parts from another. The immunologic apparatus of the recipient individual would normally reject the donated part because it is foreign. To avoid the rejection immunosuppressive drugs are given to the recipient person. The presence of immunosuppressive metabolites is tested by a modification of the MLR test, known as the one-way MLR. In this test one component of the mixture, the stimulator cells, is treated with mitomycin C to inactivate them. Within 24-48 hours the untreated cells, the responder cells begin to divide and incorporate [3H] thymidine as well as express antigens foreign to the responder T cells. The presence of an immunosuppressive metabolite is indicated by its blockage of the expression of the foreign antigens in the responder cells. An example of an immunosuppressive microbial metabolite is an analogue of the antibiotic cyclosporin, identified as FR901459 which is produced by the fungus *Stachybotris chartarum*.

28.1.1.1.3   Macrophage activation

Macrophages are white blood cells which migrate into tissues through out the body (histocytes in connective tissue, alveolar macrophages in lung, microglial in the central nervous system, mesangial in kidney, Kupffer cells in liver, and osteoclasts in bone). They engulf and digest foreign matter and are active in antigen processing and presentation and the phagocytic effectors of cell-mediated immunity and hypersensitivity. They not only destroy bacteria invading the body, but they also destroy cancer cells. Activation of macrophage cells is detected by the spreading of the macrophages as observed under the scanning electron microscope. TAN-999 is an alkaloid produced by a *Streptomyces* sp. and shown to activate macrophages.

28.1.1.2   Cardiovascular disease: Inhibition of platelet aggregation

Platelets are cell fragments from megakaryocytes. Some megakaryocytes give rise to red blood cells, while others fragment to give platelets. Blood contains 150,000 to 350,000 platelets per ml. They are important in blood clotting. When the wall of a blood vessel is damaged by disease or by a trauma such as a cut, thrombin is formed and this acts on the soluble fibrinogen present in the blood to form insoluble fibrin strands. The fibrin strands trap platelets to form blood blots. Blot clots forming within the blood vessels are dangerous, and if large enough could block blood vessels leading to the heart causing a heart attack or in blood vessels leading to the brain, a stroke. Platelet aggregation inhibitors are useful in preventing cardiovascular disorders through preventing clots. Platelet aggregation inhibition is measured against a control by a turbidometric assay of the platelets to which the metabolite and thrombin are mixed.

28.1.1.3   Cardiovascular disease: Angiogenesis inhibitors

Angiogenesis is the process of new blood vessel formation and is essential for the formation of solid tumors. Metabolites are tested for anti-angiogenesis effects by placing pellets of the metabolites dried on small cores of sterile filter paper on a five-day chick embryo. Metabolites containing anti-angiogenesis compounds prevent blood vessel formation where the paper was placed.

28.1.2 Receptor Binding Assays

Receptor binding assays have been important in drug discovery. Preparations from animal tissues have been used as membrane receptors in a wide variety of targets. The bound ligand is then separated by centrifugation or filtration and a percentage inhibition calculated on the basis of controls. Ligand receptor interactions are used to search for microbial metabolites which have potential in drug discovery for dealing with inflammatory (immunity) diseases, cancers, cardiovascular disease, and central nervous system diseases.

28.1.2.1 Receptor binding in inflammatory disease

28.1.2.1.1 Leukotriene B4 (LTB4) binding inhibitors

Leukotrienes are produced by a variety of white blood cells. Among them, LTB4 is a powerful mediator of inflammatory reactions, causing the loss of granules in granule-containing white blood cells leading to allergic reactions. LTB4 antagonists may therefore be useful in treating inflammatory diseases. In the search for LTB4 anatagonists, labeled LTB4 is incubated with a suspension of membrane from white blood cells and the test microbial metabolite. After the incubation, the bound labeled LTB4 is separated from the free ligand by filtration and centrifugation.

28.1.2.1.2 CD4 binding inhibitors

CD4 is a glycoprotein present on the surface of mature helper/inducer T white blood cells (lymphocytes) (see ch. 27). It binds to class II MHC (major Histocompatibility Complex II) and this' stabilizes the T cell receptor and its attachment, the antigen-MHCII complex. Inhibition of this interaction can have important suppressive effects on immune responses and thus it is an attractive target in the search for immunosuppressants. Additionally the CD4 molecule is an important anti-viral target because it is the cellular receptor for the HIV virus. An assay which has been used to search for anti-CD4 compounds was based on the interaction between soluble recombinant CD4 and a monoclonal antibody. The assay enabled the discovery of new anti-CD4 compounds of fungal origin.

28.1.3 Enzyme Assays

Enzymes are work horses of living things; all activities of living things are mediated through enzymes. It is therefore not surprising that enzymes are widely targeted in the search for pharmacologically active compounds. Only a few examples will be given in this section.

28.1.3.1 Inflammatory diseases

28.1.3.1.1 Cell surface sugar metabolism inhibitors

Sugars conjugated with various compounds are important in mammalian cell surfaces. They are important in cell adhesion, which is important in inflammatory disease as well as in cancer. Numerous enzymes are involved in the metabolism of sugars presented at cell surfaces. An enzyme which has been targeted is α-D-mannosidase. Broth cultures of

Streptoverticullum verticullus var. *quantum* was found to contain antagonists to the enzyme, designated mannostatins A and B.

28.1.3.1.2 Human leukocyte elastase inhibitors

Human leukocyte elastase is one of the most destructive enzymes known. It hydrolyzes several compounds found in connective tissue including elastin, proteoglycan and collagen. The enzyme is released by certain white blood cells and it may be involved with the destructive processes associated with chronic inflammatory diseases. A peptide metabolite of *Streptomyces resistomycificus* was found to antagonize the enzyme.

28.1.3.1.3 Cardiovascular disease: Inhibition of cholesterol metabolism

Cholesterol is important in cardiovascular disease because it is deposited on the walls of blood vessels, thereby decreasing the blood vessel diameter leading to high blood pressure and in some cases to occlusion of the blood vessels. This may lead to heart attacks or strokes depending on whether the occlusions occurs in a blood vessel leading to the heart or one leading to the brain. Acyl co-enzyme A cholesterol transferase (ACAT) plays an important part in atherogenesis and cholesterol adsorption from the intestine. ACAT inhibitors may therefore be useful in treating arteriosclerosis and hypercholesteremia (high cholesterol). Many new ACAT inhibitors have been identified in recent times, including some from Fungi.

28.2 NEWER METHODS OF DRUG DISCOVERY

28.2.1 Computer Aided Drug Design

The search for anti-microbial compounds and other drugs can nowadays start at the computer ie *in silico*. New drugs can be created at the computer and their efficacy determined through assessing whether or not they will bind to proteins on 'pathogenic micro-organisms' or 'disease tissues'. Drug discovery and development is immensely expensive and time-consuming. The success rate of new chemical entities selected for clinical development is approximately 20% with most failures attributed to unacceptable pharmacokinetic properties Undesirable properties, such as poor absorption, low and variable bioavailability, drug interactions may be predicted from *in vitro* and *in silico* data, thus facilitating selection of the most appropriate lead compound. The *in silico* approach is not only rapid, but it is also cost-effective. The successful *in silico* antibiotic or drug must then be tested in the wet laboratory using *in vitro* and *in vivo* methods as required by regulatory agencies discussed later in this chapter. Perhaps the best example of *in silico* drug development is the development of inhibitors of HIV-1 protease by computer-aided drug design. HIV-1 genome encodes an aspartic protease (HIV-1 PR). Inactivation of HIV-1 PR by either mutation or chemical inhibition leads to the production of immature, noninfectious viral particles thus the function of this enzyme was shown to be essential for proper virion assembly and maturation. It is not surprising, then, that HIV-1 PR was identified over a decade ago as the prime target for structure- or computer-assisted (sometimes called 'rational') drug design. The structure-assisted drug design and discovery process utilizes structural biochemical methods, such as protein crystallography, nuclear magnetic resonance (NMR), and computational biochemistry, to guide the synthesis of potential drugs. This information can, in turn, be used to help explain the basis of their activity and to improve the potency and specificity of new lead compounds. Put in another language once the structure of a target is known, the structure of a compound which will attack it can be computer-designed and then synthesized.

An aspect of *in silico* drug discovery would appear to be a process known as tethering. To facilitate the drug discovery process, many researchers are turning to fragment-based approaches to find lead molecules more efficiently. One such method, tethering, allows for the identification of small-molecule fragments that bind to specific regions of a protein target. These fragments can then be elaborated, combined with other molecules, or combined with one another to provide high-affinity drug leads.

28.2.2 Combinatorial Chemistry

The essence of combinatorial chemistry or techniques involving 'molecular diversity' is to generate enormous populations of molecules and to exploit appropriate screening techniques to isolate active components contained in these libraries. This idea has been the focus of research both in academia, but more especially in the pharmaceutical or biotechnology industry. Its developments go hand in hand with an exploding number of potential drug targets emerging from genomics and proteomics research.

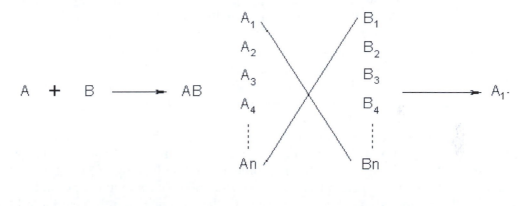

Orthodox synthesis   Combinatorial synthesis

Fig. 28.1 Illustration of Combinatorial Chemistry

Synthesis of molecules in a combinatorial fashion can quickly lead to large numbers of molecules. For example, a molecule with three points of diversity (R1, R2, and R3) can generate $NR_1 \times NR_2 \times NR_3$ possible structures, where NR_1, NR_2, and NR_3 are the number of different substituents utilized.

In this a technique a large number of structurally distinct molecules are synthesized at a time and submitted for pharmacological assay. The key of combinatorial chemistry is that a large range of analogues is synthesized using the same reaction conditions, the same reaction vessels. In this way, the chemist can synthesize many hundreds or thousands of compounds in one time instead of preparing only a few by simple methodology.

In the past, chemists have traditionally made one compound at a time. For example compound A would have been reacted with compound B to give product AB, which would have been isolated after reaction work up and purification through crystallization, distillation, or chromatography. In contrast to this approach, combinatorial chemistry offers the potential to make every combination of compound A1 to An with compound B1 to Bn (Fig. 28.1). Combinatorial chemistry has been helped by developments in automation or robotics and miniaturization of processes. These enable many hundreds of compounds to be synthesized and screened. The starting points of the compounds to be 'amplified' by combinatorial chemistry could be from plants, animals or micro-organisms. Combinatorial chemistry methods are used for discovering other drugs besides antimicrobial agents.

28.2.3 Genomic Methods in the Search for New Drugs, Including Antibiotics

The traditional approach to the development of novel antibiotics has relied on random screening (mainly of the soil) for new active molecules, using simple antibiotic activity especially death of the test organisms for primary selection. The disadvantages of using death as the main criterion for selecting anti-microbial agents are shown in Table 28.1. The result is that very few new antibiotics have been discovered over many years; on account of this progress has been achieved by modification of existing antibiotics. As a result, resistance cross-reaction across available antibiotic groups has become common.

More imaginative approaches were limited by knowledge and technology. The completion of the Human Genome Project and the availability of genome sequence data for many micro-organisms including pathogenic ones, has stimulated research on the identification of novel targets for antimicrobial compounds by providing a complete catalogue of genes which can be compared at various levels.

Recent genomic advances have enabled 'target-based' initiatives in the search for antimicrobials. Traditionally screening has been done against whole cells such as microbial cells. Such an approach has a number of disadvantages. First they are inherently insensitive and often lead to the isolation of toxic compounds. Second, the screens will only identify targets that are lethal to the bacteria under the conditions of growth in the laboratory. Third since the exact nature of the target inhibited by any new compound is unknown, rational modification of the molecule to enhance its activity and moderate toxicity is not possible. This approach has been replaced by attempts to identify effective drug targets and to design antagonists to disrupt the activity of the target.

Comparative analysis of microbial genome sequence data has revealed that: (i) the genome of each organism contains a large number of open reading frames- ie sequences which code for proteins– 20% - 40% for which the proteins are unknown and about 10% of these unknown proteins are unique to the organisms (Chapter 3); (ii) the microbial genome is a dynamic entity shaped by multiple forces including gene duplication and gene loss, genome rearrangements, and acquisition of new genes through lateral transfers; (iii) each of these mechanisms has been shown to play a major role in the evolution of pathogens and has important implications in epidemeologic studies and the spread of antibiotic resistance and pathogenicity; (iv) differences between the pathogen and non-pathogen are best explained not by the presence or absence of a gene but by

Table 28.1   Comparison of the screening strategies for novel antimicrobial compounds

Whole-cell screening	Target-based screening
(Looking directly for compounds which kill microorganisms)	(Looking for biochemical inhibitors)
Advantages	
1 Selection for compounds which penetrate cells	1 More sensitive (can detect weak or poorly penetrating compounds suitable for chemical optimization)
2 Antimicrobial properties established	2 Easy screening
3 Highly reproducible has been used successfully historically	3 Different approach can target new areas of biology facilities rational drug design
Disadvantages	
1 Insensitive	1 Need to turn an in vitro inhibitor into an antibacterial drug (complicated by penetration issues)
2 Most active compounds are toxic	
3 No rational basis for compound optimization (target unknown)	2 Genetic validation of targets (by gene knockout or reduced expression) can be misleading
4 Mixed mechanisms of action in recent years has failed to deliver	

subtle single nucleotide changes, and virulence genes have been shown to be inactivated by such single-nucleotide changes.

The ideal antimicrobial genomic target should be (i) different from the existing targets; (ii) essential for the viability of the pathogen; (iii) absent or substantially different in the human host, a parameter much easier to assess now with the availability of the complete human genome; (iv) conserved across the appropriate range of organisms; (v) easy to assay, especially in high throughput processes; (vi) easy to identify the target's inhibitors and (vii) suitable for rapid structural analysis. .

Genomics has enabled the identification of targets through (a) large-scale identification of novel potential targets through *in silico* comparison of pathogenic and non-pathogenic strains; (b) examining existing metabolic information on the organisms present in databases; (c) identifying genes necessary for bacterial growth and survival by experimental means, including transposon mutagenesis, targeted mutagenesis of conserved genes, and the expression of anti-sense RNA. These studies suggest that depending on the organism, the number of essential genes range from 150 to 500.

The potential new targets identified include aminoacyl tRNA synthetases, polypeptide deformylase, fatty acid biosynthesis, protein secretion, and cell signaling. The next step is to identify small molecular inhibitors of these proteins, often by exploiting the diversity of chemical compounds to be found in combinatorial libraries.

The target approach is however not without its shortcomings. These include the possibility that the antibacterial drug may not penetrate the cell (Table 28.1).

28.2.4 Search for Drugs Among Unculturable Microorganisms

Natural products, primarily of microbial origin, have accounted for one-third of the more than $100 billion of sales in the US and in excess of $250 billion worldwide pharmaceutical market and are also an important source of specialty chemical, agrochemical, and food or industrial processing products. Although the pharmaceutical industry appears to be spending more money in the search for new drugs, the results do not match the increased expenditure.

About $20 billion is spent currently on search for new drugs in the US, or about 20 times the figure in the 1970s. Yet only about 40 new drugs (new chemical entities, NCEs) were introduced in the mid-1990s compared to 60-70 in the 1970s. One reason for this is the diminishing returns from existing sources of search. Many of the sources of pharmaceuticals are bacteria. It is now known that culturable bacteria represented only about 5% of all bacteria. To combat the problem of diminishing returns from searching among culturable organisms, Oceanix Biosciences Corporation has developed and patented a biotechnology-based for the production of new pharmaceutical from unculturable bacteria. The procedure of the company named Combinatorial Genomics ᵀᴹ for which US Patent no 5,773,221 was granted by the US Patent and Trademark Office consists essentially isolating nonculturable micro-organisms or their high molecular weight DNA directly from environmental samples followed by the integration and expression of that genetic material in well characterized microbial host species. As to be expected, the results are unpredictable since it is based on a random and phenomenological genetic survey of unknown genetic materials.

The environmental DNA may be isolated either in a 'naked' form and subsequently encapsulated in liposomes prior to use, or may be contained in non-culturable microbial cells which are converted into spheroplasts or protoplasts prior to use. Liposomes, spheroplasts, or protoplasts containing environmental DNA are then fused, employing standard cell fusion techniques such as polyethylene glycol (PEG) mediated fusion or electrofusion, with spheroplasts or protoplasts (Chapter 9) of well characterized and easily cultured host microorganisms. Well characterized host microbes can be employed as recipient organisms including Gram-positive and Gram-negative bacteria as well as certain fungal and archaebacterial host species. Following a fusion event between a host microbe protoplast or spheroplast (auxotrophic) cell and a prepared environmental DNA sample containing liposomes, protoplasts, or spheroplasts the viable, colony-forming cells will be those in which the delivered environmental DNA is expressed.

Protoplast and liposome fusion are a versatile and well explored technique to induce genetic recombination in a variety of prokaryotic and eukaryotic microorganisms. In the presence of a fusogenic agent, such as Polyethylene glycol (PEG), or by treatment in electrofusion chambers, protoplasts and liposomes are induced to fuse and form hybrid cells. During the hybrid state, the genomes reassort and extensive genetic recombination can occur. The final, crucial step is the regeneration of viable cells from the fused protoplasts, without which no viable recombinants can be obtained. The patentees named the process Combinatorial Genomics in mimic of combinatorial chemistry where numerous compounds are prepared, in the same manner as numerous recombinations may occur between the host organisms and the unknown DNA isolated from the environment.

Growth is on agar plates and the colonies selected are tested for i) their further characterization in additional antibiotic tests employing a wider range of indicator microbial species, ii) their anti-cancer activity in a battery of malignant cell lines, and iii) their agonist or antagonist activity in a relevant central nervous system (CNS). Those bioactive agents which display promising activity as antibiotic agents, as anti-cancer agents or as pharmacologically relevant materials are then purified and subjected to chemical structural analysis. Novel bioactive agents are examined for their safety and toxicology properties.

28.4 APPROVAL OF NEW ANTIBIOTIC AND OTHER DRUGS BY THE REGULATING AGENCY

Discovering and developing safe and effective new medicines is a long, difficult and expensive process. The US system of approval for new drugs is perhaps the most rigorous in the world. In the US the regulatory agency for certifying new medicines as safe and effective is the Food and Drug Administration (FDA). In EU countries drugs are regulated by the European Medicines Agency (EMEA) which was established in 1993.

In the US which is a major producer of new drugs, it takes 12 years on average for an experimental drug to travel from the laboratory to the medicine chest. It is also an expensive process and takes on the average about $360 million to get a new medicine from the lab oratory to the medicine cupboard. Where a new medicine is a life saving one for which few or no equivalent exists it can be put on fast track and may be approved in six months.

Prior to submission for pre-clinical trial by the FDA, the firm itself would have carried out tests. Subsequently the firm submits its product for testing by the FDA. Only five in 5,000 compounds that enter preclinical testing make it to human testing. As shown in the table below, one of these five tested in people is approved. The processes of drug progress are as follows beginning with the work of the firm. (see Table 28.2)

28.4.1 Pre-Submission Work by the Pharmaceutical Firm

28.4.1.1 Synthesis and extraction

The process of identifying new molecules with the potential to produce a desired change in a biological system (e.g., to inhibit or stimulate an important enzyme, to alter a metabolic pathway, or to change cellular structure).

The process may require: 1) research on the fundamental mechanisms of disease or biological processes; 2) research on the action of known therapeutic agents; or 3) random selection and broad biological screening. New molecules can be produced through artificial synthesis or extracted from natural sources (plant, mineral, or animal). The number of compounds that can be produced based on the same general chemical structure runs into the hundreds of millions.

Table 28.2 Table showing flow chart of approval for a new drug (including antibiotics) to travel from the laboratory to the patient in the US

| | Lab studies/ Preclinical Testing | | Clinical Trials | | | | FDA | | Clinical Trials |
			Phase I	Phase II	Phase III				Phase IV
Years	3.5		1	2	3		2.5	12 Total	
Test Population	Lab and animal studies	File IND at FDA	20 to 80 healthy volunteers	100 to 300 patient volunteers	1000 to 3000 patient volunteers	File NDA at FDA			Additional Post marketing testing required by FDA (sometimes)
Purpose	Assess safety and biological activity		Determine effectiveness, safety and dosage	Evaluate effectiveness, monitor look for side effects	Verify effectiveness, adverse reactions from long-term use		Review process/ Approval		
Success Rate	5,000 compounds evaluated		5 enter trials				1 approved		

28.4.1.2 Biological screening and pharmacological testing

Studies to explore the pharmacological activity and therapeutic potential of compounds.

These tests involve the use of animals, isolated cell cultures and tissues, enzymes and cloned receptor sites as well as computer models. If the results of the tests suggest potential beneficial activity, related compounds each a unique structural modification of the original are tested to see which version of the molecule produces the highest level of pharmacological activity and demonstrates the most therapeutic promise, with the smallest number of potentially harmful biological properties.

28.4.1.3 Pharmaceutical dosage formulation and stability testing

The process of turning an active compound into a form and strength suitable for human use.

A pharmaceutical product can take any one of a number of dosage forms (e.g., liquid, tablets, capsules, ointments, sprays, patches) and dosage strengths (e.g., 50, 100, 250, 500 mg) The final formulation will include substances other than the active ingredient, called excipients. Excipients are added to improve the taste of an oral product, to allow the active ingredient to be compounded into stable tablets, to delay the drug's absorption into the body, or to prevent bacterial growth in liquid or cream preparations. The impact of each on the human body must be tested.

28.4.1.4 Toxicology and safety testing

Tests to determine the potential risk a compound poses to man and the environment. .

These studies involve the use of animals, tissue cultures, and other test systems to examine the relationship between factors such as dose level, frequency of administration, and duration of exposure to both the short- and long-term survival of living organisms. Tests provide information on the dose-response pattern of the compound and its toxic effects. Most toxicology and safety testing is conducted on new molecular entities prior to their human introduction, but companies can choose to delay long-term toxicity testing until after the therapeutic potential of the product is established.

All the above tests can take up to three and half years. If the results are promising, the firm then submits the compound and all the tests and results obtained to the FDA as an investigational new drug (IND).

28.4.2 Submission of the New Drug to the FDA

28.4.2.1 Regulatory review: Investigational new drug (IND) application

An application filed with the U.S. FDA prior to human testing.

After completing its laboratory studies, the company files an IND with FDA to begin to test the drug in people. The IND shows results of previous experiments, how, where and by whom the new studies will be conducted; the chemical structure of the compound; how it is thought to work in the body; any toxic effects found in the animal studies; and how the compound is manufactured. In addition, the IND must be reviewed and approved by the Institutional Review Board where the studies will be conducted, and progress reports on clinical trials must be submitted at least annually to FDA. The IND application is a compilation of all known information about the compound. It also includes a description of the clinical research plan for the product and the specific protocol for phase I study. Unless the FDA says no, the IND is automatically approved after 30 days and clinical tests can begin.

28.4.2.2 Clinical trials

28.4.2.2.1 Phase I Clinical Evaluation

The first testing of a new compound in human subjects, for the purpose of establishing the tolerance of healthy human subjects at different doses, defining its pharmacologic effects at anticipated therapeutic levels, and studying its absorption, distribution, metabolism, and excretion patterns in humans.

About 20 -80 healthy volunteers are used for this trail.

28.4.2.2.2 Phase II clinical evaluation

Controlled clinical trials of a compound's potential usefulness and short term risks.

A relatively small number of patients, usually no more than several hundred subjects (100 – 300), enrolled in phase II studies.

28.4.2.2.3 Phase III clinical evaluation

Controlled and uncontrolled clinical trials of a drug's safety and effectiveness in hospital and outpatient settings.

Phase III studies gather precise information on the drug's effectiveness for specific indications, determine whether the drug produces a broader range of adverse effects than those exhibited in the small study populations of phase I and II studies, and identify the best way of administering and using the drug for the purpose intended. If the drug is approved, this information forms the basis for deciding the content of the product label. Phase III studies can involve several hundred to several thousand subjects (1,000 – 3,000).

28.4.2.3   Process development for manufacturing and quality control

The firm's manufacturing capability is assessed.

Engineering and manufacturing design activities to establish a company's capacity to produce a product in large volume and development of procedures to ensure chemical stability, batch-to-batch uniformity, and overall product quality.

28.4.2.4   Bioavailability studies

The use of healthy volunteers to document the rate of absorption and excretion from the body of a compound's active ingredients.

Companies conduct bioavailability studies both at the beginning of human testing and just prior to marketing to show that the formulation used to demonstrate safety and efficacy in clinical trials is equivalent to the product that will be distributed for sale. Companies also conduct bioavailability studies on marketed products whenever they change the method used to administer the drug (e.g., from injection or oral dose form), the composition of the drug, the concentration of the active ingredient, or the manufacturing process used to produce the drug.

28.4.2.5   Regulatory review: New drug application (NDA)

The firm puts in an application for a new drug, New Drug Application (NDA)

An NDA is an application to the FDA for approval to market a new drug. All information about the drug gathered during the drug discovery and development process is assembled in the NDA Following the completion of all three phases of clinical trials, the company analyzes all of the data and files an NDA with FDA if the data successfully demonstrate safety and effectiveness. The NDA must contain all of the scientific information that the company has gathered. NDAs typically run 100,000 pages or more. By law, FDA is allowed six months to review an NDA. In almost all cases, the period between the first submission of an NDA and final FDA approval exceeds that limit; the average NDA review time for new molecular entities approved in 1992 was 29.9 months.

28.4.3   Approval

Once FDA approves the NDA, the new medicine becomes available for physicians to prescribe. The company must continue to submit periodic reports to FDA, including any cases of adverse reactions and appropriate quality-control records. For some medicines, FDA requires additional studies (Phase IV) to evaluate long-term effects.

28.4.4 Post Approval Research

Experimental studies and surveillance activities undertaken after a drug is approved for marketing.

Clinical trials conducted after a drug is marketed (referred to as phase IV studies in the United States) are an important source of information on as yet undetected adverse outcomes, especially in populations that may not have been involved the premarketing trials (e.g., children, the elderly, pregnant women) and the drug's long-term morbidity and mortality profile. Regulatory authorities can require companies to conduct Phase IV studies as a condition of market approval. Companies often conduct post-marketing studies even in the absence of a necessity to do so.

SUGGESTED READINGS

Allsop, A., Illingworth, R. 2002. The impact of genomics and related technologies on the search for new antibiotics. Journal of Applied Microbiology, 92, 7-12.

Anon, 1993. Congress of the United States, Office of Technology Assessment. Pharmaceutical R&D: Costs, Risks and Rewards: 1993; Washington, DC, USA. *pp.* 4-5.

Anon, 1999. From Test Tube to Patient: Improving Health Through Human Drugs Special Report, Center Drug Evaluation and Research. Food and Drug Administration. Rockville, MD, USA.

Austin, C. 2004. The Impact of the Completed Human Genome Sequence on the Development of Novel Therapeutics for Human Disease. Annual Review of Medicine, 55, 1–13.

Bansal, A.K. 2005. Bioinformatics in the microbial biotechnology – a mini review Microbial Cell Factories, 4, 4–19.

Beamer, L. 2002. Human BPI: One protein's journey from laboratory to clinical trials. ASM News. 68, 543-548.

Behal, V. 2000. Bioactive Products from *Streptomyces. Advances in Applied Microbiology. 47, 113–156.

Bull, A.T., Ward, A.C., Goodfellow, M. 2000. Search and Discovery Strategies For Biotechnology:. The Paradigm Shift. Microbiology and Molecular Biology Reviews, 64, 573-548.

Dale, E., Wierenga, D.E., Eaton, C.R. 2001. Processes of Product Develpoment. http://www.allpcom/drug-dev.htm. Accessed on September 28, 2005 at 12.05 pm GMT.

Debouck, C., Metcalf, B. 2000. The Impact of Genomics on Drug Discovery. Annual Review of Pharmacology and Toxicology, 40, 193–208.

Erlanson, D.A., Wells, J.A., Braisted, A.C. 2004. Tethering: Fragment-Based Drug Discovery. Annual Reviews of Biophysical and Biomolecular Structure, 33, 199–223.

Fan, F., McDevitt, D. 2002. Microbial Genomics for Antibiotic Target Discovery. In : Methods in Microbiology. Vol 33, Academic Press. Amsterdam the Netherlands, pp. 272–288.

Feling, R.H., Buchanan, G.O., Mincer, T.J., Kauffman, C.A., Jensen, P.R., Fenical, W. 2003. Salinosporamide A: a highly cytotoxic proteasome inhibitor from a novel microbial source, a marine bacterium of the new genus *Salinospora. Angewandte Chemie International Edition* 42, 355-357.

Manyak, D.M., Carlson, P.S. 1999. Combinatorial Genomics[TM]: New tools to access microbial chemical diversity In : Microbial Biosystems: New Frontiers. C.R., Bell, M. Brylinsky, P. Johnson-Green, (eds). Proceedings of the 8th International Symposium on Microbial Ecology Atlantic Canada Society for Microbial Ecology, Halifax, Canada, 1999.

Section

Waste Disposal

Treatment of
Wastes in Industry

Wastes, unwanted materials, result inevitably from industrial activities in the same way as they also do in domestic ones. If allowed to accumulate on the ground, or if dumped indiscriminately into rivers and other bodies of water, unacceptable environmental problems would result. Governments the world over usually institute legislation which regulates the handling of wastes, including those resulting from industry. In the US the Environmental Protection Agency (EPA) is the regulating agency. The EPA works to develop and enforce regulations that implement environmental laws enacted by Congress. EPA is responsible for researching and setting national standards for a variety of environmental programs, and delegates to states the responsibility for issuing permits and for monitoring and enforcing compliance.

The activities of industrial microorganisms usually occur in large volumes of water; the resulting wastes are therefore transported in aqueous medium. This chapter will examine briefly the treatment of waste water. The subject is of interest, not only from the intrinsic need to dispose of wastes in industry, but especially because the basis for ultimate waste disposal is microbial.

Waste carried in water, whether from industry or from domestic activity is known as sewage. Waste water disposal constitutes a peculiar branch of industrial microbiology. The methods to be discussed were evolved originally to handle domestic sewage, but they have been extended for use in those industries, such as the food and fermentation industries, which yield wastes degradable by microorganisms. Sewage emanating from some chemical industries especially those dealing with manmade chemicals are not only less degradable but are sometimes toxic to microorganisms and man. The processes of biological waste-water treatment to be discussed here are really an aspect of industrial microbiology within the definition of the subject adopted in this book, because they involve micro-organisms on a large scale, although there is no direct expectation of profit.

29.1 METHODS FOR THE DETERMINATION OF ORGANIC MATTER CONTENT IN WASTE WATERS

Waste waters are sampled and analyzed in order to determine the efficiency of the treatment system in use. This is particularly important at the point of the discharge of the treated waste water into rivers, streams and other natural bodies of water. If waste water discharged into a natural water is rich in degradable organic matter, large numbers of aerobic microorganisms will develop to break down the organic matter. They will use up the available oxygen and as a consequence fish and other aquatic life will die. Furthermore, anaerobic bacteria will develop following the exhaustion of oxygen; the activities of the latter will result in foul odors. Some of the methods for analyzing the organic matter content of waste waters are given below.

29.1.1   Dissolved Oxygen

Dissolved oxygen is one of the most important, though indirect, means of determining the organic matter content of waters. The heavier the amount of degradable material present in water, the greater the growth of aerobic organisms and hence the less the oxygen content. The Winkler method is widely used for determining the oxygen in water. In this method, dissolved oxygen reacts with manganous oxide to form manganic oxide. On acidification in an iodide solution, iodine is released in an amount equivalent to the oxygen reacting to form the manganic oxide. The iodine may then be titrated using thiosulphate. Membrane electrodes are now available for the same purpose. In these electrodes oxygen diffuses through the electrode and reacts with a metal to produce a current proportional to the amount of oxygen reacting with the metal.

29.1.2   The Biological or Biochemical Oxygen Demand (BOD) Tests

Due to the complexity of the organic materials introduced into water and the key role played by oxygen in supporting the aerobic bacteria which break down this organic matter, the method of the Biochemical Oxygen Demand (BOD) was developed. It is a measure of the oxygen required to stabilize or decompose the organic matter in a body of water over a five-day period at 20°C. In carrying out the test, two 250-300 ml bottles are filled with water whose BOD is to be determined. The oxygen content of one is determined immediately by the Winkler method and in the other at the end of five days incubation at 20°C. The difference between the two is the BOD.

Although it has been severely criticized, the BOD test is still widely used. Some of the criticisms are that it takes too long to obtain results and that it may infact relate only loosely to the actual organic matter content of water since it represents the overall value of the respiration of the organisms present therein. Furthermore, many industrial wastes contain materials which are either difficult to degrade or which may even be toxic to the organisms. In such cases an inoculum capable of degrading the materials must be developed by enrichment and introduced into the bottles.

29.1.3   Permanganate Value (PV) Test

This PV method determines the amount of oxygen used up by a sample in four hours from a solution of potassium permanganate in dilute H_2SO_4 in a stoppered bottle at 27°C. It gives an idea of the oxidizable materials present in water, although the actual oxidation is only 30-50% of the theoretical value. The method records the oxidation of organic materials such as phenol and aniline as well as those of sulfide, thiosulfate, and thiocyanate and would be useful in some industries. However because oxidation is incomplete it is not favored by some workers.

29.1.4   Chemical Oxygen Demand (COD)

The chemical oxygen demand is the total oxygen consumed by the chemical oxidation of that portion of organic materials in water which can be oxidized by a strong chemical oxidant. The oxidant used is a mixture of potassium dichromate and sulfuric acid and is refluxed with the sample of water being studied. The excess dichromate is titrated with ferrous ammonium sulfate. The amount of oxidizable material measured in oxygen equivalent is proportional to the dichromate used up. It is a more rapid test than BOD and since the oxidizing agents are stronger than those used in the PV test, the method can be used for a wider variety of wastes. Furthermore, when materials toxic to bacteria are present it is perhaps the best method available. Its major disadvantage is that bulky equipment and hot concentrated sulfuric acid are used.

29.1.5   Total Organic Carbon (TOC)

Total organic carbon provides a speedy and convenient way of determining the degree of organic contamination. A carbon analyzer using an infrared detection system is used to measure total organic carbon. Organic carbon is oxidized to carbon dioxide.

The CO_2 produced is carried by a 'carrier gas' into an infrared analyzer that measures the absorption wavelength of CO_2. The instrument utilizes a microprocessor that will calculate the concentration of carbon based on the absorption of light in the CO_2. The amount of carbon will be expressed in mg/L. TOC provides a more direct expression of the organic chemical content of water than BOD or COD.

29.1.6   Total Suspended Solids (TSS)

The term 'total solids' refers to matter suspended or dissolved in water or wastewater, and is related to both specific conductance and turbidity. Total solids (also referred to as total residue) is the term used for material left in a container after evaporation and drying of a water sample. Total Solids include both total suspended solids, the portion of total solids retained by a filter and total dissolved solids, the portion that passes through a filter. Total solids can be measured by evaporating a water sample in a weighed dish, and then drying the residue in an oven at 103 to 105°C. The increase in weight of the dish represents the total solids. Instead of total solids, laboratories often measure total suspended solids and/or total dissolved solids. To measure total suspended solids (TSS), the water sample is filtered through a preweighed filter. The residue retained on the filter is dried in an oven at 103 to 105°C until the weight of the filter no longer changes. The increase in weight of the filter represents the total suspended solids. TSS can also be measured by analyzing for total solids and subtracting total dissolved solids.

29.1.7   Volatile Suspended Solids (VSS)

Volatile suspended solids (VSS) are those solids (mg/liter) which can be oxidized to gas at 550°C. Most organic compounds are oxidized to CO_2 and H_2O at that temperature; inorganic compounds remain as ash.

29.2 WASTES FROM MAJOR INDUSTRIES

The composition of industrial wastes depend on the industry. Wastes from three key industries in the US are given in Table 29.1 for illustration: the oil, the pulp and paper and the food industries.

Table 29.1 Typical wastes from three industries

A	
Typical Components of an Oil Refinery Waste Water	
Handling of oil crude	Oil, sludge oil emulsions, sulfur- and nitrogen corrosion inhibitors
Crude oil distillation	Hydrocarbons, organic and inorganic acids, phenols and sulfur
Thermal cracking	Phenols, triphenols, cyanides, hydrogen sulfide
Alkylation, polymerization, isomerization processes	cid sludge, spent acid, mineral acids (sulfuric, hydrochloric), catalyst support
Refining	Hydrogen sulfide, ammonium sulfide, gums. catalyst support
Purification and extraction	Phenols, glycols, amines, spent caustic
Sweetening, stripping, fltration	Sulfur and nitrogen compounds, copper chloride, suspended matter

B		
Typical Effluent Loads from Pulp and Paper Manufacture		
Effluent	Kg/1,000 kg of product	
	Suspended solids	5-day BOD
Pulps: unbleached sulfite	10 - 20	200 - 300
Pulps: bleached sulfite	12 - 30	220 - 400
Fine paper	25 - 30	7 -20
Tissue paper	15 – 20	10 - 15

C		
Typical Effluent Loads from Food industries		
Effluent	Kg/1,000 kg of product	
	Suspended solids	5-day BOD
	Cannery wastes	
Apple canning	300 - 600	1680 - 5530
Cherries canning	200 - 600	700 - 2100
Mushrooms	50 - 240	76 - 850
	Meat Packing Industry	
Slaughter house	3000 - 930	2200 - 650
Parking house	2000 - 230	3000 - 400
Processing plant	800 - 200	800 - 200
	Poultry	
Plant waste	100 - 1500	150 - 2400

29.3   SYSTEMS FOR THE TREATMENT OF WASTES

The basic microbiological phenomenon in the treatment of wastes in aqueous environments is as follows:

(i)   The degradable organic compounds in the waste water (carbohydrates, proteins, fats, etc.) are broken down by aerobic micro-organisms mainly bacteria and to some extent, fungi. The result is an effluent with a drastically reduced organic matter content.

(ii)   The materials difficult to digest form a sludge which must be removed from time to time and which is also treated separately.

The discussion will therefore be under two headings: aerobic breakdown of raw waste-water and anaerobic breakdown of sludge.

29.3.1   Aerobic Breakdown of Raw Waste Waters

The two methods which are usually employed include the activated sludge and the trickling filter.

29.3.1.1   The activated sludge system

The activated sludge method is the most widely used method for treating waste waters. Its main features are as follows:

(a)   It uses a complex population of microorganisms of bacteria and protozoa;

(b)   This community of microorganisms has to cope with an uncontrollably diverse range of organic and inorganic compounds some of which may be toxic to the organisms.

(c)   The microorganisms occur in discreet aggregates known as flocs which are maintained in suspension in the aeration tank by mechanical agitation or during aeration or by the mixing action of bubbles from submerged aeration systems. Flocs consist of bacterial cells, extracellular polymeric substances, adsorbed organic matter, and inorganic matter. Flocs are highly variable in morphology, typically 40 to 400 µm and not easy to break apart (Fig 29.1).

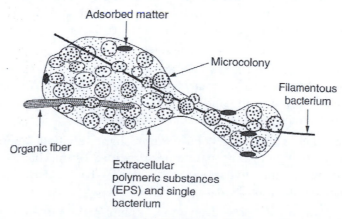

Fig. 29.1   Diagram of a Floc

(d) The flocs must have good settling properties so that separation of the biomass of microorganisms and liquid phases can occur efficiently and rapidly in the clarifier. Sometimes proper separation is not achieved giving rise to problems of bulking and foaming.

(e) Some of the settled biomass is recycled as 'returned activated sludge' or RAS to inoculate the incoming raw sewage because it contains a community of organisms adapted to the incoming sewage.

(e) The solid undigested sludge may be further treated into economically valuable products.

The advantages of the activated sludge system over the other methods to be discussed are its efficiency, economy of space and versatility. The flow diagrams of the conventional set-up and various modifications thereof are given in Fig. 29.2; others are shown in Figs 29.3, 29.4 and 29.5.

Modifications of the Activated Sludge System

(i) **The conventional activated sludge set-up**: The basic components of the conventional system are an aeration tank and a sedimentation tank. Before raw waste water enters the aeration tank it is mixed with a portion of the sludge from the sedimentation tank. The contents of the raw water are therefore broken down by organisms already adapted to the environment of the aeration tank. The incoming organisms from the sludge exist in small flocs which are maintained in suspension by the vigor of mixing in the aeration tank. It is the introduction of already adapted flocs of organisms that gave rise to the name activated sludge. Usually 25-50% of the flow through the plant is drawn off the sedimentation tank. Other modifications of the activated sludge system are given below.

(ii) **Tapered aeration**: This system takes cognizance of the heavier concentration of organic matter and hence of oxygen usage at the point where the mixture of raw sewage and the returned sludge enters the aeration tank. For this reason the aeration is heaviest at the point of entry of waste waters and diminishes towards the distal end. The diminishing aeration may be made directly into the main aeration tank (Fig. 29.2b and c) or a series of tanks with diminishing aeration may set up.

(iii) **Step aeration**: In step aeration the feed is introduced at several equally spaced points along with length of the tank thus creating a more uniform demand in the tank. As with tapered aeration the aeration may be done in a series of tanks.

(iv) **Contact stabilization**: This is used when the waste water has a high proportion of colloidal material. The colloid-rich waste waster is allowed contact with sludge for a short period of 1 - 1½ hours, in a contact basin which is aerated. After settlement in a sludge separation tank, part of the sludge is removed and part is recycled into an aeration tank from where it is mixed with the in-coming waste-water.

(v) **The Pasveer ditch**: This consists of a stadium-shaped shallow (about 3 ft) ditch in which continuous flow and oxygenation are provided by mechanical devices. It is essentially the conventional activated sludge system in which materials are circulated in ditch rather than in pipes (Fig. 29.3).

(vi) **The deep shaft process**: The deep shaft system for waste water treatment was developed by Agricultural Division of Imperial Chemical Industries (ICI) in the UK, from

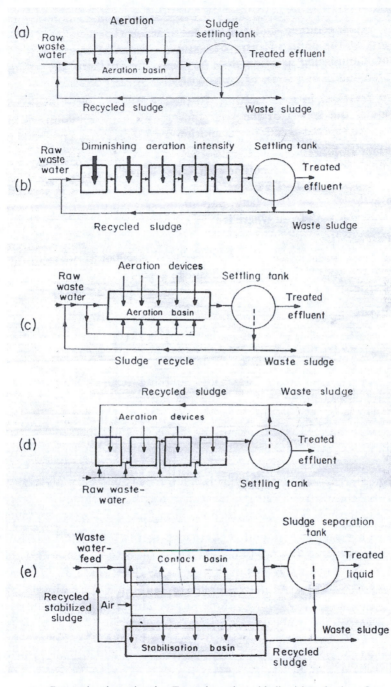

a = Conventional aeration; b = Tapered aeration with direct introduction of
raw sewage; c = Tapered aeration with tank introduction of raw sewage; d
= Step aeration; e = Contact stabilization.

Fig. 29.2 Schematic Representation of Various Modifications of the Activated Sludge Set-up

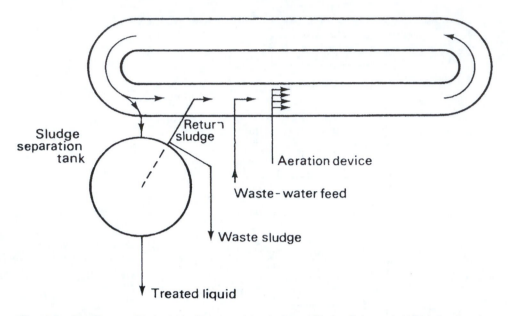

Fig. 29.3   The Pasveer Ditch: A Modification of the Activated Sludge Scheme in Which the Aeration is Done in a Basin about ft Deep in Which the Sewage Circulates

their air-lift fermentor used for the production single cell protein from methanol. It consists of an outer steel-lined concrete shaft measuring 300 ft or more installed into the ground. Waste water, and sludge recycle are injected down an inner steel tube. Compressed air is injected at a position along the center shaft deep enough to ensure that the hydrostatic weight of the water above the point of injection is high enough to force air bubbles downwards and prevent them coming upwards. The air dissolves lower down the shaft providing oxygen for the aerobic breakdown of the wastes. The water rises in the outer section of the shaft (Fig. 29.4). The system has the advantage of great rapidity in reducing the BOD and about 50% reduction in the sludge. Space is also saved.

(vii) **Enclosed tank systems and other compact systems**: Since the breakdown of waste in aerobic biological treatment is brought about by aerobic organisms, efficiency is sometimes increased by the use of oxygen or oxygen enriched air. Enclosed tanks, in which the waste water is completely mixed with the help of agitators, are used for aeration of this type. Sludge from a sedimentation tank is returned to the enclosed tank along with raw water as in the case with other systems. The advantage of the system is the absence, (or greatly reduced) obnoxious smell from the exhaust gases, and increased efficiency of waste stabilization. This system is widely used in industries the world over.

Compact activated sludge systems do not have a separate sedimentation tank. Instead sludge separation and aerobic breakdown occur in a single tank. The great advantage of such systems is the economy of space (Fig. 29.5).

29.3.1.1.1   Organisms involved in the activated sludge process

The organisms involved are bacteria and ciliates (protozoa). It was once thought that the formation of flocs which are essential for sludge formation was brought about by the

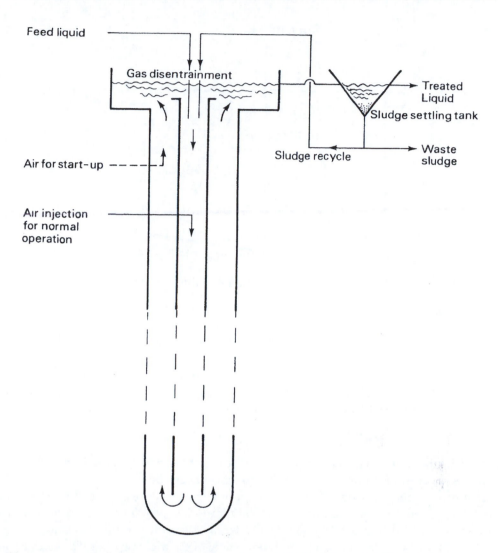

Feed liquid

Gas disentrainment

Treated
Liquid

Sludge settling tank

Air for start-up

Sludge recycle

Waste
sludge

Air injection
for normal
operation

In this system of activated sludge, the sewage is pump underground and air is injected. Because of the
depth the pressure of the air is increased causing greater dissolution of oxygen. The advantage of this
method is the saving in space use.

Fig. 29.4   The Deep Shaft Aeration System

slime-forming organism, *Zooglea ramigera*. It is now known that a wide range of bacteria
are involved, including *Pseudomonas, Achromobacter, Flavobacterium* to name a few.

29.3.1.1.2   Efficiency of activated sludge treatments

The efficiency of any system is usually determined by a reduction in the BOD of the waste
water before and after treatment. Efficiency depends on the amount of aeration, and the
contact time between the sludge and the raw waste water. Thus in conventional activated
sludge plants the contact time is about 10 hours, after which 90-95% of the BOD is
removed. When the contact time is less (in the high-rate treatment) BOD removal is 60-

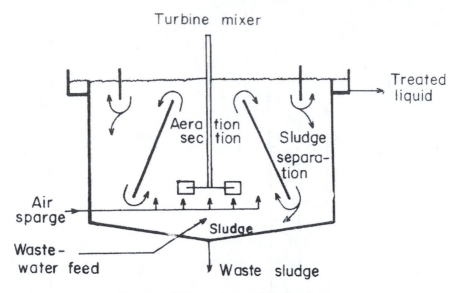

Fig. 29.5 Compact Activated Sludge System

70% and the sludge produced is more. With longer contact time, say several days, BOD reduction is over 95% and sludge extremely low.

With systems where oxygen is introduced as in the closed tank system or where there is great oxygen solubility as in the deep shaft system, contact time could be as short as 1 hour but with up to 90% BOD reduction along with substantially reduced sludge.

29.3.1.2 The trickling filter

In the trickling filter no sludge is returned to the incoming waste water. Rather the waste water is sprayed uniformly by a rotating distributor on a bed of rocks 6-10 ft deep. The rotation may be powered by an electric motor or a hydraulic impulse. The water percolates over the rocks within the bed which are 1-4 in diameter and is collected in an under-drain. The liquid is then collected from the under drain and allowed in a sedimentation tank which is an integral component of the trickling filter. The sludge from the sedimentation tank is removed from time to time. Various modifications of this basic system exist. In one modification the water may be pre-sedimented before introduction to the filter. Two filters may be placed in series and the effluent may be recycled (Fig. 29.6 and 29.7).

Microbiology of the trickling filter: A coating of microorganisms form on the stones as the waste water trickles down the filter and these organisms break-down the waste. Fungi, algae, protozoa and bacteria form on the rocks. As the filter ages the aerobic bacteria which are responsible for the breakdown of the organic matter become impeded, the system becomes inefficient and flies and obnoxious smells may result (Fig. 29.7). The microbial coating sloughs off from time to time.

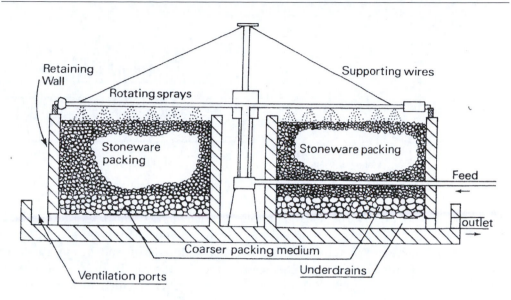

Fig. 29.6 Section through Trickling Filter Bed

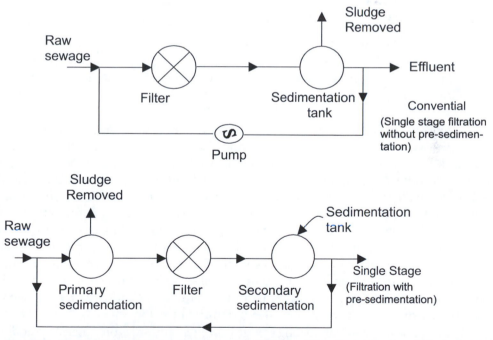

Fig. 29.7 Scheme Illustrating Two Arrangements of Trickling Filter: Conventional and Single Stage

29.3.1.3 Rotating discs

Also known as rotating biological contactors, these consist of closely packed discs about 10 ft in diameter and 1 inch apart. Discs made of plastic or metal may number up to 50 or more and are mounted on a horizontal shaft which rotates slowly, at a rate of about 0.5-

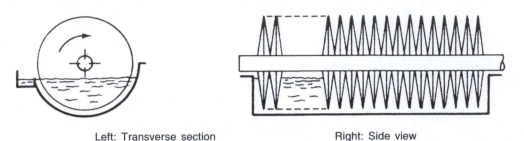

Left: Transverse section   Right: Side view

Fig. 29.8   Structure of Rotating Discs (Rotating Biological Contactor)

15 revolutions per min. During the rotation, 40-50% of the area of the discs is immersed in liquid at a time. A slime of micro-organisms, which decompose the wastes in the water, builds up on the discs. When the slime is too heavy, it sloughs off and is separated from the liquid in a clarifier. It has a short contact time and produces little sludge. The rotating disc system can be seen as a modification of the tricking filter in which the waste water is spread on rotating discs rather than on a bed of rocks.

29.4 TREATMENT OF THE SLUDGE: ANAEROBIC BREAKDOWN OF SLUDGE

As has been seen above, sludge always accompanies the aerobic breakdown of wastes in water. Its disposal is a major problem of waste treatment. Sludge consists of micro-organisms and those materials which are not readily degradable particularly cellulose. The solids in sludge form only a small percentage by weight and generally do not exceed 5%.

The goals of sludge treatment are to stabilize the sludge and reduce odors, remove some of the water and reduce volume, decompose some of the organic matter and reduce volume, kill disease causing organisms and disinfect the sludge. Untreated sludges are about 97% water. Settling the sludge and decanting off the separated liquid removes some of the water and reduces the sludge volume. Settling can result in a sludge with about 96 to 92% water. More water can be removed from sludge by using sand drying beds, vacuum filters, filter presses, and centrifuges resulting in sludges with between 80 to 50% water. This dried sludge is called a sludge cake. Anaerobic digestion is used to decompose organic matter to reduce its volume. Digestion also stabilizes the sludge to reduce odors. Caustic chemicals can be added to sludge or it may be heat treated to kill disease-causing organisms. Following treatment, liquid and cake sludges are usually spread on fields, returning organic matter and nutrients to the soil.

The commonest method of treating sludge however is by anaerobic digestion and this will be discussed below.

Anaerobic digestion consists of allowing the sludge to decompose in digesters under controlled conditions for several weeks. Digesters themselves are closed tanks with provision for mild agitation, and the introduction of sludge and release of gases. About 50% of the organic matter is broken down to gas, mostly methane. Amino acids, sugars alcohols are also produced. The broken-down sludge may then be de-watered and disposed of by any of the methods described above. Sludge so treated is less offensive and consequently easier to handle. Organisms responsible for sludge breakdown are sensitive to pH values outside 7-8, heavy metals, and detergents and these should not be introduced into digesters. Methane gas is also produced and this may sometimes be collected and used as a source of energy. Fig. 29.9 shows some anaerobic sludge digester designs.

29.5 WASTE WATER DISPOSAL IN THE PHARMACEUTICAL INDUSTRY

The treatment of wastes from a pharmaceutical industry is chosen to illustrate industrial waste treatment because the wastes are representative of a broad range of materials and include easily degradable organic materials, as well as sometimes some inorganic and even toxic compounds. Which of the various methods of disposal is used by a particular firm will depend on a number of factors foremost among which are: (a) the cost of the disposal method; (b) the location of the industry; (c) the nature of the industry and hence of its waste materials, and (d) the governmental regulations operating in the locality.

The above factors are all inter-related. For example, in siting the industry in the first place, space for, and the type of method of, waste disposal would have been considered. The cost of the disposal will be influenced not only by the nature and quantity of the waste and consequently the method adopted to handle it, but also what distance needs to be covered to have it disposed of. EPA regulations may for example dictate that the BOD of the wastes be reduced to a certain level before being discharged into a stream; any BOD reduction ultimately involves the expenditure of funds.

Nature of Wastes: The wastes from pharmaceutical firms may include easily degradable materials such as emulsion syrup, malt and tablet preparations. These contain considerable amounts of carbohydrates and hence yield wastes with high BOD.

Acids including the organic acids, acetic, formic and sulfanilic acids as well as the inorganic HCl and H_2SO_4 may be added to wastes. They have to be neutralized before being allowed into the treatment system.

Dissolved salts added in their own right or resulting from neutralization may also enter the system. Many drugs, some toxic or inhibitory to bacteria, may also be added.

Pre-treatment: Before treatment acid (or alkali) is neutralized, dissolved salts are removed usually by precipitation as calcium salts through lime addition, which also neutralizes acidity. Chloride and sulfate may be removed by ion exchange or rendered innocuous by dilution with water. Volatile compounds are stripped by pre-aeration.

Treatment: Before a routine is used within a treatment method, laboratory experiments would have been carried out to determine how much of the wastes may be efficiently handled within a given period. It may often be necessary to segregate the wastes, treating the more easily biodegradable organic forms separately from those wastes rich in inorganic materials. This is because the latter may require 'seeding' or the development of microorganisms specifically able to grow in and degrade them. Seeding is achieved by shaking a sample of the waste with a soil sample long enough for a special flora to develop.

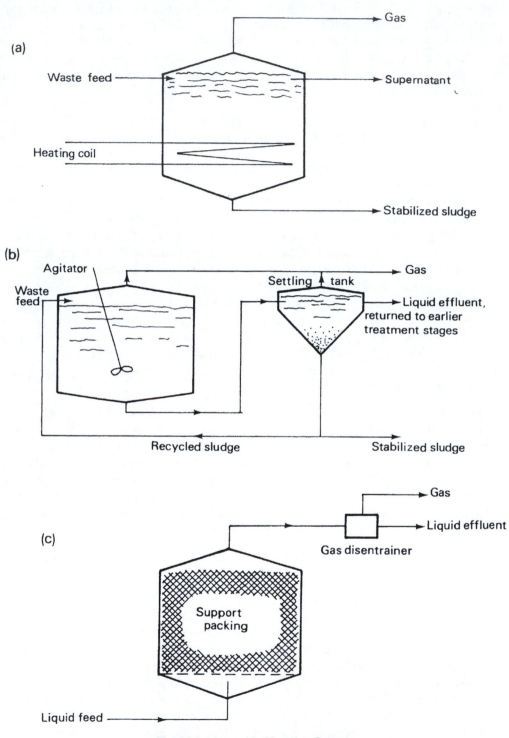

Fig. 29.9 Anaerobic Digestion Systems

SUGGESTED READINGS

Andrew, W. 1996. Biotechnology for Waste and Wastewater Treatment. Noyes Publications Westwood, N.J., USA.

Eckenfelder, W.W. 2000. Industrial water pollution control. McGraw-Hill Boston, USA.

Kosric, N., Blaszczyk, R. 1992. Industrial Effluent Processing. Encyclopedia of Microbiology. Vol 2, Academic Press. San Diego, USA. pp. 473-491.

Lindera, K.C. 2002. Activated Sludge – the Process. Encyclopedia of Environmental Microbiology Vol 1 Wiley-Interscience Publication. New York, USA. pp. 74–81.

Nielsen, P.H. 2002. Activated Sludge – the Floc. Encyclopedia of Environmental Microbiology Vol 1 Wiley-Interscience Publication. New York, USA. pp. 54–61.

Glossary

Anabolism The biochemical processes involved in the synthesis of cell constituents from simpler molecules usually requiring energy.

Anaerobic respiration When the final electron is an inorganic compound other than O_2.

Anotation The process by which useful information is added to a raw genomic DNA sequence, producing a frame work of understanding and enhancing its utility for downstream users.

Anticodon A sequence of three bases in transfer RNA which base-pairs with a codon in the messenger RNA during protein synthesis.

Antiparallel This refers to nucleic acids, where one strand runs 5' ~ 3', the other 3' ~ 5'.

Antisense mRNA, an mRNA transcript that is complementary to endogenous mRNA; it is a noncoding strand and complementary to the coding sequence of the endogenous mRNA. Introducing a transgene coding for antisense mRNA is a strategy used to block expression of a gene of interest.

Antisense RNA This is produced from a gene sequence inserted in the opposite orientation, so that the transcript is complimentary to the normal mRNA and can therefore bind to it and prevent translation.

Artificial chromosomes Cloning vectors which can carry very large inserts of foreign DNA and exist in the cell very much like a cellular chromosome. The most widely used are bacterial artificial chromosomes (BACs) and yeast artificial chromosomes (YACs).

ATP Adenosine triphosphate, the major energy carrier of the cell.

Autoradiography Detection of radioactivity in a sample labeled with a radioactive material, by placing it in contact with a photographic film; the radioactive portions will imprint on the film

Autotroph An organism able to use CO_2 as a sole source of carbon, for example plants and blue-green algae (Cyanobacteria).

Auxotroph An organism that has developed a nutritional requirement through mutation. Wthout the addition of the required material, it will not grow. The opposite is a *Prototroph* or wild type.

B lymphocyte (B cell) A lymphocyte that has immunoglobulin surface receptors, produces immunoglobulin, and may present antigens to T cells.

Bergey's Manual A compendium of an approved list of bacteria, first published in 1923. The second edition is currently being published in 5 volumes beginning in 2001 and is expected to be completed in 2007. *Bergey's Manual of Systematic Bacteriology* gets its name from Dr David H Bergey first Chairman of the Editorial Board of the Manual published by the then Society of American Bacteriologists (now called the American Society for Microbiology).

Bioinformatics The revolution in computer technology and memory storage capability has made it possible to model grand challenge problems such as large scale sequencing of genomes and management of large integrated **databases** over the Internet. This vastly improved computational capability integrated with large-scale miniaturization of biochemical techniques such as PCR, BAC, gel electrophoresis and microarray chips has delivered enormous amount of genomic and proteomic data. This integration of computation with biotechnology is Bioinformatics.

cDNA, (complementary DNA) is single-stranded DNA made in the laboratory from a messenger RNA template using the enzyme reverse transcriptase. This form of DNA is often used as a probe in the physical mapping of a chromosome; it is also used when it is desired to express a eukaryotic gene in a prokaryotic cell; for cloning into prokaryotic cell, the introns in a eukaryotic mRNA are spliced off and the intron-free mRNA converted to cDNA with reverse transcriptase.

cDNA library, A cDNA library is a collection of DNA sequences generated from mRNA sequences. This type of library contains only DNA that codes for proteins and does not include any non-coding DNA. The complete cDNA library of an organism gives an indication of the total amount of the proteins it can possibly express. The cDNA sequence also gives the genetic relationship between organisms through the similarity of their cDNA.

C'hemolithotroph An organism obtaining its energy from the oxidation of inorganic molecules.

Chemoorganotroph An organism obtaining its energy from the oxidation of organic.

Cistron A sequence of bases in DNA that specifies one polypeptide.

Clone A population of cells all descended from a single cell. Also, a number of copies of a DNA fragment obtained by allowing an inserted DNA fragment to be replicated by a phage or plasmid compounds.

Genetic Code The triplet codons that determine the types of amino acids inserted into a polypeptide chain during translation. There are 61 codons for 20 amino acids and three stop codons.

Genetic map The arrangement of genes on a chromosome.

Genome All the genes present in an organism.

Genomics The field of science that studies the entire DNA sequence of an organism's genome. The goal is to find all the genes within each genome and to use that information to develop improved medicines as well as answer scientific questions.

Histone proteins These are present in eucaryotic chromosomes; histones and DNA give structure to chromosomes in eucaryotes.

Introns Non-coding sequences within genes.

Kilo basepair 1,000 basepairs, a unit of DNA length abbreviated KB.

Knock out mice A transgenic mouse in which a gene function has been disrupted or knocked out. It is used to produce animal models for the study of human disease.

Mass spectrometry is an analytical technique used to measure the mass-to-charge ratio of ions. It is most generally used to find the composition of a physical sample by generating a mass spectrum representing the masses of sample components. The technique has several applications, including:

1. identifying unknown compounds by the mass of the compound and/or fragments thereof.
2. determining the isotopic composition of one or more elements in a compound.
3. determining the structure of compounds by observing the fragmentation of the compound.
4. quantitating the amount of a compound in a sample using carefully designed methods (mass spectrometry is not inherently quantitative).

5. studying the fundamentals of gas phase ion chemistry (the chemistry of ions and neutrals in vacuum).
6. determining other physical, chemical or even biological properties of compounds with a variety of other approaches.

Methanogenesis The biological production of methane.

Nuclear Magnetic Resonance (NMR) is a physical phenomenon based upon the magnetic property of an atom's nucleus. NMR spectroscopy is one of the principal techniques used to obtain physical, chemical, electronic and structural information about a molecule. It is the only technique that can provide detailed information on the exact three-dimensional structure of biological molecules in solution. Also, nuclear magnetic resonance is one of the techniques that has been used to build elementary quantum computers.

Operons Typically present in prokaryotes, these are clusters of genes controlled by a single operator; an operator itself is a region of an operon, close to the promoter to which a receptor protein binds.

Promoter A DNA sequence lying upstream of from the gene to which RNA polymerase binds.

Proteome The totality of the proteins present in a cell.

Proteomics The study of the study of the structure, function and regulation of the proteins in an organism.

Sense mRNA, endogenous mRNA molecules which encode functional proteins; it is a 5' to 3' mRNA molecule.

Shine-Dalgarno sequence A conserved sequence in prokaryotic mRNAs that is complementary to a sequence near the 5´ terminus of the 16S ribosomal RNA and is involved in the initiation of translation.

Site-directed mutagenesis A technique for construction a mutation in a gene in vitro by altering a base or bases in the gene.

TATA box Also called Hogness Box, an AT-rich region of the DNA with the sequence TATAT/AAT/A located before the initiation site.

Transcription factor Transcription factor is a protein that binds DNA at a specific promoter or enhancer region or site, where it regulates transcription.

Transfer RNA (tRNA) A small RNA of 75 – 85 bases that carries the anticodon and the amino acid residue required for protein synthesis.

Index